CW00518912

Spon's External Works and Landscape Price Book

2008

Spon's External Works and Landscape Price Book

Edited by

DAVIS LANGDON

in association with
LandPro Ltd
Landscape Surveyors

2008

Twenty-seventh edition

Taylor & Francis
Taylor & Francis Group
LONDON AND NEW YORK

First edition 1978
Twenty-seventh edition published 2008
by Taylor & Francis
2 Park Square, Milton Park, Abingdon, Oxon OX14 4RN

Simultaneously published in the USA and Canada
by Taylor & Francis
270 Madison Avenue, New York, NY 10016

Taylor & Francis is an imprint of the Taylor & Francis Group, an informa business

© 2008 Taylor & Francis

The right of Davis Langdon in association with LandPro Ltd Landscape Surveyors to be identified as the Author of this Work has been asserted by them in accordance with the Copyright, Designs and Patents Act 1988

Printed and bound in Great Britain by
TJ International Ltd, Padstow, Cornwall

All rights reserved. No part of this book may be reprinted or reproduced or utilised in any form or by any electronic, mechanical, or other means, now known or hereafter invented, including photocopying and recording, or in any information storage or retrieval system, without permission in writing from the publishers.

The publisher makes no representation, express or implied, with regard to the accuracy of the information contained in this book and cannot accept any legal responsibility or liability for any errors or omissions that may be made.

Publisher's note
This book has been produced from camera-ready copy supplied by the authors.

British Library Cataloguing in Publication Data
A catalogue record for this book is available from the British Library

ISBN10: 0-415-42444-5
ISBN13: 978-0-415-42444-8
ISSN: 0267-4181

Contents

Preface to the Twenty Seventh Edition

Change of Title

We remind readers who may not have updated their publication last year, of the change of title from Spon's Landscape and External Works Price Book to Spon's External Works and Landscape Price Book.

Market Conditions

Most landscape contractors continue to enjoy buoyant market conditions although competition is keen for tendered works. Domestic contractors are also benefitting from the continuing strong state of the economy and the resultant development of landscapes by both home owners and developers.

Construction price inflation in Greater London over the past year has reached 8%, according to Davis Langdon's latest Tender Price Index. Overall, new build work is forecast to rise by about 4% in real terms in 2007, and by a further 3-4% in 2008.

The surge of work as a result of the Olympic games requirements have not materialised as yet, though there is an expectant sense of preparation by contractors who are hoping to benefit from this activity. One industry source however commented that although his company is the market leader in his area, his view is that the Olympic revenues must be seen as a short term issue and that no company should rely on this factor for their long time success. It is probable that some companies who are able to negotiate their involvement rather than tender for the packages offered will increase their profits. These are the companies that over the years have shown their extreme reliability to the larger employers and therefore manage to be selected on performance rather than price. Packages are tendered but a subsequent negotiation stage follows. A trend for this practise seems to be established for prestige works.

Supplier Prices

Whereas last year many suppliers of prices to this publication held their prices, most have this year have raised prices albeit by very small percentages.
We noted during the course of our update that although there is much talk of the rise in steel prices, items with a high steel content did not rise dramatically.
The average increase in prices of material, plant and sub-contract resources in this year's edition is approximately 2.0%. Larger manufacturers have raised material prices by up to 6%.

Labour Rates

Prices for labour within the external works sector of the construction industry have remained stable or increased slightly over the last 12 months. This is, as last year, mainly due to the influx of European labour which is having a large cost impact on the construction industry. The consequence of this is that we have increased the cost of labour in this year's edition from £18.00/hour in the 2007 price book to £18.50/hour in this edition, a rise of 2.8%. Maintenance contracting remains at a lower rate price of £15.50/hour for permanently site based landscape maintenance works.
Minor works labour is based on a rate of £21.00/hour.

Tables and Memoranda

We remind readers of the information contained in this section at the back of the book. It is often overlooked but contains a wealth of information useful to external works consultants and contractors.

Profit and Overhead

Spon's External Works and Landscape prices do not allow for profit or site overhead. Readers should evaluate the market conditions in the sector or environment in which they operate and apply a percentage to these rates. Company overhead is allowed for within the labour rates used. Please refer to 'Part 1 General' for an explanation of the build-up of this year's labour rates.

Prices for Suppliers and Services

We acknowledge the support afforded to us by the suppliers of products and services who issue us with the base information used. Their contact details are published in the directory section at the front of this book. We advise that readers wishing to evaluate the cost of a product or service should approach these suppliers directly should they wish to confirm prices prior to submission of tenders or quotations.

Whilst every effort is made to ensure the accuracy of the information given in this publication, neither the editors nor the publishers in any way accept liability for loss of any kind resulting from the use made by any person of such information.

We remind readers that we would be grateful to hear from them during the course of the year with suggestions or comments on any aspect of the contents of this years book and suggestions for improvements. Our contact details are shown below.

DAVIS LANGDON
MidCity Place
71 High Holborn
London
WC1V 6QS

Tel: 020 7061 7000
e-mail: data.enquiry@davislangdon.com

SAM HASSALL
LANDPRO LTD
Landscape Surveyors
25b The Borough
Farnham
Surrey
GU9 7NJ

Tel: 01252 725513
e-mail: info@landpro.co.uk

Acknowledgements

This list has been compiled from the latest information available but as firms frequently change their names, addresses and telephone numbers due to re-organization, users are advised to check this information before placing orders.

A Plant Hire Co Ltd
102 Dalton Avenue
Birchwood Park
Warrington
Cheshire WA3 6YE
Plant hire
Tel: 01925 281000
Fax: 01925 281001
Website: www.aplant.com

ABG Geosynthetics Ltd
Unit E7
Meltham Mills Road
Meltham
West Yorkshire HD9 4DS
Geosynthetics
Tel: 01484 852096
Fax: 01484 851562
Website: www.abg-geosynthetics.com
E-mail: sales@abgltd.com

Aco Technologies Plc
Aco Business Park
Hitchin Road
Shefford
Bedfordshire SG17 5TE
Linear Drainage
Tel: 01462 816666
Fax: 01462 815895
Website: www.aco-online.co.uk
E-mail: technologies@aco.co.uk

Addagrip Surface Treatments UK Ltd
Addagrip House
Bell Lane Industrial Estate
Uckfield
East Sussex TN22 1QL
Epoxy bound surfaces
Tel: 01825 761333
Fax: 01825 768566
Website: www.addagrip.co.uk
E-mail: sales@addagrip.co.uk

ADP Netlon Turf Systems
Business Technology Centre
Radway Green
Alsager
Cheshire CW2 5PR
Erosion control, soil stabilisation, plant protection
Tel: 01270 886235
Fax: 01270 886294
Website: www.adpnetlon.co.uk
E-mail: enquiries@adp.gb.net

Alumasc Exterior Building Products Ltd
White House Works
Bold Road
Sutton
St.Helens
Merseyside WA9 4JG
Green roof systems
Tel: 01744 648400
Fax: 01744 648401
Website: www.alumasc-exterior-building-products.co.uk
E-mail: info@alumasc-exteriors.co.uk

Amberol Ltd
The Plantation
King Street
Alfreton
Derbyshire DE55 7TT
Street furniture
Tel: 01773 830930
Fax: 01773 834191
Website: www.amberol.co.uk
E-mail: sales@amberol.co.uk

Amenity and Horticultural Services Ltd
Coppards Lane
Northiam
East Sussex TN31 6QP
Horticultural composts & fertilizers
Tel: 01797 252728
Fax: 01797 252724
Website: www.ahsdirect.co.uk
E-mail: info@ahsdirect.co.uk

Amenity Land Services Ltd (ALS)
Long Lane
Wellington
Telford
Shropshire TF6 6HA
Horticultural products supplier
Tel: 01952 641949
Fax: 01952 247369
Website: www.amenity.co.uk
E-mail: sales@amenity.co.uk

Ancon Ltd
President Way
President Park
Sheffield S4 7UR
Fixings
Tel: 01142 755224
Fax: 01142 768543
Website: www.ancon.co.uk
E-mail: info@ancon.co.uk

Anderton Concrete Products Ltd
Anderton Wharf
Soot Hill
Anderton
Northwich
Cheshire CW9 6AA
Concrete fencing
Tel: 01606 79436
Fax: 01606 871590
Website: www.andertonconcrete.co.uk
E-mail: sales@andertonconcrete.co.uk

Anglo Aquarium Plant Co Ltd
Strayfield Road
Enfield
Middlesex EN2 9JE
Aquatic plants
Tel: 020 8363 8548
Fax: 020 8363 8547
Website: www.anglo-aquarium.co.uk
E-mail: sales@anglo-aquarium.co.uk

Architectural Heritage Ltd
Taddington Manor
Cutsdean
Cheltenham
Gloucestershire GL54 5RY
Ornamental stone buildings & water features
Tel: 01386 584414
Fax: 01386 584236
Website: www.architectural-heritage.co.uk
E-mail: puddy@architectural-heritage.co.uk

Artificial Lawn Co
Hartshill Nursery
Thong Lane
Gravesend
Kent DA12 4AD
Artificial grass suppliers
Tel: 01474 364320
Fax: 01474 321587
Website: www.artificiallawn.co.uk
E-mail: sales@artificiallawn.co.uk

Autopa Ltd
Cottage Leap off Butlers Leap
Rugby
Warwickshire CV21 3XP
Street furniture
Tel: 01788 550556
Fax: 01788 550265
Website: www.autopa.co.uk
E-mail: info@autopa.co.uk

Avoncrop
Weighbridge House
Station Road
Sandford
Winscombe BS25 5NX
Horticultural supplier
Tel: 01934 820363
Fax: 01934 820926
Website: www.avoncrop.co.uk
E-mail: sales@avoncrop.co.uk

AVS Fencing Supplies Ltd
Unit 1 AVS Trading Park
Chapel Lane
Milford
Surrey GU8 5HU
Fencing
Tel: 01483 410960
Fax: 01483 860867
Website: www.avsfencing.co.uk
E-mail: sales@avsfencing.co.uk

Bauder Ltd
Broughton House
Broughton Road
Ipswich
Suffolk IP1 3QR
Green roof systems
Tel: 01473 257671
Fax: 01473 230761
Website: www.bauder.co.uk
E-mail: mail@bauder.co.uk

Bayer Environmental Science
230 Cambridge Science Park
Milton Road
Cambridge CB4 0WB
Landscape chemicals
Tel: 01223 226680
Fax: 01223 226635
Website: www.bayer-escience.co.uk

Baylis Landscape Contractors Ltd
Hartshill Nursery
Thong Lane
Gravesend
Kent DA12 4AD
Sportsfield construction
Tel: 01474 569576
Fax: 01474 321587
Website: www.baylislandscapes.co.uk
E-mail: landscape@baylis.freeserve.co.uk

Boughton Loam Ltd
Telford Way Industrial Estate
Kettering
Northants NN16 8UN
Loams and topsoils
Tel: 01536 510515
Fax: 01536 510691
Website: www.boughton-loam.co.uk
E-mail: enquiries@boughton-loam.co.uk

Breedon Special Aggregates
(A division of Ennstone Johnston Ltd)
Breedon-on-the-Hill
Derby
Derbyshire DE73 1AP
Specialist gravels
Tel: 01332 862254
Fax: 01332 863149
Website: www.ennstone.co.uk
E-mail: sales@breedon.co.uk

British Seed Houses Ltd
Camp Road
Witham St Hughs
Lincoln LN6 9QJ
Grass and wildflower seed
Tel: 01522 868714
Fax: 01522 868382
Website: www.bshamenity.com
E-mail: seeds@bshlincoln.co.uk

Broxap Street Furniture
Rowhurst Industrial Estate
Chesterton
Newcastle-under-Lyme
Staffordshire ST5 6BD
Street furniture
Tel: 01782 564411
Fax: 01782 565357
Website: www.broxap.com
E-mail: sales@broxap.com

Capital Garden Products Ltd
Gibbs Reed Barn
Pashley Road
Ticehurst
East Sussex TN5 7HE
Plant containers
Tel: 01580 201092
Fax: 01580 201093
Website: www.capital-garden.com
E-mail: sales@capital-garden.com

CED Ltd
728 London Road
West Thurrock
Grays
Essex RM20 3LU
Natural stone
Tel: 01708 867237
Fax: 01708 867230
Website: www.ced.ltd.uk
E-mail: sales@ced.ltd.uk

Charcon Hard Landscaping
Hulland Ward
Ashbourne
Derbyshire DE6 3ET
Paving and street furniture
Tel: 01335 372222
Fax: 01335 370074
Website: www.aggregate.com
E-mail: ukenquiries@aggregate.com

Charles Morris (Fertilizers) Ltd
Longford House
Long Lane
Stanwell
Middlesex TW19 7AT
Topsoil
Tel: 01784 449144
Fax: 01784 449133
Website: www.charles-morris.co.uk
E-mail: sales@charles-morris.co.uk

City Electrical Factors
Eastern Road
North Lane
Aldershot
Hampshire GU12 4YD
Electrical wholesaler
Tel: 01252 327661
Fax: 01252 343089
Website: www.cef.co.uk

Colourpave Ltd
(The Bituchem Group)
Laymore
Forest Vale Industrial Estate
Cinderford
Gloucestershire GL14 2YH
Macadam contractor
Tel: 01594 826768
Fax: 01594 826948
Website: www.bituchem.com
E-mail: info@bituchem.com

Columbia Cascade Ltd
Plasycoed
Wernoleu Road
Ammanford
Carmarthenshire SA18 2JL
Street furniture
Tel: 0845 2600343
Fax: 01269 596395
Website: www.timberform.com
E-mail: columbiacascade@btconnect.com

Cooper Clarke Group Ltd
Special Products Division
Bloomfield Road
Farnworth
Bolton BL4 9LP
Wall drainage and erosion control
Tel: 01204 862222
Fax: 01204 793856
Website: www.heitonuk.com
E-mail: marketing@cooperclarke.co.uk

Coverite Ltd
Palace Gates
Bridge Road
Wood Green
London N22 4SP
Waterproofing specialist
Tel: 020 8888 7821
Fax: 020 8889 0731
Website: www.coverite.co.uk
E-mail: sales@coverite.co.uk

Crowders Nurseries
Lincoln Road
Horncastle
Lincolnshire LN9 5LZ
Plant protection
Tel: 01507 525000
Fax: 01507 524000
Website: www.crowders.co.uk
E-mail: sales@crowders.co.uk

CU Phosco Ltd
Charles House
Lower Road
Great Amwell
Ware
Hertfordshire SG12 9TA
Lighting
Tel: 01920 860600
Fax: 01920 860635
Website: www.cuphosco.co.uk
E-mail: sales@cuphosco.co.uk

Dee-Organ Ltd
5 Sandyford Road
Paisley
Renfrewshire PA3 4HP
Street furniture
Tel: 01418 897000
Fax: 01418 897764
Website: www.dee-organ.co.uk
E-mail: signs@dee-organ.co.uk

Deepdale Trees Ltd
Tithe Farm
Hatley Road
Potton
Sandy
Bedfordshire SG19 2DX
Trees, multi-stem and hedging
Tel: 01767 262636
Fax: 01767 262288
Website: www.deepdale-trees.co.uk
E-mail: mail@deepdale-trees.co.uk

DLF Trifolium Ltd
Thorn Farm
Evesham Road
Inkberrow
Worcestershire WR7 4LJ
Grass and wildflower seed
Tel: 01386 791102
Fax: 01386 792715
Website: www.dlf.co.uk
E-mail: amenity@dlf.co.uk

Duracourt (Spadeoak) Ltd
Town Lane
Wooburn Green
High Wycombe
Buckinghamshire HP10 0PD
Tennis courts
Tel: 01628 850570
Fax: 01628 810509
Website: www.duracourt.co.uk
E-mail: info@duracourt.co.uk

E.T. Clay Products
7-9 Fowler Road
Hainault
Essex IG6 3UT
Brick supplier
Tel: 0208 5012100
Fax: 0208 5009990
Website: www.etbricks.co.uk
E-mail: sales@etbricks.co.uk

Earth Anchors Ltd
15 Campbell Road
Croydon
Surrey CR0 2SQ
Anchors for site furniture
Tel: 020 8684 9601
Fax: 020 8684 2230
Website: www.earth-anchors.com
E-mail: sales@earth-anchors.com

Easy Mix Concrete Ltd
Builders Yard
Station Approach
Knebworth
Hertfordshire SG3 6AT
Concrete
Tel: 0800 1696401

Edwards Sports Products Ltd
North Mills Trading Estate
Bridport
Dorset DT6 3AH
Sports equipment
Tel: 01308 424111
Fax: 01308 455800
Website: www.edsports.co.uk
E-mail: sales@edsports.co.uk

Elliotthire
The Fen
Baston
Peterborough
Cambridgeshire PE6 9PT
Site offices
Tel: 01778 560891
Fax: 01778 560881
Website: www.elliotthire.co.uk
E-mail: hirediv@elliott-group.co.uk

English Woodlands
Burrow Nursery
Herrings Lane
Cross in Hand
Heathfield
East Sussex TN21 0UG
Plant protection
Tel: 01435 862992
Fax: 01435 867742
Website: www.ewburrownursery.co.uk
E-mail: sales@ewburrownursery.co.uk

Ensor Building Products
Blackamoor Road
Guide
Blackburn
Lancashire BB1 2LQ
Drainage suppliers
Tel: 0870 7700484
Fax: 0870 7700485
Website: www.ensorbuilding.com
E-mail: sales@ensorbuilding.com

Eve Trakway
Bramley Vale
Chesterfield
Derbyshire S44 5GA
Portable roads
Tel: 08700 767676
Fax: 08700 737373
Website: www.evetrakway.co.uk
E-mail: marketing@evetrakway.co.uk

Exclusive Leisure Ltd
28 Cannock Street
Leicester LE4 9HR
Artificial sports surfaces
Tel: 0116 233 2255
Fax: 0116 246 1561
Website: www.exclusiveleisure.co.uk
E-mail: info@exclusiveleisure.co.uk

Fairwater Ltd
Lodge Farm
Malthouse Lane
Ashington
West Sussex RH20 3BU
Water feature contractors
Tel: 01903 892228
Fax: 01903 892522
Website: www.fairwater.co.uk
E-mail: info@fairwater.co.uk

Farmura Environmental Products Ltd
Stone Hill
Egerton
Ashford
Kent TN27 9DU
Organic fertilizer suppliers
Tel: 01233 756241
Fax: 01233 756419
Website: www.farmura.com
E-mail: info@farmura.com

Fleet (Line Markers) Ltd
Fleet House
Spring Lane
Malvern
Worcestershire WR14 1AT
Sports line marking
Tel: 01684 573535
Fax: 01684 892784
Website: www.fleetlinemarkers.com
E-mail: sales@fleetlinemarkers.com

Forticrete Ltd
Hillhead Quarry
Harpur Hill
Buxton
Derbyshire SK17 9PS
Retaining wall systems
Tel: 01298 23333
Fax: 01298 23000
Website: www.forticrete.com
E-mail: enq@forticrete.com

FP McCann Ltd
Whitehill Road
Ellistown
Coalville
Leicester LE67 1ET
Precast concrete pipes
Tel: 01530 240056
Fax: 01530 240015
Website: www.fpmccann.co.uk
E-mail: info@fpmccann.co.uk

Furnitubes International Ltd
Meridian House
Royal Hill
Greenwich
London SE10 8RT
Street furniture
Tel: 020 8378 3200
Fax: 020 8378 3250
Website: www.furnitubes.com
E-mail: sales@furnitubes.com

Garden Trellis Company
Unit 1 Brunel Road
Gorse Lane Industrial Estate
Great Clacton
Essex CO15 4LU
Garden joinery specialists
Tel: 01255 688361
Fax: 01255 688362
Website: www.gardentrellis.co.uk
E-mail: info@gardentrellis.co.uk

Grace Construction Products
Ajax Avenue
Slough
Berkshire SL1 4BH
Bitu-thene tanking
Tel: 01753 692929
Fax: 01753 691623
Website: www.uk.graceconstruction.com
E-mail: uksales@grace.com

Grass Concrete Ltd
Walker House
22 Bond Street
Wakefield
West Yorkshire WF1 2QP
Grass block paving
Tel: 01924 379443
Fax: 01924 290289
Website: www.grasscrete.com
E-mail: info@grasscrete.com

Greenfix Ltd
Evesham Road
Bishops Cleeve
Cheltenham
Glos GL52 8SA
Seeded erosion control mats
Tel: 01242 679555
Fax: 01242 679855
Website: www.greenfix.co.uk
E-mail: cheltenham@greenfix.co.uk

Greenleaf Horticulture
Ivyhouse Industrial Estate
Haywood Way
Hastings TN35 4PL
Root directors
Tel: 01424 717797
Fax: 01424 205240

Grundon Waste Management Ltd
Goulds Grove
Ewelme
Wallingford
OX10 6PJ
Gravel
Tel: 0870 4438278
Website: www.grundon.com
E-mail: sales@grundon.com

HSS Hire
Unit 3
14 Wates Way
Mitcham
Surrey CR4 4HR
Tool and plant hire
Tel: 020 8685 9500
Fax: 020 8685 9600
Website: www.hss.com
E-mail: hire@hss.com

Haddonstone Ltd
The Forge House
Church Lane
East Haddon
Northampton NN6 8DB
Architectural stonework
Tel: 01604 770711
Fax: 01604 770027
Website: www.haddonstone.co.uk
E-mail: info@haddonstone.co.uk

Harrison External Display Systems
Borough Road
Darlington
Co Durham DL1 1SW
Flagpoles
Tel: 01325 355433
Fax: 01325 461726
Website: www.harrisoneds.com
E-mail: sales@harrisoneds.com

Heicom UK
4 Frog Lane
Tunbridge Wells
Kent TN1 1YT
Tel: 01892 522360
Fax: 01892 522767
Tree sand
Website: www.treesand.co.uk
E-mail: mike.q@treesand.co.uk

Hepworth
Hazelhead
Crow Edge
Sheffield S36 4HG
Drainage
Tel: 0870 4436000
Fax: 0870 4438000
Website: www.hepworthdrainage.co.uk
E-mail: info@hepworthdrainage.co.uk

Hodkin & Jones (Sheffield) Ltd
Callywhite Lane
Dronfield
Sheffield S18 6XP
Drainage
Tel: 01246 290890
Fax: 01246 290292
Website: www.hodkin-jones.co.uk
E-mail: info@hodkin-jones.co.uk

Icopal UK Ltd
Lyon Way
St Albans
Hertfordshire AL4 0LB
Lake liner
Tel: 01727 830116
Fax: 01727 868045
Website: www.monarflex.icopal.co.uk
E-mail: enq@monarflex.co.uk

Inturf
The Chestnuts
Wilberfoss
York YO41 5NT
Turf
Tel: 01759 321000
Fax: 01759 380130
Website: www.inturf.com
E-mail: info@inturf.co.uk

J Toms Ltd
7 Marley Farm
Headcorn Road
Smarden
Kent TN27 8PJ
Plant protection
Tel: 01233 770066
Fax: 01233 770055
Website: www.jtoms.co.uk
E-mail: jtoms@btopenworld.com

Jacksons Fencing
Stowting Common
Ashford
Kent TN25 6BN
Fencing
Tel: 01233 750393
Fax: 01233 750403
Website: www.jacksons-fencing.co.uk
E-mail: sales@jacksons-fencing.co.uk

James Coles & Sons (Nurseries) Ltd
The Nurseries
Uppingham Road
Thurnby
Leicester LE7 9QB
Nurseries
Tel: 0116 241 2115
Fax: 0116 243 2311
Website: www.colesnurseries.co.uk
E-mail: sales@colesnurseries.co.uk

John Anderson Hire
Unit 5 Smallford Works
Smallford Lane
St. Albans
Herts AL4 0SA
Mobile toilet facilities
Tel: 01727 822485
Fax: 01727 822886
Website: www.superloo.co.uk
E-mail: sales@superloo.co.uk

Johnsons Wellfield Quarries Ltd
Crosland Hill
Huddersfield
West Yorkshire HD4 7AB
Natural Yorkstone pavings
Tel: 01484 652311
Fax: 01484 460007
Website: www.johnsons-wellfield.co.uk
E-mail: sales@johnsons-wellfield.co.uk

Jones of Oswestry
Whittington Road
Oswestry
Shropshire SY11 1HZ
Channels, gulleys, manhole covers
Tel: 01691 653251
Fax: 01691 658222
Website: www.jonesofoswestry.com
E-mail: sales@jonesofoswestry.com

Keller Ground Engineering
Mereworth Business Centre
Danns Lane
Mereworth
Kent ME18 5LW
Hydro seeding specialist
Tel: 01622 816780
Fax: 01622 816791
Website: www.keller-ge.co.uk
E-mail: retaining@keller-ge.co.uk

Kompan Ltd
20 Denbigh Hall
Bletchley
Milton Keynes
Bucks. MK3 7QT
Play equipment
Tel: 01908 642466
Fax: 01908 270137
Website: www.kompan.com
E-mail: kompan.uk@kompan.com

Land & Water Group Ltd
3 Weston Yard
The Street
Albury
Guildford
Surrey GU5 9AF
Specialist plant hire
Tel: 01483 202733
Fax: 01483 202510
Website: www.land-water.co.uk
E-mail: enquiries@land-water.co.uk

Landline Ltd
1 Bluebridge Industrial Estate
Halstead
Essex CO9 2EX
Pond and lake installation
Tel: 01787 476699
Fax: 01787 472507
Website: www.landline.co.uk
E-mail: sales@landline.co.uk

Lappset UK Ltd
Lappset House
Henson Way
Kettering
Northants NN16 8PX
Play equipment
Tel: 01536 412612
Fax: 01536 521703
Website: www.lappset.com
E-mail: customerservicesuk@lappset.com

LDC Limited
Loampits Farm
99 Westfield Road
Woking
Surrey GU22 9QR
Willow walling
Tel: 01483 767488
Fax: 01483 764293
Website: www.ldclandscape.co.uk
E-mail: info@ldc.co.uk

Leaky Pipe Systems Ltd
Frith Farm
Dean Street
East Farleigh
Maidstone
Kent ME15 0PR
Irrigation systems
Tel: 01622 746495
Fax: 01622 745118
Website: www.leakypipe.co.uk
E-mail: sales@leakypipe.co.uk

Lister Lutyens Co Ltd
6 Alder Close
Eastbourne
East Sussex BN23 6QF
Street furniture
Tel: 01323 431177
Fax: 01323 639314
Website: www.listerteak.com
E-mail: sales@listerteak.com

Longlyf Timber Products Ltd
Grange Road
Tilford
Farnham
Surrey GU10 2DQ
Treated timber products
Tel: 01252 795042
Fax: 01252 795043
Website: www.longlyftimber.co.uk
E-mail: enquiries@longlyftimber.co.uk

Lorenz von Ehren
Maldfeldstrasse 4
D-21077 Hamburg
Germany
Topiary
Tel: 0049 40 761080
Fax: 0049 40 761081
Website: www.lve.de
E-mail: sales@lve.de

Louis Poulsen UK Ltd
Unit C 44 Barwell Business Park
Leatherhead Road
Chessington KT9 2NY
Outdoor lighting
Tel: 020 8397 4400
Fax: 020 8397 4455
Website: www.louispoulsen.co.uk
E-mail: louis.poulsen.uk@lpmail.com

Maccaferri Ltd
7400 The Quorum
Oxford Business Park North
Garsington Road
Oxford OX4 2JZ
Gabions
Tel: 01865 770555
Fax: 01865 774550
Website: www.maccaferri.co.uk
E-mail: oxford@maccaferri.co.uk

Marshalls
Landscape House
Premier Way
Lowfields Business Park
Elland HX5 9HT
Hard landscape materials and street furniture
Tel: 01422 312000
Fax: 01422 330185
Website: www.marshalls.co.uk

McArthur Group Ltd
Foundry Lane
Bristol BS5 7UE
Security fencing
Tel: 01780 761417
Fax: 01780 766630
Website: www.mcarthur-group.com
E-mail: marketing@mcarthur-group.com

Melcourt Industries Ltd
Boldridge Brake
Long Newnton
Tetbury
Gloucestershire GL8 8RT
Mulch and compost
Tel: 01666 502711
Fax: 01666 504398
Website: www.melcourt.co.uk
E-mail: mail@melcourt.co.uk

Milton Pipes Ltd
Cooks Lane
Milton Regis
Sittingbourne
Kent ME10 2QF
Soakaway rings
Tel: 01795 425191
Fax: 01795 478232
Website: www.miltonpipes.com
E-mail: sales@miltonpipes.com

Neptune Outdoor Furniture Ltd
Thompsons Lane
Marwell
Winchester
Hampshire SO21 1JH
Street furniture
Tel: 01962 777799
Fax: 01962 777723
Website: www.nofl.co.uk
E-mail: info@nofl.co.uk

Norris and Gardiner Ltd
Lime Croft Road
Knaphill
Woking
Surrey GU21 2TH
Grounds maintenance
Tel: 01483 289111
Fax: 01483 289112
E-mail: rich@norg.co.uk

Notcutts Nurseries
Woodbridge
Suffolk IP12 4AF
Nursery stock supplier
Tel: 01394 383344
Fax: 01394 445307
Website: www.notcutts.co.uk
E-mail: farmera@notcutts.co.uk

Orchard Street Furniture Ltd
119 The Street
Crowmarsh Gifford
Wallingford
Oxfordshire OX10 8EF
Street furniture
Tel: 01491 642123
Fax: 01491 642126
Website: www.orchardstreet.co.uk
E-mail: sales@orchardstreet.co.uk

Platipus Anchors Ltd
Kingsfield Business Centre
Philanthropic Road
Redhill
Surrey RH1 4DP
Tree anchors
Tel: 01737 762300
Fax: 01737 773395
Website: www.platipus-anchors.com
E-mail: info@platipus-anchors.com

RCC
Barholm Road
Tallington
Stamford
Lincolnshire PE9 4RL
Precast concrete retaining units
Tel: 01778 381000
Fax: 01778 348041
Website: www.tarmacprecast.co.uk
E-mail: precastenquiries@tarmac.co.uk

Recycled Materials Ltd
PO Box 519
Surbiton
Surrey KT6 4YL
Disposal contractors
Tel: 020 8390 7010
Fax: 020 8390 7020
Website: www.recycledmaterialsltd.co.uk
E-mail: mail@recycledmaterialsltd.co.uk

Rehavo UK Ltd
Penketh Place
West Pimbo
Skelmersdale
Lancashire WN8 9QX
Irrigation
Tel: 01695 556222
Fax: 01695 556333
Website: www.revaho.co.uk
E-mail: tomer@revaho.co.uk

Rigby Taylor Ltd
The Riverway Estate
Portsmouth Road
Peasmarsh
Guildford
Surrey GU3 1LZ
Horticultural supply
Tel: 01483 446900
Fax: 01483 534058
Website: www.rigbytaylor.com
E-mail: sales@rigbytaylor.com

RIW Ltd
Arc House
Terrace Road South
Binfield
Bracknell
Berkshire RG42 4PZ
Waterproofing products
Tel: 01344 397777
Fax: 01344 862010
Website: www.riw.co.uk
E-mail: enquiries@riw.co.uk

Road Equipment Ltd
28-34 Feltham Road
Ashford
Middlesex TW15 1DL
Plant hire
Tel: 01784 256565
Fax: 01784 240398
E-mail: roadequipment@aol.com

Rolawn Ltd
York Road
Elvington
York YO41 4XR
Industrial turf
Tel: 01904 608661
Fax: 01904 608272
Website: www.rolawn.co.uk
E-mail: info@rolawn.co.uk

RTS Ltds
UK Sales
Daisy Dene
Inglewhite Road
Goosnargh
Preston PR3 2EB
Permaloc edging
Tel: 01772 780234
Fax: 01772 780234
Website: www.rtslimited.uk.com
E-mail: rtslimited@hotmail.co.uk

Saint-Gobain Pipelines Plc
Lows Lane
Stanton-by-Dale
Ilkeston
Derbyshire DE7 4QU
Ductile iron access covers, gullies and grates
Tel: 0115 9305000
Fax: 0115 9329513
Website: www.saint-gobain-pipelines.co.uk
E-mail:
enquiries.uk.pipelines@saint-gobain.com

Scotts UK Ltd
Paper Mill Lane
Bramford
Ipswich
Suffolk IP8 4BZ
Fertilizers and chemicals
Tel: 01473 830492
Fax: 01473 830046
Website: www.scottsprofessional.co.uk
E-mail: prof.sales@scotts.com

Sleeper Supplies Ltd
PO Box 1377
Kirk Sandall
Doncaster DN3 1XT
Sleeper supplier
Tel: 0845 230 8866
Fax: 0845 230 8877
Website: www.sleeper-supplies.co.uk
E-mail: sales@sleeper-supplies.co.uk

SMP Playgrounds Ltd
Ten Acre Lane
Thorpe
Egham
Surrey TW20 8RJ
Playground equipment
Tel: 01784 489100
Fax: 01784 431079
Website: www.smp.co.uk
E-mail: sales@smp.co.uk

Southern Conveyors
Unit 2 Denton Slipways Site
Wharf Road
Gravesend DAI2 2RU
Conveyors
Tel: 01474 564145
Fax: 01474 568036

Spadeoak Construction Co Ltd
Town Lane
Wooburn Green
High Wycombe
Bucks HP10 0PD
Macadam contractors
Tel: 01628 529421
Fax: 01628 810509
Website: www.spadeoak.co.uk
E-mail: email@spadeoak.co.uk

Steelway-Fensecure Ltd
Queensgate Works
Bilston Road
Wolverhampton
West Midlands WV2 2NJ
Fencing
Tel: 01902 451733
Fax: 01902 452256
Website: www.steelway.co.uk
E-mail: sales@steelway.co.uk

Sugg Lighting Ltd
Sussex Manor Business Park
Gatwick Road
Crawley
West Sussex RH10 9GD
Lighting
Tel: 01293 540111
Fax: 01293 540114
Website: www.sugglighting.co.uk
E-mail: sales@sugglighting.co.uk

Swan Plant Ltd
Haig Road
Parkgate Industrial Estate
Knutsford
Cheshire WA16 8DX
Landscape maintenance plant hire
Tel: 01565 650567
Fax: 01565 650769
Website: www.swan-services.co.uk
E-mail: info@swan-services.co.uk

Tarmac Southern Ltd
London Sales Office
24 Park Royal Road
London NW10 7JW
Ready mixed concrete
Tel: 020 8965 1864
Fax: 020 8965 1863

Tensar International
New Wellington Street
Blackburn
Lancashire BB2 4PJ
Erosion control, soil stabilisation
Tel: 01254 266862
Fax: 01254 266868
Website: www.tensar-international.com
E-mail: sales@tensar.co.uk

Terranova Lifting Ltd
Terranova House
Bennett Road
Reading
Berks RG2 OQX
Crane hire
Tel: 0118 931 2345
Fax: 0118 931 4114
Website: www.terranova-lifting.co.uk
E-mail: cranes@terranovagroup.co.uk

Thompson Landscapes
610 Goffs Lane
Goffs Oak
Herts EN7 5EP
Swimming pools
Tel: 01707 873444
Fax: 01707 873447
Website: www.thompsonlandscapes.com
E-mail: info@thompsonlandscapes.com

Townscape Products Ltd
Fulwood Road South
Sutton-in-Ashfield
Nottinghamshire NG17 2JZ
Hard landscaping
Tel: 01623 513355
Fax: 01623 440267
Website: www.townscape-products.co.uk
E-mail: sales@townscape-products.co.uk

Tubex Ltd
Aberaman Park
Aberaman
South Wales CF44 6DA
Plant protection, tree guards
Tel: 01685 888000
Fax: 01685 888001
Website: www.tubex.com
E-mail: plantcare@tubex.com

Turf Management Systems
Dromenach Farm
Seven Hills Road
Iver Heath
Buckinghamshire SL0 0PA
Specialist turf systems
Tel: 01895 834411
Fax: 01895 834892
Website: www.boughton-loam.co.uk
E-mail: enquiries@boughton-loam.co.uk

Wade International Ltd
Third Avenue
Halstead
Essex CO9 2SX
Stainless steel drainage
Tel: 01787 475151
Fax: 01787 475579
Website: www.wadedrainage.co.uk
E-mail: sales@wade.eu

Wavin Plastics Ltd
Parsonage Way
Chippenham
Wiltshire SN15 5PN
Drainage products
Tel: 01249 766600
Fax: 01249 443286
Website: www.wavin.co.uk
E-mail: info@wavin.co.uk

White Horse Contractors Ltd
Blakes Oak Farm
Lodge Hill
Abingdon
Oxfordshire OX14 2JD
Drainage
Tel: 01865 736272
Fax: 01865 326176
Website: www.whitehorsecontractors.co.uk
E-mail: whc@whitehorsecontractors.co.uk

Wicksteed Leisure Ltd
Digby Street
Kettering
Northamptonshire NN16 8YJ
Play equipment
Tel: 01536 517028
Fax: 01536 410633
Website: www.wicksteed.co.uk
E-mail: sales@wicksteed.co.uk

Woodscape Ltd
Church Works
Church Street
Church
Lancashire BB5 4JT
Street furniture
Tel: 01254 383322
Fax: 01254 381166
Website: www.woodscape.co.uk
E-mail: sales@woodscape.co.uk

Wybone Ltd
Mason Way
Platts Common Industrial Estate
Hoyland
Barnsley
South Yorkshire S74 9TF
Street furniture
Tel: 01226 744010
Fax: 01226 350105
Website: www.wybone.co.uk
E-mail: sales@wybone.co.uk

Yeoman Aggregates Ltd
(Foster Yeoman Ltd)
Marston House
Frome
Somerset BA11 5DU
Aggregates
Tel: 020 8896 6820
Fax: 020 8896 6829
Website: www.foster-yeoman.co.uk
E-mail: sales@foster-yeoman.co.uk

SPON'S PRICEBOOKS 2008

Spon's Architects' and Builders' Price Book 2008

Davis Langdon

The most detailed, professionally relevant source of construction price information currently available anywhere.

The 2008 edition provides an extensive overhaul of prices in line with buoyant market conditions. It includes an expanded Approximate Estimates section and a new section on Sustainability. It also includes the following new items: Additional high quality balustrades, PVC wall linings, Rainscreen cladding, Resin-bonded gravel, SAS suspended ceilings, Self-levelling screeds, 'Speedfit' pipes and fittings, Tapered insulation boards, Waterproofing membranes, Window refurbishment

HB & electronic package * £135
978 0-415-42448-6 1000pp approx.

Spon's External Works and Landscape Price Book 2008

Davis Langdon

The only comprehensive source of information for detailed external works and landscape costs.

The 2008 edition gives an expanded Approximate Estimates section, and new items including Professional and amenity sports pitch construction, Costs of tender production, Small unit concrete kerbs, Ornamental trellises, Borehole drilling, irrigation sensors, and an irrigation water recycling system. And more on excavation and filling, landscape chemicals prices and applications, and linear drainage.

HB & electronic package * £105
978 0-415-42444-8 424pp approx.

Spon's Mechanical and Electrical Services Price Book 2008

Davis Langdon, Mott Green Wall

The only comprehensive and best annual services engineering price book available for the UK.

As well as an extensive overhaul of prices, updates to the 2008 edition include an expanded Approximate Estimate section, updated coverage of sustainability and of communication and data, along with additions to both the ductwork and lifts sections. High efficiency plant has been refreshed and the CHP and Biomass plant has been reviewed.

HB & electronic package * £135
978 0-415-42442-4 640pp approx.

Spon's Civil Engineering and Highway Works Price Book 2008

Davis Langdon

The 21st edition, in its easy-to-read format, incorporates a general review throughout, updated in line with the latest CIJC wage agreement, and includes a restructuring of the highway works unit costs section to align with the latest revisions to the *Manual of Contract Documents for Highway Works.*

HB & electronic package * £145
978 0-415-42446-2 800pp approx.

Receive a free electronic version when you order any Spon 2008 Price Book with estimating software to help you produce tender documents, customise data, perform word searches and simple calculations.
Visit www.pricebooks.co.uk

To Order: Tel: +44 (0) 8700 768853, or +44 (0) 1264 343071 Fax: +44 (0) 1264 343005, or
Post: Taylor and Francis Customer Services, Thomson Publishing Services, Cheriton House, Andover, Hants, SP10 5BE, UK Email: book.orders@tandf.co.uk

For a complete listing of all our titles visit www.tandf.co.uk

Taylor & Francis
Taylor & Francis Group

How to Use this Book

INTRODUCTION

First-time users of *Spon's External Works and Landscape Price Book* and others who may not be familiar with the way in which prices are compiled may find it helpful to read this section before starting to calculate the costs of landscape works.
The cost of an item of landscape construction or planting is made up of many components:

- the cost of the product;
- the labour and additional materials needed to carry out the job;
- the cost of running the contractor's business.

These are described more fully below.

IMPORTANT NOTES ON THE PROFIT ELEMENT OF RATES IN THIS BOOK

The rates shown in the Approximate Estimates and Measured Works sections of this book do not contain a profit element unless the rate has been provided by a sub-contractor. Prices are shown at cost. This is the cost including the costs of company overhead required to run perform this task or project. For reference please see the tables on pages 8 - 9.

Analysed Rates Versus Sub-contractor Rates

As a general rule if a rate is shown as an analysed rate in the Measured Works section, i.e. it has figures shown in the columns other than the "Unit" and "Total Rate" column, it can be assumed that this has no profit or overhead element and that this calculation is shown as a direct labour/material supply/plant task being performed by a contractor.

On the other hand if a rate is shown as a Total Rate only, this would normally be a sub-contractor's rate and would contain the profit and overhead for the sub-contractor.

The foregoing applies for the most part to the Approximate Estimates section, however in some items there may be an element of sub-contractor rates within direct works build-ups.

As an example of this, to excavate, lay a base and place a macadam surface, a general landscape contractor would normally perform the earthworks and base installation. The macadam surfacing would be performed by a specialist sub-contractor to the landscape contractor.

The Approximate Estimate for this item uses rates from the Measured Works section to combine the excavation, disposal, base material supply and installation with all associated plant. There is no profit included on these elements. The cost of the surfacing, however, as supplied to us in the course of our annual price enquiry from the macadam sub-contractor, would include the sub-contractor's profit element.

The landscape contractor would add this to his section of the work but would normally apply a mark up at a lower percentage to the sub-contract element of the composite task.
Users of this book should therefore allow for the profit element at the prevailing rates. Please see the worked examples below and the notes on overheads for further clarification.

INTRODUCTORY NOTES ON PRICING CALCULATIONS USED IN THIS BOOK

There are two pricing sections to this book:

Major Works - Average Contract Value £100 000.00 - £300 000.00

Minor Works - Average Contract Value £10 000.00 - £70 000.00

Typical Project Profiles Used in this Book

	Major Works	Minor Works
Contract value	100 000.00 - £300 000.00	£10 000.00 - £70 000.00
Labour rate (see page 5)	£18.50 per hour	£21.00 per hour
Labour rate for maintenance contracts	£15.50 per hour	-
Number of site staff	18	6 - 9
Project area	6000 m^2	1200 m^2
Project location	Outer London	Outer London
Project components	50% hard landscape 50% soft landscape and planting	20% hard landscape 80% soft landscape and planting
Access to works areas	Very good	Very good
Contract	Main contract	Main contract
Delivery of materials	Full loads	Part loads

As explained in more detail later the prices are generally based on wage rates and material costs current at spring 2007. They do not allow for preliminary items, which are dealt with in the Preliminaries section of this book, or for any Value Added Tax which may be payable.

Adjustments should be made to standard rates for time, location, local conditions, site constraints and any other factors likely to affect the costs of a specific scheme.

Term contracts for general maintenance of large areas should be executed at rates somewhat lower than those given in this section.

There is now a facility available to readers that enables a comparison to be made between the level of prices given and those for projects carried out in regions other than outer London; this is dealt with on page 21.

Units of Measurement

The units of measurement have been varied to suit the type of work and care should be taken when using any prices to ascertain the basis of measurement adopted.

The prices per unit of area for executing various mechanical operations are for work in the following areas and under the following conditions:

Prices per m^2	relate to areas not exceeding 100 m^2 (any plan configuration)
Prices per 100 m^2	relate to areas exceeding 100 m^2 but not exceeding 1/4 ha (generally clear areas but with some subdivision)
Prices per ha	relate to areas over 1/4 ha (clear areas suitable for the use of tractors and tractor-operated equipment)

The prices per unit area for executing various operations by hand generally vary in direct proportion to the change in unit area.

Approximate Estimates

These are combined Measured Work prices which give an approximate cost for a complete section of landscape work. For example, the construction of a car park comprises excavation, levelling, road-base, surfacing and marking parking bays. Each of these jobs is priced separately in the Measured Works section, but a comprehensive price is given in the Approximate Estimates section, which is intended to provide a quick guide to the cost of the job. It will be seen that the more items that go to make up an approximate estimate, the more possibilities there are for variations in the PC prices and the user should ensure that any PC price included in the estimate corresponds to his specification.

Worked example

In many instances a modular format has been use in order to enable readers to build up a rate for a required task. The following table describes the trench excavation pipe laying and backfilling operations as contained in section R12 of this book.

Pipe laying 100 m

	£
Excavate for drain 150 mm wide x 450 mm; excavated material to spoil heaps on site by machine	434.84
Lay flexible plastic pipe 100 mm wide	136.63
Backfilling with gravel rejects, blinding with sand and topping with 150 mm topsoil	229.27
Total	**800.54**

The figures given in Spon's External Works and Landscape Price Book are intended for general guidance, and if a significant variation appears in any one of the cost groups, the Price for Measured Work should be re-calculated.

Measured Works

Prime Cost: Commonly known as the 'PC'. Prime Cost is the actual price of the material item being addressed such as paving, shrubs, bollards or turves, as sold by the supplier. Prime Cost is given 'per square metre', 'per 100 bags' or 'each' according to the way the supplier sells his product. In researching the material prices for the book we requested that the suppliers price for full loads of their product delivered to a site close to the M25 in London. Spon's rates do not include VAT. Some companies may be able to obtain greater discounts on the list prices than those shown in the book. Prime Cost prices for those products and plants which have a wide cost range will be found under the heading of Market Prices in the main sections of this book, so that the user may select the product most closely related to his specification.

Materials: The PC material plus the additional materials required to fix the PC material. Every job needs materials for its completion besides the product bought from the supplier. Paving needs sand for bedding, expansion joint strips and cement pointing; fencing needs concrete for post setting and nails or bolts; tree planting needs manure or fertilizer in the pit, tree stakes, guards and ties. If these items were to be priced out separately, Spon's External Works and Landscape Price Book (and the Bill of Quantities) would be impossibly unwieldy, so they are put together under the heading of Materials.

Labour: This figure covers the cost of planting shrubs or trees, laying paving, erecting fencing etc. and is calculated on the wage rate (skilled or unskilled) and the time needed for the job. Extras such as highly skilled craft work, difficult access, intermittent working and the need for labourers to back up the craftsman all add to the cost. Large regular areas of planting or paving are cheaper to install than smaller intricate areas, since less labour time is wasted moving from one area to another.

Plant, consumable stores and services: This rather impressive heading covers all the work required to carry out the job which cannot be attributed exactly to any one item. It covers the use of machinery ranging from JCB's to shovels and compactors, fuel, static plant, water supply (which is metered on a construction site), electricity and rubbish disposal. The cost of transport to site is deemed to be included elsewhere and should be allowed for elsewhere or as a preliminary item. Hired plant is calculated on an average of 36 hours working time per week.

Sub-contract rates: Where there is no analysis against an item, this is deemed to be a rate supplied by a sub-contractor. In most cases these are specialist items where most external works contractors would not have the expertise or the equipment to carry out the task described. It should be assumed that sub-contractor rates include for the sub-contractor's profit. An example of this may be found for the tennis court rates in section Q26.

Labour Rates Used in this Edition

The rates for labour used in this edition have been based on surveys carried out on a cross section of external works contractors. These rates include for company overheads such as employee administration, transport insurance, and on costs such as National Insurance. Please see the calculation of these rates shown on page 8. The rates do not include for profit.

Overheads: An allowance for this is included in the labour rates which are described above. The general overheads of the contract such as insurance, site huts, security, temporary roads and the statutory health and welfare of the labour force are not directly assignable to each item, so they are distributed as a percentage on each, or as a separate preliminary cost item. The contractor's and sub-contractor's profits are not included in this group of costs. Site overheads, which will vary from contract to contract according to the difficulties of the site, labour shortages, inclement weather or involvement with other contractors, have not been taken into account in the build up of these rates, while overhead (or profit) may have to take into account losses on other jobs and the cost to the contractor of unsuccessful tendering.

ADJUSTMENT AND VARIATION OF THE RATES IN THIS BOOK

It will be appreciated that a variation in any one item in any group will affect the final Measured Works price. Any cost variation must be weighed against the total cost of the contract. A small variation in Prime Cost where the items are ordered in thousands, may have more effect on the total cost than a large variation on a few items. A change in design that necessitates the use of earth-moving equipment which must be brought to the site for that one job will cause a dramatic rise in the contract cost. Similarly, a small saving on multiple items will provide a useful reserve to cover unforeseen extras.

Worked examples

A variation in the Prime Cost of an item can arise from a specific quotation.

For example:
The PC of 900 x 600 x 50 mm precast concrete flags is given as £3.08 each which equates to £5.70/m^2, and to which the costs of bedding and pointing are added to give a total material cost of £9.52/m^2.
Further costs for labour and mechanical plant give a resultant price of £16.92/m^2.
If a quotation of £6.00/m^2 was received from a supplier of the flags the resultant price would be calculated as £16.92 less the original cost (£5.70) plus the revised cost (£6.00) to give £17.22/m^2.

A variation will also occur if, for example, the specification changes from 50 light standard trees to 50 extra heavy standard trees. In this case the Prime Cost will increase due to having to buy extra heavy standard trees instead of light standard stock.

Original price: The prime cost of an Acer platanoides 8 -10 cm bare root tree is given as £12.52, to which the cost of handling and planting, etc. is added to give an overall cost of £18.15. Further costs for labour and mechanical plant for mechanical excavation of a 600 x 600 x 600 mm pit (£3.39), a single tree stake (£7.71), importing "Topgrow" compost in 75 litre bags (£1.59), backfilling with imported topsoil (£4.76) and disposing of the excavated material (£4.71), give a resultant price of £40.31. Therefore the total cost of 50 light standard trees is £2015.50.

Revised price: The prime cost of an Acer platanoides root ball 14 -16 cm is given as £31.50, however, taking into account the additional staking, extra labour for handling and planting, (£27.25) a larger tree pit, 900 x 900 x 600 mm (£7.59) a larger stake (£11.11) and more topsoil,(£10.78) and the disposal (£10.60) this cost gives a total unit price of £98.83. This is significantly more than the light standard version. Therefore the total cost of 50 extra heavy trees is £4941.50 and the additional cost is £2926.00.

This example of the effects of changing the tree size illustrates that caution is needed when revising tender prices, as merely altering the prime cost of an item will not accurately reflect the total cost of the revised item.

HOW THIS BOOK IS UPDATED EACH YEAR

The basis for this book is a database of Material, Labour, Plant and Sub-contractor resources each with its own area of the database.

Material, Plant and Sub-contractor Resources

Each year the suppliers of each material, plant or sub-contract item are approached and asked to update their prices to those that will prevail in September of that year.
These resource prices are individually updated in the database. Each resource is then linked to one or many tasks. The tasks in the task library section of the database is automatically updated by changes in the resource library. A quantity of the resource is calculated against the task. The calculation is generally performed once and the links remain in the database.
On occasions where new information or method or technology are discovered or suggested, these calculations would be revisited. A further source of information are simple time and production observations made during the course of the last year.

Labour Resource Update

Most tasks, except those shown as sub-contractor rates (see above), employ an element of labour. The Data Department at Davis Langdon conducts ongoing research into the costs of labour in various parts of the country. Tasks or entire sections are then re-examined and recalculated. Comments on the rates published in this book are welcomed and may be submitted to the contact addresses shown in the preface.

PART 1

General

This part of the book contains the following sections:

FAST ESTIMATE
IT'S ABOUT TIME

Advanced pricing tools for Spon Price Book users

Everything you need to price work quickly and accurately using your own custom rates together with your Spon Price libraries.

- **Resource Library** - Keep your costs up-to-date and in one place
- **Suppliers** - Allocate resources to individual suppliers
- **Price Lists** - Access to the latest product prices from the UK's largest merchant
- **Composite Resources** - A real time saver for mortar; labour gangs etc.
- **Price Book Options** - Access industry standard Price books from Spon Press
- **Item Library** - Maintain your own rate library (with full build ups)
- **Composite Items** - Save time by selecting a single item covering multiple tasks
- **BQ Preparation** - Deliver professional documents to clients and colleagues
- **Resource Requirements** - Specify exactly what you need to complete the job
- **Purchase Ordering -** Convert job requirements into enquiries and orders
- **Flexibility** - Works for builders, contractors, sub-contractors and specialists
- **Price Excel® and CITE Bills** - Import, price and export bills supplied in these industry standard formats.
- **Output to Microsoft® Applications** - Export data to Word®, Excel®, Project®, etc.
- **Subcontractor Bills** - Generate bills from selected resources or tagged items.

Fast Estimate shares the same clear pricing processes as Spon Estimating, but with more powerful pricing tools to give you outstanding control over your tender documents.

Try it today - Download your FREE trial copy:

www.estek.co.uk

Common Arrangement of Work Sections

The main work sections relevant to landscape work and their grouping

A Preliminaries/general conditions

- A10 Project particulars
- A11 Tender and contract documents
- A12 The site/existing buildings
- A13 Description of the work
- A20 The contract/sub-contract
- A30 Employer's requirements: tendering/sub-letting/supply
- A31 Employer's requirements: provision, content and use of documents
- A32 Employer's requirements: management of the works
- A33 Employer's requirements: quality standards/control
- A34 Employer's requirements: security/safety/protection
- A35 Employer's requirements: specific limitations on method/sequence/timing/ use of site
- A36 Employer's requirements: facilities/temporary works/ services
- A37 Employer's requirements: operation/maintenance of the finished building
- A40 Contractor's general cost items: management and staff
- A41 Contractor's general cost items: site accommodation
- A42 Contractor's general cost items: services and facilities
- A43 Contractor's general cost items: mechanical plant
- A44 Contractor's general cost items: temporary works
- A50 Works/products by/on behalf of the employer
- A51 Nominated sub-contractors
- A52 Nominated suppliers
- A53 Work by statutory authorities/ undertakers
- A54 Provisional work
- A55 Dayworks
- A60 Preliminaries/general conditions for demolition contract
- A61 Preliminaries/general conditions for investigation/survey contract
- A62 Preliminaries/general conditions for piling/embedded retaining wall contract
- A63 Preliminaries/general conditions for landscape contract
- A70 General specification requirements for work package

B Complete buildings/structures/ units

- B10 Prefabricated buildings/structures
- B11 Prefabricated building units

C Existing site/buildings/services

- C10 Site survey
- C11 Ground investigation
- C12 Underground services survey

D Groundwork

- D11 Soil stabilization
- D20 Excavating and filling
- D41 Crib walls/gabions/reinforced earth

E In situ concrete/large precast concrete

- E05 In situ concrete construction generally
- E10 Mixing/casting/curing in situ concrete
- E20 Formwork for in situ concrete
- E30 Reinforcement for in situ concrete

F Masonry

- F10 Brick/block walling
- F20 Natural stone rubble walling

- F21 Natural stone ashlar walling/dressings
- F22 Cast stone ashlar walling/dressings
- F30 Accessories/sundry items for brick/block/stone walling
- F31 Precast concrete sills/lintels/copings/features

G Structural/carcassing metal/timber

- G31 Prefabricated timber unit decking

H Cladding/covering

- H51 Natural stone slab cladding/features
- H52 Cast stone slab cladding/features

J Waterproofing

- J10 Specialist waterproof rendering
- J20 Mastic asphalt tanking/damp proofing
- J21 Mastic asphalt roofing/insulation/finishes
- J22 Proprietary roof decking with asphalt finish
- J30 Liquid applied tanking/damp proofing
- J31 Liquid applied waterproof roof coatings
- J40 Flexible sheet tanking/damp proofing
- J44 Sheet linings for pools/lakes/waterways

M Surface finishes

- M10 Cement:sand/concrete screeds
- M20 Plastered/rendered/roughcast coatings
- M40 Stone/concrete/quarry/ceramic tiling/mosaic
- M60 Painting/clear finishing

P Building fabric sundries

- P30 Trenches/pipeways/pits for buried engineering services

Q Paving/planting/fencing/site furniture

- Q10 Kerbs/edgings/channels/paving accessories
- Q20 Granular sub-bases to roads/pavings
- Q21 In situ concrete roads/pavings
- Q22 Coated macadam/asphalt roads/pavings
- Q23 Gravel/hoggin/woodchip roads/pavings
- Q24 Interlocking brick/block roads/pavings
- Q25 Slab/brick/sett/cobble pavings
- Q26 Special surfacings/pavings for sport/general amenity
- Q30 Seeding/turfing
- Q31 Planting
- Q32 Planting in special environments
- Q35 Landscape maintenance
- Q40 Fencing
- Q50 Site/street furniture/equipment

R Disposal systems

- R12 Drainage belowground
- R13 Land drainage

S Piped supply systems

- S10 Cold water
- S14 Irrigation
- S15 Fountains/water features

V Electrical supply/power/lighting systems

- V41 Street/area/flood lighting

Free Updates

with three easy steps…

1. Register today on www.pricebooks.co.uk/updates

2. We'll alert you by email when new updates are posted on our website

3. Then go to www.pricebooks.co.uk/updates and download.

All four Spon Price Books – *Architects' and Builders'*, *Civil Engineering and Highway Works*, *External Works and Landscape* and *Mechanical and Electrical Services* – are supported by an updating service. Three updates are loaded on our website during the year, in November, February and May. Each gives details of changes in prices of materials, wage rates and other significant items, with regional price level adjustments for Northern Ireland, Scotland and Wales and regions of England. The updates terminate with the publication of the next annual edition.

As a purchaser of a Spon Price Book you are entitled to this updating service for the 2008 edition – free of charge. Simply register via the website www.pricebooks.co.uk/updates and we will send you an email when each update becomes available.

If you haven't got internet access we can supply the updates by an alternative means. Please write to us for details: Spon Price Book Updates, Taylor & Francis Marketing Department, 2 Park Square, Milton Park, Abingdon, Oxfordshire, OX14 4RN.

Find out more about spon books
Visit www.tandfbuiltenvironment.com for more details.

New books from Spon

The following books can be ordered directly from Taylor & Francis or from your nearest good bookstore

Representing Landscape Architecture, M Treib
hbk 9780415700429, pbk 9780415700436
Landscape and Sustainability, J Benson
hbk 9780415404433
Project Management Demystified 3rd edition, G Reiss
pbk 9780415421638
Principles of Project Finance, W Tan
hbk 9780415415767, pbk 9780415415774
Cultured Landscape, S Harvey *et al.*
hbk 0419250301, pbk 0419250409
Garden History, T Turner
hbk 0415317487
Spon's Estimating Costs Guide to Minor Landscaping, Gardening and External Works, B Spain
pbk 0415344107
Effective Press Relations for the Built Environment, H Elias
hbk 0415348668, pbk 0415348676
Dynamic Landscape, N Dunnett
hbk 0415256208
Outdoor Lighting Guide, Institution of Lighting Engineers
hbk 0415370078
Urban Sound Environments, J Kang
hbk 0415358574
Elements of Visual Design in the Landscape 2nd edition, S Bell
hbk 041532517X, pbk 0415325188
Contemporary Landscapes of Contemplation, R Krinke *et al.*
hbk 041570068X, pbk 0415700698
Spon's First Stage Estimating Handbook 2nd edition, B Spain
pbk 0415386195
Spon's Irish Construction Price Book 2nd edition, Franklin + Andrews
hbk 0415344093
Urban Drainage 2nd edition, D Butler *et al.*
hbk 041530606X, pbk 0415306078
Predicting Outdoor Sound, K Attenborough *et al.*
hbk 0419235108

Please send your order to:
Marketing Department, Taylor & Francis, 2 Park Square, Milton Park Abingdon, Oxfordshire, OX14 4RN.
or visit www.tandfbuiltenvironment.com for more details.

Labour Rates Used in this Edition

Based on surveys carried out on a cross section of external works contractors, the following rates for labour have been used in this edition.
These rates include for company overheads such as employee administration, transport insurance, and on-costs such as National Insurance.
The rates do not include for profit.

The rates for all labour used in this edition are as follows:

Major Works

General contracting	£18.50/hour
Maintenance contracting	£15.50/hour

Minor Works

General contracting	£21.00/hour

eBooks – at www.eBookstore.tandf.co.uk

A library at your fingertips!

eBooks are electronic versions of printed books. You can store them on your PC/laptop or browse them online.

They have advantages for anyone needing rapid access to a wide variety of published, copyright information.

eBooks can help your research by enabling you to bookmark chapters, annotate text and use instant searches to find specific words or phrases. Several eBook files would fit on even a small laptop or PDA.

NEW: Save money by eSubscribing: cheap, online access to any eBook for as long as you need it.

Annual subscription packages

We now offer special low-cost bulk subscriptions to packages of eBooks in certain subject areas. These are available to libraries or to individuals.

For more information please contact
webmaster.ebooks@tandf.co.uk

We're continually developing the eBook concept, so keep up to date by visiting the website.

www.eBookstore.tandf.co.uk

Computation of Labour Rates Used in this Edition

Different organizations will have varying views on rates and costs, which will in any event be affected by the type of job, availability of labour and the extent to which mechanical plant can be used. However this information should assist the reader to:

(1) compare the prices to those used in his own organisation
(2) calculate the effect of changes in wage rates or prices of materials
(3) calculate prices for work similar to the examples given

From 25 June 2007 basic weekly rates of pay for craft and general operatives are £379.08 and £285.09 respectively; to these rates have been added allowances for the items below in accordance with the recommended procedure of the Chartered Institute of Building in its "Code of Estimating Practice". The resultant hourly rates are £13.47 and £10.09 for craft operatives and general operatives respectively.

The items for which allowances have been made are:

- Lost time
- Construction Industry Training Board Levy
- Holidays with pay
- Accidental injury, retirement and death benefits scheme
- Sick pay
- National Insurance
- Severance pay and sundry costs
- Employer's liability and third party insurance

The tables that follow illustrate how the above hourly rates have been calculated. Productive time has been based on a total of 1801.8 hours worked per year for daywork calculations above and for 1953.54 hours worked per year for the all-in labour rates (including 5 hours per week average overtime) for the all in labour rates below.

How the Labour Rate has been Calculated

A survey of typical landscape/external works companies indicates that they are currently paying above average wages for multi skilled operatives regardless of specialist or supervisory capability. In our current overhaul of labour constants and costs we have departed from our previous policy of labour rates based on national awards and used instead a gross rate per hour for all rate calculations. Estimators can readily adjust the rate if they feel it is inappropriate for their work.

The productive labour of any organisation must return their salary plus the cost of the administration which supports the labour force.

The labour rate used in this edition is calculated as follows:

Team size
A three man team with annual salaries as shown and allowances for National Insurance, uniforms, site tools and a company vehicle, returns a basic labour rate of £13.76 per hour.

Basic labour rate calculation for Spon's 2008 Major Works

Standard Working Hours per year (2007/8) = 1802.5 (figure from calculation on the previous page)

Actual Working Hours per year (2007/8) = 1750.5 (allows for 52 hours sick)

	Number of	**Labour Team**	**NI**	**Nett Cost**	**Uniforms**	**Site Tools**	**Total**
Foreman	1	21500.00	2558.50	24058.50	100.00	100.00	
Craftsman	1	18250.00	2171.75	20421.75	100.00	100.00	
Labourer	1	14250.00	1695.75	15945.75	100.00	100.00	
Overtime Hours							
nominal hours /annum	50.00	18.42	2.19	1030.78			
at 1.5 times normal	50.00	15.64	1.86	874.97			
standard rate	50.00	12.21	1.45	683.19			
Total Staff in Team	3.00			63014.95	300.00	300.00	**63,614.95**
Vehicle Costs	**Working Days**	**£/Day**					
Inclusive of fuel, insurances, etc.	240.00	36.00					8,640.00
						TOTAL	**£72,254.95**

The basic average labour rate excluding overhead costs (below) is

$$\frac{£72254.95}{\text{(Total staff in team)} \times \text{(Working hours per year)}}$$

= £13.76

Overhead costs

Add to the above basic rate the company overhead costs for a small company as per the table below. These costs are absorbed by the number of working men multiplied by the number of working hours supplied in the table below. This then generates an hourly overhead rate which is added to the Nett cost rate above.

Illustrative overhead for small company employing 9 - 18 landscape operatives

Cost Centre	Number of	Cost	Total
MD: Salary only: excludes profits	1	40000.00	40000.00
Contracts managers incl. vehicle.	1	40000.00	40000.00
Secretary	1.00	15000.00	15000.00
Book Keeper	1.00	15000.00	15000.00
Rental	12.00	500.00	6000.00
Insurances	1.00	5000.00	5000.00
Telephone and Mobiles	12.00	250.00	3000.00
Office Equipment	1.00	1000.00	1000.00
Stationary	12.00	50.00	600.00
Advertising	12.00	200.00	2400.00
Other vehicles not allocated to contract teams	1.00	9000.00	9000.00
Other consultants	1.00	6000.00	6000.00
Accountancy	1.00	2500.00	2500.00
Lights heating water	12.00	200.00	2400.00
Other	1.00	10000.00	10000.00
TOTAL OFFICE OVERHEAD			157900.00

The final labour rate used is then generated as follows:

$$\text{Nett labour rate} + \frac{\text{Total Office Overhead}}{\text{(Working hours per year x Labour resources employed)}}$$

Total nr of site staff	Admin Cost per hour	Total Rate per Man Hour
6	13.65	£28.79
9	9.10	£23.78
12	6.83	£21.28
15	5.46	£19.77
18	4.55	£18.77

Hourly labour rates used for Spon's External Works and Landscape 2008 (Major Works)

General contracting	£18.50
Maintenance contracting	£15.50

Minor Works

- The minor works labour calculation is based on the above but the administration cost of the organisation is £97000.00 per annum
- The owner of the business does not carry out site works and is assumed to be paid a salary of £35000.00 per annum; this figure constitutes part of the £97000.00 shown
- There are between 6 and 9 site operatives

Hourly labour rates used for Spon's External Works and Landscape 2008 (Minor Works)

General contracting	£21.00

Example of Computation using National Wage Awards - Building Craft and General Operatives
(not used for rate calculations in this book)

			Craft Operatives		General Operatives	
Effective from 25 June 2007			£	£	£	£
Wages at standard basic rate						
Productive time	44.3	wks	379.08	16793.24	285.09	12629.49
Lost time allowance	0.9	wks	379.09	341.17	285.09	256.58
Non-productive overtime	5.8	wks	568.62	3298.00	427.64	2480.28
				20432.41		15366.35
National Working Rules	45.2	wks		-		-
Sick pay	1	wk		-		-
CITB allowance (0.50% of payroll)	1	year		115.27		86.69
Holiday pay	4.2	wks	452.04	1898.57	339.96	1427.83
Public Holiday	1.6	wks	452.04	723.26	339.96	543.94
Travelling allowance (say 30 km)	226	dys	2.47		2.47	
Fare allowance (say 30 km)	226	dys	6.25		6.25	
Employer's contributions to						
EasyBuild Stakeholder Pension (death and accident cover is provided free)	52	wks	5.00	260.00	5.00	260.00
National Insurance (average weekly payment)	48	wks	41.69	2001.12	28.18	1352.64
				25460.63		19037.45
Severance pay and sundry costs	Plus		1.5%	381.46	1.5%	285.56
				25812.09		19323.01
Employer's liability and third-party insurance	Plus		2.0%	516.24	2.0%	386.46
Total cost per annum				26328.33		19709.47
Total cost per hour				**13.47**		**10.09**

Notes:

1. Absence due to sickness has been assumed to be for periods not exceeding 3 days for which no payment is due (Working Rule 20.7.3).
2. EasyBuild Stakeholder Pension effective from 1 July 2002. Death and accident benefit cover is provided free of charge. Taken as £4.00/week average as range increased for 2006/09 wage award.
3. All N.I. Payments are at not-contracted out rates applicable from April 2007. National Insurance is paid for 48 complete weeks (52 wks - 4.2 wks) based on employer making regular monthly payments into the Template holiday pay scheme and by doing so the employer achieves National Insurance savings on holiday wages.

Calculation of Annual Hours Worked

Number of actual hours worked yearly				
Normal working hours (46.2 wks x 39 hrs)				1801.80
Less sick				
1 week				-39.00
				1762.80
Less time lost for inclement weather @ 2%				-35.26
				1727.54
Overtime hours				
52 wks x 5 hrs (overtime)			260.00	
Less				
Annual holidays	4.2 wks x 5 hrs	21.00		
Public holidays	1.6 wks x 5 hrs	8.00		
Sickness	1 week x 5 hrs	5.00	-34.00	
				226.00
Number of actual hours worked yearly				1953.54

Computation of the Cost of Materials

Percentages of default waste are placed against material resources within the supplier database. These range from 2.5% (for bricks) to 20% for topsoil.
An allowance for the cost of unloading, stacking etc. should be added to the cost of materials.

The following are typical hours of labour for unloading and stacking some of the more common building materials.

Material	Unit	Labourer hour
Cement	tonne	0.67
Lime	tonne	0.67
Common bricks	1000	1.70
Light engineering bricks	1000	2.00
Heavy engineering bricks	1000	2.40

ESSENTIAL READING FROM TAYLOR AND FRANCIS

Spon's Estimating Costs Guide to Small Groundworks, Landscaping and Gardening

Second Edition

Especially written for contractors and small businesses carrying out small works, Spon's Estimating Costs Guide to Small Groundworks, Landscaping and Gardening contains accurate information on thousands of rates, each broken down to labour, material overheads and profit.

This is the first book to focus on garden maintenance work on blocks of flats and individual houses, schools and sports fields, garden makeovers, laying patios and paths, small land drainage schemes on farmland, and small-scale local authority maintenance work.

More than just a price book, it gives easy-to-read, professional advice on setting up and running a business including help on producing estimates faster, and keeping estimates accurate and competitive.

Suitable for any size firm from one-man-band to established business, this book contains valuable commercial and cost information that contractors can't afford to be without.

December 2007: 216x138 mm: 304 pages
PB: 978-0-415-43442-3: £27.99

To Order: Tel: +44 (0) 1264 343071 Fax: +44 (0) 1264 343005, or
Post: Taylor and Francis Customer Services, Thomson Publishing Services, Cheriton House, Andover, Hants, SP10 5BE, UK Email: book.orders@tandf.co.uk

For a complete listing of all our titles visit:
www.sponpress.com

Taylor & Francis
Taylor & Francis Group

Computation of Mechanical Plant Costs

Plant used within resource calculations in this book have been used according to the following principles:

1. Plant which is non-pedestrian is fuelled by the plant hire company.
2. Operators of excavators above the size of 5 tonnes are provided by the plant hire company and no labour calculation is allowed for these operators within the rate.
3. Three to five tonne excavators and dumpers are operated by the landscape contractor and additional labour is shown within the calculation of the rate.
4. Small plant is operated by the landscape contractor.
5. No allowance for delivery or collection has been made within the rates shown.
6. Downtime is shown against each plant type either for non-productive time or mechanical down time.

The following tables lists the plant resources used in this year's book.

Prices shown reflect 32 working hours per week.

Plant Supplied by Road Equipment Ltd, Tel: 01784 256565

Description	Unit	Supply Quantity	Rate Used £	Down Time %
Excavator 360 Tracked 21 tonne *fuelled and operated*	hour	1	41.25	20
Excavator 360 Tracked 5 tonne *fuelled only*	hour	1	10.93	20
Excavator 360 Tracked 7 tonne *fuelled only*	hour	1	13.13	20
Mini Excavator JCB 803 Rubber Tracks Self Drive *fuelled only*	hour	1	8.75	20
Mini Excavator JCB 803 Steel Tracks Self Drive *fuelled only*	hour	1	7.50	20
JCB 3Cx 4x4 Sitemaster, *fuelled and operated*	hour	1	30.00	20
JCB 3Cx 4x4 Sitemaster + Breaker *fuelled and operated*	hour	1	31.25	20
Dumper 3 tonne, Thwaites Self Drive *fuelled only*	hour	1	3.44	20
Dumper 5 tonne, Thwaites Self Drive *fuelled only*	hour	1	4.69	20
Dumper 6 tonne, Thwaites Self Drive *fuelled only*	hour	1	5.00	20
Manitou Telehandler *fuelled only*	hour	1	18.75	40

Plant Supplied by HSS Hire

Description	Unit	Supply Quantity	Supply Cost £	Rate Used £	Down Time %
Post Hole Borer, 2 man, weekly rate (rate based on 25 productive hours)	hour	40	96.00	3.84	37.5
Petrol Poker Vibrator + 50 mm head (rate based on 25 productive hours)	hour	40	43.00	1.72	37.5
Vibrating Plate Compactor (rate based on 25 productive hours)	hour	40	60.00	2.40	37.5
Petrol Masonry Saw Bench, 350 mm (rate based on 25 productive hours)	hour	40	112.00	4.48	37.5
Diamond Blade Consumable, 230 mm	mm	1	19.00	19.00	-
Alloy Access Tower, 5.2 m	hour	40	138.50	27.70	-
Oxy-acetylene Cutting Kit (rate based on 16 productive hours)	hour	40	90.00	5.63	80
Heavy-duty Petrol Breaker, 5 hrs/day (rate based on 25 productive hours)	hour	40	138.00	5.52	37.5
Cultivator, 110 kg (rate based on 25 productive hours)	hour	40	178.00	7.12	37.5
Rotavator + Tractor + Operator (rate based on 32 productive hours)	hour	40	394.50	12.33	-
Vibrating Roller, 136 kg/10.1 Kn, 4 hrs/day (rate based on 20 productive hours)	hour	40	122.00	6.10	25
Vibrating Roller, 470 kg/18.7 Kn, 4 hrs/day (rate based on 20 productive hours)	hour	40	150.00	7.50	25

Landfill Tax

Waste Liable at the Lower Rate

Group	Description of material	Conditions	
1	Rocks and soils	Naturally occurring	includes clay, sand, gravel, sandstone, limestone, crushed stone, china clay, construction stone, stone from the demolition of buildings or structures, slate, topsoil, peat, silt and dredgings glass includes fritted enamel, but excludes glass fibre and glass reinforced plastics.
2	Ceramic or concrete materials		ceramics includes bricks, bricks and mortar, tiles, clay ware, pottery, china and refractories concrete includes reinforced concrete, concrete blocks, breeze blocks and aircrete blocks, but excludes concrete plant washings
3	Minerals	Processed or prepared, not used	moulding sands excludes sands containing organic binders clays includes moulding clays and clay absorbents, including Fuller's earth and bentonite man-made mineral fibres includes glass fibres, but excludes glass-reinforced plastic and asbestos silica, mica and mineral abrasives
4	Furnace slags		vitrified wastes and residues from thermal processing of minerals where, in either case, the residue is both fused and insoluble slag from waste incineration
5	Ash		comprises only bottom ash and fly ash from wood, coal or waste combustion excludes fly ash from municipal, clinical, and hazardous waste incinerators and sewage sludge incinerators
6	Low activity inorganic compound		comprises only titanium dioxide, calcium carbonate, magnesium carbonate, magnesium oxide, magnesium hydroxide, iron oxide, ferric hydroxide, aluminium oxide, aluminium hydroxide & zirconium dioxide
7	Calcium sulphate	Disposed of either at a site not licensed to take putrescible waste or in a containment cell which takes only calcium sulphate	includes gypsum and calcium sulphate based plasters, but excludes plasterboard
8	Calcium hydroxide and brine	Deposited in brine cavity	
9	Water	Containing other qualifying material in suspension	

Volume to Weight Conversion Factors

Waste category	Typical waste types	Cubic metres to tonne - multiply by:	Cubic yards to tonne - multiply by:
Inactive or inert waste	Largely water insoluble and non or very slowly biodegradable: e.g. sand, subsoil, concrete, bricks, mineral fibres, fibreglass etc.	1.5	1.15

Notes:

The Landfill tax came into operation on 1 October 1996. It is levied on operators of licensed landfill sites at the following rates with effect from 1 April 2007:

£2 per tonne - Inactive or inert wastes.
Included are soil, stones, brick, plain and reinforced concrete, plaster and glass.

£24 per tonne - All other taxable wastes.
Included are timber, paint and other organic wastes generally found in demolition work, builders skips etc.

From 1 April 2008 the rate of for "all other taxable wastes" will be increased by £8 to £32 per tonne. The rate for "inactive or inert wastes" will increase to £2.50 per tonne.

Mixtures containing wastes not classified as inactive or inert will not qualify for the lower rate of tax unless the amount of non-qualifying material is small and there is no potential for pollution. Water can be ignored and the weight discounted.

Calculating the Weight of Waste
There are two options:

- If licensed sites have a weighbridge, tax will be levied on the actual weight of waste.
- If licensed sites do not have a weighbridge, tax will be levied on the permitted weight of the lorry based on an alternative method of calculation based on volume to weight factors for various categories of waste.

For further information contact the National Advisory Service, Telephone: 0845 010 9000

PART 2

Approximate Estimates-Major Works

This part of the book contains the following sections:

ESSENTIAL READING FROM TAYLOR AND FRANCIS

Project Management Demystified

Third Edition

Geoff Reiss

Concise, practical and entertaining to read, this excellent introduction to project management is an indispensable book for both professionals and students working in or studying project management in business, engineering or the public sector.

Approachable and written in an easy-to-use style, it shows readers how, where and when to use the various project management techniques, demonstrating how to achieve efficient management of human, material and financial resources to make major contributions to projects and be an appreciated and successful project manager.

This new edition contains expanded sections on programme management, portfolio management, and the public sector. An entirely new chapter covers the evaluation, analysis and management of risks and issues. A much expanded section explores the rise and utilisation of methodologies like Prince2.

Contents: Introduction. Setting the Stage. Getting the Words in the Right Order. Nine Steps to a Successful Project. The Scope of the Project and its Objectives. Project Planning. A Fly on the Wall. Resource Management. Progress Monitoring and Control. Advanced Critical-Path Topics. The People Issues. Risk and Issue Management. Terminology

June 2007: 234x156mm: 240 pages
PB: 978-0-415-42163-8: £19.99

To Order: Tel: +44 (0) 1264 343071 Fax: +44 (0) 1264 343005, or
Post: Taylor and Francis Customer Services, Thomson Publishing Services, Cheriton House, Andover, Hants, SP10 5BE, UK Email: book.orders@tandf.co.uk

For a complete listing of all our titles visit:
www.tandf.co.uk

Taylor & Francis
Taylor & Francis Group

Cost Indices

The purpose of this section is to show changes in the cost of carrying out landscape work (hard surfacing and planting) since 1990. It is important to distinguish between costs and tender prices: the following table reflects the change in cost to contractors but does not necessarily reflect changes in tender prices. In addition to changes in labour and material costs, which are reflected in the indices given below, tender prices are also affected by factors such as the degree of competition at the time of tender and in the particular area where the work is to be carried out, the availability of labour and materials, and the general economic situation. This can mean that in a period when work is scarce tender prices may fall despite the fact that costs are rising, and when there is plenty of work available, tender prices may increase at a faster rate than costs.

The Constructed Cost Index

A Constructed Cost Index based on PSA Price Adjustment Formulae for Construction Contracts (Series 2). Cost indices for the various trades employed in a building contract are published monthly by HMSO and are reproduced in the technical press.

The indices comprise 49 Building Work indices plus seven 'Appendices' and other specialist indices. The Building Work indices are compiled by monitoring the cost of labour and materials for each category and applying a weighting to these to calculate a single index.

Although the PSA indices are prepared for use with price-adjustment formulae for calculating reimbursement of increased costs during the course of a contract, they also present a time series of cost indices for the main components of landscaping projects. They can therefore be used as the basis of an index for landscaping costs.

The method used here is to construct a composite index by allocating weightings to the indices representing the usual work categories found in a landscaping contract, the weightings being established from an analysis of actual projects. These weightings totalled 100 in 1976 and the composite index is calculated by applying the appropriate weightings to the appropriate PSA indices on a monthly basis, which is then compiled into a quarterly index and rebased to 1976 = 100.

Constructed Landscaping (Hard Surfacing and Planting) Cost Index

Based on approximately 50% soft landscaping area and 50% hard external works.
1976 = 100

Year	First Quarter	Second Quarter	Third Quarter	Fourth Quarter	Annual Average
1990	330	334	353	357	344
1991	356	356	364	364	360
1992	364	365	376	378	371
1993	380	381	383	384	382
1994	387	388	394	397	391
1995	399	404	413	413	407
1996	417	419	428	430	424
1997	430	432	435	442	435
1998	442	445	466	466	455
1999	466	468	488	490	478
2000	493	496	513	514	504
2001	512	514	530	530	522
2002	531	540	573	574	555
2003	577	582	603	602	591
2004	602	610	639	639	623
2005	641	648	684	683	664
2006	688	692	710	710	710
2007	711*				

* *Provisional*

This index is updated every quarter in Spon's Price Book Update. The updating service is available, free of charge, to all purchasers of Spon's Price Books (complete the reply-paid card enclosed).

Regional Variations

Prices in Spon's External Works and Landscape Price Book are based upon conditions prevailing for a competitive tender in the outer London area. For the benefit of readers, this edition includes regional variation adjustment factors which can be used for an assessment of price levels in other regions.

Special further adjustment may be necessary when considering city centre or very isolated locations.

Region	Adjustment Factor
Outer London	1.00
Inner London	1.06
East Anglia	0.91
East Midlands	0.89
Northern	0.94
Northern Ireland	0.72
North West	0.89
Scotland	0.94
South East	0.94
South West	0.90
Wales	0.87
West Midlands	0.90
Yorkshire and Humberside	0.93

The following example illustrates the adjustment of prices for regions other than outer London, by use of regional variation adjustment factors.

		£
A.	Value of items priced using Spon's External Works and Landscape Price Book	100 000
B.	Adjustment to value of A. to reflect Northern Region price level 100 000 x 0.94	94 000

ESSENTIAL READING FROM TAYLOR AND FRANCIS

2nd Edition

Spon's Irish Construction Price Book

Franklin + Andrews

This new edition of *Spon's Irish Construction Price Book*, edited by Franklin + Andrews, is the only complete and up-to-date source of cost data for this important market.

- All the materials costs, labour rates, labour constants and cost per square metre are based on current conditions in Ireland
- Structured according to the new Agreed Rules of Measurement (second edition)
- 30 pages of Approximate Estimating Rates for quick pricing

This price book is an essential aid to profitable contracting for all those operating in Ireland's booming construction industry.

Franklin + Andrews, Construction Economists, have offices in 100 countries and in-depth experience and expertise in all sectors of the construction industry.

April 2004: 246x174 mm: 448 pages
HB: 0-415-34409-3: £135.00

To Order: Tel: +44 (0) 1264 343071 Fax: +44 (0) 1264 343005, or
Post: Taylor and Francis Customer Services, Thomson Publishing Services, Cheriton House, Andover, Hants, SP10 5BE, UK Email: book.orders@tandf.co.uk

For a complete listing of all our titles visit:
www.sponpress.com

Taylor & Francis
Taylor & Francis Group

Approximate Estimates

Prices in this section are based upon the Prices for Measured Works, but allow for incidentals which would normally be measured separately in a Bill of Quantities. They do not include for Preliminaries which are priced elsewhere in this book.

Items shown as sub-contract or specialist rates would normally include the specialist's overhead and profit. All other items which could fall within the scope of works of general landscape and external works contractors would not include profit.

Based on current commercial rates, profits of 15% to 35% may be added to these rates to indicate the likely "with profit" values of the tasks below. The variation quoted above is dependent on the sector in which the works are taking place - domestic, public or commercial.

ESSENTIAL READING FROM TAYLOR AND FRANCIS

Spon's Estimating Cost Guide to Minor Works, Alterations and Repairs to Fire, Flood, Gale and Theft Damage

Third Edition

Bryan Spain

All the cost data you need to keep your estimating accurate, competitive and profitable.

Do you work on jobs between £50 and £50,000? Then this book is for you.

Specially written for contractors, quantity surveyors and clients carrying out small works, this book contains accurate information on thousands of rates each broken down to labour, material overheads and profit.More than just a price book, it gives easy-to-read professional advice on setting up and running a business.

December 2005: 216x138 mm: 320 pages
PB: 0-415-38213-0: £29.99

To Order: Tel: +44 (0) 1264 343071 Fax: +44 (0) 1264 343005, or
Post: Taylor and Francis Customer Services, Thomson Publishing Services, Cheriton House, Andover, Hants, SP10 5BE, UK Email: book.orders@tandf.co.uk

For a complete listing of all our titles visit:
www.tandf.co.uk

Taylor & Francis
Taylor & Francis Group

PRELIMINARIES

Item Excluding site overheads and profit	Unit	**Total rate £**
PRELIMINARIES		
CONTRACT ADMINISTRATION AND MANAGEMENT		
Prepare and maintain Health and Safety file, prepare risk and Coshh assessments, method statements, works programmes before the start of works on site; project value:		
£35000.00	nr	**240.00**
£75000.00	nr	**360.00**
£100000.00	nr	**480.00**
£200000.00 - £500000.00	nr	**1200.00**
Contract management		
Site management by site surveyor carrying out supervision and administration duties only; inclusive of on costs		
full time management	week	**1200.00**
managing one other contract	week	**600.00**
Parking		
Parking expenses where vehicles do not park on the site area; per vehicle		
Metropolitan area; city centre	week	**200.00**
Metropolitan area; outer areas	week	**160.00**
Suburban restricted parking areas	week	**40.00**
Congestion charging		
London only	week	**40.00**
SITE SETUP AND SITE ESTABLISHMENT		
Site fencing; supply and erect temporary protective fencing and remove at completion of works		
Cleft chestnut paling		
1.20 m high 75 larch posts at 3 m centres	100 m	**997.00**
1.50 m high 75 larch posts at 3 m centres	100 m	**1225.00**
"Heras" fencing		
2 week hire period	100 m	**728.00**
4 week hire period	100 m	**1008.00**
8 week hire period	100 m	**1568.00**
12 week hire period	100 m	**2128.00**
16 week hire period	100 m	**2688.00**
24 week hire period	100 m	**3808.00**
Site compounds		
Establish site administration area on reduced compacted Type 1 base; 150 mm thick; allow for initial setup of site office and welfare facilities for site manager and maximum 8 site operatives		
temporary base or hard standing; 20 m x 10 m	200 m^2	**2155.00**
To hardstanding area above; erect office and welfare accommodation; allow for furniture, telephone connection and power; restrict access using temporary safety fencing to the perimeter of the site administration area		
Establishment; delivery and collection costs only		
Site office and chemical toilet; 20 m of site fencing	nr	**743.00**
Site office and toilet; 40 m of site fencing	nr	**808.00**
Site office and toilet; 60 m of site fencing	nr	**873.00**
Weekly hire rates of site compound equipment		
Site office 2.4 x 3.6 m; chemical toilet; 20 m of site fencing	week	**87.00**
Site office 4.8 x 2.4 m; chemical toilet; 20 m of site fencing; armoured store 2.4 x 3.6 m	week	**103.00**
Site office 2.4 x 3.6 m; chemical toilet; 40 m of site fencing	week	**115.00**
Site office 2.4 x 3.6 m; chemical toilet; 60 m of site fencing	week	**143.00**

PRELIMINARIES

Item Excluding site overheads and profit	Unit	**Total rate £**
SITE SETUP AND SITE ESTABLISHMENT - cont'd		
Removal of site compound		
Remove base; fill with topsoil and reinstate to turf on completion of site works	200 m^2	**1272.00**
Surveying and setting out		
Note in all instances a quotation should be obtained from a surveyor. The following cost projections are based on the costs of a surveyor using EDM (electronic distance measuring equipment) where an average of 400 different points on site are recorded in a day. The surveyed points are drawn in a CAD application and the assumption used that there is one day of drawing up for each day of on-site surveying.		
Survey existing site using laser survey equipment and produce plans of existing site conditions		
Site up to 1000 to 10000 m^2 with sparse detail and features; boundary and level information only; average 5 surveying stations	nr	**750.00**
Site up to 1000 m^2 with dense detail such as buildings, paths, walls and existing vegetation; average 5 survey stations	nr	**1500.00**
Site up to 4000 m^2 with dense detail such as buildings, paths, walls and existing vegetation; average 10 survey stations	nr	**2250.00**

DEMOLITION AND SITE CLEARANCE

Item Excluding site overheads and profit	Unit	**Total rate £**
DEMOLITION AND SITE CLEARANCE		
Clear existing scrub vegetation, including shrubs and hedges and burn on site; spread ash on specified area		
By machine		
light scrub and grasses	100 m²	**18.00**
undergrowth brambles and heavy weed growth	100 m²	**39.00**
small shrubs	100 m²	**39.00**
As above but remove vegetation to licensed tip	100 m²	**39.00**
By hand	100 m²	**82.00**
As above but remove vegetation to licensed tip	100 m²	**59.00**
Fell trees on site; grub up roots; all by machine; remove debris to licensed tip		
trees 600 girth	each	**215.00**
trees 1.5 - 3.00 m girth	each	**622.00**
trees over 3.00 m girth	each	**1415.00**
Demolish existing surfaces		
Break up plain concrete slab; remove to licensed tip		
150 thick	m²	**9.00**
200 thick	m²	**11.00**
300 thick	m²	**17.00**
Break up plain concrete slab and preserve arisings for hardcore; break down to maximum size of 200 mm x 200 mm and transport to location maximum distance 50 m		
150 thick	m²	**6.00**
200 thick	m²	**7.00**
300 thick	m²	**11.00**
Break up reinforced concrete slab and remove to licensed tip		
150 thick	m²	**13.00**
200 thick	m²	**17.00**
300 thick	m²	**25.00**
Break out existing surface and associated 150 mm thick granular base load to 20 tonne muck away vehicle for removal off site		
Macadam 70 mm thick	m²	**7.70**
Macadam surface 150 thick	m²	**11.00**
Block paving 50 thick	m²	**7.30**
Block paving 80 thick	m²	**8.00**
Demolish existing free standing walls; grub out foundations; remove arisings to tip		
Brick wall; 112 mm thick		
300 mm high	m	**9.60**
500 mm high	m	**10.60**
Brick wall; 225 mm thick		
300 mm high	m	**12.00**
500 mm high	m	**17.80**
1.00 m high	m	**23.80**
1.20 m high	m	**25.80**
1.50 m high	m	**32.60**
1.80 m high	m	**36.00**
Site clearance - generally		
Clear away light fencing and gates (chain link, chestnut paling, light boarded fence or similar) and remove to licensed tip	100 m	**119.00**

GROUNDWORK

Item Excluding site overheads and profit	Unit	Total rate £
GROUNDWORK		
Cut and strip by machine turves 50 thick		
Load to dumper or palette by hand, stack on site not exceeding 100 m travel to stack	100 m²	**50.00**
As above but all by hand, including barrowing	100 m²	**432.00**
Prices for excavating and reducing to levels are for work in light or medium soils; multiplying factors for other soils are as follows		
clay	1.5	-
compact gravel	1.2	-
soft chalk	2.0	-
hard rock	3.0	-
Excavating topsoil for preservation		
Remove topsoil average depth 300 mm; deposit in spoil heaps not exceeding 100 m travel to heap; treat once with Paraquat-Diquat weedkiller; turn heap once during storage		
by machine	m³	**11.00**
by hand	m³	**109.00**
Excavate to reduce levels; remove spoil to dump not exceeding 100 m travel; all by machine		
0.25 m deep average	m²	**2.00**
0.30 m deep	m²	**3.00**
1.0 m deep using 21 tonne 360 tracked excavator	m²	**5.00**
1.0 m deep using JCB	m²	**7.00**
Excavate to reduce levels; remove spoil to dump not exceeding 100 m travel; all by hand		
0.10 m deep average	m²	**10.00**
0.20 m deep average	m²	**20.00**
0.30 m deep average	m²	**30.00**
1.0 m deep average	m²	**100.00**
Extra for carting spoil to licensed tip offsite (20 tonnes)	m³	**22.00**
Extra for carting spoil to licensed tip offsite (by skip - machine loaded)	m³	**45.00**
Extra for carting spoil to licensed tip offsite (by skip - hand loaded)	m³	**99.00**
Spread excavated material to levels in layers not exceeding 150 mm; using scraper blade or bucket		
Average thickness 100 mm	m²	**1.00**
Average thickness 100 mm but with imported topsoil	m²	**3.00**
Average thickness 200 mm	m²	**1.00**
Average thickness 200 mm but with imported topsoil	m²	**6.00**
Average thickness 250 mm	m²	**1.00**
Average thickness 250 mm but with imported topsoil	m²	**7.00**
Extra for work to banks exceeding 30 slope	30%	-
Rip subsoil using approved sub-soiling machine to a depth of 600 mm below topsoil at 600 mm centres; all tree roots and debris over 150 x 150 mm to be removed; cultivate ground where shown on drawings to depths as shown		
100 mm	100 m²	**8.00**
200 mm	100 m²	**8.00**
300 mm	100 m²	**9.00**
400 mm	100 m²	**9.00**
Form landform from imported material; mounds to falls and grades to receive turf or seeding treatments; material dumped by 20 tonne lorry to location; volume 100 m³ spread over a 200 m² area; thickness varies from 500 mm to 150 mm		
Imported recycled topsoil		
by large excavator	100 m³	**2257.26**
by 5 tonne excavator	100 m³	**2400.26**
by 3 tonne excavator	100 m³	**2511.26**

GROUNDWORK

Item Excluding site overheads and profit	Unit	**Total rate £**
TRENCHES		
Excavate trenches for foundations; trenches 225 mm deeper than specified foundation thickness; pour plain concrete foundations GEN 1 10 N/mm and to thickness as described; disposal of excavated material off site; dimensions of concrete foundations		
250 mm wide x 250 mm deep	m	**12.00**
300 mm wide x 300 mm deep	m	**15.00**
400 mm wide x 600 mm deep	m	**35.00**
600 mm wide x 400 mm deep	m	**37.00**
Excavate trenches for foundations; trenches 225 mm deeper than specified foundation thickness; pour plain concrete foundations GEN 1 10 N/mm and to thickness as described; disposal of excavated material off site; dimensions of concrete foundations		
250 mm wide x 250 mm deep	m^3	**218.00**
300 mm wide x 300 mm deep	m^3	**157.00**
600 mm wide x 400 mm deep	m^3	**151.00**
400 mm wide x 600 mm deep	m^3	**135.00**
Excavate trenches for services; grade bottoms of excavations to required falls; remove 20% of excavated material off site		
Trenches 300 wide		
600 mm deep	m	**5.89**
750 mm deep	m	**7.55**
900 mm deep	m	**9.06**
1.00 m deep	m	**9.97**
Ha-ha		
Excavate ditch 1200 mm deep x 900 mm wide at bottom battered one side 45° slope; excavate for foundation to wall and place GEN 1 concrete foundation 150 x 500 mm; construct one and a half brick wall (brick PC £300.00/1000) battered 10° from vertical, laid in cement:lime:sand (1:1:6) mortar in English garden wall bond; precast concrete coping weathered and throated set 150 mm above ground on high side; rake bottom and sides of ditch and seed with low maintenance grass at 35 g/m²		
wall 600 high	m	**242.00**
wall 900 high	m	**302.00**
wall 1200 high	m	**376.00**
Excavate ditch 1200 mm deep x 900 mm wide at bottom battered one side 45° slope; place deer or cattle fence 1.20 m high at centre of excavated trench; rake bottom and sides of ditch and seed with low maintenance grass at 35 g/m²		
fence 1200 high	m	**22.00**

GROUND STABILIZATION

Item Excluding site overheads and profit	Unit	Total rate £
GROUND STABILIZATION		
Excavate and grade banks to grade to receive stabilization treatments below; removal of excavated material not included		
By 21 tonne excavator	m^3	0.28
By 7 tonne excavator	m^3	1.67
By 5 tonne excavator	m^3	1.71
By 3 tonne excavator	m^3	2.82
Excavate sloped bank to vertical to receive retaining or stabilization treatment priced below; allow 1.00 m working space at top of bank; backfill working space and remove balance of arisings off site		
Bank height 1.00; excavation by 5 tonne excavator		
10 degree	m	93.00
30 degree	m	32.00
45 degree	m	20.00
60 degree	m	20.00
Bank height 1.00; excavation by 5 tonne excavator arisings moved to stockpile on site		
10 degree	m	16.00
30 degree	m	8.00
45 degree	m	6.00
60 degree	m	5.00
Bank height 2.00; excavation by 5 tonne excavator		
10 degree	m	365.00
30 degree	m	120.00
45 degree	m	72.00
60 degree	m	44.00
Bank height 2.00; excavation by 5 tonne excavator but arisings moved to stockpile on site		
10 degree	m	56.00
30 degree	m	23.00
45 degree	m	17.00
60 degree	m	14.00
Bank height 3.00; excavation by 21 tonne excavator		
20 degree	m	378.00
30 degree	m	240.00
45 degree	m	136.00
60 degree	m	80.00
As above but arisings moved to stockpile on site		
20 degree	m	41.00
30 degree	m	28.00
45 degree	m	13.00
60 degree	m	9.00
Bank stabilization; erosion control mats		
Excavate fixing trench for erosion control mat 300 x 300 mm; backfill after placing mat selected below	m	6.00
To anchor trench above, place mat to slope to be retained; allow extra over to the slope for the area of matting required to the anchor trench		
Rake only bank to final grade; lay erosion control solution; sow with low maintenance grass at 35 g/m^2; spread imported topsoil to BS3882 25 mm thick incorporating medium grade sedge peat at 3 kg/m^2 and fertilizer at 30 g/m^2; water lightly		
unseeded "Eromat Light"	m^2	3.00
seeded "Covamat Standard"	m^2	6.00
Lay Greenfix biodegradable pre-seeded erosion control mat, fixed with 6 x 300 mm steel pegs at 1.0 m centres	m^2	6.00
"Tensar" open textured erosion mat	m^2	5.00
Excavate trench and lay foundation concrete 1:3:6 600 x 300 mm deep		
By machine	m	27.00
By hand	m	38.00

GROUND STABILIZATION

Item Excluding site overheads and profit	Unit	**Total rate £**
Concrete block retaining walls; Stepoc		
Excavate trench 750 mm deep and lay concrete foundation 600 wide x 600 deep; construct Forticrete precast hollow concrete block wall with 450 mm below ground laid all in accordance with manufacturer's instructions; fix reinforcing bar 12 mm as work proceeds; fill blocks with concrete 1:3:6 as work proceeds		
Walls 1.00 m high		
Type 256; 400 x 225 x 256 mm	m	**191.00**
Type 190; 400 x 225 x 190 mm	m	**166.00**
As above but 1.50 m high		
Type 256; 400 x 225 x 256 mm	m	**248.00**
Type 190; 400 x 225 x 190 mm	m	**215.00**
Walls 1.50 m high as above but with foundation 1.80 m wide x 600 mm deep		
Type 256; 400 x 225 x 256 mm	m	**296.00**
Type 190; 400 x 225 x 190 mm	m	**260.00**
Walls 1.80 m high		
Type 256; 400 x 225 x 256 mm	m	**320.00**
Type 190; 400 x 225 x 190 mm	m	**281.00**
On foundation measured above, supply and RCC precast concrete "L" shaped units, constructed all in accordance with manufacturer's instructions; backfill with approved excavated material compacted as the work proceeds		
1250 mm high x 1000 mm wide	m	**155.00**
2400 mm high x 1000 mm wide	m	**242.00**
3750 mm high x 1000 mm wide	m	**283.00**
Grass concrete; bank revetments; excluding bulk earthworks		
Bring bank to final grade by machine; lay regulating layer of Type 1; lay filter membrane; lay 100 mm drainage layer of broken stone or approved hardcore 28 - 10 mm size; blind with 25 mm sharp sand; lay grass concrete surface (price does not include edgings or toe beams); fill with approved topsoil and fertilizer at 35 g/m²; seed with dwarf rye based grass seed mix		
Grasscrete in situ reinforced concrete surfacing GC 1, 100 thick	m²	**74.00**
Grasscrete in situ reinforced concrete surfacing GC 2, 150 thick	m²	**82.00**
Grassblock 103 open matrix blocks 406 x 406 x 103	m²	**75.00**
Gabion walls		
Construct revetment of Maccaferri Ltd Reno mattress gabions laid on firm level ground, tightly packed with broken stone or concrete and securely wired; all in accordance with manufacturers' instructions		
gabions 6 m x 2 m x 0.17 m; one course	m²	**67.00**
gabions 6 m x 2 m x 0.17 m; two courses	m²	**90.00**
Timber Log Retaining Walls		
Excavate trench 300 wide to one third of the finished height of the retaining walls below; lay 100 mm hardcore; fix machine rounded logs set in concrete 1:3:6; remove excavated material from site; fix geofabric to rear of timber logs; backfill with previously excavated material set aside in position; all works by machine		
100 mm diameter logs		
500 mm high (constructed from 1.80 m lengths)	m	**77.00**
1.20 mm high (constructed from 1.80 m lengths)	m	**114.00**
1.60 mm high (constructed from 2.40 m lengths)	m	**166.00**
2.00 mm high (constructed from 3.00 m lengths)	m	**223.00**
As above but 150 mm diameter logs		
1.20 mm high (constructed from 1.80 m lengths)	m	**179.00**
1.60 mm high (constructed from 2.40 m lengths)	m	**220.00**
2.00 mm high (constructed from 3.00 m lengths)	m	**289.00**
2.40 mm high (constructed from 3.60 m lengths)	m	**373.00**
Timber crib wall		
Excavate trench to receive foundation 300 mm deep; place plain concrete foundation 150 thick in 11.50 N/mm² concrete (sulphate-resisting cement); construct timber crib retaining wall and backfill with excavated spoil behind units		
Timber crib wall system, average 4.2 m high	m	**206.00**
Timber crib wall system, average 2.2 m high	m	**153.00**

GROUND STABILIZATION

Item Excluding site overheads and profit	Unit	**Total rate £**
GROUND STABILIZATION - cont'd		
Geogrid soil reinforcement (Note: measured per metre run at the top of the bank)		
Backfill excavated bank; lay Tensar geogrid; cover with excavated material as work proceeds		
2.00 m high bank; geogrid at 1.00 m vertical lifts		
10 degree slope	m	**166.00**
20 degree slope	m	**90.00**
30 degree slope	m	**64.00**
2.00 m high bank; geogrid at 0.50 m vertical lifts		
45 degree slope	m	**63.00**
60 degree slope	m	**50.00**
3.00 m high bank		
10 degree slope	m	**166.00**
20 degree slope	m	**90.00**
30 degree slope	m	**64.00**
45 degree slope	m	**150.00**

IN SITU CONCRETE

Item Excluding site overheads and profit	Unit	Total rate £
IN SITU CONCRETE		
Mix concrete on site; aggregates delivered in 20 tonne loads; deliver mixed concrete to location by mechanical dumper distance 25 m		
1:3:6	m^3	**81.70**
1:2:4	m^3	**92.10**
As above but ready mixed concrete		
10 N/mm^2	m^3	**99.50**
15 N/mm^2	m^3	**106.00**
Mix concrete on site; aggregates delivered in 20 tonne loads; deliver mixed concrete to location by barrow distance 25 m		
1:3:6	m^3	**125.00**
1:2:4	m^3	**136.00**
As above but aggregates delivered in 1 tonne bags		
1:3:6	m^3	**153.00**
1:2:4	m^3	**163.00**
As above but ready mixed concrete		
10 N/mm^2	m^3	**143.00**
15 N/mm^2	m^3	**149.00**
As above but concrete discharged directly from ready mix lorry to required location		
10 N/mm^2	m^3	**99.00**
15 N/mm^2	m^3	**105.00**
Excavate foundation trench mechanically; remove spoil offsite; lay 1:3:6 site mixed concrete foundations; distance from mixer 25 m; depth of trench to be 225 mm deeper than foundation to allow for 3 underground brick courses priced separately		
Foundation size		
200 mm deep x 400 mm wide	m	**19.00**
300 mm deep x 500 mm wide	m	**27.00**
400 mm deep x 400 mm wide	m	**27.00**
400 mm deep x 600 mm wide	m	**40.00**
600 mm deep x 600 mm wide	m	**57.00**
As above but hand excavation and disposal to spoil heap 25 m by barrow and off site by grab lorry		
200 mm deep x 400 mm wide	m	**38.00**
300 mm deep x 500 mm wide	m	**62.00**
400 mm deep x 400 mm wide	m	**62.00**
400 mm deep x 600 mm wide	m	**69.00**
600 mm deep x 600 mm wide	m	**128.00**
Excavate foundation trench mechanically; remove spoil offsite; lay ready mixed concrete GEN1 discharged directly from delivery lorry to location; depth of trench to be 225 mm deeper than foundation to allow for 3 underground brick courses priced separately; foundation size		
200 mm deep x 400 mm wide	m	**18.00**
300 mm deep x 500 mm wide	m	**27.00**
400 mm deep x 400 mm wide	m	**26.00**
400 mm deep x 600 mm wide	m	**39.00**
600 mm deep x 600 mm wide	m	**55.00**
Reinforced concrete wall to foundations above (site mixed concrete)		
up to 1.00 m high x 200 thick	m^2	**130.00**
up to 1.00 m high x 300 thick	m^2	**148.00**
Reinforced concrete wall to foundations above (ready mix concrete RC35)		
1.00 m high x 200 thick	m^2	**124.00**
1.00 m high x 300 thick	m^2	**136.00**

IN SITU CONCRETE

Item Excluding site overheads and profit	Unit	Total rate £
IN SITU CONCRETE - cont'd		
Grade ground to levels and falls; excavate to reduce levels for concrete slab; lay waterproof membrane, 100 mm hardcore and form concrete slab reinforced with A 142 mesh in 1:2:4 site mixed concrete to thickness; remove excavated material off site		
Concrete 1:2:4 site mixed		
100 thick	m^2	**63.04**
150 thick	m^2	**70.23**
250 thick	m^2	**84.61**
300 thick	m^2	**91.79**
Concrete ready mixed GEN 2		
100 thick	m^2	**61.25**
150 thick	m^2	**67.55**
250 thick	m^2	**80.13**
300 thick	m^2	**86.42**

BRICK/BLOCK WALLING

Item Excluding site overheads and profit	Unit	**Total rate £**
BRICK/BLOCK WALLING		
Excavate foundation trench 500 deep; remove spoil to dump off site; lay site mixed concrete foundations 1:3:6 350 x 150 thick; construct half brick wall with one brick piers at 2.0 m centres; laid in cement:lime:sand (1:1:6) mortar with flush joints; fair-face one side; DPC two courses underground; engineering brick in cement:sand (1:3) mortar; coping of headers on end		
Wall 900 high above DPC		
in engineering brick (class B) - £260.00/1000	m	**157.00**
in sandfaced facings - £300.00/1000	m	**179.00**
in reclaimed bricks - £800.00/1000	m	**212.00**
Excavate foundation trench 400 deep; remove spoil to dump off site; lay GEN 1 concrete foundations 450 wide x 250 thick; construct one brick wall with one and a half brick piers at 3.0 m centres; all in English garden wall bond; laid in cement:lime:sand (1:1:6) mortar with flush joints, fair-face one side; DPC two courses engineering brick in cement:sand (1:3) mortar; precast concrete coping 152 x 75 mm		
Wall 900 high above DPC		
in engineering brick (class B) - £260.00/1000	m	**264.00**
in sandfaced facings - £300.00/1000	m	**272.00**
in reclaimed bricks PC £800.00/1000	m	**402.00**
Wall 1200 high above DPC		
in engineering brick (class B) - £260.00/1000	m	**294.00**
in sandfaced facings - £300.00/1000	m	**304.00**
in reclaimed bricks PC £800.00/1000	m	**398.00**
Wall 1800 high above DPC		
in engineering brick (class B) - £260.00/1000	m	**427.00**
in sandfaced facings - £300.00/1000	m	**442.00**
in reclaimed bricks PC £800.00/1000	m	**583.00**
Excavate foundation trench 450 deep; remove spoil to dump off site; lay GEN 1 concrete foundations 600 x 300 thick; construct one and a half brick wall with two thick brick piers at 3.0 m centres; all in English garden wall bond; laid in cement:lime:sand (1:1:6) mortar with flush joints; fair-face one side; DPC two courses engineering brick in cement:sand (1:3) mortar; coping of headers on edge		
Wall 900 high above DPC		
in engineering brick (class B) - £260.00/1000	m	**533.00**
in sandfaced facings - £300.00/1000	m	**279.00**
in rough stocks - £450.00/1000	m	**295.00**
Wall 1200 high above DPC		
in engineering brick (class B) - £260.00/1000	m	**302.00**
in sandfaced facings - £300.00/1000	m	**311.00**
in rough stocks - £450.00/1000	m	**383.00**
Wall 1800 high above DPC		
in engineering brick (class B) - £260.00/1000	m	**374.00**
in sandfaced facings - £300.00/1000	m	**449.00**
in rough stocks - £450.00/1000	m	**478.00**
Excavate foundation trench 450 deep; remove spoil to dump off site; lay GEN 1 concrete foundations 600 x 300 thick; construct wall of concrete blocks		
Solid blocks 7 N/mm^2		
100 mm thick	m^2	**56.00**
140 mm thick	m^2	**63.00**
100 mm blocks laid "on flat"		
215 mm thick	m^2	**110.00**
Hollow blocks filled with concrete		
215 mm thick	m^2	**89.00**
Hollow blocks but with steel bar cast into the foundation		
215 mm thick	m^2	**110.00**
Excavate foundation trench 450 deep; remove spoil to dump off site; lay GEN 1 concrete foundations 600 x 300 thick; construct wall of concrete block with brick face 112.5 thick to stretcher bond; place stainless steel ties at 4 nr/m^2 of wall face		
Solid blocks 7 N/mm^2 with strecher bond brick face; bricks PC £500.00/1000		
100 mm thick	m^2	**171.00**
140 mm thick	m^2	**177.00**

ROADS AND PAVINGS

Item Excluding site overheads and profit	Unit	Total rate £
ROADS AND PAVINGS		
BASES FOR PAVING		
Excavate ground and reduce levels to receive 38 mm thick slab and 25 mm mortar bed; dispose of excavated material off site; treat substrate with total herbicide		
Lay granular fill Type 1 150 thick laid to falls and compacted		
all by machine	m²	**13.00**
all by hand except disposal by grab	m²	**44.00**
Lay 1:2:4 concrete base 150 thick laid to falls		
all by machine	m²	**25.00**
all by hand except disposal by grab	m²	**61.00**
150 mm hardcore base with concrete base 150 deep		
concrete 1:2:4 site mixed	m²	**33.00**
concrete PAV 1 35 ready mixed	m²	**31.00**
Hardcore base 150 deep; concrete reinforced with A142 mesh		
site mixed concrete 1:2:4; 150 mm deep	m²	**47.00**
site mixed concrete 1:2:4; 250 mm deep	m²	**62.00**
concrete PAV 1 35 N/mm ready mixed; 150 mm deep	m²	**37.00**
concrete PAV 1 35 N/mm ready mixed; 250 mm deep	m²	**52.00**
KERBS AND EDGINGS		
Note: excavation is by machine unless otherwise mentioned		
Excavate trench and construct concrete foundation 300 mm wide x 150 mm deep; lay precast concrete kerb units bedded in semi-dry concrete; slump 35 mm maximum; haunching one side; disposal of arisings off site		
Kerbs laid straight		
125 mm high x 125 mm thick bullnosed type BN	m	**25.00**
125 x 255 mm; ref HB2; SP	m	**27.00**
150 x 305 mm; ref HB1	m	**31.00**
Excavate and construct concrete foundation 450 mm wide x 150 mm deep; lay precast concrete kerb units bedded in semi-dry concrete; slump 35 mm maximum; haunching one side; lay channel units bedded in 1:3 sand mortar; jointed in cement:sand mortar 1:3		
Kerbs; 125 mm high x 125 mm thick bullnosed type BN		
dished channel; 125 x 225; Ref CS	m	**45.00**
square channel; 125 x 150; Ref CS2	m	**43.00**
bullnosed channel; 305 x 150; Ref CBN	m	**51.00**
Granite Kerbs		
straight; 125 x 250	m	**47.00**
curved; 125 x 250	m	**50.00**
Brick channel; class B engineering bricks		
Excavate and construct concrete foundation 600 mm wide x 200 mm deep; lay channel to depths and falls bricks to be laid as headers along the channel; bedded in 1:3 lime:sand mortar bricks close jointed in cement:sand mortar 1:3		
3 courses wide	m	**63.00**
Precast concrete edging on concrete foundation 100 x 150 deep and haunching one side 1:2:4 including all necessary excavation disposal and formwork		
Rectangular chamfered or bullnosed; to one side of straight path		
50 x 150	m	**21.00**
50 x 200	m	**26.00**
50 x 250	m	**27.00**
Rectangular chamfered or bullnosed as above but to both sides of straight paths		
50 x 150	m	**45.00**
50 x 200	m	**52.00**
50 x 250	m	**54.00**
Timber edgings; softwood		
Straight		
150 x 38	m	**7.00**
150 x 50	m	**7.00**

ROADS AND PAVINGS

Item Excluding site overheads and profit	Unit	**Total rate £**
Curved; over 5 m radius		
150 x 38	m	**9.00**
150 x 50	m	**9.00**
Curved; 4 - 5 m radius		
150 x 38	m	**9.00**
150 x 50	m	**10.00**
Curved; 3 - 4 m radius		
150 x 38	m	**10.00**
150 x 50	m	**11.00**
Curved; 1.00 - 3.00 m radius		
150 x 38	m	**12.00**
150 x 50	m	**13.00**
Timber edgings; hardwood		
150 x 38	m	**12.00**
150 x 50	m	**14.00**
Brick or concrete block edge restraint; excavate for groundbeam; lay concrete 1:2:4 200 wide x 150 mm deep; on 50 mm thick sharp sand bed; lay blocks or bricks inclusive of haunching one side		
Blocks 200 x 100 x 60; PC £6.74/m^2; butt jointed		
header course	m	**15.00**
stretcher course	m	**12.00**
Bricks 215 x 112.5 x 50; PC £300.00/1000; with mortar joints		
header course	m	**17.00**
stretcher course	m	**16.00**
Sawn yorkstone edgings; excavate for groundbeam; lay concrete 1:2:4 150 mm deep x 33.3% wider than the edging; on 35 mm thick mortar bed; inclusive of haunching one side		
Yorkstone 50 mm thick		
100 mm wide x random lengths	m	**23.00**
100 mm x 100 mm	m	**27.00**
100 mm wide x 200 long	m	**35.00**
250 mm wide x random lengths	m	**36.00**
500 mm wide x random lengths	m	**60.00**
Granite edgings; excavate for groundbeam; lay concrete 1:2:4 150 mm deep x 33.3% wider than the edging; on 35 mm thick mortar bed; inclusive of haunching one side		
Granite 50 mm thick		
100 mm wide x random lengths	m	**23.00**
100 mm x 100 mm	m	**27.00**
100 mm wide x 200 long	m	**35.00**
250 mm wide x random lengths	m	**36.00**
500 mm wide x random lengths	m	**60.00**
ROADS		
Excavate weak points of excavation by hand and fill with well rammed Type 1 granular material		
100 thick	m^2	**10.00**
200 thick	m^2	**13.00**
Reinforced concrete roadbed 150 thick		
Excavate road bed and dispose excavated material off site; bring to grade; lay reinforcing mesh A142 lapped and joined; lay 150 mm thick in situ reinforced concrete roadbed 21 N/mm^2; with 15 impregnated fibreboard expansion joints and polysulphide based sealant at 50 m centres; on 150 hardcore blinded with ash or sand; kerbs 155 x 255 to both sides; including foundations haunched one side in 11.5 N/mm^2 concrete with all necessary formwork; Falls, crossfalls and cambers not exceeding 15 degrees		
4.90 m wide	m	**230.00**
6.10 m wide	m	**312.00**
7.32 m wide	m	**[illegible]**

ROADS AND PAVINGS

Item Excluding site overheads and profit	Unit	Total rate £
ROADS - cont'd		
Macadam roadway over 1000 m^2		
Excavate 350 mm for pathways or roadbed; bring to grade; lay 100 well-rolled hardcore; lay 150 mm Type 1 granular material; lay precast edgings kerbs 50 x 150 on both sides; including foundations haunched one side in 11.5 N/mm² concrete with all necessary formwork; machine lay surface of 90 mm macadam 60 mm base course and 30 mm wearing course; all disposal off site		
1.50 m wide	m	**143.00**
2.00 m wide	m	**175.00**
3.00 m wide	m	**241.00**
4.00 m wide	m	**307.00**
5.00 m wide	m	**357.00**
6.00 m wide	m	**439.00**
7.00 m wide	m	**504.00**
Macadam roadway between 400 and 1000 m^2		
Excavate 350 mm for pathways or roadbed; bring to grade; lay 100 well-rolled hardcore; lay 150 mm Type 1 granular material; lay precast edgings kerbs 50 x 150 on both sides; including foundations haunched one side in 11.5 N/mm² concrete with all necessary formwork; machine lay surface of 90 mm macadam 60 mm base course and 30 mm wearing course; all disposal off site		
1.50 m wide	m	**148.00**
2.00 m wide	m	**181.00**
3.00 m wide	m	**250.00**
4.00 m wide	m	**319.00**
5.00 m wide	m	**388.00**
6.00 m wide	m	**457.00**
7.00 m wide	m	**526.00**
CAR PARKS		
Excavate 350 for pathways or roadbed to receive surface 100 mm thick priced separately; bring to grade; lay 100 well-rolled hardcore; lay 150 mm Type 1 granular material; lay kerbs BS 7263; 125 x 255 on both sides; including foundations haunched one side in 11.5 N/mm² concrete with all necessary formwork		
Work to falls, crossfalls and cambers not exceeding 15 degrees		
1.50 m wide	m	**125.00**
2.00 m wide	m	**152.00**
3.00 m wide	m	**202.00**
4.00 m wide	m	**253.00**
5.00 m wide	m	**304.00**
6.00 m wide	m	**354.00**
7.00 m wide	m	**405.00**
As above but excavation 450 mm deep and base of Type 1 at 250 mm thick		
1.50 m wide	m	**194.00**
2.00 m wide	m	**241.00**
3.00 m wide	m	**336.00**
4.00 m wide	m	**431.00**
5.00 m wide	m	**527.00**
6.00 m wide	m	**622.00**
7.00 m wide	m	**717.00**
To excavated and prepared base above, lay roadbase of 40 size dense bitumen macadam to BS 4987 70 thick; lay wearing course of 10 size dense bitumen macadam 30 thick; mark out car parking bays 5.0 m x 2.4 m with thermoplastic road paint; surfaces all mechanically laid		
per bay 5.0 m x 2.4 m	each	**211.00**
gangway	m^2	**15.00**
As above but stainless steel road studs 100 x 100 to BS 873:pt.4; two per bay in lieu of thermoplastic paint		
per bay 5.0 m x 2.4 m	each	**213.00**

ROADS AND PAVINGS

Item Excluding site overheads and profit	Unit	**Total rate £**
Car park as above but with interlocking concrete blocks 200 x 100 x 80 mm; grey		
per bay 5.0 m x 2.4 m	each	**450.00**
gangway	m²	**38.00**
Car park as above but with interlocking concrete blocks 200 x 100 x 80 mm; colours		
per bay 5.0 m x 2.4 m	each	**462.00**
gangway	m²	**39.00**
Car park as above but with interlocking concrete blocks 200 x 100 x 60 mm; grey		
per bay 5.0 m x 2.4 m	each	**440.00**
gangway	m²	**37.00**
Car park as above but with interlocking concrete blocks 200 x 100 x 60 mm; colours		
per bay 5.0 m x 2.4 m	each	**447.00**
gangway	m²	**38.00**
Car park as above but with laying of grass concrete Grasscrete in situ continuously reinforced cellular surfacing; including expansion joints at 10 m centres; fill with topsoil and peat (5:1) and fertilizer at 35 g/m²; seed with dwarf rye grass at 35 g/m²		
GC2; 150 thick for HGV traffic including dust carts		
per bay 5.0 m x 2.4 m	each	**435.00**
gangway	m²	**37.00**
GC1; 100 thick for cars and light traffic		
per bay 5.0 m x 2.4 m	each	**338.00**
gangway	m²	**29.00**
CAR PARKING FOR DISABLED PEOPLE		
To excavated and prepared base above, lay roadbase of 20 size dense bitumen macadam to BS 4987 80 thick; lay wearing course of 10 size dense bitumen macadam 30 thick; mark out car parking bays with thermoplastic road paint		
per bay 5.80 m x 3.25 m (ambulant)	each	**273.00**
per bay 6.55 m x 3.80 m (wheelchair)	each	**360.00**
gangway	m²	**15.00**
Car park as above but with interlocking concrete blocks 200 x 100 x 60 mm; grey; but mark out bays		
per bay 5.80 m x 3.25 m (ambulant)	each	**3600.00**
per bay 6.55 m x 3.80 m (wheelchair)	each	**4190.00**
gangway	m²	**37.00**
PLAY AREA FOR BALL GAMES		
Excavate for playground; bring to grade; lay 225 consolidated hardcore; blind with 100 type 1; lay macadam base of 40 size dense bitumen macadam to 75 thick; lay wearing course 30 thick		
Over 1000 m²	100 m²	**3860.00**
400 m² - 1000 m²	100 m²	**3966.00**
Excavate for playground; bring to grade; lay 100 consolidated hardcore; blind with 100 type 1; lay macadam base of dense bitumen macadam to 50 thick; lay wearing course 20 thick		
Over 1000 m²	100 m²	**2865.00**
400 m² - 1000 m²	100 m²	**3604.00**
Excavate for playground; bring to grade; lay 100 consolidated hardcore; blind with 100 type 1; lay macadam base of dense bitumen macadam to 50 thick; lay wearing course of Addagrip resin coated aggregate 3 mm diameter 6 mm thick		
Over 1000 m²	100 m²	**4939.00**
400 m² -1000 m²	100 m²	**5535.00**
Safety surfaced play area		
Excavate for playground; bring to grade; lay with 150 type 1; lay macadam base of dense bitumen macadam to 50 thick		
lay safety surface 35 mm; coloured	100 m²	**9275.00**

ROADS AND PAVINGS

Item Excluding site overheads and profit	Unit	**Total rate £**
INTERLOCKING BLOCK PAVING		
Edge restraint to block paving; excavate for groundbeam, lay concrete 1:2:4 200 wide x 150 mm deep; on 50 mm thick sharp sand bed; inclusive of haunching one side		
Blocks 200 x 100 x 60; PC £6.74/m^2; butt jointed		
header course	m	**15.00**
stretcher course	m	**12.00**
Bricks 215 x 112.5 x 50; PC £300.00/1000; with mortar joints		
header course	m	**17.00**
stretcher course	m	**16.00**
Excavate ground; treat substrate with total herbicide; supply and lay granular fill Type 1 150 thick laid to falls and compacted; supply and lay block pavers; laid on 50 compacted sharp sand; vibrated; joints filled with loose sand; excluding edgings or kerbs measured separately		
Concrete blocks		
200 x 100 x 60	m^2	**50.00**
200 x 100 x 80	m^2	**52.00**
Reduce levels; lay 150 granular material Type 1; lay vehicular block paving to 90 degree herringbone pattern; on 50 compacted sand bed; vibrated; jointed in sand and vibrated; excavate and lay precast concrete edging 50 x 150 to BS 7263; on concrete foundation 1:2:4		
Blocks 200 x 100 x 60 mm		
1.0 m wide clear width between edgings	m	**92.00**
1.5 m wide clear width between edgings	m	**117.00**
2.0 m wide clear width between edgings	m	**142.00**
Blocks 200 x 100 x 60 mm but blocks laid 45 degree herringbone pattern including cutting edging blocks		
1.0 m wide clear width between edgings	m	**98.00**
1.5 m wide clear width between edgings	m	**123.00**
2.0 m wide clear width between edgings	m	**148.00**
Blocks 200 x 100 x 60 mm but all excavation by hand disposal off site by grab		
1.0 m wide clear width between edgings	m	**187.00**
1.5 m wide clear width between edgings	m	**223.00**
2.0 m wide clear width between edgings	m	**261.00**
BRICK PAVING		
WORKS BY MACHINE		
Excavate and lay base Type 1 150 thick remove arisings; all by machine; lay clay brick paving		
200 x 100 x 50 thick; butt jointed on 50 mm sharp sand bed		
PC £300.00/1000	m^2	**63.00**
PC £600.00/1000	m^2	**78.00**
200 x 100 x 50 thick; 10 mm mortar joints on 35 mm mortar bed		
PC £300.00/1000	m^2	**76.00**
PC £600.00/1000	m^2	**89.00**
Excavate and lay base Type 1 250 thick all by machine; remove arisings; lay clay brick paving		
200 x 100 x 50 thick; 10 mm mortar joints on 35 mm mortar bed		
PC £300.00/1000	m^2	**83.00**
Excavate and lay 150 mm readymix concrete base reinforced with A393 mesh; all by machine; remove arisings; lay clay brick paving		
200 x 100 x 50 thick; 10 mm mortar joints on 35 mm mortar bed; running or stretcher bond		
PC £300.00/1000	m^2	**96.00**
PC £600.00/1000	m^2	**109.00**
200 x 100 x 50 thick; 10 mm mortar joints on 35 mm mortar bed; butt jointed; herringbone bond		
PC £300.00/1000	m^2	**83.00**
PC £600.00/1000	m^2	**98.00**

ROADS AND PAVINGS

Item Excluding site overheads and profit	Unit	**Total rate £**
Excavate and lay base readymix concrete base 150 mm thick reinforced with A393 mesh; all by machine; remove arisings; lay clay brick paving		
215 x 102.5 x 50 thick; 10 mm mortar joints on 35 mm mortar bed		
PC £300.00/1000; herringbone	m²	**98.00**
PC £600.00/1000; herringbone	m²	**111.00**
WORKS BY HAND		
Excavate and lay base Type 1 150 thick by hand; arisings barrowed to spoil heap maximum distance 25 m and removal off site by grab; lay clay brick paving		
200 x 100 x 50 thick; butt jointed on 50 mm sharp sand bed		
PC £300.00/1000	m²	**97.90**
PC £600.00/1000	m²	**125.00**
Excavate and lay 150 mm concrete base; 1:3:6: site mixed concrete reinforced with A393 mesh; remove arisings to stockpile and then off site by grab; lay clay brick paving		
215 x 102.5 x 50 thick; 10 mm mortar joints on 35 mm mortar bed		
PC £300.00/1000	m²	**95.60**
PC £600.00/1000	m²	**123.00**
FLAG PAVING TO PEDESTRIAN AREAS		
Prices are inclusive of all mechanical excavation and disposal		
Supply and lay precast concrete flags; excavate ground and reduce levels; treat substrate with total herbicide; supply and lay granular fill Type 1 150 thick laid to falls and compacted		
Standard precast concrete flags to BS 7263 bedded and jointed in lime:sand mortar (1:3)		
450 x 450 x 70 chamfered	m²	**38.00**
450 x 450 x 50 chamfered	m²	**35.00**
600 x 300 x 50	m²	**34.00**
600 x 450 x 50	m²	**33.00**
600 x 600 x 50	m²	**31.00**
750 x 600 x 50	m²	**31.00**
900 x 600 x 50	m²	**30.00**
Coloured flags bedded and jointed in lime:sand mortar (1:3)		
600 x 600 x 50	m²	**33.00**
450 x 450 x 70 chamfered	m²	**42.00**
600 x 600 x 50	m²	**33.00**
400 x 400 x 65	m²	**43.00**
750 x 600 x 50	m²	**32.00**
900 x 600 x 50	m²	**31.00**
Marshalls Saxon; textured concrete flags; reconstituted yorkstone in colours; butt jointed bedded in lime:sand mortar (1:3)		
300 x 300 x 35	m²	**63.00**
450 x 450 x 50	m²	**54.00**
600 x 300 x 35	m²	**50.00**
600 x 600 x 50	m²	**47.00**
Tactile flags; Marshalls blister tactile pavings; red or buff; for blind pedestrian guidance laid to designed pattern		
450 x 450	m²	**45.00**
400 x 400	m²	**48.00**
PEDESTRIAN DETERRENT PAVING		
Excavate ground and bring to levels; treat substrate with total herbicide; supply and lay granular fill Type 1 150 mm thick laid to falls and compacted; supply and lay precast deterrent paving units bedded in lime:sand mortar (1:3) and jointed in lime:sand mortar (1:3)		
Marshalls Mono		
Lambeth pyramidal paving 600 x 600 x 75	m²	**38.00**
Townscape		
Abbey square cobble pattern pavings; reinforced 600 x 600 x 60	m²	**58.00**
Geoset chamfered studs 600 x 600 x 60	m²	**54.00**

ROADS AND PAVINGS

Item Excluding site overheads and profit	Unit	**Total rate £**
IMITATION YORKSTONE PAVINGS		
Excavate ground and bring to levels, treat substrate with total herbicide, supply and lay granular fill Type 1 150 mm thick laid to falls and compacted, supply and lay imitation yorkstone paving laid to coursed patterns bedded in lime:sand mortar (1:3) and jointed in lime:sand mortar (1:3)		
Marshalls Heritage square or rectangular		
300 x 300 x 38	m²	**66.00**
600 x 300 x 30	m²	**52.00**
600 x 450 x 38	m²	**51.00**
450 x 450 x 38	m²	**49.00**
600 x 600 x 38	m²	**46.00**
As above but laid to random rectangular patterns		
various sizes selected from the above	m²	**57.00**
Imitation yorkstone laid random rectangular as above but on concrete base 150 thick		
by machine	m²	**69.00**
by hand	m²	**100.00**
NATURAL STONE SLAB PAVING		
WORKS BY MACHINE		
Excavate ground by machine and reduce levels, to receive 65 mm thick slab and 35 mm mortar bed; dispose of excavated material off site; treat substrate with total herbicide; lay granular fill Type 1 150 thick laid to falls and compacted; lay to random rectangular pattern on 35 mm mortar bed		
New riven slabs		
laid random rectangular	m²	**120.00**
New riven slabs; but to 150 mm plain concrete base		
laid random rectangular	m²	**123.00**
New riven slabs; disposal by grab		
laid random rectangular	m²	**120.00**
Reclaimed Cathedral grade riven slabs		
laid random rectangular	m²	**145.00**
Reclaimed Cathedral grade riven slabs; but to 150 mm plain concrete base		
laid random rectangular	m²	**156.00**
Reclaimed Cathedral grade riven slabs; disposal by grab		
laid random rectangular	m²	**149.00**
New slabs sawn 6 sides		
laid random rectangular	m²	**110.00**
3 sizes, laid to coursed pattern	m²	**113.00**
New slabs sawn 6 sides; but to 150 mm plain concrete base		
laid random rectangular	m²	**120.00**
3 sizes, laid to coursed pattern	m²	**125.00**
New slabs sawn 6 sides; disposal by grab		
laid random rectangular	m²	**112.00**
3 sizes, laid to coursed pattern	m²	**117.00**
WORKS BY HAND		
Excavate ground by hand and reduce levels, to receive 65 mm thick slab and 35 mm mortar bed; barrow all materials and arisings 25 m; dispose of excavated material off site by grab; treat substrate with total herbicide; lay granular fill Type 1 150 thick laid to falls and compacted; lay to random rectangular pattern on 35 mm mortar bed		
New riven slabs		
laid random rectangular	m²	**147.00**
New riven slabs laid random rectangular; but to 150 mm plain concrete base		
laid random rectangular	m²	**162.00**
New riven slabs; but disposal to skip		
laid random rectangular	m²	**161.00**
Reclaimed Cathedral grade riven slabs		
laid random rectangular	m²	**182.00**
Reclaimed Cathedral grade riven slabs; but to 150 mm plain concrete base		
laid random rectangular	m²	**197.00**

ROADS AND PAVINGS

Item Excluding site overheads and profit	Unit	**Total rate £**
Reclaimed Cathedral grade riven slabs; but disposal to skip		
laid random rectangular	m²	**196.00**
New slabs sawn 6 sides		
laid random rectangular	m²	**143.00**
3 sizes, sawn 6 sides laid to coursed pattern	m²	**148.00**
New slabs sawn 6 sides; but to 150 mm plain concrete base		
laid random rectangular	m²	**158.00**
3 sizes, sawn 6 sides laid to coursed pattern	m²	**163.00**
GRANITE SETT PAVING - PEDESTRIAN		
Excavate ground and bring to levels; lay 100 hardcore to falls; compacted with 5 tonne roller; blind with compacted Type 1 50 mm thick; lay 100 concrete 1:2:4; supply and lay granite setts 100 x 100 x 100 mm bedded in cement:sand mortar (1:3) 25 mm thick minimum; close butted and jointed in fine sand; all excluding edgings or kerbs measured separately		
Setts laid to bonded pattern and jointed		
new setts 100 x 100 x 100	m²	**134.00**
second-hand cleaned 100 x 100 x 100	m²	**150.00**
Setts laid in curved pattern		
new setts 100 x 100 x 100	m²	**140.00**
second-hand cleaned 100 x 100 x 100	m²	**156.00**
GRANITE SETT PAVING - TRAFFICKED AREAS		
Excavate ground and bring to levels; lay 100 hardcore to falls; compacted with 5 tonne roller; blind with compacted Type 1 50 mm thick; lay 150 site mixed concrete 1:2:4 reinforced with steel fabric to BS 4483 ref: A 142; supply and lay granite setts 100 x 100 x 100 mm bedded in cement:sand mortar (1:3) 25 mm thick minimum; close butted and jointed in fine sand; all excluding edgings or kerbs measured separately		
Site mixed concrete		
new setts 100 x 100 x 100	m²	**142.00**
second-hand cleaned 100 x 100 x 100	m²	**159.00**
Ready mixed concrete		
new setts 100 x 100 x 100	m²	**138.00**
second-hand cleaned 100 x 100 x 100	m²	**155.00**
CONCRETE PAVING		
Pedestrian areas		
Excavate to reduce levels; lay 100 mm Type 1; lay PAV 1 air entrained concrete; joints at max width of 6.0m cut out and sealed with sealant to BS 5212; inclusive of all formwork and stripping		
100 mm thick	m²	**66.00**
Trafficked areas		
Excavate to reduce levels; lay 150 mm Type 1 lay PAV 2 40 N/mm² air entrained concrete; reinforced with steel mesh to BS 4483 200 x 200 square at 2.22 kg/m²; joints at max width of 6.0m cut out and sealed with sealant to BS 5212		
150 mm thick	m²	**86.00**
BEACH COBBLE PAVING		
Excavate ground and bring to levels and fill with compacted Type 1 fill 100 thick; lay GEN 1 concrete base 100 thick; supply and lay cobbles individually laid by hand bedded in cement:sand mortar (1:3) 25 thick minimum; dry grout with 1:3 cement:sand grout; brush off surplus grout and water in; sponge off cobbles as work proceeds; all excluding formwork edgings or kerbs measured separately		
Scottish beach cobbles 200 - 100 mm	m²	**84.00**
Kidney flint cobbles 100 - 75 mm	m²	**89.00**
Scottish beach cobbles 75 - 50 mm	m²	**98.60**

ROADS AND PAVINGS

Item Excluding site overheads and profit	Unit	Total rate £
CONCRETE SETT PAVING		
Excavate ground and bring to levels; supply and lay 150 granular fill Type 1 laid to falls and compacted; supply and lay setts bedded in 50 sand and vibrated; joints filled with dry sand and vibrated; all excluding edgings or kerbs measured separately		
Marshalls Mono; Tegula precast concrete setts		
random sizes 60 thick	m²	**45.00**
single size 60 thick	m²	**43.00**
random size 80 thick	m²	**49.00**
single size 80 thick	m²	**59.00**
GRASS CONCRETE PAVING		
Reduce levels; lay 150 mm Type 1 granular fill compacted; supply and lay precast grass concrete blocks on 20 sand and level by hand; fill blocks with 3 mm sifted topsoil and pre-seeding fertilizer at 50g/m²; sow with perennial ryegrass/chewings fescue seed at 35 g/m²		
Grass concrete		
GB103 406 x 406 x 103	m²	**37.00**
GB83 406 x 406 x 83	m²	**35.00**
extra for geotextile fabric underlayer	m²	**1.00**
Firepaths		
Excavate to reduce levels; lay 300 mm well rammed hardcore; blinded with 100 mm type 1; supply and lay Marshalls Mono "Grassguard 180" precast grass concrete blocks on 50 sand and level by hand; fill blocks with sifted topsoil and pre-seeding fertilizer at 50 g/m²		
firepath 3.8 m wide	m	**234.00**
firepath 4.4 m wide	m	**271.00**
firepath 5.0 m wide	m	**309.00**
turning areas	m²	**62.00**
Charcon Hard Landscaping Grassgrid; 366 x 274 x 100	m²	**31.00**
SLAB/BRICK PATHS		
Stepping stone path inclusive of hand excavation and mechanical disposal, 100 mm Type 1 and sand blinding; slabs to comply with BS 7263 laid 100 mm apart to existing turf		
600 wide; 600 x 600 x 50 slabs		
natural finish	m	**47.00**
coloured	m	**47.00**
exposed aggregate	m	**55.00**
900 wide; 600 x 900 x 50 slabs		
natural finish	m	**51.00**
coloured	m	**53.00**
exposed aggregate	m	**64.00**
Pathway inclusive of mechanical excavation and disposal; 100 mm Type 1 and sand blinding; slabs to comply with BS 7263 close butted		
Straight butted path 900 wide; 600 x 900 x 50 slabs		
natural finish	m	**20.00**
coloured	m	**20.00**
exposed aggregate	m	**28.00**
Straight butted path 1200 wide; double row; 600 x 900 x 50 slabs, laid stretcher bond		
natural finish	m	**34.00**
coloured	m	**35.00**
exposed aggregate	m	**52.00**
Straight butted path 1200 wide; one row of 600 x 600 x 50, one row 600 x 900 x 50 slabs		
natural finish	m	**33.00**
coloured	m	**34.00**
exposed aggregate	m	**59.00**

ROADS AND PAVINGS

Item Excluding site overheads and profit	Unit	Total rate £
Straight butted path 1500 wide; slabs of 600 x 900 x 50 and 600 x 600 x 50; laid to bond		
natural finish	m	**41.00**
coloured	m	**42.00**
exposed aggregate	m	**59.00**
Straight butted path 1800 wide; two rows of 600 x 900 x 50 slabs; laid bonded		
natural finish	m	**45.00**
coloured	m	**53.00**
exposed aggregate	m	**78.00**
Straight butted path 1800 wide; three rows of 600 x 900 x 50 slabs; laid stretcher bond		
natural finish	m	**49.00**
coloured	m	**57.00**
exposed aggregate	m	**83.00**
Brick paved paths; bricks 215 x 112.5 x 65 with mortar joints; prices are inclusive of excavation, disposal off site, 100 hardcore, 100 1:2:4 concrete bed; edgings of brick 215 wide, haunched; all jointed in cement:lime:sand mortar (1:1:6)		
Path 1015 wide laid stretcher bond; edging course of headers		
rough stocks - £400.00/1000	m	**118.00**
engineering brick - £280.00/1000	m	**113.00**
Path 1015 wide laid stack bond		
rough stocks - £400.00/1000	m	**123.00**
engineering brick - £280.00/1000	m	**118.00**
Path 1115 wide laid header bond		
rough stocks - £400.00/1000	m	**129.00**
engineering brick - £280.00/1000	m	**125.00**
Path 1330 wide laid basketweave bond		
rough stocks - £400.00/1000	m	**151.00**
engineering brick - £280.00/1000	m	**145.00**
Path 1790 wide laid basketweave bond		
rough stocks - £400.00/1000	m	**200.00**
engineering brick - £280.00/1000	m	**145.00**
Brick paved paths; brick paviors 200 x 100 x 50 chamfered edge with butt joints; prices are inclusive of excavation, 100 mm Type 1, and 50 mm sharp sand; jointing in kiln dried sand brushed in; exclusive of edge restraints		
Path 1000 wide laid stretcher bond; edging course of headers		
rough stocks - £400.00/1000	m	**110.00**
engineering brick - £280.00/1000	m	**113.00**
Path 1330 wide laid basketweave bond		
rough stocks - £400.00/1000	m	**151.00**
engineering brick - £280.00/1000	m	**145.00**
Path 1790 wide laid basketweave bond		
rough stocks - £400.00/1000	m	**200.00**
engineering brick - £280.00/1000	m	**145.00**
GRAVEL PATHS		
Reduce levels and remove spoil to dump on site; lay 150 hardcore well rolled; lay 25 mm sand blinding and geofabric; fix timber edge 150 x 38 to both sides of straight paths; lay 50 mm Cedec gravel watered and rolled;		
Lay Cedec gravel 50 thick; watered and rolled		
1.0 m wide	m	**29.00**
1.5 m wide	m	**24.00**
2.0 m wide	m	**47.00**
Lay Breedon gravel 50 thick; watered and rolled		
1.0 m wide	m	**28.00**
1.5 m wide	m	**37.00**
2.0 m wide	m	**47.00**

ROADS AND PAVINGS

Item Excluding site overheads and profit	Unit	**Total rate £**
BARK PAVING		
Excavate to reduce levels; remove all topsoil to dump on site; treat area with herbicide; lay 150 mm clean hardcore; blind with sand to BS 882 Grade C; lay 0.7 mm geotextile filter fabric water flow 50 l/m²/sec; supply and fix treated softwood edging boards 50 x 150 mm to bark area on hardcore base extended 150 mm beyond the bark area; boards fixed with galvanized nails to treated softwood posts 750 x 50 x 75 mm driven into firm ground at 1.0 m centres; edging boards to finish 25 mm above finished bark surface; tops of posts to be flush with edging boards and once weathered		
Supply and lay 100 mm Melcourt conifer walk chips 10 - 40 mm size		
1.00 m wide	m	**65.00**
2.00 m wide	m	**108.00**
3.00 m wide	m	**154.00**
4.00 m wide	m	**196.00**
Extra over for wood fibre	m^2	**-1.52**
Extra over for hardwood chips	m^2	**-1.36**
FOOTPATHS		
Excavate footpath to reduce level; remove arisings to tip on site maximum distance 25 m; lay Type 1 granular fill 100 thick; lay base course of 28 size dense bitumen macadam 50 thick; wearing course of 10 size dense bitumen macadam 20 thick; timber edge 150 x 38 mm		
Areas over 1000 m^2		
1.0 m wide	m	**31.00**
1.5 m wide	m	**43.00**
2.0 m wide	m	**54.00**
Areas 400 m^2 - 1000 m^2		
1.0 m wide	m	**34.00**
1.5 m wide	m	**48.00**
2.0 m wide	m	**73.00**

SPECIAL SURFACES FOR SPORT/PLAYGROUNDS

Item Excluding site overheads and profit	Unit	**Total rate £**
SPECIAL SURFACES FOR SPORT/PLAYGROUNDS		
BARK PLAY AREA		
Excavate playground area to 450 depth; lay 150 broken stone or clean hardcore; lay filter membrane; lay bark surface		
Melcourt Industries		
Playbark 10/50 - 300 thick	100 m²	**3957.00**
Playbark 8/25 - 300 thick	100 m²	**3766.00**
BOWLING GREEN CONSTRUCTION; BAYLIS LANDS		
Bowling Green; complete		
Excavate 300 mm deep and grade to level; excavate and install 100 mm land drain to perimeter and backfill with shingle; install 60 mm land drain to surface at 4.5 m centres backfilled with shingle; install 50 m non perforated pipe 50 m long; install 100 mm compacted and levelled drainage stone, blind with grit and sand; spread 150 mm imported 70:30 sand:soil accurately compacted and levelled; lay bowling green turf and top dress twice luted into surface; exclusive of perimeter ditches and bowls protection		
6 rink green 38.4 x 38.4 m	each	**53573.00**
install "Toro" automatic bowling green irrigation system with pump, tank, controller, electrics, pipework and 8 nr "Toro 780" sprinklers; excluding pumphouse (optional)	each	**8319.00**
Supply and install to perimeter of green, "Sportsmark" preformed bowling green ditch channels		
"Ultimate Design 99" steel reinforced concrete channel, 600 mm long section	each	**7895.00**
"Ultimate GRC" glass reinforced concrete channel, 1.2 m long section	each	**12012.00**
"Ultimate Design 2001" medium density rotational moulded channel; 1 m long section with integral bowls protection	each	**12012.00**
Supply and fit bowls protection material to rear hitting face of "Ultimate" channels 1 and 2 above		
"Curl Grass" artificial grass, 0.45 m wide	each	**2219.00**
"Astroturf" artificial grass, 0.45 m wide	each	**2013.00**
50 mm rubber bowls bumper (2 rows)	each	**4572.00**
bowls protection ditch liner laid loose, 300 mm wide	each	**1353.00**
JOGGING TRACK		
Excavate track 250 deep; treat substrate with Casoron G4; lay filter membrane; lay 100 depth gravel waste or similar; lay 100 mm compacted gravel	100 m²	**6306.00**
Extra for treated softwood edging 50 x 150 on 50 x 50 posts both sides	m	**9.00**
PLAYGROUNDS		
Excavate playground area and dispose of arisings to tip; lay Type 1 granular fill; lay macadam surface two coat work 80 thick; base course of 28 size dense bitumen macadam 50 thick; wearing course of 10 size dense bitumen macadam 30 thick		
Areas over 1000 m²		
Excavation 180 mm base 100 mm thick	m²	**25.00**
Excavation 225 mm base 150 mm thick	m²	**29.00**
Areas 400 m² - 1000 m²		
Excavation 180 mm base 100 mm thick	m²	**28.00**
Excavation 225 mm base 150 mm thick	m²	**32.00**
Excavate playground area to given levels and falls; remove soil off site and backfill with compacted Type 1 granular fill; lay ready mixed concrete to fall 2% in all directions		
base 150 mm thick; surface 100 mm thick	m²	**25.00**
base 150 mm thick; surface 150 mm thick	m²	**31.00**

SPECIAL SURFACES FOR SPORT/PLAYGROUNDS

Item Excluding site overheads and profit	Unit	**Total rate £**
SAFETY SURFACING; BAYLIS LANDSCAPE CONTRACTORS		
Excavate ground and reduce levels to receive safety surface; dispose of excavated material off site; treat substrate with total herbicide; lay granular fill Type 1 150 thick laid to falls and compacted; lay macadam base 40 mm thick; supply and lay "Rubaflex" wet pour safety system to thicknesses as specified		
All by machine except macadam by hand		
Black		
15 mm thick	100 m²	**5060.00**
35 mm thick	100 m²	**6530.00**
60 mm thick	100 m²	**7890.00**
Coloured		
15 mm thick	100 m²	**8790.00**
35 mm thick	100 m²	**9360.00**
60 mm thick	100 m²	**10300.00**
All by hand except disposal by grab		
Black		
15 mm thick	100 m²	**8350.00**
35 mm thick	100 m²	**9910.00**
60 mm thick	100 m²	**11400.00**
Coloured		
15 mm thick	100 m²	**12100.00**
35 mm thick	100 m²	**12800.00**
60 mm thick	100 m²	**13800.00**
SPORTSGROUND CONSTRUCTION; BAYLIS LANDSCAPE CONTRACTORS		
Plain sports pitches; site clearance, grading and drainage not included		
Cultivate ground and grade to levels; apply pre-seeding fertilizer at 900 kg/ha; apply pre-seeding selective weedkiller; seed in two operations with sports pitch type grass seed at 350 kg/ha; harrow and roll lightly; including initial cut; size		
Association football, senior 114 m x 72 m	each	**3525.00**
Association football, junior 106 m x 58 m	each	**2843.00**
Rugby union pitch 156 m x 81 m	each	**5173.00**
Rugby league pitch 134 m x 60 m	each	**3411.00**
Hockey pitch 95 m x 60 m	each	**2615.00**
Shinty pitch 186 m x 96 m	each	**7277.00**
Men's lacrosse pitch 100 m x 55 m	each	**2502.00**
Women's lacrosse pitch 110 m x 73 m	each	**3639.00**
Target archery ground 150 m x 50 m	each	**2956.00**
Cricket outfield 160 m x 142 m	each	**9209.00**
Cycle track outfield 160 m x 80 m	each	**5685.00**
Polo ground 330 m x 220 m	each	**27172.00**
Cricket square; excavate to depth of 150 mm; pass topsoil through 6 mm screen; return and mix evenly with imported marl or clay loam; bring to accurate levels; apply pre-seeding fertilizer at 50 g/m; apply selective weedkiller; seed with cricket square type g grass seed at 50 g/m²; rake in and roll lightly; erect and remove temporary protective chestnut fencing; allow for initial cut and watering three times		
22.8 m x 22.8 m	each	**14780.00**

PREPARATION FOR PLANTING/TURFING

Item Excluding site overheads and profit	Unit	**Total rate £**
PREPARATION FOR PLANTING/TURFING		
SURFACE PREPARATION BY MACHINE		
Treat area with systemic non selective herbicide 1 month before starting cultivation operations; rip up subsoil using subsoiling machine to a depth of 250 below topsoil at 1.20 m centres in light to medium soils; rotavate to 200 deep in two passes; cultivate with chain harrow; roll lightly; clear stones over 50 mm		
By tractor	100 m²	**8.80**
As above but carrying out operations in clay or compacted gravel	100 m²	**9.40**
As above but ripping by tractor rotavation by pedestrian operated rotavator, clearance and raking by hand, herbicide application by knapsack sprayer	100 m²	**41.00**
As above but carrying out operations in clay or compacted gravel	100 m²	**54.00**
Spread and lightly consolidate topsoil brought from spoil heap not exceeding 100 m; in layers not exceeding 150; grade to specified levels; remove stones over 25; treat with Paraquat-Diquat weedkiller; all by machine		
100 mm thick	100 m²	**66.00**
150 mm thick	100 m²	**100.00**
300 mm thick	100 m²	**199.00**
450 mm thick	100 m²	**298.00**
Extra to above for imported topsoil PC £18.50 m^3 allowing for 20% settlement		
100 mm thick	100 m²	**222.00**
150 mm thick	100 m²	**333.00**
300 mm thick	100 m²	**666.00**
450 mm thick	100 m²	**999.00**
500 mm thick	100 m²	**1110.00**
600 mm thick	100 m²	**1332.00**
750 mm thick	100 m²	**1665.00**
1.00 m thick	100 m²	**2220.00**
Extra for incorporating mushroom compost at 50 mm/m² into the top 150 mm of topsoil (compost delivered in 20 m^3) loads		
manually spread, mechanically rotavated	100 m²	**151.00**
mechanically spread and rotavated	100 m²	**110.00**
Extra for incorporating manure at 50 mm/m² into the top 150 mm of topsoil loads		
manually spread, mechanically rotavated 20 m^3 loads	100 m²	**251.00**
mechanically spread and rotavated 60 m^3 loads	100 m²	**161.00**
SURFACE PREPARATION BY HAND		
Spread only and lightly consolidate topsoil brought from spoil heap in layers not exceeding 150; grade to specified levels; remove stones over 25 mm; treat with Paraquat-Diquat weedkiller; all by hand		
100 mm thick	100 m²	**371.00**
150 mm thick	100 m²	**556.00**
300 mm thick	100 m²	**1111.00**
450 mm thick	100 m²	**1666.00**
As above but inclusive of loading to location by barrow maximum distance 25 m; finished topsoil depth		
100 mm thick	100 m²	**815.00**
150 mm thick	100 m²	**1222.00**
300 mm thick	100 m²	**2443.00**
450 mm thick	100 m²	**3664.00**
500 mm thick	100 m²	**4069.00**
600 mm thick	100 m²	**4879.00**
As above but loading to location by barrow maximum distance 100 m; finished topsoil depth		
100 mm thick	100 m²	**833.00**
150 mm thick	100 m²	**1249.00**
300 mm thick	100 m²	**2498.00**
450 mm thick	100 m²	**3747.00**
500 mm thick	100 m²	**4162.00**
600 mm thick	100 m²	**4990.00**

SPECIAL SURFACES FOR SPORT/PLAYGROUNDS

Item Excluding site overheads and profit	Unit	Total rate £
SURFACE PREPARATION BY HAND - cont'd		
Extra to the above for incorporating mushroom compost at 50 mm/m² into the top 150 mm of topsoil; by hand	100 m²	**178.00**
Extra to above for imported topsoil PC £18.50 m³ allowing for 20% settlement		
100 mm thick	100 m²	**222.00**
150 mm thick	100 m²	**333.00**
300 mm thick	100 m²	**666.00**
450 mm thick	100 m²	**999.00**
500 mm thick	100 m²	**1110.00**
600 mm thick	100 m²	**1332.00**
750 mm thick	100 m²	**1665.00**
1.00 m thick	100 m²	**2220.00**

SEEDING AND TURFING

Item Excluding site overheads and profit	Unit	Total rate £
SEEDING AND TURFING		
Bring top 200 of topsoil to a fine tilth using tractor drawn implements; remove stones over 25 by mechanical stone rake and bring to final tilth by harrow; apply pre-seeding fertilizer at 50 g/m² and work into top 50 during final cultivation; seed with certified grass seed in two operations; roll seedbed lightly after sowing		
General amenity grass at 35 g/m²; BSH A3	100 m²	**32.00**
General amenity grass at 35 g/m²; Johnsons Taskmaster	100 m²	**28.00**
Shaded areas; BSH A6 at 50 g/m²	100 m²	**41.00**
Motorway and road verges; Perryfields Pro 120 25 - 35 g/m²	100 m²	**31.00**
Bring top 200 of topsoil to a fine tilth using pedestrian operated rotavator; remove stones over 25 mm; apply pre-seeding fertilizer at 50 g/m² and work into top 50 during final hand cultivation; seed with certified grass seed in two operations; rake and roll seedbed lightly after sowing		
General amenity grass at 35 g/m²; BSH A3	100 m²	**100.00**
General amenity grass at 35 g/m²; Johnsons Taskmaster	100 m²	**97.00**
Shaded areas; BSH A6 at 50 g/m²	100 m²	**110.00**
Motorway and road verges; Perryfields Pro 120 25-35 g/m²	100 m²	**99.00**
Extra to above for using imported topsoil spread by machine		
100 minimum depth	100 m²	**251.00**
150 minimum depth	100 m²	**376.00**
Extra for slopes over 30	50 %	**-**
Extra to above for using imported topsoil spread by hand; maximum distance for transporting soil 100 m		
100 minimum depth	m²	**593.00**
150 minimum depth	m²	**889.00**
Extra for slopes over 30	50 %	**-**
Extra for mechanically screening top 25 of topsoil through 6 mm screen and spreading on seedbed, debris carted to dump on site not exceeding 100 m	m³	**6.05**
Bring top 200 of topsoil to a fine tilth; remove stones over 50; apply pre-seeding fertilizer at 50 g/m² and work into top 50 of topsoil during final cultivation; seed with certified grass seed in two operations; harrow and roll seedbed lightly after sowing		
Areas inclusive of fine levelling by specialist machinery		
outfield grass at 350 kg/ha - Perryfields Pro 40	ha	**3744.00**
outfield grass at 350 kg/ha - Perryfields Pro 70	ha	**3495.00**
sportsfield grass at 300 kg/ha - Johnsons Sportsmaster	ha	**3594.00**
Seeded areas prepared by chain harrow		
low maintenance grass at 350 kg/ha	ha	**4280.00**
verge mixture grass at 150 kg/ha	ha	**3539.00**
Extra for wild flora mixture at 30 kg/ha		
BSH WSF 75 kg/ha	ha	**3225.00**
Extra for slopes over 30	50%	**-**
Cut existing turf to 1.0 x 1.0 x 0.5 m turves, roll up and move to stack not exceeding 100 m		
by pedestrian operated machine, roll up and stack by hand	100 m²	**70.00**
all works by hand	100 m²	**201.00**
Extra for boxing and cutting turves	100 m²	**4.00**
Bring top 200 of topsoil to a fine tilth in 2 passes; remove stones over 25 by and bring to final tilth; apply pre-seeding fertilizer at 50 g/m² and work into top 50 during final cultivation; roll turf bed lightly		
Using tractor drawn implements and mechanical stone rake	m²	**1.00**
Cultivation by pedestrian rotavator, all other operations by hand	m²	**1.00**
As above but bring turf from stack not exceeding 100 m; lay turves to stretcher bond using plank barrow runs; firm turves using wooden turf beater		
using tractor drawn implements and mechanical stone rake	m²	**2.00**
cultivation by pedestrian rotavator, all other operations by hand	m²	**3.00**
As above but including imported turf; Rolawn Medallion		
using tractor drawn implements and mechanical stone rake	m²	**4.00**
cultivation by pedestrian rotavator, all other operations by hand	m²	**5.00**

SEEDING AND TURFING

Item Excluding site overheads and profit	Unit	**Total rate £**
SEEDING AND TURFING - cont'd		
Extra over to all of the above for watering on two occasions and carrying out initial cut		
by ride on triple mower	100 m^2	**2.00**
by pedestrian mower	100 m^2	**5.00**
by pedestrian mower; box cutting	100 m^2	**6.00**
Extra for using imported topsoil spread and graded by machine		
25 minimum depth	m^2	**1.00**
75 minimum depth	m^2	**2.00**
100 minimum depth	m^2	**3.00**
150 minimum depth	m^2	**4.00**
Extra for using imported topsoil spread and graded by hand; distance of barrow run 25 m		
25 minimum depth	m^2	**2.00**
75 minimum depth	m^2	**4.00**
100 minimum depth	m^2	**5.00**
150 minimum depth	m^2	**8.00**
Extra for work on slopes over 30 including pegging with 200 galvanized wire pins	m^2	**2.00**
Inturf Big Roll		
Supply, deliver in one consignment, fully prepare the area and install in Big Roll format Inturf 553, a turfgrass comprising dwarf perennial ryegrass, smooth stalked meadowgrass and fescues; installation by tracked machine		
Preparation by tractor drawn rotavator	m^2	**3.00**
Cultivation by pedestrian rotavator, all other operations by hand	m^2	**4.00**
Erosion control		
On ground previously cultivated, bring area to level, treat with herbicide; lay 20 mm thick open texture erosion control mat with 100 mm laps, fixed with 8 x 400 mm steel pegs at 1.0 m centres; sow with low maintenance grass suitable for erosion control on slopes at 35 g/m^2; spread imported topsoil 25 mm thick and fertilizer at 35 g/m^2; water lightly using sprinklers on two occasions	m^2	**7.00**
As above but hand-watering by hose pipe maximum distance from mains supply 50 m	m^2	**7.00**

PLANTING

Item Excluding site overheads and profit	Unit	**Total rate £**
PLANTING		
HEDGE PLANTING		
Works by machine; excavate trench for hedge 300 wide x 450 deep; deposit spoil alongside and plant hedging plants in single row at 200 centres; backfill with excavated material incorporating organic manure at 1 m³ per 5 m³; carry out initial cut; including delivery of plants from nursery		
Bare root hedging plants		
PC - £0.30	100 m	**611.00**
PC - £0.60	100 m	**761.00**
PC - £1.00	100 m	**961.00**
PC - £1.20	100 m	**1061.00**
PC - £1.50	100 m	**1211.00**
As above but two rows of hedging plants at 300 centres staggered rows		
PC - £0.30	100 m	**797.00**
PC - £0.60	100 m	**997.00**
PC - £1.00	100 m	**1264.00**
PC - £1.50	100 m	**1597.00**
Works by hand; excavate trench for hedge 300 wide x 450 deep; deposit spoil alongside and plant hedging plants in single row at 200 centres; backfill with excavated material incorporating organic manure at 1 m³ per 5 m³; carry out initial cut; including delivery of plants from nursery		
Two rows of hedging plants at 300 centres staggered rows		
PC - £0.30	100 m	**1010.00**
PC - £0.60	100 m	**1201.00**
PC - £1.00	100 m	**1468.00**
PC - £1.50	100 m	**1801.00**
TREE PLANTING		
Excavate tree pit by hand; fork over bottom of pit; plant tree with roots well spread out; backfill with excavated material, incorporating tree planting compost at 1 m³ per 3 m³ of soil, one tree stake and two ties; tree pits square in sizes shown		
Light standard bare root tree in pit; PC £8.40		
600 x 600 deep	each	**37.00**
900 x 900 deep	each	**51.00**
Standard bare root tree in pit; PC £10.75		
600 x 600 deep	each	**43.00**
900 x 600 deep	each	**52.00**
900 x 900 deep	each	**63.00**
Standard root balled tree in pit; PC £25.75		
600 x 600 deep	each	**60.00**
900 x 600 deep	each	**55.00**
900 x 900 deep	each	**79.00**
Selected standard bare root tree, in pit; PC £16.90		
900 x 900 deep	each	**70.00**
1.00 m x 1.00 m deep	each	**80.00**
Selected standard root ball tree in pit; PC £31.90		
900 x 900 deep	each	**85.00**
1.00 m x 1.00 m deep	each	**115.00**
Heavy standard bare root tree, in pit; PC £28.00		
900 x 900 deep	each	**86.00**
1.00 m x 1.00 m deep	each	**110.00**
1.20 m x 1.00 m deep	each	**123.00**
Heavy standard root ball tree, in pit; PC £43.00		
900 x 900 deep	each	**100.00**
1.00 m x 1.00 m deep	each	**115.00**
1.20 m x 1.00 m deep	each	**137.00**
Extra heavy standard bare root tree in pit; PC £33.50		
1.00 m x 1.00 deep	each	**112.00**
1.20 m x 1.00 m deep	each	**164.00**
1.50 m x 1.00 m deep	each	**173.00**

PLANTING

Item Excluding site overheads and profit	Unit	Total rate £
TREE PLANTING - cont'd		
Extra heavy standard root ball tree in pit; PC £51.50		
1.00 m x 1.00 deep	each	**131.00**
1.20 m x 1.00 m deep	each	**183.00**
1.50 m x 1.00 m deep	each	**192.00**
Excavate tree pit by machine; fork over bottom of pit; plant tree with roots well spread out; backfill with excavated material, incorporating organic manure at 1 m³ per 3 m³ of soil; one tree stake and two ties; tree pits square in sizes shown		
Light standard bare root tree in pit; PC £8.40		
600 x 600 deep	each	**30.00**
900 x 900 deep	each	**42.00**
Standard bare root tree in pit; PC £10.75		
600 x 600 deep	each	**37.00**
900 x 600 deep	each	**39.00**
900 x 900 deep	each	**45.00**
Standard root balled tree in pit; PC £25.75		
600 x 600 deep	each	**53.00**
900 x 600 deep	each	**56.00**
900 x 900 deep	each	**61.00**
Selected standard bare root tree, in pit; PC £16.90		
900 x 900 deep	each	**54.00**
1.00 m x 1.00 m deep	each	**64.00**
Selected standard root ball tree in pit; PC £31.90		
900 x 900 deep	each	**85.00**
1.00 m x 1.00 m deep	each	**99.00**
Heavy standard bare root tree, in pit; PC £28.00		
900 x 900 deep	each	**86.00**
1.00 m x 1.00 m deep	each	**74.00**
1.20 m x 1.00 m deep	each	**100.00**
Heavy standard root ball tree, in pit; PC £43.00		
900 x 900 deep	each	**100.00**
1.00 m x 1.00 m deep	each	**99.00**
1.20 m x 1.00 m deep	each	**113.00**
Extra heavy standard bare root tree in pit; PC £33.50		
1.00 m x 1.00 deep	each	**96.00**
1.20 m x 1.00 m deep	each	**110.00**
1.50 m x 1.00 m deep	each	**137.00**
Extra heavy standard root ball tree in pit; PC £51.50		
1.00 m x 1.00 deep	each	**114.00**
1.20 m x 1.00 m deep	each	**129.00**
1.50 m x 1.00 m deep	each	**156.00**
SEMI MATURE TREE PLANTING		
Excavate tree pit deep by machine; fork over bottom of pit; plant rootballed tree Acer platanoides "Emerald Queen" using telehandler where necessary; backfill with excavated material, incorporating Melcourt Topgrow bark/manure mixture at 1 m³ per 3 m³ of soil; Platipus underground guying system; tree pits 1500 x 1500 x 1500 mm deep inclusive of Platimats; excavated material not backfilled to tree pit spread to surrounding area		
16 - 18 cm girth - £80.00	each	**240.00**
18 - 20 cm girth - £93.00	each	**258.00**
20 - 25 cm girth - £115.00	each	**348.00**
25 - 30 cm girth - £210.00	each	**499.00**
30 - 35 cm girth - £340.00	each	**729.00**
As above but treepits 2.00 x 2.00 x 1.5 m deep		
40 - 45 cm girth - £680.00	each	**1210.00**
45 - 50 cm girth - £970.00	each	**1521.00**
55 - 60 cm girth - £1390.00	each	**1998.00**
67 - 70 cm girth - £2200.00	each	**2979.00**
75 - 80 cm girth - £4600.00	each	**5426.00**

PLANTING

Item Excluding site overheads and profit	Unit	**Total rate £**
Extra to the above for imported topsoil moved 25 m from tipping area and disposal off site of excavated material		
Tree pits 1500 x 1500 x 1500 m deep		
16 - 18 cm girth	each	**171.00**
18 - 20 cm girth	each	**163.00**
20 - 25 cm girth	each	**155.00**
25 - 30 cm girth	each	**149.00**
30 - 35 cm girth	each	**143.00**
TREE PLANTING WITH MOBILE CRANES		
Excavate tree pit 1.50 x 1.50 x 1.00 m deep; supply and plant semi mature trees delivered in full loads; trees lifted by crane; inclusive of backfilling tree pit with imported topsoil, compost, fertilizers and underground guying using "Platipus" anchors		
Self managed lift; local authority applications, health and safety, traffic management or road closures not included; tree size and distance of lift; 35 tonne crane		
25 - 30 cm; max 25 m distance	each	**616.00**
30 - 35 cm; max 25 m distance	each	**763.00**
35 - 40 cm; max 25 m distance	each	**1170.00**
55 - 60 cm; max 15 m distance	each	**1981.00**
80 - 90 cm; max 10 m distance	each	**6297.00**
Managed lift; inclusive of all local authority applications, health and safety, traffic management or road closures all by crane hire company; tree size and distance of lift; 35 tonne crane		
25 - 30 cm; max 25 m distance	each	**626.00**
30 - 35 cm; max 25 m distance	each	**781.00**
35 - 40 cm; max 25 m distance	each	**1190.00**
55 - 60 cm; max 15 m distance	each	**2025.00**
80 - 90 cm; max 10 m distance	each	**6448.00**
Self managed lift; local authority applications, health and safety, traffic management or road closures not included; tree size and distance of lift; 80 tonne crane		
25 - 30 cm; max 40 m distance	each	**620.00**
30 - 35 cm; max 40 m distance	each	**771.00**
35 - 40 cm; max 40 m distance	each	**1180.00**
55 - 60 cm; max 33 m distance	each	**2000.00**
80 - 90 cm; max 23 m distance	each	**6362.00**
Managed lift; inclusive of all local authority applications, health and safety, traffic management or road closures all by crane hire company; tree size and distance of lift; 35 tonne crane		
25 - 30 cm; max 40 m distance	each	**630.00**
30 - 35 cm; max 40 m distance	each	**789.00**
35 - 40 cm; max 40 m distance	each	**1200.00**
55 - 60 cm; max 33 m distance	each	**2045.00**
80 - 90 cm; max 23 m distance	each	**6517.00**
SHRUBS, GROUNDCOVERS AND BULBS		
Excavate planting holes on 250 x 250 mm x 300 deep to area previously ripped and rotavated; excavated material left alongside planting hole		
By mechanical auger		
250 mm centres (16 plants per m^2)	m^2	**11.00**
300 mm centres (11.11 plants per m^2)	m^2	**8.00**
400 mm centres (6.26 plants per m^2)	m^2	**5.00**
450 mm centres (4.93 plants per m^2)	m^2	**4.00**
500 mm centres (4 plants per m^2)	m^2	**3.00**
600 mm centres (2.77 plants per m^2)	m^2	**2.00**
750 mm centres (1.77 plants per m^2)	m^2	**2.00**
900 mm centres (1.23 plants per m^2)	m^2	**1.00**
1.00 m centres (1 plants per m^2)	m^2	**1.00**
1.50 m centres (0.44 plants per m^2)	m^2	**1.00**

PLANTING

Item Excluding site overheads and profit	Unit	**Total rate £**
SHRUBS, GROUNDCOVERS AND BULBS - cont'd		
As above but excavation by hand		
250 mm centres (16 plants per m^2)	m^2	**11.00**
300 mm centres (11.11 plants per m^2)	m^2	**8.00**
400 mm centres (6.26 plants per m^2)	m^2	**5.00**
450 mm centres (4.93 plants per m^2)	m^2	**4.00**
500 mm centres (4 plants per m^2)	m^2	**3.00**
600 mm centres (2.77 plants per m^2)	m^2	**2.00**
750 mm centres (1.77 plants per m^2)	m^2	**2.00**
900 mm centres (1.23 plants per m^2)	m^2	**1.00**
1.00 m centres (1 plants per m^2)	m^2	**1.00**
1.50 m centres (0.44 plants per m^2)	m^2	**1.00**
Clear light vegetation from planting area and remove to dump on site, dig planting holes, plant whips with roots well spread out, backfill with excavated topsoil, including one 38 x 38 treated softwood stake, two tree ties, and mesh guard 1.20 m high; planting matrix 1.5 m x 1.5 m; allow for beating up once at 10% of original planting, cleaning and weeding round whips once, applying weedkiller once at 35 gm/m², applying fertilizer once at 35 gm/m², using the following mix of whips, bare rooted		
Plant bare root plants average price £0.27 each to a required matrix plant mix as above	100 m²	**246.00**
Plant bare root plants average price £0.75 each to a required matrix plant mix as above	100 m²	**256.00**
Cultivate and grade shrub bed; bring top 300 mm of topsoil to a fine tilth, incorporating mushroom compost at 50 mm and Enmag slow release fertilizer; rake and bring to given levels; remove all stones and debris over 50 mm; dig planting holes average 300 x 300x 300 mm deep; supply and plant specified shrubs in quantities as shown below; backfill with excavated material as above; water to field capacity and mulch 50 mm bark chips 20 - 40 mm size; water and weed regularly for 12 months and replace failed plants		
Shrubs - 3 L PC £2.80; ground covers - 9 cm PC £1.50		
100% shrub area 300 centres		
300 mm centres	100 m^2	**5763.00**
400 mm centres	100 m^2	**3440.00**
500 mm centres	100 m^2	**2366.00**
600 mm centres	100 m^2	**1782.00**
100% groundcovers		
200 mm centres	100 m^2	**7240.00**
300 mm centres	100 m^2	**3469.00**
400 mm centres	100 m^2	**2151.00**
500 mm centres	100 m^2	**1540.00**
Groundcover 30%/shrubs 70% at the distances shown below		
200 mm/300 mm	100 m^2	**6141.00**
300 mm/400 mm	100 m^2	**3418.00**
300 mm/500 mm	100 m^2	**2685.00**
400 mm/500 mm	100 m^2	**2281.00**
Groundcover 50%/shrubs 50% at the distances shown below		
200 mm/300 mm	100 m^2	**6506.00**
300 mm/400 mm	100 m^2	**3450.00**
300 mm/500 mm	100 m^2	**2925.00**
400 mm/500 mm	100 m^2	**2253.00**
Cultivate ground by machine and rake to level; plant bulbs as shown; bulbs PC £25.00/100		
15 bulbs per m²	100 m²	**657.00**
25 bulbs per m²	100 m²	**1075.00**
50 bulbs per m²	100 m²	**2120.00**

PLANTING

Item Excluding site overheads and profit	Unit	Total rate £
Cultivate ground by machine and rake to level; plant bulbs as shown; bulbs PC £13.00/100		
15 bulbs per m²	100 m²	**456.00**
25 bulbs per m²	100 m²	**740.00**
50 bulbs per m²	100 m²	**1450.00**
Form holes in grass areas and plant bulbs using bulb planter, backfill with organic manure and turf plug; bulbs PC £13.00/100		
15 bulbs per m²	100 m²	**658.00**
25 bulbs per m²	100 m²	**1096.00**
50 bulbs per m²	100 m²	**2192.00**
BEDDING		
Spray surface with glyphosate; lift and dispose of turf when herbicide action is complete; cultivate new area for bedding plants to 400 mm deep; spread compost 100 deep and chemical fertilizer "Enmag" and rake to fine tilth to receive new bedding plants; remove all arisings to skip		
Existing turf area		
disposal to skip	100 m²	**583.00**
disposal to compost area on site; distance 25 m	100 m²	**581.00**
Plant bedding to existing planting area; bedding planting PC £0.25 each		
Clear existing bedding; cultivate soil to 230 mm deep; incorporate compost 75 mm and rake to fine tilth; collect bedding from nursery and plant at 100 mm ccs; irrigate on completion; maintain weekly for 12 weeks		
mass planted; 100 mm ccs	m²	**33.00**
to patterns; 100 mm ccs	m²	**37.00**
mass planted; 150 mm ccs	m²	**19.00**
to patterns; 150 mm ccs	m²	**23.00**
mass planted; 200 mm ccs	m²	**19.00**
to patterns; 200 mm ccs	m²	**23.00**
Extra for watering by hand held hose pipe		
Flow rate 25 litres/minute		
10 litres/m²	100 m²	**0.11**
15 litres/m²	100 m²	**0.17**
20 litres/m²	100 m²	**0.23**
25 litres/m²	100 m²	**0.28**
Flow rate 40 litres/minute		
10 litres/m²	100 m²	**0.07**
15 litres/m²	100 m²	**0.11**
20 litres/m²	100 m²	**0.14**
25 litres/m²	100 m²	**0.18**
PLANTING PLANTERS		
To brick planter; coat insides with 2 coats RIW liquid asphaltic composition; fill with 50 mm shingle and cover with geofabric; fill with screened topsoil incorporating 25% Topgrow compost and Enmag		
Planters 1.00 m deep		
1.00 x 1.00	each	**110.00**
1.00 x 2.00	each	**172.00**
1.00 x 3.00	each	**243.00**
Planters 1.50 m deep		
1.00 x 1.00	each	**150.00**
1.00 x 2.00	each	**222.00**
1.00 x 3.00	each	**363.00**

PLANTING

Item Excluding site overheads and profit	Unit	**Total rate £**
PLANTING PLANTERS - cont'd		
Container planting; fill with 50 mm shingle and cover with geofabric; fill with screened topsoil incorporating 25% Topgrow compost and Enmag		
Planters 1.00 m deep		
400 x 400 x 400 mm deep	each	**8.00**
400 x 400 x 600 mm deep	each	**10.00**
1.00 x 400 wide x 400 deep	each	**16.00**
1.00 x 600 wide x 600 deep	each	**25.00**
1.00 x 100 wide x 400 deep	each	**26.00**
1.00 x 100 wide x 600 deep	each	**38.00**
1.00 x 100 wide x 1.00 deep	each	**61.00**
1.00 m diameter x 400 deep	each	**21.00**
1.00 m diameter x 1.00 m deep	each	**49.00**
2.00 m diameter x 1.00 m deep	each	**189.00**

LANDSCAPE MAINTENANCE

Item Excluding site overheads and profit	Unit	**Total rate £**
LANDSCAPE MAINTENANCE		
MAINTENANCE OF GRASSED AREAS		
Maintenance executed as part of a landscape construction contract		
Grass cutting		
Grass cutting; fine turf, using pedestrian guided machinery; arisings boxed and disposed of off site		
per occasion	100 m²	**6.00**
per annum 26 cuts	m²	**2.00**
per annum 18 cuts	m²	**1.00**
Grass cutting; standard turf, using self propelled 3 gang machinery		
per occasion	100 m²	**1.00**
per annum 26 cuts	m²	**1.00**
per annum 18 cuts	m²	**1.00**
Maintenance for one year; recreation areas, parks, amenity grass areas, using tractor drawn machinery		
per occasion	ha	**29.00**
per annum 26 cuts	ha	**737.00**
per annum 18 cuts	ha	**510.00**
Aeration of turfed areas		
Aerate ground with spiked aerator; apply spring/summer fertilizer once; apply autumn/winter fertilizer once; cut grass, 16 cuts; sweep up leaves twice		
as part of a landscape contract, defects liability	ha	**1379.00**
as part of a long term maintenance contract	ha	**1063.00**
PLANTED AREA MAINTENANCE		
Vegetation control; native planting - roadside railway or forestry planted areas		
Post planting maintenance; control of weeds and grass; herbicide spray applications; maintain weed free circles 1.00 m diameter to planting less than 5 years old in roadside, rail or forestry planting environments and the like; strim grass to 50 - 75 mm; prices per occasion (3 applications of each operation normally required)		
Knapsack spray application; glyphosate; planting at		
1.50 mm centres	ha	**994.00**
1.75 mm centres	ha	**897.00**
2.00 mm centres	ha	**873.00**
Maintain planted areas; control of weeds and grass; maintain weed free circles 1.00 m diameter to planting less than 5 years old in roadside, rail or forestry planting environments and the like; strim surrounding grass to 50 - 75 mm; prices per occasion (3 applications of each operation normally required)		
Herbicide spray applications; CDA (controlled droplet application) glyphosate and strimming; plants planted at the following centres		
1.50 mm centres	ha	**933.00**
1.75 mm centres	ha	**897.00**
2.00 mm centres	ha	**873.00**
Post planting maintenance; control of weeds and grass; herbicide spray applications; CDA (controlled droplet application); "Xanadu" glyphosate/diuron; maintain weed free circles 1.00 m diameter to planting less than 5 years old in roadside, rail or forestry planting environments and the like; strim grass to 50 - 75 mm; prices per occasion (1.5 applications of herbicide and 3 strim operations normally required)		
Plants planted at the following centres		
1.50 mm centres	ha	**973.00**
1.75 mm centres	ha	**926.00**
2.00 mm centres	ha	**896.00**

LANDSCAPE MAINTENANCE

Item Excluding site overheads and profit	Unit	**Total rate £**
PLANTED AREA MAINTENANCE - cont'd		
Ornamental shrub beds		
Hand weed ornamental shrub bed during the growing season; planting less than 2 years old		
Mulched beds; weekly visits; planting centres		
600 ccs	100 m^2	**10.00**
400 ccs	100 m^2	**12.00**
300 ccs	100 m^2	**15.00**
ground covers	100 m^2	**19.00**
Mulched beds; monthly visits; planting centres		
600 ccs	100 m^2	**14.00**
400 ccs	100 m^2	**19.00**
300 ccs	100 m^2	**23.00**
ground covers	100 m^2	**28.00**
Non-mulched beds; weekly visits; planting centres		
600 ccs	100 m^2	**14.00**
400 ccs	100 m^2	**15.00**
300 ccs	100 m^2	**28.00**
ground covers	100 m^2	**37.00**
Non-mulched beds; monthly visits; planting centres		
600 ccs	100 m^2	**19.00**
400 ccs	100 m^2	**23.00**
300 ccs	100 m^2	**28.00**
ground covers	100 m^2	**47.00**
Remulch planting bed at the start of the planting season; top up mulch 25 mm thick; Melcourt Ltd		
Larger areas maximum distance 25 m; 80 m^3 loads		
Ornamental bark mulch	100 m^2	**123.00**
Melcourt Bark Nuggets	100 m^2	**112.00**
Amenity Bark	100 m^2	**87.00**
Forest biomulch	100 m^2	**79.00**
Smaller areas; maximum distance 25 m; 25 m^3 loads		
Ornamental bark mulch	100 m^2	**156.00**
Melcourt Bark Nuggets	100 m^2	**145.00**
Amenity Bark	100 m^2	**119.00**
Forest biomulch	100 m^2	**111.00**

FENCING

Item Excluding site overheads and profit	Unit	**Total rate £**
FENCING		
CHAIN LINK AND WIRE FENCING		
Chain link fencing; supply and erect chain link fencing; form post holes and erect concrete posts and straining posts with struts at 50 m centres all set in 1:3:6 concrete; fix line wires		
3 mm galvanized wire 50 mm chain link fencing		
900 mm high	m	**24.00**
1200 mm high	m	**27.00**
1800 mm high	m	**29.00**
2400 mm high	m	**56.00**
Plastic coated 3.15 gauge galvanized wire mesh		
900 mm high	m	**22.00**
1200 mm high	m	**25.00**
1800 mm high	m	**28.00**
Extra for additional concrete straining posts with 1 strut set in concrete		
900 high	each	**67.00**
1200 high	each	**70.00**
1400 high	each	**78.00**
1800 high	each	**83.00**
2400 high	each	**109.00**
Extra for additional concrete straining posts with 2 struts set in concrete		
900 high	each	**96.00**
1200 high	each	**100.00**
1400 high	each	**122.12**
1800 high	each	**126.89**
2400 high	each	**127.08**
Chain link fencing 3 mm galvanized as above but with mild steel angle posts and straining posts instead of concrete posts		
900 mm high	m	**14.00**
1200 mm high	m	**17.00**
1800 mm high	m	**22.00**
2400 mm high	m	**28.00**
Extra for additional angle iron straining posts with 1 strut set in concrete		
900 high	each	**64.00**
1200 high	each	**84.00**
1400 high	each	**96.00**
1800 high	each	**110.00**
2400 high	each	**125.00**
Extra for additional angle iron straining posts with 2 struts set in concrete		
900 high	each	**95.00**
1200 high	each	**124.00**
1400 high	each	**143.00**
1800 high	each	**170.00**
2400 high	each	**206.00**
TIMBER FENCING		
Erect chestnut pale fencing, cleft chestnut pales, two lines galvanized wire, galvanized tying wire, treated softwood posts at 3.0 m centres and straining posts and struts at 50 m centres driven into firm ground		
900 high; posts 75 dia. x 1200 long	m	**9.00**
1200 high; posts 75 dia. x 1500 long	m	**10.00**
Construct timber rail, horizontal hit and miss type, rails 150 x 25, posts 100 x 100 at 1.8 m centres, twice stained with coloured wood preservative, including excavation for posts and concreting into ground (C7P)		
In treated softwood		
1800 mm high	m	**59.00**
In primed softwood		
1800 mm high	m	**65.00**

FENCING

Item Excluding site overheads and profit	Unit	Total rate £
TIMBER FENCING - cont'd		
Erect close boarded timber fence in treated softwood, pales 89 x 19 lapped, 152 x 25 gravel boards, two 76 x 38 rectangular rails		
Concrete posts 100 x 100 at 3.0 m centres set into ground in 1:3:6 concrete		
900 mm high	m	**25.00**
1350 mm high	m	**33.00**
1800 mm high	m	**35.00**
As above but with oak posts		
1350 mm high	m	**32.00**
1800 mm high	m	**36.00**
As above but with softwood posts		
1350 mm high	m	**30.00**
1650 mm high	m	**31.00**
1800 mm high	m	**32.00**
Erect post and rail fence, three horizontal rails 100 x 38, fixed with galvanized nails to 100 x 100 posts driven into firm ground		
In treated softwood		
1200 mm high	m	**22.00**
In oak or chestnut		
1200 mm high	m	**37.00**
Construct post and rail fence to BS 1722: Part 7 Type MPR 13/4 1300 mm high with five rails 87 x 38 mm; rails morticed into treated softwood posts 2100 75 x 150 mm at 3.0 m centres set 700 mm into firm ground with intermediate prick posts 1800 x 87 x 38		
1300 high	m	**23.00**
Construct cleft oak rail fence, with rails 300 mm minimum girth tennoned both ends; 125 x 100 mm treated softwood posts double mortised for rails; corner posts 125 x 125 mm, driven into firm ground at 2.5 m centres		
3 rails	m	**25.00**
4 rails	m	**29.00**
DEER STOCK RABBIT FENCING		
Construct rabbit-stop fencing; erect galvanized wire netting, mesh 31, 900 above ground, 150 below ground turned out and buried, on 75 dia. treated timber posts 1.8 m long driven 700 into firm ground at 4.0 m centres; netting clipped to top and bottom straining wires 2.63 mm diameter; straining post 150 mm dia. x 2.3 m long driven into firm ground at 50 m intervals		
Turned in 150 mm	100 m	**870.00**
Buried 150 mm	100 m	**950.00**
Deer fence		
Construct deer-stop fencing, erect 5 no. 4 dia. plain galvanized wires and 5 no. 2 ply galvanized barbed wires at 150 spacing, on 45 x 45 x 5 mm angle iron posts 2.4 m long driven into firm ground at 3.0 m centres, driven into firm ground at 3.0 m centres, with timber droppers 25 x 38 x 1.0 m long at 1.5 m centres	100 m	**2077.00**
Forestry fencing		
Supply and erect forestry fencing of three lines of 3 mm plain galvanized wire tied to 1700 x 65 mm dia. angle iron posts at 2750 m centres with 1850 x 100 mm dia. straining posts and 1600 x 80 mm dia. struts at 50.0 m centres driven into firm ground		
1800 high; 3 wires	100 m	**1456.00**
1800 high; 3 wires including cattle fencing	100 m	**1758.00**

FENCING

Item Excluding site overheads and profit	Unit	**Total rate £**
CONCRETE FENCING		
Supply and erect precast concrete post and panel fence in 2 m bays, panels to be shiplap profile, aggregate faced one side, posts set 600 mm in ground in concrete		
1500 mm high	m	**26.00**
1800 mm high	m	**31.00**
2100 mm high	m	**35.00**
SECURITY FENCING		
Supply and erect chain link fence, 51 mm x 3 mm mesh, with line wires and stretcher bars bolted to concrete posts at 3.0 m centres and straining posts at 10 m centres; posts set in concrete 450 x 450 x 33% of height of post deep; fit straight extension arms of 45 x 45 x 5 mm steel angle with three lines of barbed wire and droppers; all metalwork to be factory hot-dip galvanized for painting on site		
900 high	m	**28.00**
1200 high	m	**32.00**
1800 high	m	**44.00**
As above but with PVC coated 3.15 mm mesh (diameter of wire 2.5 mm)		
900 high	m	**26.00**
1200 high	m	**30.00**
1800 high	m	**40.00**
Add to fences above for base of fence to be fixed with hairpin staples cast into concrete ground beam 1:3:6 site mixed concrete; mechanical excavation disposal to on site spoil heaps		
125 x 225 mm deep	m	**5.00**
Add to fences above for straight extension arms of 45 x 45 x 5 mm steel angle with three lines of barbed wire and droppers	m	**4.00**
Supply and erect palisade security fence Jacksons "Barbican" 2500 mm high with rectangular hollow section steel pales at 150 mm centres on three 50 x 50 x 6 mm rails; rails bolted to 80 x 60 mm posts set in concrete 450 x 450 x 750 mm deep at 2750 mm centres, tops of pales to points and set at 45 degree angle; all metalwork to be hot-dip factory galvanized for painting on site	m	**90.00**
Supply and erect single gate to match above complete with welded hinges and lock		
1000 mm wide	each	**807.00**
4.0 m wide	each	**868.00**
8.0 m wide	pair	**1768.00**
Supply and erect Orsogril proprietary welded steel mesh panel fencing on steel posts set 750 mm deep in concrete foundations 600 x 600 mm; supply and erect proprietary single gate 2.0 m wide to match fencing		
930 high	100 m	**11800.00**
1326 high	100 m	**15100.00**
1722 high	100 m	**19800.00**
RAILINGS		
Erect ms bar railing of 19 mm balusters at 115 mm centres welded to ms top and bottom rails 40 x 10; bays 2.0 m long, bolted to 51 x 51 ms hollow section posts set in C15P concrete; all metal work galvanized after fabrication		
900 mm high	m	**63.00**
1200 mm high	m	**86.00**
1500 mm high	m	**94.00**
Supply and erect mild steel pedestrian guard rail Class A to BS 3049; panels 1000 mm high x 2000 mm wide with 150 mm toe space and 200 mm visibility gap at top; rails to be rectangular hollow sealed section 50 x 30 x 2.5 mm, vertical support 25 x 19 mm central between intermediate and top rail; posts to be set 300 mm into paving base; all components factory welded and factory primed for painting on site	m	**86.00**

FENCING

Item Excluding site overheads and profit	Unit	**Total rate £**
BALLSTOP FENCING		
Supply and erect plastic coated 30 x 30 mm netting fixed to 60.3 diameter 12 mm solid bar lattice galvanized dual posts; top, middle and bottom rails with 3 horizontal rails on 60.3 mm dia. nylon coated tubular steel posts at 3.0 m centres and 60.3 mm dia. straining posts with struts at 50 m centres set 750 mm into FND2 concrete footings 300 x 300 x 600 mm deep; include framed chain link gate 900 x 1800 mm high to match complete with hinges and locking latch		
4500 high	100 m	**7595.00**
5000 high	100 m	**8689.00**
6000 high	100 m	**9588.00**
CATTLE GRID		
Cattle grid Excavate pit 4.0 m x 3.0 m x 500; dispose of spoil on site; lay concrete base (C7P) 100 thick on 150 hardcore; excavate trench and lay 100 dia. clay agricultural drain outlet; form concrete sides to pit 150 thick (C15P) including all formwork; construct supporting walls one brick thick in engineering brick laid in cement mortar (1:3) at 400 mm centres; install cattle grid by Jacksons Fencing; erect side panels on steel posts set in concrete (1:3:6); erect 2 nr warning signs standard DoT pattern		
3.66 x 2.55 m	each	**2568.00**
Erect stock gate, ms tubular field gate, diamond braced 1.8 m high hung on tubular steel posts set in concrete (C7P), complete with ironmongery, all galvanized		
Width 3.00 m	each	**294.00**
Width 4.20 m	each	**320.00**
TRIP RAILS		
Steel trip rail Erect trip rail of galvanized steel tube 38 internal dia. with sleeved joint fixed to 38 dia. steel posts as above 700 long, set in C7P concrete at 1.20 m centres; metalwork primed and painted two coats metal preservative paint	m	**122.00**
Birdsmouth fencing; timber		
600 high	m	**18.00**
900 high	m	**19.00**

STREET FURNITURE

Item Excluding site overheads and profit	Unit	**Total rate £**
STREET FURNITURE		
BENCHES, SEATS AND BOLLARDS		
Bollards and access restriction		
Supply and install powder coated steel parking posts and bases to specified colour; fold down top locking type, complete with keys and instructions; posts and bases set in concrete footing 200 x 200 x 300 mm deep	each	**179.00**
Supply and install 10 no. cast iron "Doric" bollards 920 mm high above ground x 170 dia. bedded in concrete base 400 dia. x 400 mm deep	10 nr	**1497.00**
Benches and seating		
In grassed area excavate for base 2500 x 1575 mm and lay 100 mm hardcore, 100 mm concrete, brick pavers in stack bond bedded in 25 mm cement:lime:sand mortar; supply and fix where shown on drawing proprietary seat, hardwood slats on black powder coated steel frame, bolted down with 4 no. 24 x 90 mm recessed hex-head stainless steel anchor bolts set into concrete	set	**1403.00**
Cycle stand		
Supply and fix cycle stand 1250 m long x 550 mm long of 60.3 mm black powder coated hollow steel sections to BS 4948:Part 2, one-piece with rounded top corners, set 250 mm into paving	each	**304.00**
Street planters		
Supply and locate in position precast concrete planters; fill with topsoil placed over 50 mm shingle and Terram; plant with assorted 5 litre and 3 litre shrubs at to provide instant effect 970 diameter x 470 high; white exposed aggregate finish	each	**40.00**

DRAINAGE

Item Excluding site overheads and profit	Unit	**Total rate £**
DRAINAGE		
SURFACE WATER DRAINAGE		
Clay gully Excavate hole; supply and set in concrete (C10P) vitrified clay trapped mud (dirt) gully with rodding eye to BS 65, complete with galvanized bucket and cast iron hinged locking grate and frame, flexible joint to pipe; connect to drainage system with flexible joints	each	**190.00**
Concrete road gully Excavate hole and lay 100 concrete base (1:3:6) 150 x150 to suit given invert level of drain; supply and connect trapped precast concrete road gully 450 dia x 1.07 m deep with 160 outlet to BS 5911; set in concrete surround; connect to vitrified clay drainage system with flexible joints; supply and fix straight bar dished top cast iron grating and frame; bedded in cement:sand mortar (1:3)	each	**320.00**
Gullies PVC-u Excavate hole and lay 100 concrete (C20P) base 150 x 150 to suit given invert level of drain; connect to drainage system; backfill with DoT Type 1 granular fill; install gully; complete with cast iron grate and frame		
trapped PVC-u gully	each	**81.00**
bottle gully 228 x 228 x 317 deep	each	**67.00**
bottle gully 228 x 228 x 642 deep	each	**81.00**
yard gully 300 dia. x 600 deep	each	**226.00**
Inspection chambers; brick manhole; excavate pit for inspection chamber including earthwork support and disposal of spoil to dump on site not exceeding 100 m; lay concrete (1:2:4) base 1500 dia. x 200 thick; 110 vitrified clay channels; benching in concrete (1:3:6) allowing one outlet and two inlets for 110 dia pipe; construct inspection chamber 1 brick thick walls of engineering brick Class B; backfill with excavated material; complete with 2 no. cast iron step irons		
1200 x 1200 x 1200 mm		
cover slab of precast concrete	each	**868.00**
access cover; Group 2; 600 x 450 mm	each	**871.00**
1200 x 1200 x 1500 mm		
access cover; Group 2; 600 x 450 mm	each	**1039.00**
recessed cover 5 tonne load; 600 x 450 mm; filled with block paviors	each	**1225.00**
Cast iron inspection chamber **Excavate pit for inspection chamber; supply and install cast iron inspection chamber unit; bedded in Type 1 granular material**		
650 mm deep; 100 x 150 mm; one branch each side	each	**285.00**
Interceptor trap Supply and fix vitrified clay interceptor trap 100 mm inlet, 100 mm outlet; to manhole bedded in 10 aggregate concrete (C10P), complete with brass stopper and chain	each	**151.00**
Connect to drainage system with flexible joints	each	**24.00**
Pipe laying Excavate trench by excavator 600 deep; lay Type 2 bedding; backfill to 150 mm above pipe with gravel rejects; lay non woven geofabric and fill with topsoil to ground level		
160 PVC-U drainpipe	100 m	**2090.00**
110 PVC-U drainpipe	100 m	**1247.00**
150 vitrified clay	100 m	**1948.00**
100 vitrified clay	100 m	**1324.00**
Linear drainage to design sensitive areas Excavate trench by machine; Lay Aco Brickslot channel drain on concrete base and surround to falls; all to manufacturers specifications		
paving surround to both sides of channel	m	**106.51**

DRAINAGE

Item Excluding site overheads and profit	Unit	Total rate £
Linear drainage to pedestrian area		
Excavate trench by machine; lay Aco NK100 channel drain on concrete base and surround to falls; all to manufacturers specifications; paving surround to channel with brick paving PC £300.00/1000		
slotted galvanised grating; paving surround to one side of channel	m	**147.77**
slotted galvanised grating; paving surround to both sides of channel	m	**157.37**
stainless steel grating; paving surround to one side of channel	m	**278.87**
slotted galvanised grating; paving surround to both sides of channel	m	**288.47**
Linear drainage to light vehicular area		
Excavate trench by machine; Lay Aco NK100 channel drain on concrete base and surround to falls; all to manufacturers specifications; paving surround to channel with brick paving PC £300.00/1000		
slotted galvanised grating; paving surround to one side of channel	m	**147.77**
slotted galvanised grating; paving surround to both sides of channel	m	**157.37**
'Heelguard™' composite grating; paving surround to one side of channel	m	**112.87**
'Heelguard™' ductile grating; paving surround to both sides of channel	m	**114.71**
Accessories for channel drain		
Sump unit with sediment bucket	nr	**79.09**
End cap; inlet/outlet	nr	**16.92**
AGRICULTURAL DRAINAGE		
Excavate and form ditch and bank with 45 degree sides in light to medium soils; all widths taken at bottom of ditch		
300 wide x 600 deep	100 m	**110.00**
600 wide x 900 deep	100 m	**133.00**
1.20 m wide x 900 deep	100 m	**735.00**
1.50 m wide x 1.20 m deep	100 m	**1238.00**
Clear and bottom existing ditch average 1.50 m deep, trim back vegetation and remove debris to licensed tip not exceeding 13 km, lay jointed concrete pipes; to BS 5911 pt.1 class S; including bedding, haunching and topping with 150 mm concrete; 11.50 N/mm^2 - 40 mm aggregate; backfill with approved spoil from site		
pipes 300 dia.	100 m	**5603.00**
pipes 450 dia.	100 m	**7048.00**
pipes 600 dia.	100 m	**9278.00**
Clay land drain		
Excavate trench by excavator to 450 deep; lay 100 vitrified clay drain with butt joints, bedding Class B; backfill with excavated material screened to remove stones over 40, backfill to be laid in layers not exceeding 150; top with 150 mm topsoil remove surplus material to approved dump on site not exceeding 100m; final level of fill to allow for settlement	100 m	**1686.00**
SUB SOIL DRAINAGE; BY MACHINE		
Main drain; remove 150 topsoil and deposit alongside trench, excavate drain trench by machine and lay flexible perforated drain, lay bed of gravel rejects 100 mm; backfill with gravel rejects or similar to within 150 of finished ground level; complete fill with topsoil; remove surplus spoil to approved dump on site		
Main drain 160 mm in supplied in 35 m lengths		
450 mm deep	100 m	**549.00**
600 mm deep	100 m	**620.00**
900 mm deep	100 m	**976.70**
Extra for couplings	each	**3.00**
As above but with 100 mm main drain supplied in 100 m lengths		
450 mm deep	100 m	**846.00**
600 mm deep	100 m	**1109.00**
900 mm deep	100 m	**1819.00**
Extra for couplings	each	**2.00**

DRAINAGE

Item Excluding site overheads and profit	Unit	**Total rate £**
SUB SOIL DRAINAGE; BY MACHINE - cont'd		
Laterals to mains above; herringbone pattern; excavation and backfilling as above; inclusive of connecting lateral to main drain		
160 mm pipe to 450 mm deep trench		
laterals at 1.0 m centres	100 m²	**549.00**
laterals at 2.0 m centres	100 m²	**275.00**
laterals at 3.0 m centres	100 m²	**181.00**
laterals at 5.0 m centres	100 m²	**110.00**
laterals at 10.0 m centres	100 m²	**55.00**
160 mm pipe to 600 mm deep trench		
laterals at 1.0 m centres	100 m²	**620.00**
laterals at 2.0 m centres	100 m²	**310.00**
laterals at 3.0 m centres	100 m²	**205.00**
laterals at 5.0 m centres	100 m²	**124.00**
laterals at 10.0 m centres	100 m²	**62.00**
160 mm pipe to 900 mm deep trench		
laterals at 1.0 m centres	100 m²	**976.70**
laterals at 2.0 m centres	100 m²	**489.00**
laterals at 3.0 m centres	100 m²	**323.00**
laterals at 5.0 m centres	100 m²	**196.00**
laterals at 10.0 m centres	100 m²	**97.70**
Extra for 160/160 mm couplings connecting laterals to main drain		
laterals at 1.0 m centres	10 m	**78.00**
laterals at 2.0 m centres	10 m	**39.00**
laterals at 3.0 m centres	10 m	**26.00**
laterals at 5.0 m centres	10 m	**16.00**
laterals at 10.0 m centres	10 m	**8.00**
100 mm pipe to 450 mm deep trench		
laterals at 1.0 m centres	100 m²	**846.00**
laterals at 2.0 m centres	100 m²	**423.00**
laterals at 3.0 m centres	100 m²	**280.00**
laterals at 5.0 m centres	100 m²	**170.00**
laterals at 10.0 m centres	100 m²	**85.00**
100 mm pipe to 600 mm deep trench		
laterals at 1.0 m centres	100 m²	**1109.00**
laterals at 2.0 m centres	100 m²	**555.00**
laterals at 3.0 m centres	100 m²	**366.00**
laterals at 5.0 m centres	100 m²	**222.00**
laterals at 10.0 m centres	100 m²	**111.00**
100 mm pipe to 900 mm deep trench		
laterals at 1.0 m centres	100 m²	**1819.00**
laterals at 2.0 m centres	100 m²	**910.00**
laterals at 3.0 m centres	100 m²	**601.00**
laterals at 5.0 m centres	100 m²	**364.00**
laterals at 10.0 m centres	100 m²	**182.00**
80 mm pipe to 450 mm deep trench		
laterals at 1.0 m centres	100 m²	**805.00**
laterals at 2.0 m centres	100 m²	**403.00**
laterals at 3.0 m centres	100 m²	**266.00**
laterals at 5.0 m centres	100 m²	**161.00**
laterals at 10.0 m centres	100 m²	**81.00**
80 mm pipe to 600 mm deep trench		
laterals at 1.0 m centres	100 m²	**1068.00**
laterals at 2.0 m centres	100 m²	**534.00**
laterals at 3.0 m centres	100 m²	**353.00**
laterals at 5.0 m centres	100 m²	**214.00**
laterals at 10.0 m centres	100 m²	**107.00**
80 mm pipe to 900 mm deep trench		
laterals at 1.0 m centres	100 m²	**1778.00**
laterals at 2.0 m centres	100 m²	**889.00**
laterals at 3.0 m centres	100 m²	**587.00**
laterals at 5.0 m centres	100 m²	**356.00**
laterals at 10.0 m centres	100 m²	**178.00**

DRAINAGE

Item Excluding site overheads and profit	Unit	**Total rate £**
Extra for 100/80 mm junctions connecting laterals to main drain		
laterals at 1.0 m centres	10 m	**36.00**
laterals at 2.0 m centres	10 m	**18.00**
laterals at 3.0 m centres	10 m	**20.00**
laterals at 5.0 m centres	10m	**8.00**
laterals at 10.0 m centres	10m	**4.00**
SUB SOIL DRAINAGE; BY HAND		
Main drain; remove 150 topsoil and deposit alongside trench; excavate drain trench by machine and lay flexible perforated drain; lay bed of gravel rejects 100 mm; backfill with gravel rejects or similar to within 150 of finished ground level; complete fill with topsoil; remove surplus spoil to approved dump on site		
Main drain 160 mm in supplied in 35 m lengths		
450 mm deep	100 m	**1731.00**
600 mm deep	100 m	**2158.00**
900 mm deep	100 m	**2870.00**
Extra for couplings	each	**3.00**
As above but with 100 mm main drain supplied in 100 m lengths		
450 mm deep	100 m	**1144.00**
600 mm deep	100 m	**1439.00**
900 mm deep	100 m	**2005.00**
Extra for couplings	each	**2.00**
Laterals to mains above; herringbone pattern; excavation and backfilling as above; inclusive of connecting lateral to main drain		
160 mm pipe to 450 mm deep trench		
laterals at 1.0 m centres	100 m²	**1731.00**
laterals at 2.0 m centres	100 m²	**866.00**
laterals at 3.0 m centres	100 m²	**572.00**
laterals at 5.0 m centres	100 m²	**347.00**
laterals at 10.0 m centres	100 m²	**174.00**
160 mm pipe to 600 mm deep trench		
laterals at 1.0 m centres	100 m²	**2158.00**
laterals at 2.0 m centres	100 m²	**1079.00**
laterals at 3.0 m centres	100 m²	**712.00**
laterals at 5.0 m centres	100 m²	**432.00**
laterals at 10.0 m centres	100 m²	**216.00**
160 mm pipe to 900 mm deep trench		
laterals at 1.0 m centres	100 m²	**2870.00**
laterals at 2.0 m centres	100 m²	**1435.00**
laterals at 3.0 m centres	100 m²	**947.00**
laterals at 5.0 m centres	100 m²	**574.00**
laterals at 10.0 m centres	100 m²	**287.00**
Extra for 160/160 mm couplings connecting laterals to main drain		
laterals at 1.0 m centres	10 m	**78.00**
laterals at 2.0 m centres	10 m	**39.00**
laterals at 3.0 m centres	10 m	**26.00**
laterals at 5.0 m centres	10 m	**16.00**
laterals at 10.0 m centres	10 m	**8.00**
100 mm pipe to 450 mm deep trench		
laterals at 1.0 m centres	100 m²	**1144.00**
laterals at 2.0 m centres	100 m²	**572.00**
laterals at 3.0 m centres	100 m²	**378.00**
laterals at 5.0 m centres	100 m²	**229.00**
laterals at 10 m centres	100 m²	**115.00**
100 mm pipe to 600 mm deep trench		
laterals at 1.0 m centres	100 m²	**1439.00**
laterals at 2.0 m centres	100 m²	**720.00**
laterals at 3.0 m centres	100 m²	**475.00**
laterals at 5.0 m centres	100 m²	**288.00**
laterals at 10.0 m centres	100 m²	**144.00**

DRAINAGE

Item Excluding site overheads and profit	Unit	**Total rate £**
SUB SOIL DRAINAGE; BY HAND - cont'd		
Laterals to mains above; herringbone pattern - cont'd		
100 mm pipe to 900 mm deep trench		
laterals at 1.0 m centres	100 m²	**2005.00**
laterals at 2.0 m centres	100 m²	**1010.00**
laterals at 3.0 m centres	100 m²	**662.00**
laterals at 5.0 m centres	100 m²	**401.00**
laterals at 10.0 m centres	100 m²	**201.00**
80 mm pipe to 450 mm deep trench		
laterals at 1.0 m centres	100 m²	**1102.00**
laterals at 2.0 m centres	100 m²	**551.00**
laterals at 3.0 m centres	100 m²	**364.00**
laterals at 5.0 m centres	100 m²	**221.00**
laterals at 10.0 m centres	100 m²	**111.00**
80 mm pipe to 600 mm deep trench		
laterals at 1.0 m centres	100 m²	**1398.00**
laterals at 2.0 m centres	100 m²	**699.00**
laterals at 3.0 m centres	100 m²	**462.00**
laterals at 5.0 m centres	100 m²	**280.00**
laterals at 10.0 m centres	100 m²	**140.00**
80 mm pipe to 900 mm deep trench		
laterals at 1.0 m centres	100 m²	**1964.00**
laterals at 2.0 m centres	100 m²	**982.00**
laterals at 3.0 m centres	100 m²	**648.00**
laterals at 5.0 m centres	100 m²	**393.00**
laterals at 10.0 m centres	100 m²	**197.00**
Extra for 100/80 mm couplings connecting laterals to main drain		
laterals at 1.0 m centres	10 m	**36.00**
laterals at 2.0 m centres	10 m	**18.00**
laterals at 3.0 m centres	10 m	**20.00**
laterals at 5.0 m centres	10 m	**8.00**
laterals at 10.0 m centres	10 m	**4.00**
SOAKAWAYS		
Construct soakaway from perforated concrete rings; excavation, casting in situ concrete ring beam base; filling with gravel 250 mm deep; placing perforated concrete rings; surrounding with geofabric and backfilling with 250 mm granular surround and excavated material; step irons and cover slab; inclusive of all earthwork retention and disposal offsite of surplus material		
900 mm diameter		
1.00 m deep	nr	**475.00**
2.00 m deep	nr	**784.00**
1200 mm diameter		
1.00 m deep	nr	**643.00**
2.00 m deep	nr	**1080.00**
2400 mm diameter		
1.00 m deep	nr	**1596.00**
2.00 m deep	nr	**2663.00**

IRRIGATION

Item Excluding site overheads and profit	Unit	**Total rate £**
IRRIGATION		
LANDSCAPE IRRIGATION		
Automatic irrigation; Revaho UK Ltd		
Large garden consisting of 24 stations; 7000 m^2 irrigated area		
turf only	nr	**11000.00**
70/30% - turf/shrub beds	nr	**13000.00**
Medium sized garden consisting of 12 stations; 3500 m^2 irrigated area		
turf only	nr	**7200.00**
70/30% - turf/shrub beds	nr	**9900.00**
Medium sized garden consisting of 6 stations of irrigated area; 1000 m^2		
turf only	nr	**4900.00**
70/30% - turf/shrub beds	nr	**5400.00**
50/50% - turf/shrub beds	nr	**5900.00**
Leaky pipe irrigation; Leaky Pipe Ltd		
Works by machine; main supply and connection to laterals; excavate trench for main or ring main 450 mm deep; supply and lay pipe; backfill and lightly compact trench		
20 mm LDPE	100 m	**432.00**
16 mm LDPE	100 m	**403.00**
Works by hand; main supply and connection to laterals; excavate trench for main or ring main 450 mm deep; supply and lay pipe; backfill and lightly compact trench		
20 mm LDPE	100 m	**1317.00**
16 mm LDPE	100 m	**1288.00**
Turf area irrigation; laterals to mains; to cultivated soil; excavate trench 150 deep using hoe or mattock; lay moisture leaking pipe laid 150 mm below ground at centres of 350 mm		
Low leak	100 m^2	**513.00**
High leak	100 m^2	**464.00**
Landscape area irrigation; laterals to mains; moisture leaking pipe laid to the surface of irrigated areas at 600 mm centres		
Low leak	100 m^2	**264.00**
High leak	100 m^2	**235.00**
Landscape area irrigation; laterals to mains; moisture leaking pipe laid to the surface of irrigated areas at 900 mm centres		
Low leak	100 m^2	**177.00**
High leak	100 m^2	**157.00**
Multistation controller	each	**369.00**
Solenoid valves connected to automatic controller	each	**408.00**

WATER FEATURES

Item Excluding site overheads and profit	Unit	**Total rate £**
WATER FEATURES		
LAKES AND PONDS		
FAIRWATER LTD		
Excavate for small pond or lake maximum depth 1.00 m; remove arisings off site; grade and trim to shape; lay 75 mm sharp sand; line with 0.75 mm butyl liner 75 mm sharp sand and geofabric; cover over with sifted topsoil; anchor liner to anchor trench; install balancing tank and automatic top-up system		
Pond or lake of organic shape		
100 m^2; perimeter 50 m	each	**5300.00**
250 m^2; perimeter 90 m	each	**7900.00**
500 m^2; perimeter 130 m	each	**18353.00**
1000 m^2; perimeter 175 m	each	**36761.00**
Excavate for lake average depth 1.0 m, allow for bringing to specified levels; reserve topsoil; remove spoil to approved dump on site; remove all stones and debris over 75; lay polythene sheet including welding all joints and seams by specialist; screen and replace topsoil 200 mm thick		
Prices are for lakes of regular shape		
500 micron sheet	1000 m	**9213.80**
1000 micron sheet	1000 m	**10738.00**
Extra for removing spoil to tip	m^3	**22.00**
Extra for 25 sand blinding to lake bed	100 m^2	**108.00**
Extra for screening topsoil	m^2	**2.00**
Extra for spreading imported topsoil	100 m^2	**543.00**
Plant aquatic plants in lake topsoil		
Aponogeton distachyum - £280.00/100	100	**354.00**
Acorus calamus - £196.00/100	100	**270.00**
Butomus umbellatus - £196.00/100	100	**270.00**
Typha latifolia - £196.00/100	100	**270.00**
Nymphaea - £815.00/100	100	**889.00**
Formal water features		
Excavate and construct water feature of regular shape; lay 100 hardcore base and 150 concrete 1:2:4 site mixed; line base and vertical face with butyl liner 0.75 micron and construct vertical sides of reinforced blockwork; rendering 2 coats; anchor the liner behind blockwork; install pumps balancing tanks and all connections to mains supply		
1.00 x 1.00 x 1.00 m deep	nr	**2289.00**
2.00 x 1.00 x 1.00 m deep	nr	**2893.00**

TIMBER DECKING

Item Excluding site overheads and profit	Unit	**Total rate £**
TIMBER DECKING		
WYCKHAM-BLACKWELL LTD		
Timber decking; support structure of timber joists for decking laid on blinded base (measured separately)		
joists 50 x 150 mm	10 m^2	**377.00**
joists 50 x 200 mm	10 m^2	**323.00**
joists 50 x 250 mm	10 m^2	**303.00**
As above but inclusive of timber decking boards in yellow cedar 142 mm wide x 42 mm thick		
joists 50 x 150 mm	10 m^2	**817.00**
joists 50 x 200 mm	10 m^2	**763.00**
joists 50 x 250 mm	10 m^2	**743.00**
As above but boards 141 x 26 mm thick		
joists 50 x 200 mm	10 m^2	**323.00**
joists 50 x 250 mm	10 m^2	**303.00**
As above but decking boards in red cedar 131 mm wide x 42 mm thick		
joists 50 x 150 mm	10 m^2	**774.00**
joists 50 x 200 mm	10 m^2	**720.00**
joists 50 x 250 mm	10 m^2	**700.00**
Add to all of the above for handrails fixed to posts 100 x 100 x 1370 high		
square balusters at 100 centres	m	**74.00**
square balusters at 300 centres	m	**46.00**
turned balusters at 100 centres	m	**96.00**
turned balusters at 300 centres	m	**53.00**

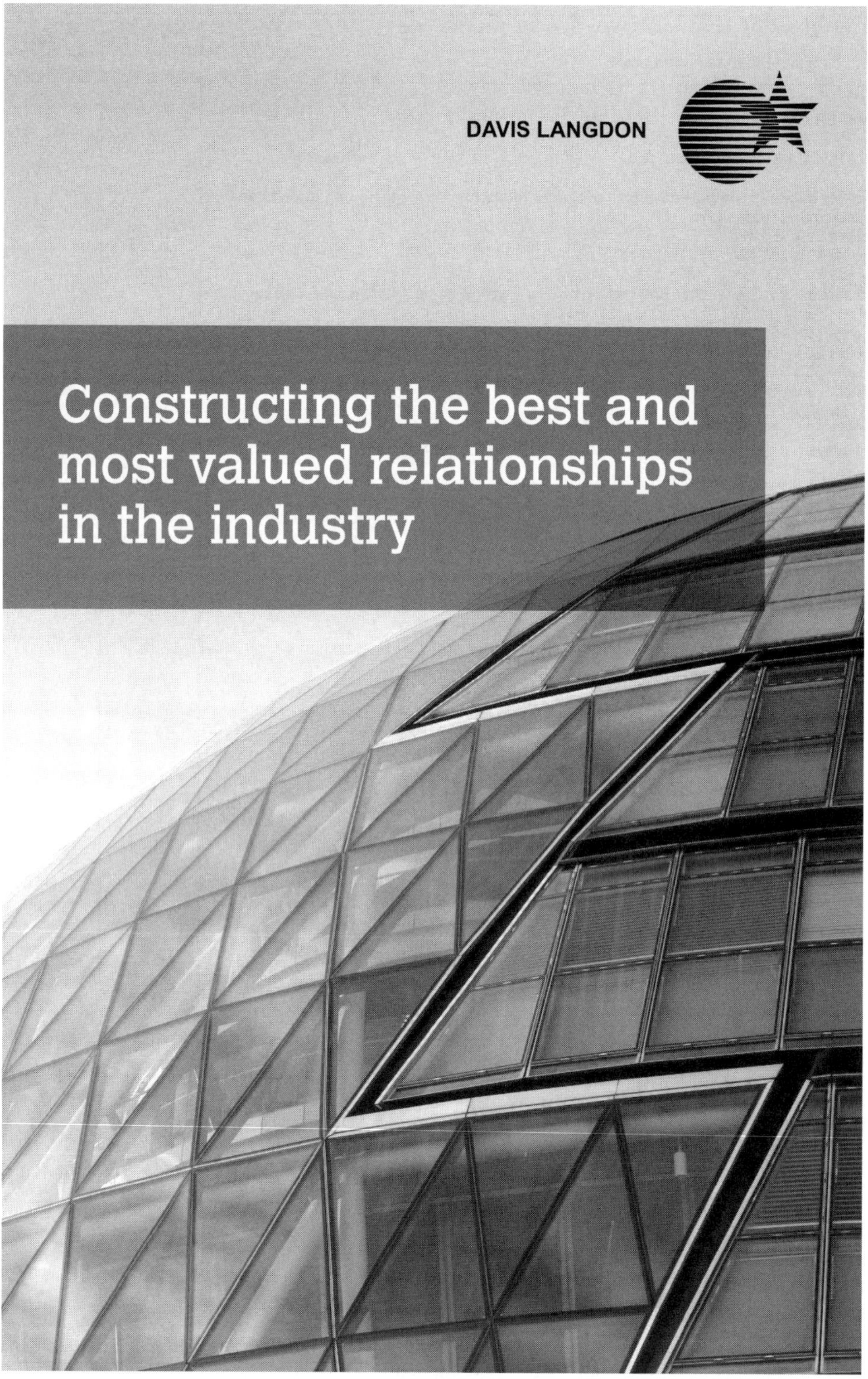
DAVIS LANGDON
Constructing the best and most valued relationships in the industry

www.davislangdon.com

Offices in Europe & the Middle East, Africa, Asia Pacific, Australasia and the USA

Project Management | Cost Management | Management Consulting | Legal Support
Specification Consulting | Engineering Services | Property Tax & Finance

ESSENTIAL READING FROM TAYLOR AND FRANCIS

Spon's Estimating Costs Guide to Electrical Works

Third Edition

Bryan Spain

All the cost data you need to keep your estimating accurate, competitive and profitable.

Do you work on jobs between £50 and £25,000? Then this book is for you.

Specially written for contractors and small businesses carrying out small works, Spon's Estimating Costs Guide to Electrical Works provides accurate information on thousands of rates, each broken down to labour, material overheads and profit for residential, retail and light industrial premises. It is the first book to include typical project costs for new installations, stripping out, rewiring and upgrading for flats and houses.

In addition, vital information and advice is given on setting up and running a business, employing staff, tax, VAT and CIS4's.

For the cost of approximately two hours of your charge-out rate (or less), this book will help you to:

produce estimates faster
keep your estimates accurate and competitive
run your business more effectively
save time.
No matter how big your firm is - from one-man-band to an established business - this book contains valuable commercial and cost information that you can't afford to be without.

December 2005: 216x138 mm: 304 pages
PB: 0-415-38614-4: £29.99

To Order: Tel: +44 (0) 1264 343071 Fax: +44 (0) 1264 343005, or
Post: Taylor and Francis Customer Services, Thomson Publishing Services, Cheriton House, Andover, Hants, SP10 5BE, UK Email: book.orders@tandf.co.uk

For a complete listing of all our titles visit:
www.tandf.co.uk

Taylor & Francis
Taylor & Francis Group

PART 3

Prices for Measured Works - Major Works

INTRODUCTION

Typical Project Profile

Contract value	£100 000.00 - £300 000.00
Labour rate (see page 4)	£18.50 per hour
Labour rate for maintenance contracts	£15.50 per hour
Number of site staff	18
Project area	6000 m^2
Project location	Outer London
Project components	50% hard landscape 50% soft landscape and planting
Access to works areas	Very good
Contract	Main contract
Delivery of materials	Full loads
Profit and site overheads	Excluded

ESSENTIAL READING FROM TAYLOR AND FRANCIS

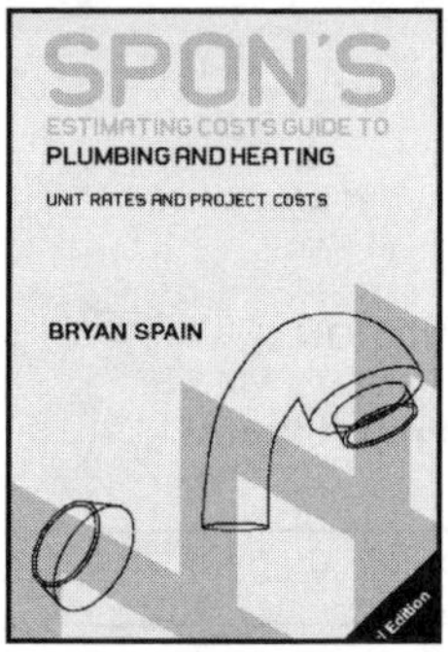

Spon's Estimating Cost Guide to Plumbing and Heating

Unit Rates and Project Costs

Third Edition

Bryan Spain

All the cost data you need to keep your estimating accurate, competitive and profitable.
Do you work on jobs between £50 and £25,000? Then this book is for you.
Specially written for contractors and small businesses carrying out small works, Spon's Estimating Costs Guide to Plumbing and Heating contains accurate information on thousands of rates, each broken down into labour, material overheads and profit.
The first book to include typical project costs for:
* rainwater goods installations
* bathrooms
* external waste systems
* central heating systems
* hot and cold water systems

More than just a price book, it gives easy-to-read professional advice on setting up and running a business including:
* taxation
* book-keeping, VAT and CIS4's
* legal obligations

For the cost of approximately two hours of your charge-out rate (or less), this book will help you to:
* produce estimates faster
* keep your estimates accurate and competitive
* run your business more effectively
* help save you time

No matter how big your firm is - from one-man-band to an established business - this book contains valuable commercial and cost information that you can't afford to be without.

December 2005: 216x138 mm: 264 pages
PB: 0-415-38618-7: £29.99

To Order: Tel: +44 (0) 1264 343071 Fax: +44 (0) 1264 343005, or
Post: Taylor and Francis Customer Services, Thomson Publishing Services, Cheriton House, Andover, Hants, SP10 5BE, UK Email: book.orders@tandf.co.uk

For a complete listing of all our titles visit:
www.tandf.co.uk

Taylor & Francis
Taylor & Francis Group

NEW ITEMS FOR THIS EDITION

Item Excluding site overheads and profit	PC £	Labour hours	Labour £	Plant £	Material £	Unit	**Total rate £**
A30 EMPLOYER'S REQUIREMENTS: TENDERING/SUB-LETTING/SUPPLY							
Tendering costs for an employed estimator for the acquisition of a landscape main or sub-contract; inclusive of measurement, sourcing of materials, and suppliers, cost and area calculations, pre and post tender meetings, bid compliance method statements and submission documents							
Remeasureable contract from prepared bills; contract value							
Value £25,000.00	-	-	-	-	-	nr	**180.00**
Value £50,000.00	-	-	-	-	-	nr	**300.00**
Value £100,000.00	-	-	-	-	-	nr	**720.00**
Value £200,000.00	-	-	-	-	-	nr	**1200.00**
Value £500,000.00	-	-	-	-	-	nr	**1560.00**
Value £1,000,000.00	-	-	-	-	-	nr	**2400.00**
Lump sum contract from specifications and drawings only; contract value							
Value £25,000.00	-	-	-	-	-	nr	**480.00**
Value £50,000.00	-	-	-	-	-	nr	**600.00**
Value £100,000.00	-	-	-	-	-	nr	**1200.00**
Value £200,000.00	-	-	-	-	-	nr	**2160.00**
Value £500,000.00	-	-	-	-	-	nr	**3000.00**
Value £1,000,000.00	-	-	-	-	-	nr	**4800.00**
D20 EXCAVATING AND FILLING							
Grading operations; surface previously excavated to reduce levels to prepare to receive subsequent treatments; grading to accurate levels and falls 20 mm tolerances							
Clay or heavy soils or hardcore							
5 tonne excavator	-	0.04	0.74	0.22	-	m^2	**0.96**
JCB Sitemaster 3CX	-	0.02	0.37	0.66	-	m^2	**1.03**
21 tonne 360 tracked excavator	-	0.01	0.19	0.41	-	m^2	**0.60**
by hand	-	0.10	1.85	-	-	m^2	**1.85**
Loamy topsoils							
5 tonne excavator	-	0.03	0.49	0.15	-	m^2	**0.64**
JCB Sitemaster 3CX	-	0.01	0.25	0.44	-	m^2	**0.68**
21 tonne 360 tracked excavator	-	0.01	0.14	0.31	-	m^2	**0.44**
by hand	-	0.05	0.93	-	-	m^2	**0.93**
Sand or graded granular materials							
5 tonne excavator	-	0.02	0.37	0.11	-	m^2	**0.48**
JCB Sitemaster 3CX	-	0.01	0.19	0.33	-	m^2	**0.52**
21 tonne 360 tracked excavator	-	0.01	0.11	0.24	-	m^2	**0.34**
by hand	-	0.03	0.62	-	-	m^2	**0.62**
Excavating; by hand							
Topsoil for preservation; loading to barrows							
average depth 100 mm	-	0.24	4.44	-	-	m^2	**4.44**
average depth 150 mm	-	0.36	6.66	-	-	m^2	**6.66**
average depth 200 mm	-	0.58	10.66	-	-	m^2	**10.66**
average depth 250 mm	-	0.72	13.32	-	-	m^2	**13.32**
average depth 300 mm	-	0.86	15.98	-	-	m^2	**15.98**

NEW ITEMS FOR THIS EDITION

Item Excluding site overheads and profit	PC £	Labour hours	Labour £	Plant £	Material £	Unit	**Total rate £**
D20 EXCAVATING AND FILLING - cont'd							
Grading operations; surface recently filled to raise levels to prepare to receive subsequent treatments; grading to accurate levels and falls 20 mm tolerances							
Clay or heavy soils or hardcore							
5 tonne excavator	-	0.03	0.56	0.16	-	m^2	**0.72**
JCB Sitemaster 3CX	-	0.02	0.28	0.50	-	m^2	**0.77**
21 tonne 360 tracked excavator	-	0.01	0.12	0.28	-	m^2	**0.40**
by hand	-	0.08	1.54	-	-	m^2	**1.54**
Loamy topsoils							
5 tonne excavator	-	0.02	0.38	0.11	-	m^2	**0.49**
JCB Sitemaster 3CX	-	0.01	0.19	0.34	-	m^2	**0.53**
21 tonne 360 tracked excavator	-	0.01	0.12	0.28	-	m^2	**0.40**
by hand	-	0.04	0.74	-	-	m^2	**0.74**
Sand or graded granular materials							
5 tonne excavator	-	0.02	0.28	0.08	-	m^2	**0.36**
JCB Sitemaster 3CX	-	0.01	0.14	0.25	-	m^2	**0.39**
21 tonne 360 tracked excavator	-	0.01	0.09	0.21	-	m^2	**0.30**
by hand	-	0.03	0.53	-	-	m^2	**0.53**
Q10 KERBS/EDGINGS/CHANNELS/PAVING ACCESSORIES							
Marshalls Mono; small element precast concrete kerb system; Keykerb Large (KL) upstand of 100 - 125 mm; on 150 mm concrete foundation including haunching with in situ concrete 1:3:6 1 side							
Bullnosed or half battered; 100 x 127 x 200							
laid straight	9.43	0.80	14.80	-	10.27	m	**25.07**
radial blocks laid to curve; 8 blocks/1/4 circle - 500 mm radius	7.25	1.00	18.50	-	8.10	m	**26.60**
radial blocks laid to curve; 8 radial blocks, alternating 8 standard blocks/1/4 circle - 1000 mm radius	11.97	1.25	23.13	-	12.81	m	**35.94**
radial blocks, alternating 16 standard blocks/1/4 circle - 1500 mm radius	11.01	1.50	27.75	-	11.86	m	**39.61**
internal angle 90 degree	3.98	0.20	3.70	-	3.98	nr	**7.68**
external angle	3.98	0.20	3.70	-	3.98	nr	**7.68**
drop crossing kerbs; KL half battered to KL splay; LH and RH	16.66	1.00	18.50	-	17.50	pair	**36.00**
drop crossing kerbs; KL half battered to KS bullnosed	16.20	1.00	18.50	-	17.04	pair	**35.54**
Marshalls Mono; small element precast concrete kerb system; Keykerb Small (KS) upstand of 25 - 50 mm; on 150 mm concrete foundation including haunching with in situ concrete 1:3:6: 1 side							
Half battered							
laid straight	6.62	0.80	14.80	-	7.46	m	**22.26**
radial blocks laid to curve; 8 blocks/1/4 circle - 500 mm radius	5.44	1.00	18.50	-	6.28	m	**24.78**
radial blocks laid to curve; 8 radial blocks, alternating 8 standard blocks/1/4 circle - 1000 mm radius	8.75	1.25	23.13	-	9.59	m	**32.72**
radial blocks, alternating 16 standard blocks/1/4 circle - 1500 mm radius	9.37	1.50	27.75	-	10.21	m	**37.96**
internal angle 90 degree	3.98	0.20	3.70	-	3.98	nr	**7.68**
external angle	3.98	0.20	3.70	-	3.98	nr	**7.68**

NEW ITEMS FOR THIS EDITION

Item Excluding site overheads and profit	PC £	Labour hours	Labour £	Plant £	Material £	Unit	Total rate £
Q26 SPECIAL SURFACINGS/PAVINGS FOR SPORT/GENERAL AMENITY							
Sports Pitches							
Natural sports pitches; White Horse Contractors Ltd; sports pitches; gravel raft construction							
Professional standard							
Football pitch 6500 m^2	-	-	-	-	-	nr	**150,000**
Cricket wicket - county standard	-	-	-	-	-	nr	**40,000**
Playing fields; Sport England compliant							
Football pitch 6500 m^2	-	-	-	-	-	nr	**80,000**
Cricket wicket; playing field standard	-	-	-	-	-	nr	**25,000**
Track and infield	-	-	-	-	-	nr	**125,000**
Artificial sports pitches; White Horse Contractors Ltd; 3rd generation rubber crumb							
Sports pitches to respective national governing body standards; inclusive of all excavation, drainage; lighting, fencing etc.							
Football; 106 m x 71 m	-	-	-	-	-	nr	**450,000**
Rugby union; multi-use pitch; 120 m x 75 m	-	-	-	-	-	nr	**550,000**
Hockey; water based; 101.4 m x 63 m	-	-	-	-	-	nr	**460,000**
Athletic track; 8 lane; International amateur athletic federation	-	-	-	-	-	nr	**750,000**
Multi-sport pitches; inclusive of all excavation, drainage; lighting, fencing etc;							
Sand dressed; 101.4 m x 63 m	-	-	-	-	-	nr	**365,000**
Sand filled; 101.4 m x 63 m	-	-	-	-	-	nr	**340,000**
Polymeric (synthetic bound rubber)	-	-	-	-	-	m^2	**70.00**
Macadam	-	-	-	-	-	m^2	**42.00**
Q30 SEEDING/TURFING							
Regional Environmental Mixes							
"British Seed Houses"; ref RE2 (Lowland Meadow); 5 g/m^2	-	-	-	-	33.75	100 m^2	-
"British Seed Houses"; ref RE3 (Riverflood Plain/Water Meadow); 5 g/m^2	-	-	-	-	33.25	100 m^2	-
"British Seed Houses"; ref RE4 (Lowland Limestone); 5 g/m^2	-	-	-	-	33.25	100 m^2	-
"British Seed Houses"; ref RE5 (Calcareous Sub-mountain Restoration); 5 g/m^2	-	-	-	-	37.10	100 m^2	-
"British Seed Houses"; ref RE7 (Acid Sub-mountain Restoration); 5 g/m^2	-	-	-	-	32.25	100 m^2	-
"British Seed Houses"; ref RE8 (Coastal Reclamation); 5 g/m^2	-	-	-	-	41.45	100 m^2	-
"British Seed Houses"; ref RE9 (Farmland Mixture); 5 g/m^2	-	-	-	-	28.85	100 m^2	-
"British Seed Houses"; ref RE10 (Marginal Land); 5 g/m^2	-	-	-	-	42.75	100 m^2	-
"British Seed Houses"; ref RE11 (Heath Scrubland); 5 g/m^2	-	-	-	-	12.30	100 m^2	-
"British Seed Houses"; ref RE12 (Drought Land); 5 g/m^2	-	-	-	-	25.95	100 m^2	-

NEW ITEMS FOR THIS EDITION

Item Excluding site overheads and profit	PC £	Labour hours	Labour £	Plant £	Material £	Unit	**Total rate £**
Q30/35 SEEDING/TURFING/LANDSCAPE MAINTENANCE							
Market prices of landscape chemicals at suggested application rates							
Residual herbicides							
Embargo G; Rigby Taylor Ltd; residual pre and post emergent herbicide							
80 - 125 kg/ha; annual meadow-grass, black grass, charlock, common chickweed, common mouse-ear, common orache, common poppy, corn marigold, corn spurrey, fat-hen, groundsel, hedge mustard, scentless mayweed, small nettle, sow-thistle, stinking chamomile, wild-oat	-	-	-	-	0.48	100 m^2	-
Premiere; Scotts Horticulture; granular weedkiller approved for use in new plantings of trees and shrubs							
Application rate 1 kg/100 m^2 (100 kg/ha)	-	-	-	-	30.38	100 m^2	-
Casoron G; Scotts Horticulture; control of broadleaved weeds; existing weeds and germinating weeds; a wide range of annual and perennial weeds							
Selective; 560 g/100 m^2: control of germinating annual and perennial weeds and light to moderate infestation of established annuals	-	-	-	-	2.53	100 m^2	-
Selective; 1 kg/100 m^2: control more persistent weeds in the above category	-	-	-	-	2.53	100 m^2	-
Selective; 1.25 kg/100 m^2: control established weed	-	-	-	-	5.65	100 m^2	-
Selective; 1 kg/280 m^2	-	-	-	-	1.46	100 m^2	-
Contact herbicides; Scotts Horticulture							
Dextrone X (Diquat Paraquat)							
8.5 L/ha	-	-	-	-	0.77	100 m^2	-
3.0 L/ha	-	-	-	-	0.28	100 m^2	-
Speedway 2 (Diquat Paraquat)							
1220 m^2/4 kg	-	-	-	-	2.25	100 m^2	-
3333 m^2/4 kg	-	-	-	-	0.82	100 m^2	-
Selective herbicides							
Rigby Taylor; junction for broadleaf weeds in turf grasses							
Application rate; 1.2 L/ha	-	-	-	-	0.45	100 m^2	-
Rigby Taylor; Greenor systemic selective treatment of weeds in turfgrass							
Application rate; 4 L/ha	-	-	-	-	0.88	100 m^2	-
Rigby Taylor Bastion -T; control of weeds in both established and newly seeded turf							
Application rate	-	-	-	-	0.88	100 m^2	-
Scotts; Intrepid 2; for broadleaf weed control in grass							
Application rate; 7700 m^2/5 L	-	-	-	-	0.52	100 m^2	-
Scotts; Re-Act; selective herbicide; established managed amenity turf, and newly seeded grass; controls many annual and perennial weeds; will not vaporise in hot conditions							
Application rate; 9090 - 14,285 m^2/5 L	-	-	-	-	0.45	100 m^2	-
Bayer Environmental Science; Asulox; post-emergence translocated herbicide for the control of bracken and docks							
Application rate; docks; 17857 m^2/5 L	-	-	-	-	0.42	100 m^2	-
Application rate; bracken; 4545 m^2/5 L	-	-	-	-	1.65	100 m^2	-

NEW ITEMS FOR THIS EDITION

Item Excluding site overheads and profit	PC £	Labour hours	Labour £	Plant £	Material £	Unit	**Total rate £**
Bayer Environmental Science; Clovotox; selective herbicide for the control of clover and other difficult broadleaved weeds in turf							
Application rate; 4545 m^2/5 L	-	-	-	-	0.77	100 m^2	-
Dicotox Extra; Bayer Environmental Science; showerproof herbicide for the control of certain broadleaved weeds on turf and for the control of heather and woody weeds in forestry situations; also controls ragwort							
Application rate; 2.8 L/ha; plantains; hawksweed buttercups, etc.	-	-	-	-	0.22	100 m^2	-
Application rate; 5.6 L/ha; birdsfoot clovers mouse-ear etc. (1 application)	-	-	-	-	0.43	100 m^2	-
Application rate; 5.6 L/ha; pearlwort, yarrow (2 applications)	-	-	-	-	0.87	100 m^2	-
Application rate; 13 L/ha; woody weeds and heather in established trees and shrub areas	-	-	-	-	1.01	100 m^2	-
Dormone; Bayer Environmental Science; selective herbicide for the control of broadleaved weeds in amenity grassland situations; may also be used in situations in or near water							
Application rate; 2.8 L/ha	-	-	-	-	0.18	100 m^2	-
Application rate; 5.6 L/ha	-	-	-	-	0.36	100 m^2	-
Near water; application rate; 4.5 L/ha	-	-	-	-	0.29	100 m^2	-
Near water; application rate; 9 L/ha	-	-	-	-	0.58	100 m^2	-
Spearhead; Bayer Environmental Science; selective herbicide for the control of broadleaved weeds in established turf							
Application rate; 4.5 L/ha	-	-	-	-	0.84	100 m^2	-
Total herbicides (Note: these application rates will vary dependent on season)							
Rigby Taylor Ltd; Gallup Biograde amenity; Glyphosate 360 g/L formulation							
General use; woody weeds; ash beech bracken bramble; 3 L/ha	-	-	-	-	0.27	100 m^2	-
Annual and perennial grasses; heather (peat soils); 4 L/ha	-	-	-	-	0.36	100 m^2	-
Pre-planting; general clearance; 5 L/ha	-	-	-	-	0.45	100 m^2	-
Heather; mineral soils; 6 L/ha	-	-	-	-	0.54	100 m^2	-
Rhododendron; 10 L/ha	-	-	-	-	0.90	100 m^2	-
Gallup Hi-aktiv Amenity; 490 g/L formulation							
General use; woody weeds; ash, beech, bracken, bramble; 2.2 L/ha	-	-	-	-	0.23	100 m^2	-
Annual and perennial grasses; heather (peat soils); 2.9 L/ha	-	-	-	-	0.31	100 m^2	-
Pre-planting; general clearance; 3.7 L/ha	-	-	-	-	0.39	100 m^2	-
Heather; mineral soils; 4.4 L/ha	-	-	-	-	0.47	100 m^2	-
Rhododendron; 7.3 L/ha	-	-	-	-	0.77	100 m^2	-
Roundup Pro Biactive; Scotts							
Application rate; 5 L/ha	-	-	-	-	0.53	100 m^2	-
Casoron G; Scotts							
Total; 2.25 kg/100 m^2	-	-	-	-	10.17	100 m^2	-
Kerb							
"Kerb Flowable (Propyzamide)"	-	-	-	-	2.16	100 m^2	-
"Kerb Granules" (2 x 2000 tree pack)	-	-	-	-	0.06	tree	-

NEW ITEMS FOR THIS EDITION

Item Excluding site overheads and profit	PC £	Labour hours	Labour £	Plant £	Material £	Unit	Total rate £
Q30/35 SEEDING/TURFING/LANDSCAPE MAINTENANCE - cont'd							
Woody weed herbicides; Rigby Taylor Ltd							
Timbrel; summer applied; water based Bayer Environmental Science; selective scrub and brushwood herbicide							
2.0 L/ha; bramble, briar, broom, gorse, nettle	-	-	-	-	0.98	100 m^2	-
4.0 L/ha; alder, birch, blackthorn, dogwood, elder, poplar, rosebay willowherb, sycamore	-	-	-	-	1.97	100 m^2	-
6.0 L/ha; beech, box, buckthorn, elm, hazel, hornbeam, horse chestnut, lime, maple, privet, rowan, Spanish chestnut, willow, wild pear	-	-	-	-	2.95	100 m^2	-
8.0 L/ha; ash, oak, rhododendron	-	-	-	-	3.93	100 m^2	-
Timbrel; winter applied; paraffin or diesel based; Bayer Environmental Science; selective scrub and brushwood herbicide							
3.0 L/ha; bramble, briar, broom, gorse, nettle	-	-	-	-	1.47	100 m^2	-
6.0 L/ha; alder, ash, beech, birch, blackthorn, box, buckthorn, dogwood, elder, elm, hazel, hornbeam, horse chestnut, lime, maple, oak, poplar, privet, rowan, Spanish chestnut, sycamore, willow, wild pear	-	-	-	-	2.95	100 m^2	-
10.0 L/ha; hawthorn, laurel, rhododendron	-	-	-	-	4.92	100 m^2	-
Aquatic herbicides							
Casoron G; control of broad-leaved weeds; existing weeds and germinating weeds; a wide range of annual and perennial weeds							
Aquatic: 1,665 m^2/25 kg	-	-	-	-	6.78	100 m^2	-
Aquatic: 5,555 m^2/25 kg	-	-	-	-	2.03	100 m^2	-
Fungicides							
Rigby Taylor; Masalon; systemic							
Application rate; 8 L/ha	-	-	-	-	7.18	100 m^2	-
Scotts UK Daconil turf							
Application rate; 333 m^2/L	-	-	-	-	5.13	100 m^2	-
Moss control							
Enforcer; Scotts; surface biocide and mosskiller for the control of mosses in turf or on external hard surfaces							
Turf areas; application rate 1 L/200 m^2	-	-	-	-	6.13	100 m^2	-
Hard surfaces; moss and fungi control; application rate 1 L/588 m^2	-	-	-	-	2.11	100 m^2	-
Insect control							
Merit Turf; insecticide for the control of chafer grubs and leatherjackets; Bayer Environmental Science							
Application rate; 10 kg/3333 m^2	-	-	-	-	5.20	100 m^2	-
Mildothane Turf Liquid; casting worm control and fungicide; Bayer Environmental Science							
Application rate; 5 L/6667 m^2	-	-	-	-	0.07	100 m^2	-
Spray mark indicator							
Rigby Taylor; Trail Blazer; mixed with chemicals; price per 1000 L applied	-	-	-	-	19.75	1000 L	-

NEW ITEMS FOR THIS EDITION

Item Excluding site overheads and profit	PC £	Labour hours	Labour £	Plant £	Material £	Unit	**Total rate £**
Q40 FENCING							
Traditional trellis panels; The Garden Trellis Company; bespoke trellis panels for decorative, screening or security applications; timber planed all round; height of trellis 1800 mm							
Freestanding panels; posts 70 x 70 set in concrete; timber frames mitred and grooved 45 x 34 mm; heights and widths to suit; slats 32 mm x 10 mm joinery quality tanalised timber at 100 mm ccs; elements fixed by galvanised staples; capping rail 70 x 34							
HV68 horizontal and vertical slats; softwood	63.00	1.00	18.50	-	69.16	m	**87.66**
D68 diagonal slats; softwood	72.00	1.00	18.50	-	78.16	m	**96.66**
HV68 horizontal and vertical slats; hardwood iroko	126.00	1.00	18.50	-	132.16	m	**150.66**
D68 diagonal slats; hardwood iroko or Western Red Cedar	144.00	1.00	18.50	-	150.16	m	**168.66**
Trellis panels fixed to face of existing wall or railings; timber frames mitred and grooved 45 x 34 mm; heights and widths to suit; slats 32 mm x 10 mm joinery quality tanalised timber at 100 mm ccs; elements fixed by galvanised staples; capping rail 70 x 34							
HV68 horizontal and vertical slats; softwood	57.60	0.50	9.25	-	59.52	m	**68.77**
D68 diagonal slats; softwood	68.40	0.50	9.25	-	70.32	m	**79.57**
HV68 horizontal and vertical slats; iroko or Western Red Cedar	117.00	0.50	9.25	-	118.92	m	**128.17**
D68 diagonal slats; iroko or Western Red Cedar	135.00	0.50	9.25	-	136.92	m	**146.17**
Contemporary style trellis panels; The Garden Trellis Company; bespoke trellis panels for decorative, screening or security applications; timber planed all round							
Freestanding panels 30/15; posts 70 x 70 set in concrete with 90 x 30 top capping; slats 30 x 14 mm with 15 mm gaps; vertical support at 450 mm ccs							
joinery treated softwood	81.00	1.00	18.50	-	87.16	m	**105.66**
hardwood iroko or Western Red Cedar	162.00	1.00	18.50	-	168.16	m	**186.66**
Panels 30/15 face fixed to existing wall or fence; posts 70 x 70 set in concrete with 90 x 30 top capping; slats 30 x 14 mm with 15 mm gaps; vertical support at 450 mm ccs							
joinery treated softwood	72.00	0.50	9.25	-	73.92	m	**83.17**
hardwood iroko or Western Red Cedar	153.00	0.50	9.25	-	154.92	m	**164.17**
Integral arches to ornamental trellis panels; The Garden Trellis Company							
Arches to trellis panels in 45 mm x 34 mm grooved timbers to match framing; fixed to posts							
R450 1/4 circle; joinery treated softwood	35.00	1.50	27.75	-	41.16	nr	**68.91**
R450 1/4 circle; hardwood iroko or Western Red Cedar	40.00	1.50	27.75	-	46.16	nr	**73.91**
Arches to span 1800 wide							
joinery treated softwood	45.00	1.50	27.75	-	51.16	nr	**78.91**
hardwood iroko or Western Red Cedar	60.00	1.50	27.75	-	66.16	nr	**93.91**
Painting or staining of trellis panels; high quality coatings							
microporous opaque paint or spirit based stain	-	-	-	-	15.00	m^2	-

NEW ITEMS FOR THIS EDITION

Item Excluding site overheads and profit	PC £	Labour hours	Labour £	Plant £	Material £	Unit	**Total rate £**
R12 DRAINAGE BELOW GROUND							
Linear Drainage							
Loadings for slot drains							
A15 - 1.5 tonne - pedestrian							
B125 - 12.5 tonne domestic use							
C250 - 25 tonne; car parks, supermarkets, industrial units							
D400 - 40 tonne; highways							
E600 - 60 tonne; forklifts							
Linear drainage; channels; Aco Building Products; laid to concrete bed C25 on compacted granular base on 200 mm deep concrete bed; haunched with 200 mm concrete surround; all in 750 x 430 wide trench with compacted 200 granular base surround (excavation and sub-base not included)							
Slot drains							
ACO brickslot; load class C250; discreet slot drainage system ideal for visually sensitive paved areas	69.06	0.33	6.17	-	70.48	m	**76.65**
Extra over for laying of paving (PC £300.00/1000) adjacent to channel on epoxy mortar bed; stretcher bond	0.36	0.33	6.17	-	3.44	m	**9.60**
Channel footpath drain							
Aco channel drain to pedestrian footpaths; connection to existing kerbdrain; laying on 150 mm concrete base and surround; excavation sub-base not included	23.56	2.00	37.00	-	32.31	m	**69.31**
NK100; to C250; 100 mm bore; polymer concrete drain units; tapered to falls; on 150 mm concrete with 150 concrete surround							
Constant or pre-sloped invert	33.67	0.80	14.80	-	44.41	m	**59.21**
Accessories for NK100							
Sump unit - complete with sediment bucket	60.59	1.00	18.50	-	60.59	nr	**79.09**
End cap - closing piece	-	-	-	-	7.94	nr	-
End cap - inlet/outlet	-	-	-	-	16.92	nr	-
Step connectors	-	-	-	-	12.97	nr	-
Aco Building Products; Extra over for grating covers for NK100							
Load class A15 (pedestrian areas)							
Slotted galvanised steel grating. Ref: 423QL/424QL	-	-	-	-	34.15	m	-
Slotted stainless steel grating. Ref: 470QL	-	-	-	-	99.70	m	-
1433 Load Class C250 (Cars and light vans)							
Slotted galvanised steel grating. Ref: 405/406.	-	-	-	-	34.15	m	-
Galvanised steel mesh grating. Ref: 413/414.	-	-	-	-	25.60	m	-
Heelguard™' composite 522QL	-	-	-	-	33.40	m	-
'Heelguard™' ductile iron grating. Ref: 508QL.	-	-	-	-	35.24	m	-
Wade Ltd; stainless steel channel drains; specialised applications; bespoke manufacture							
Drain in stainless steel; c/w tie in lugs and inbuilt falls to 100 mm spigot outlet; stainless steel gratings							
"NE" channel; Ref 12430 secured gratings	252.00	1.20	22.20	-	262.74	m	**284.94**

NEW ITEMS FOR THIS EDITION

Item Excluding site overheads and profit	PC £	Labour hours	Labour £	Plant £	Material £	Unit	Total rate £
R13 LAND DRAINAGE							
Fin drains; Cooper Clarke Ltd; Geofin shallow drain; drainage of sports fields and grassed landscaped areas							
Geofin shallow drain; to trenches; excavation by trenching machine; backfilled with single size aggregate 20 mm and covered with sharp sand rootzone 200 thick							
Geofin 25 mm thick x 150 deep	2.85	0.05	0.93	0.70	11.03	m	**12.65**
Geofin 25 mm thick x 450 deep	3.09	0.07	1.23	0.93	15.00	m	**17.16**
Geofin 25 mm thick x 900 deep	3.80	0.10	1.85	0.93	22.40	m	**25.18**
Fin drains; Cooper Clarke Ltd; Geofin composite fin drain laid to slabs							
Geofin fin drain laid horizontally to slab or blinded ground; covered with 20 mm shingle 200 mm thick							
Geofin 25 mm thick x 900 wide laid flat to falls	4.22	0.02	0.37	0.14	10.88	m^2	**11.39**
S10 COLD WATER							
Borehole drilling; White Horse Contractors Ltd							
Drilling operations to average 100 m depth; lining with 150 mm diameter sieve							
Drilling to 100 m	-	-	-	-	-	nr	**12000.00**
Rate per m over 100 m	-	-	-	-	-	m	**120.00**
Pump, rising main, cable and kiosk	-	-	-	-	-	nr	**5000.00**
S14 IRRIGATION							
Revaho Ltd; water recycling							
Fully automatic self cleaning water recycle system for use with domestic grey water including sub surface tank self flushing filter pump and control system; excludes plumbing of the drainage system from the house into the tank	-	-	-	-	-	nr	**2500.00**
Revaho Ltd; EZ Pro Max moisture sensing control system							
Large garden comprising 7000 m^2 and 24 stations with 4 moisture sensors							
Turf only	-	-	-	-	-	nr	**12000.00**
Turf/shrub beds 70/30	-	-	-	-	-	nr	**7000.00**
Medium garden comprising 3500 m^2 and 12 stations with 2 moisture sensors							
Turf only	-	-	-	-	-	nr	**8000.00**
Turf/shrub beds 70/30	-	-	-	-	-	nr	**10400.00**
Smaller gardens garden comprising 1000 m^2 and 6 stations with 1 moisture sensor							
Turf only	-	-	-	-	-	nr	**4900.00**
Turf/shrub beds 70/30	-	-	-	-	-	nr	**5400.00**
Turf/shrub beds 50/50	-	-	-	-	-	nr	**5800.00**
S15 FOUNTAINS/WATER FEATURES							
Natural swimming pools; Fairwater Ltd							
Natural swimming pool of 50 m^2 area; shingle regeneration zone 50 m^2 planted with marginal and aquatic plants	-	-	-	-	-	nr	**55000.00**

NEW ITEMS FOR THIS EDITION

Item Excluding site overheads and profit	PC £	Labour hours	Labour £	Plant £	Material £	Unit	**Total rate £**
S15 FOUNTAINS/WATER FEATURES - cont'd							
Swimming pools; Thompson Landscapes							
Conventional swimming pool construction inclusive of excavation reinforced gunnite construction, heating, lighting, pumping, filtration and mosaic lined							
Rectangular construction 14 x 6 m	-	-	-	-	-	nr	**51000.00**
Duraglide cover inclusive of all electrical works	-	-	-	-	-	nr	**22000.00**
Automatic cleaning system with main drains	-	-	-	-	-	nr	**13000.00**
Water features; Fairwater Ltd							
Natural stream; butyl lined level changes of 1.00 M; water pumped from lower pond (not included) via balancing tank; level changes via Purbeck stone water falls							
20 m long stream	-	-	-	-	-	nr	**18000.00**
Bespoke freestanding water wall; steel or glass panel; self contained recirculation system and reservoir							
water wall; 2.00 high x 850 mm wide	-	-	-	-	-	nr	**8500.00**

A PRELIMINARIES

Item Excluding site overheads and profit	PC £	Labour hours	Labour £	Plant £	Material £	Unit	**Total rate £**
A11 TENDER AND CONTRACT DOCUMENTS							
Health and Safety							
Produce health and safety file including preliminary meeting and subsequent progress meetings with external planning officer in connection with health and safety; project value:							
£35,000	-	8.00	240.00	-	-	nr	**240.00**
£75000	-	12.00	360.00	-	-	nr	**360.00**
£100000	-	16.00	480.00	-	-	nr	**480.00**
£200,000 to £500,000	-	40.00	1200.00	-	-	nr	**1200.00**
Maintain health and safety file for project duration							
£35,000	-	4.00	120.00	-	-	week	**120.00**
£75000	-	4.00	120.00	-	-	week	**120.00**
£100000	-	8.00	240.00	-	-	week	**240.00**
£200,000 to £500,000	-	8.00	240.00	-	-	week	**240.00**
Produce written risk assessments on all areas of operations within the scope of works of the contract; project value:							
£35,000	-	2.00	60.00	-	-	week	**60.00**
£75000	-	3.00	90.00	-	-	week	**90.00**
£100000	-	5.00	150.00	-	-	week	**150.00**
£200,000 to £500,000	-	8.00	240.00	-	-	week	**240.00**
Produce Coshh assessments on all substances to be used in connection with the contract; project value:							
£35,000	-	1.50	45.00	-	-	nr	**45.00**
£75000	-	2.00	60.00	-	-	nr	**60.00**
£100000	-	3.00	90.00	-	-	nr	**90.00**
£200,000 to £500,000	-	3.00	90.00	-	-	nr	**90.00**
Method Statements							
Provide detailed method statements on all aspects of the works; project value:							
£ 30,000	-	2.50	75.00	-	-	nr	**75.00**
£ 50,000	-	3.50	105.00	-	-	nr	**105.00**
£ 75,000	-	4.00	120.00	-	-	nr	**120.00**
£ 100,000	-	5.00	150.00	-	-	nr	**150.00**
£ 200,000	-	6.00	180.00	-	-	nr	**180.00**
A30 EMPLOYER'S REQUIREMENTS: TENDERING/SUB-LETTING/SUPPLY							
Tendering costs for an employed estimator for the acquisition of a landscape main or sub-contract; inclusive of measurement, sourcing of materials and suppliers, cost and area calculations, pre and post tender meetings, bid compliance method statements and submission documents							
Remeasureable contract from prepared bills; contract value							
Value £25,000.00	-	-	-	-	-	nr	**180.00**
Value £50,000.00	-	-	-	-	-	nr	**300.00**
Value £100,000.00	-	-	-	-	-	nr	**720.00**
Value £200,000.00	-	-	-	-	-	nr	**1200.00**
Value £500,000.00	-	-	-	-	-	nr	**1560.00**
Value £1,000,000.00	-	-	-	-	-	nr	**2400.00**

A PRELIMINARIES

Item Excluding site overheads and profit	PC £	Labour hours	Labour £	Plant £	Material £	Unit	**Total rate £**
A30 EMPLOYER'S REQUIREMENTS: TENDERING/SUB-LETTING/SUPPLY - cont'd							
Tendering costs - cont'd							
Lump sum contract from specifications and drawings only; contract value							
Value £25,000.00	-	-	-	-	-	nr	**480.00**
Value £50,000.00	-	-	-	-	-	nr	**600.00**
Value £100,000.00	-	-	-	-	-	nr	**1200.00**
Value £200,000.00	-	-	-	-	-	nr	**2160.00**
Value £500,000.00	-	-	-	-	-	nr	**3000.00**
Value £1,000,000.00	-	-	-	-	-	nr	**4800.00**
A32 EMPLOYER'S REQUIREMENTS: MANAGEMENT OF THE WORKS							
Programmes							
Allow for production of works programmes prior to the start of the works; project value:							
£ 30,000	-	3.00	90.00	-	-	nr	**90.00**
£ 50,000	-	6.00	180.00	-	-	nr	**180.00**
£ 75,000	-	8.00	240.00	-	-	nr	**240.00**
£ 100,000	-	10.00	300.00	-	-	nr	**300.00**
£ 200,000	-	14.00	420.00	-	-	nr	**420.00**
Allow for updating the works programme during the course of the works; project value:							
£ 30,000	-	1.00	30.00	-	-	nr	**30.00**
£ 50,000	-	1.50	45.00	-	-	nr	**45.00**
£ 75,000	-	2.00	60.00	-	-	nr	**60.00**
£ 100,000	-	3.00	90.00	-	-	nr	**90.00**
£ 200,000	-	5.00	150.00	-	-	nr	**150.00**
Setting out							
Setting out for external works operations comprising hard and soft works elements; placing of pegs and string lines to Landscape Architect's drawings; surveying levels and placing level pegs; obtaining approval from the Landscape Architect to commence works; areas of entire site							
1,000 m^2	-	5.00	92.50	-	6.00	nr	**98.50**
2,500 m^2	-	8.00	148.00	-	12.00	nr	**160.00**
5,000 m^2	-	8.00	148.00	-	12.00	nr	**160.00**
10,000 m^2	-	32.00	592.00	-	30.00	nr	**622.00**
A34 EMPLOYER'S REQUIREMENTS: SECURITY/SAFETY/PROTECTION							
Jacksons Fencing; Express fence; framed mesh unclimbable fencing; including precast concrete supports and couplings							
Weekly hire							
2.0 m high; weekly hire rate	-	-	-	1.40	-	m	**1.40**
erection of fencing; labour only	-	0.10	1.85	-	-	m	**1.85**
removal of fencing loading to collection vehicle	-	0.07	1.23	-	-	m	**1.23**
delivery charge	-	-	-	0.80	-	m	**0.80**
return haulage charge	-	-	-	0.60	-	m	**0.60**

A PRELIMINARIES

Item Excluding site overheads and profit	PC £	Labour hours	Labour £	Plant £	Material £	Unit	**Total rate £**
A41 CONTRACTOR'S GENERAL COST ITEMS: SITE ACCOMMODATION							
General							
The following items are instances of the commonly found preliminary costs associated with external works contracts. The assumption is made that the external works contractor is sub-contracted to a main contractor.							
Elliotthire; erect temporary accommodation and storage on concrete base measured separately							
Prefabricated office hire; jackleg; open plan							
3.6 x 2.4 m	-	-	-	-	-	week	**35.00**
4.8 x 2.4 m	-	-	-	-	-	week	**38.00**
Armoured store; erect temporary secure storage container for tools and equipment							
3.0 x 2.4 m	-	-	-	-	-	week	**13.00**
3.6 x 2.4 m	-	-	-	-	-	week	**13.00**
Delivery and collection charges on site offices							
delivery charge	-	-	-	-	-	load	**120.00**
collection charge	-	-	-	-	-	load	**120.00**
Toilet facilities							
John Anderson Ltd.; serviced self-contained toilet delivered to and collected from site; maintained by toilet supply company							
single chemical toilet including wash hand basin & water	-	-	-	-	-	week	**24.00**
delivery and collection; each way	-	-	-	-	-	nr	**19.00**
A43 CONTRACTOR'S GENERAL COST ITEMS: MECHANICAL PLANT							
Southern Conveyors; moving only of granular material by belt conveyor; conveyors fitted with troughed belts, receiving hopper, front and rear undercarriages and driven by electrical and air motors; support work for conveyor installation (scaffolding), delivery, collection and installation all excluded							
Conveyor belt width 400 mm; mechanically loaded and removed at offload point; conveyor length							
up to 5 m	-	1.50	27.75	6.94	-	m^3	**34.69**
10 m	-	1.50	27.75	7.71	-	m^3	**35.46**
12.5 m	-	1.50	27.75	8.05	-	m^3	**35.80**
15 m	-	1.50	27.75	8.48	-	m^3	**36.23**
20 m	-	1.50	27.75	9.26	-	m^3	**37.01**
25 m	-	1.50	27.75	9.95	-	m^3	**37.70**
30 m	-	1.50	27.75	10.72	-	m^3	**38.47**
Conveyor belt width 600 mm; mechanically loaded and removed at offload point; conveyor length							
up to 5 m	-	1.00	18.50	5.03	-	m^3	**23.53**
10 m	-	1.00	18.50	5.66	-	m^3	**24.16**
12.5 m	-	1.00	18.50	6.00	-	m^3	**24.50**
15 m	-	1.00	18.50	6.35	-	m^3	**24.85**
20 m	-	1.00	18.50	7.03	-	m^3	**25.54**
25 m	-	1.00	18.50	7.69	-	m^3	**26.20**
30 m	-	1.00	18.50	8.35	-	m^3	**26.85**

A PRELIMINARIES

Item Excluding site overheads and profit	PC £	Labour hours	Labour £	Plant £	Material £	Unit	**Total rate £**
A43 CONTRACTOR'S GENERAL COST ITEMS: MECHANICAL PLANT - cont'd							
Southern Conveyors - cont'd							
Conveyor belt width 400 mm; mechanically loaded and removed by hand at offload point; conveyor length							
up to 5 m	-	3.19	59.02	5.19	-	m^3	**64.21**
10 m	-	3.19	59.02	6.22	-	m^3	**65.24**
12.5 m	-	3.19	59.02	6.68	-	m^3	**65.70**
15 m	-	3.19	59.02	7.25	-	m^3	**66.27**
20 m	-	3.19	59.02	8.28	-	m^3	**67.30**
25 m	-	3.19	59.02	9.20	-	m^3	**68.22**
30 m	-	3.19	59.02	10.23	-	m^3	**69.25**
Terranova Cranes Ltd; materials handling and lifting; telescopic cranes supply and management; exclusive of roadway management or planning applications; prices below illustrate lifts of 1 tonne at maximum crane reach; lift cycle of 0.25 hours; mechanical filling of material skip if appropriate							
35 tonne mobile crane; lifting capacity of 1 tonne at 26 m; C.P.A hire (all management by hirer)							
granular materials including concrete	-	0.75	13.88	15.86	-	tonne	**29.73**
palletised or packed materials	-	0.50	9.25	13.13	-	tonne	**22.38**
35 tonne mobile crane; lifting capacity of 1 tonne at 26 m; contract lift							
granular materials or concrete	-	0.25	4.63	43.36	-	tonne	**47.98**
palletised or packed materials	-	-	-	40.63	-	tonne	**40.63**
50 tonne mobile crane; lifting capacity of 1 tonne at 34 m; C.P.A hire (all management by hirer)							
granular materials including concrete	-	0.75	13.88	19.92	-	tonne	**33.80**
palletised or packed materials	-	0.50	9.25	17.19	-	tonne	**26.44**
50 tonne mobile crane; lifting capacity of 1 tonne at 34 m; contract lift							
granular materials or concrete	-	0.25	4.63	49.55	-	tonne	**54.18**
palletised or packed materials	-	-	-	46.82	-	tonne	**46.82**
80 tonne mobile crane; lifting capacity of 1 tonne at 44 m; C.P.A. hire (all management by hirer)							
granular materials including concrete	-	0.75	13.88	27.73	-	tonne	**41.61**
palletised or packed materials	-	0.50	9.25	25.00	-	tonne	**34.25**
80 tonne mobile crane; lifting capacity of 1 tonne at 44 m; contract lift							
granular materials or concrete	-	0.25	4.63	55.86	-	tonne	**60.48**
palletised or packed materials	-	-	-	53.13	-	tonne	**53.13**
A44 CONTRACTOR'S GENERAL COST ITEMS: TEMPORARY WORKS							
Eve Trakway; portable roadway systems							
Temporary roadway system laid directly onto existing surface or onto PVC matting to protect existing surface. Most systems are based on a weekly hire charge with transportation, installation and recovery charges included.							
Heavy Duty Trakpanel - per panel (3.05 m x 2.59 m) per week	-	-	-	-	-	m^2	**5.06**
Outrigger Mats for use in conjunction with Heavy Duty Trakpanels - per set of 4 mats per week	-	-	-	-	-	set	**82.40**
Medium Duty Trakpanel - per panel (2.44 m x 3.00 m) per week	-	-	-	-	-	m^2	**5.46**
LD20 Eveolution - Light Duty Trakway - roll out system minimum delivery 50 m	-	-	-	-	-	m^2	**4.86**
Terraplas Walkways - turf protection system - per section (1 m x 1 m) per week	-	-	-	-	-	m^2	**5.00**

B COMPLETE BUILDINGS/STRUCTURES/UNITS

Item Excluding site overheads and profit	PC £	Labour hours	Labour £	Plant £	Material £	Unit	**Total rate £**
B10 PREFABRICATED BUILDINGS/STRUCTURES							
Cast stone buildings; Haddonstone Ltd; ornamental garden buildings in Portland Bath or terracotta finished cast stone; prices for stonework and facades only; excavations, foundations, reinforcement, concrete infill, roofing and floors all priced separately							
Pavilion Venetian Folly L9400; Tuscan columns, pedimented arch, quoins and optional balustrading							
4184 mm high x 4728 mm wide x 3147 mm deep	9605.00	175.00	3237.50	193.90	9705.09	nr	**13136.49**
Pavilion L9300; Tuscan columns							
3496 mm high x 3634 mm wide	5038.00	144.00	2664.00	166.20	5138.09	nr	**7968.29**
Small Classical Temple L9250; 6 column with fibreglass lead effect finish dome roof							
overall height 3610 mm, diameter 2540 mm	5832.00	130.00	2405.00	138.50	5932.09	nr	**8475.59**
Large Classical Temple L9100; 8 column with fibreglass lead effect finish dome roof							
overall height 4664 mm, diameter 3190 mm	9580.00	165.00	3052.50	138.50	9680.09	nr	**12871.09**
Stepped floors to temples							
Single step; Large Classical Temple	1106.00	26.00	481.00	-	1161.04	nr	**1642.04**
Single step; Small Classical Temple	781.00	24.00	444.00	-	826.87	nr	**1270.87**
Stone structures; Architectural Heritage Ltd; hand carved from solid natural limestone with wrought iron domed roof, decorated frieze and base and integral seats; supply and erect only; excavations and concrete bases priced separately							
'The Park Temple', five columns; 3500 mm high x 1650 mm diameter	9800.00	96.00	1776.00	60.00	9800.00	nr	**11636.00**
'The Estate Temple', six columns; 4000 mm high x 2700 mm diameter	16800.00	120.00	2220.00	120.00	16800.00	nr	**19140.00**
Stone structures; Architectural Heritage Ltd; hand carved from solid natural limestone with solid oak trelliage; prices for stonework and facades only; supply and erect only; excavations and concrete bases priced separately							
'The Pergola', 2240 mm high x 2640 mm wide x 6990 mm long	11000.00	96.00	1776.00	240.00	11030.54	nr	**13046.54**
Ornamental timber buildings; Architectural Heritage Ltd; hand built in English oak with fire retardant wheat straw thatch; hand painted, fired lead glass windows							
'The Thatched Edwardian Summer House', 3710 high x 3540 wide x 2910 deep overall, 2540 x 1910 internal	19500.00	32.00	592.00	-	19500.00	nr	**20092.00**
Ornamental stone structures; Architectural Heritage Ltd; setting to bases or plinths (not included)							
'The Obelisk', classic natural stone obelisk, tapering square form on panelled square base, 1860 high x 360 square	1200.00	2.00	37.00	-	1200.00	1	**1237.00**
'The Narcissus Column', natural limestone column on pedestal surmounted by bronze sculpture, overall height 3440 with 440 square base	3200.00	2.00	37.00	-	3200.00	1	**3237.00**
'The Armillary Sundial', artificial stone baluster pedestal base surmounted by verdigris copper armillary sphere calibrated sundial	2200.00	1.00	18.50	-	2200.00	1	**2218.50**

D GROUNDWORK

Item Excluding site overheads and profit	PC £	Labour hours	Labour £	Plant £	Material £	Unit	Total rate £
D11 SOIL STABILIZATION							
Soil stabilization - General							
Preamble - Earth-retaining and stabilizing materials are often specified as part of the earth-forming work in landscape contracts, and therefore this section lists a number of products specially designed for large-scale earth control. There are two types: rigid units for structural retention of earth on steep slopes; and flexible meshes and sheets for control of soil erosion where structural strength is not required. Prices for these items depend on quantity, difficulty of access to the site and availability of suitable filling material: estimates should be obtained from the manufacturer when the site conditions have been determined.							
Crib walls; Keller Ground Engineering							
"Timbercrib" timber crib walling system; machine filled with crushed rock inclusive of reinforced concrete footing cribfill stone, rear wall land drain and rear wall drainage/separation membrane; excluding excavation							
ref 600/38; for retaining walls up to 2.20 m high	-	-	-	-	-	m^2	**126.00**
ref 750/38; for retaining walls up to 3.10 m high	-	-	-	-	-	m^2	**131.00**
ref 900/48; for retaining walls up to 3.60 m high	-	-	-	-	-	m^2	**163.00**
ref 1050/48; for retaining walls up to 4.40 m high	-	-	-	-	-	m^2	**173.00**
ref 1200/48; for retaining walls up to 5.20 m high	-	-	-	-	-	m^2	**195.00**
ref 1500/48; for retaining walls up to 6.50 m high	-	-	-	-	-	m^2	**215.00**
ref 1800/48; for retaining walls up to 8.20 m high	-	-	-	-	-	m^2	**242.00**
Retaining walls; Keller Ground Engineering							
"Textomur" reinforced soil system embankments							
reinforced soil slopes at angles of 60 - 70 degrees to the horizontal; as an alternative to reinforced concrete or gabion solutions; finishing with excavated material/grass or shrubs	-	-	-	-	-	m^2	**95.00**
Retaining walls; RCC							
Retaining walls of units with plain concrete finish; prices based on 24 tonne loads but other quantities available (excavation, temporary shoring, foundations and backfilling not included)							
1000 wide x 1250 mm high	117.00	1.27	23.50	7.22	119.50	m	**150.22**
1000 wide x 1750 mm high	133.00	1.27	23.50	7.22	136.31	m	**167.03**
1000 wide x 2400 mm high	198.00	1.27	23.50	7.22	202.68	m	**233.40**
1000 wide x 2690 mm high	245.00	1.27	23.50	43.31	251.21	m	**318.03**
1000 wide x 3000 mm high	254.00	1.47	27.20	54.14	262.12	m	**343.46**
1000 wide x 3750 mm high	163.00	1.57	29.14	64.97	176.11	m	**270.21**
Retaining walls; Maccaferri Ltd							
Wire mesh gabions; galvanized mesh 80 mm x 100 mm; filling with broken stones 125 mm - 200 mm size; wire down securely to manufacturer's instructions; filling front face by hand							
2 x 1 x 0.50 m	13.00	2.00	37.00	4.81	80.83	nr	**122.64**
2 x 1 x 1.00 m	21.50	4.00	74.00	9.63	157.16	nr	**240.79**
PVC coated gabions							
2 x 1 x 0.5 m	20.95	2.00	37.00	4.81	88.78	nr	**130.59**
2 x 1 x 1.00 m	29.52	4.00	74.00	9.63	165.18	nr	**248.81**
"Reno" mattress gabions							
6 x 2 x 0.17 m	59.97	3.00	55.50	7.22	198.34	nr	**261.06**
6 x 2 x 0.23 m	63.95	4.50	83.25	9.63	251.16	nr	**344.04**
6 x 2 x 0.30 m	75.10	6.00	111.00	10.83	319.29	nr	**441.12**

D GROUNDWORK

Item Excluding site overheads and profit	PC £	Labour hours	Labour £	Plant £	Material £	Unit	**Total rate £**
Retaining walls; Tensar International							
"Tensar" retaining wall system; modular dry laid concrete blocks; 220 mm x 400 mm long x 150 mm high connected to "Tensar RE" geogrid with proprietary connectors; geogrid laid horizontally within the fill at 300 mm centres; on 150 x 450 concrete foundation; filling with imported granular material							
1.00 m high	80.00	3.00	55.50	8.25	89.00	m^2	**152.75**
2.00 m high	80.00	4.00	74.00	8.25	89.00	m^2	**171.25**
3.00 m high	80.00	4.50	83.25	8.25	89.00	m^2	**180.50**
Retaining walls; Grass Concrete Ltd							
"Betoflor" precast concrete landscape retaining walls including soil filling to pockets (excavation, concrete foundations, backfilling stones to rear of walls and planting not included)							
"Betoflor" interlocking units; 250 mm long x 250 mm x 200 mm modular deep; in walls 250 mm wide	-	-	-	-	-	m^2	**75.76**
extra over "Betoflor" interlocking units for colours	-	-	-	-	-	m^2	**7.28**
"Betoatlas" interlocking units; 250 mm long x 500 mm wide x 200 mm modular deep; in walls 500 mm wide	-	-	-	-	-	m^2	**113.63**
extra over "Betoatlas" interlocking units for colours	-	-	-	-	-	m^2	**13.86**
ABG Ltd "Webwall"; honeycomb structured HDPE polymer strips with interlocking joints; laid to prepared terraces and backfilled with excavated material to form raked wall for planting							
500 mm high units laid and backfilled horizontally within the face of the slope to produce stepped gradient; measured as bank face area; wall height:							
up to 1.00 m high	-	-	-	-	-	m^2	**80.00**
up to 2.00 m high	-	-	-	-	-	m^2	**100.00**
over 2.00 m high	-	-	-	-	-	m^2	**110.00**
250 mm high units laid and backfilled horizontally within the face of the slope to produce stepped gradient; measured as bank face area; wall height:							
up to 1.00 m high	-	-	-	-	-	m^2	**90.00**
up to 2.00 m high	-	-	-	-	-	m^2	**110.00**
over 2.00 m high	-	-	-	-	-	m^2	**120.00**
Setting out; grading and levelling; compacting bottoms of excavations	-	0.13	2.47	0.33	-	m	**2.79**
Extra over "Betoflor" retaining walls for C7P concrete foundations							
700 mm x 300 mm deep	15.98	0.27	4.94	-	15.98	m	**20.92**
Retaining walls; Forticrete Ltd							
"Keystone" pc concrete block retaining wall; geogrid included for walls over 1.00 m high; excavation, concrete foundation, stone backfill to rear of wall all measured separately							
1.0 m high	70.00	2.40	44.40	-	71.75	m^2	**116.15**
2.0 m high	70.00	2.40	44.40	6.92	86.62	m^2	**137.94**
3.0 m high	70.00	2.40	44.40	9.23	96.95	m^2	**150.58**
4.0 m high	70.00	2.40	44.40	11.08	104.95	m^2	**160.43**

D GROUNDWORK

Item Excluding site overheads and profit	PC £	Labour hours	Labour £	Plant £	Material £	Unit	Total rate £
D11 SOIL STABILIZATION - cont'd							
Retaining walls; Forticrete Ltd							
Stepoc blocks; interlocking blocks; 10 mm reinforcing laid loose horizontally to preformed notches and vertical reinforcing nominal size 10 mm fixed to starter bars; infilling with concrete; foundations and starter bars measured separately							
Type 325 400 x 225 x 325 mm	40.00	1.40	25.90	-	58.93	m^2	**84.83**
Type 256 400 x 225 x 256 mm	38.00	1.20	22.20	-	57.27	m^2	**79.47**
Type 190 400 x 225 x 190 mm	28.00	1.00	18.50	-	43.67	m^2	**62.17**
Embankments; Tensar International							
Embankments; reinforced with "Tensar Uniaxial Geogrid"; ref 40 RE; 40 kN/m width; Geogrid laid horizontally within fill to 100% of vertical height of slope at 1.00 m centres; filling with excavated material							
slopes less than 45 degrees	2.60	0.19	3.43	1.80	2.73	m^3	**7.96**
slopes exceeding 45 degrees	5.59	0.19	3.43	2.72	5.87	m^3	**12.02**
Embankments; reinforced with "Tensar Uniaxial Geogrid"; ref 55 RE; 55 kN/m width; Geogrid laid horizontally within fill to 100% of vertical height of slope at 1.00 m centres; filling with excavated material							
slopes less than 45 degrees	3.15	0.19	3.43	1.80	3.31	m^3	**8.54**
slopes exceeding 45 degrees	6.77	0.19	3.43	2.72	7.11	m^3	**13.26**
Extra for "Tensar Mat"; erosion control mats; to faces of slopes of 45 degrees or less; filling with 20 mm fine topsoil; seeding	3.50	0.02	0.37	0.17	4.57	m^2	**5.11**
Embankments; reinforced with "Tensar Uniaxial Geogrid" ref "55 RE"; 55 kN/m width; Geogrid laid horizontally within fill to 50% of horizontal length of slopes at specified centres; filling with imported fill PC £16.00/m^3							
300 mm centres; slopes up to 45 degrees	5.23	0.23	4.32	2.06	24.69	m^3	**31.07**
600 mm centres; slopes up to 45 degrees	2.61	0.19	3.43	2.72	21.95	m^3	**28.10**
Extra for "Tensar Mat"; erosion control mats; to faces of slopes of 45 degrees or less; filling with 20 mm fine topsoil; seeding	3.50	0.02	0.37	0.17	4.57	m^2	**5.11**
Embankments; reinforced with "Tensar Uniaxial Geogrid" ref "55 RE"; 55 kN/m width; Geogrid laid horizontally within fill; filling with excavated material; wrapping around at faces							
300 mm centres; slopes exceeding 45 degrees	10.49	0.19	3.43	1.80	11.01	m^3	**16.25**
600 mm centres; slopes exceeding 45 degrees	5.23	0.19	3.43	1.80	5.49	m^3	**10.72**
Extra over embankments for bagwork face supports for slopes exceeding 45 degrees	-	0.50	9.25	-	13.00	m^2	**22.25**
Extra over embankments for seeding of bags	-	0.10	1.85	-	0.20	m^2	**2.05**
Extra over embankments for "Bodkin" joints	-	-	-	-	1.00	m	
Extra over embankments for temporary shuttering to slopes exceeding 45 degrees	-	0.20	3.70-	-	10.00	m^2	**13.70**
Anchoring systems; surface stabilization; Platipus Anchors Ltd; centres and depths shown should be verified with a design engineer; they may vary either way dependent on circumstances							
Concrete revetments; hard anodised stealth anchors with stainless steel accessories installed to a depth of 1 m							
"SO4" anchors to a depth of 1.00 - 1.50 m at 1.00 m ccs	24.15	0.33	6.17	1.38	24.15	m^2	**31.70**

D GROUNDWORK

Item Excluding site overheads and profit	PC £	Labour hours	Labour £	Plant £	Material £	Unit	**Total rate £**
Loadlocking and crimping tool; for stressing and crimping the anchors							
purchase price	367.50	-	-	-	367.50	nr	-
hire rate per week	63.00	-	-	-	63.00	nr	-
Surface erosion; geotextile anchoring on slopes; aluminium alloy Stealth anchors							
"SO4", "SO6" to a depth of 1.0 - 2.0 m at 2.00 m ccs	11.03	0.20	3.70	1.38	11.03	m^2	**16.11**
Brick block or in situ concrete, distressed retaining walls; anchors installed in two rows at varying tensions between 18 kN - 50 kN along the length of the retaining wall; core drilling of wall not included							
"SO8", "BO6", "BO8" bronze Stealth and Bat anchors installed in combination with stainless steel accessories to a depth of 6 m; average cost base on 1.50 m ccs; long term solution	346.50	0.50	9.25	1.38	346.50	m^2	**357.13**
"SO8", "BO6", "BO8" Cast SG Iron Stealth and Bat anchors installed in combination with galvanised steel accessories to a depth of 6 m; average cost base on 1.50 m ccs; short term solution	148.05	0.50	9.25	1.38	148.05	m^2	**158.68**
Timber retaining wall support; anchors installed 19 kN along the length of the retaining wall							
"SO6", "SO8", bronze Stealth anchors installed in combination with stainless steel accessories to a depth of 3.0 m. 1.50 centres; average cost	51.98	0.50	9.25	1.10	51.98	m^2	**62.33**
Gabions; anchors for support and stability to moving, overturning or rotating gabion retaining walls							
"SO8", "BO6" anchors including base plate at 1.00 m centres to a depth of 4.00 m	262.50	0.33	6.17	1.38	262.50	m^2	**270.05**
Soil nailing of geofabrics; anchors fixed through surface of fabric or erosion control surface treatment (not included)							
"SO8", "BO6" anchors including base plate at 1.50 m centres to a depth of 3.00 m	105.00	0.17	3.08	0.92	105.00	m^2	**109.00**
Embankments; Grass Concrete Ltd							
"Grasscrete"; in situ reinforced concrete surfacing; to 20 mm thick sand blinding layer (not included); including soiling and seeding							
ref GC1; 100 mm thick	-	-	-	-	-	m^2	**28.12**
ref GC2; 150 mm thick	-	-	-	-	-	m^2	**36.18**
"Grassblock 103"; solid matrix precast concrete blocks; to 20 mm thick sand blinding layer; excluding edge restraint; including soiling and seeding							
406 mm x 406 mm x 103 mm; fully interlocking	-	-	-	-	-	m^2	**28.98**
Grass reinforcement; Farmura Environmental Ltd							
"Matrix" grass paver; recycled polyethylene and polypropylene mixed interlocking erosion control and grass reinforcement system laid to rootzone prepared separately and filled with screened topsoil and seeded with grass seed; green							
640 x 330 x38 mm	12.50	0.13	2.31	-	14.82	m^2	**17.13**
330 x 330 x 38 mm	12.50	0.17	3.08	-	14.82	m^2	**17.91**
extra over for coloured material	19.00	-	-	-	19.00	m^2	-

D GROUNDWORK

Item Excluding site overheads and profit	PC £	Labour hours	Labour £	Plant £	Material £	Unit	Total rate £
D11 SOIL STABILIZATION - cont'd							
Embankments; Cooper Clarke Group Ltd							
"Ecoblock" polyethylene; 925 mm x 310 mm x 50 mm; heavy duty for car parking and fire paths							
to firm sub-soil (not included)	14.15	0.04	0.74	-	15.10	m^2	**15.84**
to 100 mm granular fill and "Geotextile"	14.15	0.08	1.48	0.13	19.99	m^2	**21.60**
to 250 mm granular fill and "Geotextile"	14.15	0.20	3.70	0.22	26.20	m^2	**30.12**
Extra for filling "Ecoblock" with topsoil; seeding with rye grass at 50 g/m^2	0.94	0.02	0.37	0.18	1.12	m^2	**1.68**
"Geoweb" polyethylene soil-stabilizing panels; to soil surfaces brought to grade (not included); filling with excavated material							
panels; 2.5 x 8.0 m x 100 mm deep	6.00	0.10	1.85	2.55	6.12	m^2	**10.52**
panels; 2.5 x 8.0 m x 200 mm deep	11.95	0.13	2.47	3.05	12.19	m^2	**17.71**
"Geoweb" polyethylene soil-stabilizing panels; to soil surfaces brought to grade (not included); filling with ballast							
panels; 2.5 x 8.0 m x 100 mm deep	6.00	0.11	2.06	2.55	8.03	m^2	**12.63**
panels; 2.5 x 8.0 m x 200 mm deep	11.95	0.15	2.85	3.05	16.00	m^2	**21.89**
"Geoweb" polyethylene soil-stabilizing panels; to soil surfaces brought to grade (not included); filling with ST2 10 N/mm^2 concrete							
panels; 2.5 x 8.0 m x 100 mm deep	6.00	0.16	2.96	-	13.68	m^2	**16.64**
panels; 2.5 x 8.0 m x 200 mm deep	11.95	0.20	3.70	-	27.31	m^2	**31.01**
Extra over for filling "Geoweb" with imported topsoil; seeding with rye grass at 50 g/m^2	1.94	0.07	1.23	0.30	2.32	m^2	**3.85**
Timber log retaining walls; Longlyf Timber Products Ltd							
Machine rounded softwood logs to trenches priced separately; disposal of excavated material priced separately; inclusive of 75 mm hardcore blinding to trench and backfilling trench with site mixed concrete 1:3:6; geofabric pinned to rear of logs; heights of logs above ground							
500 mm (constructed from 1.80 m lengths)	19.70	1.50	27.75	-	28.82	m	**56.57**
1.20 m (constructed from 1.80 m lengths)	39.40	1.30	24.05	-	60.44	m	**84.49**
1.60 m (constructed from 2.40 m lengths)	51.30	2.50	46.25	-	77.63	m	**123.88**
2.00 m (constructed from 3.00 m lengths)	72.60	3.50	64.75	-	104.22	m	**168.97**
As above but with 150 mm machine rounded timbers							
500 mm	28.52	2.50	46.25	-	37.64	m	**83.89**
1.20 m	94.97	1.75	32.38	-	116.02	m	**148.39**
1.60 m	95.07	3.00	55.50	-	121.40	m	**176.90**
As above but with 200 mm machine rounded timbers							
1.80 m (constructed from 2.40 m lengths)	128.60	4.00	74.00	-	160.22	m	**234.22**
2.40 m (constructed from 3.60 m lengths)	192.50	4.50	94.50	-	229.64	m	**324.14**
Railway sleeper walls; Sleeper Supplies Ltd; retaining wall from railway sleepers; fixed with steel galvanised pins 12 mm driven into the ground; sleepers laid flat							
Grade 1 softwood; 2590 x 250 x 150 mm							
150 mm; 1 sleeper high	7.02	0.50	9.25	-	7.46	m	**16.71**
300 mm; 2 sleepers high	14.05	1.00	18.50	-	14.93	m	**33.43**
450 mm; 3 sleepers high	20.90	1.50	27.75	-	22.06	m	**49.81**
600 mm; 4 sleepers high	28.16	2.00	37.00	-	29.31	m	**66.31**
Grade 1 softwood as above but with 2 nr galvanised angle iron stakes set into concrete internally and screwed to the inside face of the sleepers							
750 mm; 5 sleepers high	34.84	2.50	46.25	-	71.71	m^2	**117.96**
900 mm; 6 sleepers high	41.70	2.75	50.88	-	80.58	m^2	**131.45**

D GROUNDWORK

Item Excluding site overheads and profit	PC £	Labour hours	Labour £	Plant £	Material £	Unit	**Total rate £**
Grade 1 hardwood; 2590 x 250 x 150 mm							
150 mm; 1 sleeper high	5.65	0.50	9.25	-	6.09	m	**15.34**
300 mm; 2 sleepers high	11.30	1.00	18.50	-	12.18	m	**30.68**
450 mm; 3 sleepers high	16.81	1.50	27.75	-	17.97	m	**45.72**
600 mm; 4 sleepers high	22.65	2.00	37.00	-	23.81	m	**60.81**
Grade 1 hardwood as above but with 2 nr galvanised angle iron stakes set into concrete internally and screwed to the inside face of the sleepers							
750 mm; 5 sleepers high	28.02	2.50	46.25	-	64.90	m^2	**111.15**
900 mm; 6 sleepers high	33.54	2.75	50.88	-	72.42	m^2	**123.30**
New pine softwood; 2500 x 245 x 120 mm							
120 mm; 1 sleeper high	7.33	0.50	9.25	-	7.77	m	**17.02**
240 mm; 2 sleepers high	14.66	1.00	18.50	-	15.54	m	**34.04**
360 mm; 3 sleepers high	21.98	1.50	27.75	-	23.14	m	**50.89**
480 mm; 4 sleepers high	29.31	2.00	37.00	-	30.47	m	**67.47**
New pine softwood as above but with 2 nr galvanised angle iron stakes set into concrete internally and screwed to the inside face of the sleepers							
600 mm; 5 sleepers high	36.64	2.50	46.25	-	73.51	m^2	**119.76**
720 mm; 6 sleepers high	43.97	2.75	50.88	-	82.85	m^2	**133.72**
New oak hardwood 2500 x 200 x 130 mm							
130 mm; 1 sleeper high	8.33	0.50	9.25	-	8.77	m	**18.02**
260 mm; 2 sleepers high	16.66	1.00	18.50	-	17.54	m	**36.04**
390 mm; 3 sleepers high	24.98	1.50	27.75	-	26.14	m	**53.89**
520 mm; 4 sleepers high	33.31	2.00	37.00	-	34.47	m	**71.47**
New oak hardwood as above but with 2 nr galvanised angle iron stakes set into concrete internally and screwed to the inside face of the sleepers							
640 mm; 5 sleepers high	41.64	2.50	46.25	-	78.51	m^2	**124.76**
760 mm; 6 sleepers high	49.97	2.75	50.88	-	88.85	m^2	**139.72**
Excavate foundation trench; set railway sleepers vertically on end in concrete 1:3:6 continuous foundation to 33.3% of their length to form retaining wall							
Grade 1 hardwood; finished height above ground level							
300 mm	10.02	3.00	55.50	1.88	19.46	m	**76.83**
500 mm	14.52	3.00	55.50	1.88	26.21	m	**83.58**
600 mm	20.18	3.00	55.50	1.88	38.68	m	**96.06**
750 mm	21.78	3.50	64.75	1.88	41.08	m	**107.70**
1.00 m	29.77	3.75	69.38	1.88	46.41	m	**117.66**
Excavate and place vertical steel universal beams 165 wide in concrete base at 2.590 centres; fix railway sleepers set horizontally between beams to form horizontal fence or retaining wall							
Grade 1 hardwood; bay length 2.590 m							
0.50 m high; (2 sleepers)	13.66	2.50	46.25	-	45.60	bay	**91.85**
0.75 m high; (3 sleepers)	20.08	2.60	48.10	-	67.66	bay	**115.76**
1.00 m high; (4 sleepers)	26.91	3.00	55.50	-	87.16	bay	**142.66**
1.25 m high; (5 sleepers)	34.36	3.50	64.75	-	112.59	bay	**177.34**
1.50 m high; (6 sleepers)	40.37	3.00	55.50	-	133.92	bay	**189.42**
1.75 m high; (7 sleepers)	46.99	3.00	55.50	-	140.54	bay	**196.04**
Willow walling to riverbanks; LDC Ltd							
Woven willow walling as retention to riverbanks; driving or concreting posts in at 2 m centres; intermediate posts at 500 mm centres							
1.20 m high	-	-	-	-	-	m	**106.72**
1.60 m high	-	-	-	-	-	m	**136.07**

D GROUNDWORK

Item Excluding site overheads and profit	PC £	Labour hours	Labour £	Plant £	Material £	Unit	**Total rate £**
D11 SOIL STABILIZATION - cont'd							
Flexible sheet materials; Tensar International							
"Tensar Mat"; erosion mats; 3 m - 4.50 m wide; securing with "Tensar" pegs; lap rolls 100 mm; anchors at top and bottom of slopes; in trenches	3.63	0.02	0.31	-	3.99	m^2	**4.30**
Topsoil filling to "Tensar Mat"; including brushing and raking	0.46	0.02	0.31	0.10	0.58	m^2	**0.99**
"Tensar Bi-axial Geogrid"; to graded compacted base; filling with 200 mm granular fill; compacting (turf or paving to surfaces not included); 400 mm laps							
ref SS20; 39 mm x 39 mm mesh	1.70	0.01	0.25	0.14	4.58	m^2	**4.97**
ref SS30; 39 mm x 39 mm mesh	2.20	0.01	0.25	0.14	5.08	m^2	**5.47**
Flexible sheet materials; Terram Ltd							
"Terram" synthetic fibre filter fabric; to graded base (not included)							
"Terram 1000"; 0.70 mm thick; mean water flow 50 litre/m^2/s	0.41	0.02	0.31	-	0.41	m^2	**0.72**
"Terram 2000"; 1.00 mm thick; mean water flow 33 litre/m^2/s	0.91	0.02	0.31	-	1.09	m^2	**1.40**
"Terram Minipack"	0.53	-	0.06	-	0.53	m^2	**0.59**
Flexible sheet materials; Greenfix Ltd							
"Greenfix"; erosion control mats; 10 mm - 15 mm thick; fixing with 4 nr crimped pins in accordance with manufacturer's instructions; to graded surface (not included)							
unseeded "Eromat 1"; 2.40 m wide	168.54	2.00	37.00	-	239.39	100 m^2	**276.39**
unseeded "Eromat 2"; 2.40 m wide	187.70	2.00	37.00	-	260.47	100 m^2	**297.47**
unseeded "Eromat 3"; 2.40 m wide	223.98	2.00	37.00	-	300.38	100 m^2	**337.38**
seeded "Covamat 1"; 2.40 m wide	410.78	2.00	37.00	-	485.32	100 m^2	**522.32**
seeded "Covamat 2"; 2.40 m wide	428.78	2.00	37.00	-	504.22	100 m^2	**541.22**
seeded "Covamat 3"; 2.40 m wide	470.78	2.00	37.00	-	548.32	100 m^2	**585.32**
"Bioroll" 300 mm diameter to river banks and revetments	19.00	0.13	2.31	-	20.48	m	**22.80**
Extra over "Greenfix" erosion control mats for fertilizer applied at 70 g/m^2	4.32	0.20	3.70	-	4.32	100 m^2	**8.02**
Extra over "Greenfix" erosion control mats for Geojute; fixing with steel pins	2.06	0.02	0.41	-	2.29	m^2	**2.70**
Extra over "Greenfix" erosion control mats for laying to slopes exceeding 30 degrees	-	-	-	-	-	25%	**-**
Extra for the following operations							
Spreading 25 mm approved topsoil							
by machine	0.46	0.01	0.11	0.10	0.56	m^2	**0.77**
by hand	0.46	0.04	0.65	-	0.56	m^2	**1.20**
Grass seed; PC £3.00/kg; spreading in two operations; by hand							
35 g/m^2	10.50	0.17	3.08	-	10.50	100 m^2	**13.58**
50 g/m^2	15.00	0.17	3.08	-	15.00	100 m^2	**18.08**
70 g/m^2	21.00	0.17	3.08	-	21.00	100 m^2	**24.08**
100 g/m^2	30.00	0.20	3.70	-	30.00	100 m^2	**33.70**
125 g/m^2	37.50	0.20	3.70	-	37.50	100 m^2	**41.20**
Extra over seeding by hand for slopes over 30 degrees (allowing for the actual area but measured in plan)							
35 g/m^2	1.56	-	0.07	-	1.56	100 m^2	**1.63**
50 g/m^2	2.25	-	0.07	-	2.25	100 m^2	**2.32**
70 g/m^2	3.15	-	0.07	-	3.15	100 m^2	**3.22**
100 g/m^2	4.50	-	0.08	-	4.50	100 m^2	**4.58**
125 g/m^2	5.61	-	0.08	-	5.61	100 m^2	**5.69**

D GROUNDWORK

Item Excluding site overheads and profit	PC £	Labour hours	Labour £	Plant £	Material £	Unit	**Total rate £**
Grass seed; PC £3.00/kg; spreading in two operations; by machine							
35 g/m²	10.50	-	-	0.51	10.50	100 m²	**11.01**
50 g/m²	15.00	-	-	0.51	15.00	100 m²	**15.51**
70 g/m²	21.00	-	-	0.51	21.00	100 m²	**21.51**
100 g/m²	30.00	-	-	0.51	30.00	100 m²	**30.51**
125 kg/ha	375.00	-	-	50.95	375.00	ha	**425.95**
150 kg/ha	450.00	-	-	50.95	450.00	ha	**500.95**
200 kg/ha	600.00	-	-	50.95	600.00	ha	**650.95**
250 kg/ha	750.00	-	-	50.95	750.00	ha	**800.95**
300 kg/ha	900.00	-	-	50.95	900.00	ha	**950.95**
350 kg/ha	1050.00	-	-	50.95	1050.00	ha	**1100.95**
400 kg/ha	1200.00	-	-	50.95	1200.00	ha	**1250.95**
500 kg/ha	1500.00	-	-	50.95	1500.00	ha	**1550.95**
700 kg/ha	2100.00	-	-	50.95	2100.00	ha	**2150.95**
1400 kg/ha	4200.00	-	-	50.95	4200.00	ha	**4250.95**
Extra over seeding by machine for slopes over 30 degrees (allowing for the actual area but measured in plan)							
35 g/m²	10.50	-	-	0.08	1.57	100 m²	**1.65**
50 g/m²	15.00	-	-	0.08	2.25	100 m²	**2.33**
70 g/m²	21.00	-	-	0.08	3.15	100 m²	**3.23**
100 g/m²	30.00	-	-	0.08	4.50	100 m²	**4.58**
125 kg/ha	375.00	-	-	7.64	56.25	ha	**63.89**
150 kg/ha	450.00	-	-	7.64	67.50	ha	**75.14**
200 kg/ha	600.00	-	-	7.64	90.00	ha	**97.64**
250 kg/ha	750.00	-	-	7.64	112.50	ha	**120.14**
300 kg/ha	900.00	-	-	7.64	135.00	ha	**142.64**
350 kg/ha	1050.00	-	-	7.64	157.50	ha	**165.14**
400 kg/ha	1200.00	-	-	7.64	180.00	ha	**187.64**
500 kg/ha	1500.00	-	-	7.64	225.00	ha	**232.64**
700 kg/ha	2100.00	-	-	7.64	315.00	ha	**322.64**
1400 kg/ha	4200.00	-	-	7.64	630.00	ha	**637.64**
D20 EXCAVATING AND FILLING							
MACHINE SELECTION TABLE							
Road Equipment Ltd; machine volumes for excavating/filling only and placing excavated material alongside or to a dumper; no bulkages are allowed for in the material volumes; these rates should be increased by user-preferred percentages to suit prevailing site conditions; the figures in the next section for "Excavation; mechanical" and filling; allow for the use of banksmen within the rates shown below							
1.5 tonne excavators;: digging volume							
1 cycle/minute; 0.04 m³	-	0.42	7.71	2.39	-	m³	**10.10**
2 cycles/minute; 0.08 m³	-	0.21	3.85	1.41	-	m³	**5.26**
3 cycles/minute; 0.12 m³	-	0.14	2.57	1.09	-	m³	**3.66**
3 tonne excavators; digging volume							
1 cycles/minute; 0.13 m³	-	0.13	2.37	1.27	-	m³	**3.64**
2 cycles/minute; 0.26 m³	-	0.06	1.19	1.63	-	m³	**2.82**
3 cycles/minute; 0.39 m³	-	0.04	0.79	0.75	-	m³	**1.54**
5 tonne excavators; digging volume							
1 cycle/minute; 0.28 m³	-	0.06	1.10	1.50	-	m³	**2.60**
2 cycles/minute; 0.56 m³	-	0.03	0.55	1.16	-	m³	**1.71**
3 cycles/minute; 0.84 m³	-	0.02	0.37	1.17	-	m³	**1.54**
7 tonne excavators; with operator; digging volume							
1 cycle/minute; 0.28 m³	-	0.06	1.10	2.28	-	m³	**3.38**
2 cycles/minute; 0.56 m³	-	0.03	0.55	1.12	-	m³	**1.67**
3 cycles/minute; 0.84 m³		0.02	0.37	1.22	-	m³	**1.59**

D GROUNDWORK

Item Excluding site overheads and profit	PC £	Labour hours	Labour £	Plant £	Material £	Unit	Total rate £
D20 EXCAVATING AND FILLING - cont'd							
Road Equipment Ltd; machine volumes - cont'd							
21 tonne excavators; supplied with operator; digging volume							
1 cycle/minute; 1.21 m^3	-	-	-	0.68	-	m^3	**0.68**
2 cycles/minute; 2.42 m^3	-	-	-	0.28	-	m^3	**0.28**
3 cycles/minute; 3.63 m^3	-	-	-	0.19	-	m^3	**0.19**
Backhoe loader; excavating; JCB 3 CX rear bucket capacity 0.28 m^3							
1 cycle/minute; 0.28 m^3	-	-	-	1.78	-	m^3	**1.78**
2 cycles/minute; 0.56 m^3	-	-	-	0.89	-	m^3	**0.89**
3 cycles/minute; 0.84 m^3	-	-	-	0.59	-	m^3	**0.59**
Backhoe loader; loading from stockpile; JCB 3 CX front bucket capacity 1.00 m^3							
1 cycle/minute; 1.00 m^3	-	-	-	0.50	-	m^3	**0.50**
2 cycles/minute; 2.00 m^3	-	-	-	0.25	-	m^3	**0.25**
Note: All volumes below are based on excavated "Earth" moist at 1997 kg/m^3 solid or 1598 kg/m^3 loose; a 25% bulkage factor has been used; the weight capacities below exceed the volume capacities of the machine in most cases; see the memorandum section at the back of this book for further weights of materials.							
Dumpers; Road Equipment Ltd							
1 tonne high tip skip loader; volume 0.485 m^3 (775 kg)							
5 loads per hour	-	0.41	7.63	1.56	-	m^3	**9.19**
7 loads per hour	-	0.29	5.45	1.15	-	m^3	**6.60**
10 loads per hour	-	0.21	3.81	0.84	-	m^3	**4.66**
3 tonne dumper; max. volume 2.40 m^3 (3.38 t); available volume 1.9 m^3							
4 loads per hour	-	0.14	2.57	0.50	-	m^3	**3.07**
5 loads per hour	-	0.11	2.06	0.40	-	m^3	**2.46**
7 loads per hour	-	0.08	1.47	0.30	-	m^3	**1.77**
10 loads per hour	-	0.06	1.03	0.22	-	m^3	**1.25**
6 tonne dumper; max. volume 3.40 m^3 (5.4 t); available volume 3.77 m^3							
4 loads per hour	-	0.07	1.23	0.35	-	m^3	**1.58**
5 loads per hour	-	0.05	0.98	0.28	-	m^3	**1.27**
7 loads per hour	-	0.04	0.70	0.21	-	m^3	**0.91**
10 loads per hour	-	0.03	0.49	0.16	-	m^3	**0.65**
Market prices of topsoil; prices shown include for 20% settlement							
Multiple source screened topsoil	-	-	-	-	22.20	m^3	-
Single source topsoil; British Sugar PLC	-	-	-	-	31.14	m^3	-
"P30" high grade topsoil; Charles Morris (Fertilizers) Ltd	-	-	-	-	48.00	m^3	-
Site preparation							
Felling and removing trees off site							
girth 600 mm - 1.50 m (95 mm - 240 mm trunk diameter)	-	5.00	92.50	19.04	-	nr	**111.54**
girth 1.50 m - 3.00 m (240 mm - 475 mm trunk diameter)	-	17.00	314.50	76.15	-	nr	**390.65**
girth 3.00 m - 4.00 girth (475 mm - 630 mm trunk diameter)	-	48.53	897.77	121.83	-	nr	**1019.60**
Removing tree stumps							
girth 600 mm - 1.50 m	-	2.00	37.00	66.00	-	nr	**103.00**
girth 1.50 m - 3.00 m	-	7.00	129.50	101.06	-	nr	**230.56**
girth over 3.00 m	-	12.00	222.00	173.25	-	nr	**395.25**

D GROUNDWORK

Item Excluding site overheads and profit	PC £	Labour hours	Labour £	Plant £	Material £	Unit	**Total rate £**
Stump grinding; disposing to spoil heaps							
girth 600 mm - 1.50 m	-	2.00	37.00	32.20	-	nr	**69.20**
girth 1.50 m - 3.00 m	-	2.50	46.25	64.41	-	nr	**110.66**
girth over 3.00 m	-	4.00	74.00	56.35	-	nr	**130.35**
Clearing site vegetation							
mechanical clearance	-	0.25	4.63	11.56	-	100 m^2	**16.19**
hand clearance	-	2.00	37.00	-	-	100 m^2	**37.00**
Lifting turf for preservation							
machine lift and stack	-	0.75	13.88	9.48	-	100 m^2	**23.36**
hand lift and stack	-	8.33	154.16	-	-	100 m^2	**154.16**
Note: The figures in this section relate to the machine capacities shown earlier in this section. The figures below however allow for dig efficiency based on depth. The figures below also allow for a banksman. The figures below allow for bulkages of 25% on loamy soils; adjustments should be made for different soil types.							
Excavating; mechanical; topsoil for preservation							
3 tonne tracked excavator (bucket volume 0.13 m^3)							
average depth 100 mm	-	2.40	44.40	67.20	-	100 m^2	**111.60**
average depth 150 mm	-	3.36	62.16	94.08	-	100 m^2	**156.24**
average depth 200 mm	-	4.00	74.00	112.00	-	100 m^2	**186.00**
average depth 250 mm	-	4.40	81.40	123.20	-	100 m^2	**204.60**
average depth 300 mm	-	4.80	88.80	134.40	-	100 m^2	**223.20**
JCB Sitemaster 3CX (bucket volume 0.28 m^3)							
average depth 100 mm	-	1.00	18.50	33.00	-	100 m^2	**51.50**
average depth 150 mm	-	1.25	23.13	41.25	-	100 m^2	**64.38**
average depth 200 mm	-	1.78	32.93	58.74	-	100 m^2	**91.67**
average depth 250 mm	-	1.90	35.15	62.70	-	100 m^2	**97.85**
average depth 300 mm	-	2.00	37.00	66.00	-	100 m^2	**103.00**
Excavating; mechanical; to reduce levels							
5 tonne excavator (bucket volume 0.28 m^3)							
maximum depth not exceeding 0.25 m	-	0.14	2.59	0.77	-	m^3	**3.36**
maximum depth not exceeding 1.00 m	-	0.10	1.76	0.52	-	m^3	**2.28**
JCB Sitemaster 3CX (bucket volume 0.28 m^3)							
maximum depth not exceeding 0.25 m	-	0.07	1.29	2.31	-	m^3	**3.60**
maximum depth not exceeding 1.00 m	-	0.06	1.10	1.96	-	m^3	**3.06**
21 tonne 360 tracked excavator (bucket volume 1.21 m^3)							
maximum depth not exceeding 1.00 m	-	0.01	0.20	0.45	-	m^3	**0.66**
maximum depth not exceeding 2.00 m	-	0.02	0.31	0.68	-	m^3	**0.99**
Pits; 3 tonne tracked excavator							
maximum depth not exceeding 0.25 m	-	0.33	6.17	2.50	-	m^3	**8.67**
maximum depth not exceeding 1.00 m	-	0.50	9.25	3.75	-	m^3	**13.00**
maximum depth not exceeding 2.00 m	-	0.60	11.10	4.50	-	m^3	**15.60**
Trenches; width not exceeding 0.30 m; 3 tonne excavator							
maximum depth not exceeding 0.25 m	-	1.33	24.67	6.00	-	m^3	**30.67**
maximum depth not exceeding 1.00 m	-	0.69	12.67	3.08	-	m^3	**15.76**
maximum depth not exceeding 2.00 m	-	0.60	11.10	2.70	-	m^3	**13.80**
Trenches; width exceeding 0.30 m; 3 tonne excavator							
maximum depth not exceeding 0.25 m	-	0.60	11.10	2.70	-	m^3	**13.80**
maximum depth not exceeding 1.00 m	-	0.50	9.25	2.25	-	m^3	**11.50**
maximum depth not exceeding 2.00 m	-	0.38	7.12	1.73	-	m^3	**8.85**

D GROUNDWORK

Item Excluding site overheads and profit	PC £	Labour hours	Labour £	Plant £	Material £	Unit	Total rate £
D20 EXCAVATING AND FILLING - cont'd							
Excavating; mechanical; to reduce levels - cont'd							
Extra over any types of excavating irrespective of depth for breaking out existing materials; JCB with breaker attachment							
hard rock	-	0.50	9.25	62.50	-	m^3	**71.75**
concrete	-	0.50	9.25	23.44	-	m^3	**32.69**
reinforced concrete	-	1.00	18.50	42.50	-	m^3	**61.00**
brickwork, blockwork or stonework	-	0.25	4.63	23.44	-	m^3	**28.06**
Extra over any types of excavating irrespective of depth for breaking out existing hard pavings; JCB with breaker attachment							
concrete; 100 mm thick	-	-	-	1.72	-	m^2	**1.72**
concrete; 150 mm thick	-	-	-	2.86	-	m^2	**2.86**
concrete; 200 mm thick	-	-	-	3.44	-	m^2	**3.44**
concrete; 300 mm thick	-	-	-	5.16	-	m^2	**5.16**
reinforced concrete; 100 mm thick	-	0.08	1.54	2.27	-	m^2	**3.81**
reinforced concrete; 150 mm thick	-	0.08	1.39	2.98	-	m^2	**4.36**
reinforced concrete; 200 mm thick	-	0.10	1.85	3.97	-	m^2	**5.82**
reinforced concrete; 300 mm thick	-	0.15	2.77	5.95	-	m^2	**8.73**
tarmacadam; 75 mm thick	-	-	-	1.72	-	m^2	**1.72**
tarmacadam and hardcore; 150 mm thick	-	-	-	2.75	-	m^2	**2.75**
Extra over any types of excavating irrespective of depth for taking up							
precast concrete paving slabs	-	0.07	1.23	0.55	-	m^2	**1.79**
natural stone paving	-	0.10	1.85	0.83	-	m^2	**2.68**
cobbles	-	0.13	2.31	1.03	-	m^2	**3.35**
brick paviors	-	0.13	2.31	1.03	-	m^2	**3.35**
Excavating; hand							
Topsoil for preservation; loading to barrows							
average depth 100 mm	-	0.24	4.44	-	-	m^2	**4.44**
average depth 150 mm	-	0.36	6.66	-	-	m^2	**6.66**
average depth 200 mm	-	0.58	10.66	-	-	m^2	**10.66**
average depth 250 mm	-	0.72	13.32	-	-	m^2	**13.32**
average depth 300 mm	-	0.86	15.98	-	-	m^2	**15.98**
Excavating; hand							
Topsoil to reduce levels							
maximum depth not exceeding 0.25 m	-	2.40	44.40	-	-	m^3	**44.40**
maximum depth not exceeding 1.00 m	-	3.12	57.72	-	-	m^3	**57.72**
Pits							
maximum depth not exceeding 0.25 m	-	2.67	49.33	-	-	m^3	**49.33**
maximum depth not exceeding 1.00 m	-	3.47	64.13	-	-	m^3	**64.13**
maximum depth not exceeding 2.00 m (includes earthwork support)	-	6.93	128.27	-	-	m^3	**183.34**
Trenches; width not exceeding 0.30 m							
maximum depth not exceeding 0.25 m	-	2.86	52.86	-	-	m^3	**52.86**
maximum depth not exceeding 1.00 m	-	3.72	68.84	-	-	m^3	**68.84**
maximum depth not exceeding 2.00 m (includes earthwork support)	-	3.72	68.84	-	-	m^3	**96.37**
Trenches; width exceeding 0.30 m wide							
maximum depth not exceeding 0.25 m	-	2.86	52.86	-	-	m^3	**52.86**
maximum depth not exceeding 1.00 m	-	4.00	74.00	-	-	m^3	**74.00**
maximum depth not exceeding 2.00 m (includes earthwork support)	-	6.00	111.00	-	-	m^3	**166.07**
Extra over any types of excavating irrespective of depth for breaking out existing materials; hand held pneumatic breaker							
rock	-	5.00	92.50	27.60	-	m^3	**120.10**
concrete	-	2.50	46.25	13.80	-	m^3	**60.05**
reinforced concrete	-	4.00	74.00	27.70	-	m^3	**101.70**
brickwork, blockwork or stonework	-	1.50	27.75	8.28	-	m^3	**36.03**

D GROUNDWORK

Item Excluding site overheads and profit	PC £	Labour hours	Labour £	Plant £	Material £	Unit	**Total rate £**
Filling to make up levels; mechanical (JCB rear bucket)							
Arising from the excavations							
average thickness not exceeding 0.25 m	-	0.07	1.36	2.21	-	m^3	**3.56**
average thickness less than 500 mm	-	0.06	1.16	1.88	-	m^3	**3.03**
average thickness 1.00 m	-	0.05	0.86	1.39	-	m^3	**2.25**
Obtained from on site spoil heaps; average 25 m distance; multiple handling							
average thickness less than 250 mm	-	0.07	1.32	2.71	-	m^3	**4.03**
average thickness less than 500 mm	-	0.06	1.09	2.23	-	m^3	**3.32**
average thickness 1.00 m	-	0.05	0.93	1.90	-	m^3	**2.83**
Obtained off site; planting quality topsoil PC £18.50/m^3							
average thickness less than 250 mm	-	0.07	1.32	2.71	22.20	m^3	**26.23**
average thickness less than 500 mm	-	0.06	1.09	2.23	22.20	m^3	**25.52**
average thickness 1.00 m	-	0.05	0.93	1.90	22.20	m^3	**25.02**
Obtained off site; hardcore; PC £18.00/m^3							
average thickness less than 250 mm	-	0.08	1.54	3.16	21.60	m^3	**26.31**
average thickness less than 500 mm	-	0.06	1.16	2.38	21.60	m^3	**25.13**
average thickness 1.00 m	-	0.05	0.97	2.00	21.60	m^3	**24.57**
Filling to make up levels; hand							
Arising from the excavations							
average thickness exceeding 0.25 m; depositing in layers 150 mm maximum thickness	-	0.60	11.10	-	-	m^3	**11.10**
Obtained from on site spoil heaps; average 25 m distance; multiple handling							
average thickness exceeding 0.25 m thick; depositing in layers 150 mm maximum thickness	-	1.00	18.50	-	-	m^3	**18.50**
Disposal							
Note: Most commercial site disposal is carried out by 20 tonne, 8 wheeled vehicles. It has been customary to calculate disposal from construction sites in terms of full 15 m^3 loads. Spon's research has found that based on weights of common materials such as clean hardcore and topsoil, vehicles could not load more than 12 m^3 at a time. Most hauliers do not make it apparent that their loads are calculated by weight and not by volume. The rates below reflect these lesser volumes which are limited by the 20 tonne limit. The volumes shown below are based on volumes "in the solid". Weight and bulking factors have been applied. For further information please see the weights of typical materials in the Earthworks section of the Memoranda at the back of this book.							
Disposal; mechanical; Recycled Materials Ltd							
Light soils and loams (bulking factor - 1.25); 40 tonnes (2 loads per hour)							
Excavated material; off site; to tip; mechanically loaded (JCB)							
inert	-	0.04	0.77	1.25	19.80	m^3	**21.82**
clean concrete	-	0.04	0.77	1.25	5.40	m^3	**7.42**
soil (sandy and loam) dry	-	0.04	0.77	1.25	19.80	m^3	**21.82**
soil (sandy and loam) wet	-	0.04	0.77	1.25	21.00	m^3	**23.02**
broken out compacted materials such as road bases and the like	-	0.04	0.77	1.25	22.50	m^3	**24.52**
soil (clay) dry	-	0.04	0.77	1.25	25.70	m^3	**27.72**
soil (clay) wet	-	0.04	0.77	1.25	26.70	m^3	**28.72**
slightly contaminated	-	0.04	0.77	1.25	24.75	m^3	**26.77**
rubbish	-	0.04	0.77	1.25	29.70	m^3	**31.72**

D GROUNDWORK

Item Excluding site overheads and profit	PC £	Labour hours	Labour £	Plant £	Material £	Unit	Total rate £
D20 EXCAVATING AND FILLING - cont'd							
Disposal; mechanical - cont'd							
As above but allowing for 3 loads (36 m^3 - 60 tonne) per hour removed							
Inert material	-	0.03	0.51	0.83	19.80	m^3	**21.15**
Excavated material; off site; to tip; mechanically loaded by grab; capacity of load 7.25 m^3							
inert material	-	-	-	-	-	m^3	**37.95**
soil (sandy and loam) dry	-	-	-	-	-	m^3	**37.95**
soil (sandy and loam) wet	-	-	-	-	-	m^3	**40.25**
broken out compacted materials such as road bases and the like	-	-	-	-	-	m^3	**42.17**
soil (clay) dry	-	-	-	-	-	m^3	**49.26**
soil (clay) wet	-	-	-	-	-	m^3	**51.18**
Disposal by skip; 6 yd^3 (4.6 m^3)							
Excavated material loaded to skip							
by machine	-	-	-	44.46	-	m^3	**44.46**
by hand	-	3.00	55.50	42.79	-	m^3	**98.29**
Excavated material; on site							
In spoil heaps							
average 25 m distance	-	0.02	0.42	0.88	-	m^3	**1.30**
average 50 m distance	-	0.04	0.72	1.52	-	m^3	**2.24**
average 100 m distance	-	0.06	1.15	2.42	-	m^3	**3.56**
average 200 m distance	-	0.12	2.22	4.68	-	m^3	**6.90**
Spreading on site							
average 25 m distance	-	0.02	0.42	4.06	-	m^3	**4.48**
average 50 m distance	-	0.04	0.72	4.58	-	m^3	**5.30**
average 100 m distance	-	0.06	1.15	5.61	-	m^3	**6.76**
average 200 m distance	-	0.12	2.22	5.34	-	m^3	**7.56**
Disposal; hand							
Excavated material; on site; in spoil heaps							
average 25 m distance	-	2.40	44.40	-	-	m^3	**44.40**
average 50 m distance	-	2.64	48.84	-	-	m^3	**48.84**
average 100 m distance	-	3.00	55.50	-	-	m^3	**55.50**
average 200 m distance	-	3.60	66.60	-	-	m^3	**66.60**
Excavated material; spreading on site							
average 25 m distance	-	2.64	48.84	-	-	m^3	**48.84**
average 50 m distance	-	3.00	55.50	-	-	m^3	**55.50**
average 100 m distance	-	3.60	66.60	-	-	m^3	**66.60**
average 200 m distance	-	4.20	77.70	-	-	m^3	**77.70**
Cultivating							
Ripping up subsoil; using approved subsoiling machine; minimum depth 250 mm below topsoil; at 1.20 m centres; in							
gravel or sandy clay	-	-	-	2.40	-	100 m^2	**2.40**
soil compacted by machines	-	-	-	2.80	-	100 m^2	**2.80**
clay	-	-	-	3.00	-	100 m^2	**3.00**
chalk or other soft rock	-	-	-	6.01	-	100 m^2	**6.01**
Extra for subsoiling at 1 m centres	-	-	-	0.60	-	100 m^2	**0.60**
Breaking up existing ground; using pedestrian operated tine cultivator or rotavator; loam or sandy soil							
100 mm deep	-	0.22	4.07	1.90	-	100 m^2	**5.97**
150 mm deep	-	0.28	5.09	2.37	-	100 m^2	**7.46**
200 mm deep	-	0.37	6.78	3.16	-	100 m^2	**9.95**
As above but in heavy clay or wet soils							
100 mm deep	-	0.44	8.14	3.80	-	100 m^2	**11.94**
150 mm deep	-	0.66	12.21	5.70	-	100 m^2	**17.91**
200 mm deep	-	0.82	15.26	7.12	-	100 m^2	**22.38**

D GROUNDWORK

Item Excluding site overheads and profit	PC £	Labour hours	Labour £	Plant £	Material £	Unit	**Total rate £**
Breaking up existing ground; using tractor drawn tine cultivator or rotavator							
100 mm deep	-	-	-	0.27	-	100 m^2	**0.27**
150 mm deep	-	-	-	0.34	-	100 m^2	**0.34**
200 mm deep	-	-	-	0.45	-	100 m^2	**0.45**
400 mm deep	-	-	-	1.36	-	100 m^2	**1.36**
Cultivating ploughed ground; using disc, drag, or chain harrow							
4 passes	-	-	-	1.63	-	100 m^2	**1.63**
Rolling cultivated ground lightly; using self-propelled agricultural roller	-	0.06	1.03	0.60	-	100 m^2	**1.63**
Importing and storing selected and approved topsoil; to BS 3882; 20 tonne load = 11.88 m^3 average							
20 tonne loads	18.50	-	-	-	22.50	m^3	-
Grading operations; surface previously excavated to reduce levels to prepare to receive subsequent treatments; grading to accurate levels and falls 20 mm tolerances							
Clay or heavy soils or hardcore							
5 tonne excavator	-	0.04	0.74	0.22	-	m^2	**0.96**
JCB Sitemaster 3CX	-	0.02	0.37	0.66	-	m^2	**1.03**
21 tonne 360 tracked excavator	-	0.01	0.19	0.41	-	m^2	**0.60**
by hand	-	0.10	1.85	-	-	m^2	**1.85**
Loamy topsoils							
5 tonne excavator	-	0.03	0.49	0.15	-	m^2	**0.64**
JCB Sitemaster 3CX	-	0.01	0.25	0.44	-	m^2	**0.68**
21 tonne 360 tracked excavator	-	0.01	0.14	0.31	-	m^2	**0.44**
by hand	-	0.05	0.93	-	-	m^2	**0.93**
Sand or graded granular materials							
5 tonne excavator	-	0.02	0.37	0.11	-	m^2	**0.48**
JCB Sitemaster 3CX	-	0.01	0.19	0.33	-	m^2	**0.52**
21 tonne 360 tracked excavator	-	0.01	0.11	0.24	-	m^2	**0.34**
by hand	-	0.03	0.62	-	-	m^2	**0.62**
Excavating; by hand							
Topsoil for preservation; loading to barrows							
average depth 100 mm	-	0.24	4.44	-	-	m^2	**4.44**
average depth 150 mm	-	0.36	6.66	-	-	m^2	**6.66**
average depth 200 mm	-	0.58	10.66	-	-	m^2	**10.66**
average depth 250 mm	-	0.72	13.32	-	-	m^2	**13.32**
average depth 300 mm	-	0.86	15.98	-	-	m^2	**15.98**
Grading operations; surface recently filled to raise levels to prepare to receive subsequent treatments; grading to accurate levels and falls 20 mm tolerances							
Clay or heavy soils or hardcore							
5 tonne excavator	-	0.03	0.56	0.16	-	m^2	**0.72**
JCB Sitemaster 3CX	-	0.02	0.28	0.50	-	m^2	**0.77**
21 tonne 360 tracked excavator	-	0.01	0.12	0.28	-	m^2	**0.40**
by hand	-	0.08	1.54	-	-	m^2	**1.54**
Loamy topsoils							
5 tonne excavator	-	0.02	0.38	0.11	-	m^2	**0.49**
JCB Sitemaster 3CX	-	0.01	0.19	0.34	-	m^2	**0.53**
21 tonne 360 tracked excavator	-	0.01	0.12	0.28	-	m^2	**0.40**
by hand	-	0.04	0.74	-	-	m^2	**0.74**
Sand or graded granular materials							
5 tonne excavator	-	0.02	0.28	0.08	-	m^2	**0.36**
JCB Sitemaster 3CX	-	0.01	0.14	0.25	-	m^2	**0.39**
21 tonne 360 tracked excavator	-	0.01	0.09	0.21	-	m^2	**0.30**
by hand	-	0.03	0.53	-	-	m^2	**0.53**

D GROUNDWORK

Item Excluding site overheads and profit	PC £	Labour hours	Labour £	Plant £	Material £	Unit	**Total rate £**
D20 EXCAVATING AND FILLING - cont'd							
Surface treatments							
Compacting							
bottoms of excavations	-	0.01	0.09	0.02	-	m^2	**0.12**
Surface preparation							
Trimming surfaces of cultivated ground to final levels, removing roots stones and debris exceeding 50 mm in any direction to tip offsite; slopes less than 15 degrees							
clean ground with minimal stone content	-	0.25	4.63	-	-	100 m^2	**4.63**
slightly stony - 0.5 kg stones per m^2	-	0.33	6.16	-	-	100 m^2	**6.16**
very stony - 1.0 - 3.00 kg stones per m^2	-	0.50	9.25	-	0.01	100 m^2	**9.26**
clearing mixed slightly contaminated rubble inclusive of roots and vegetation	-	0.50	9.25	-	0.04	100 m^2	**9.29**
clearing brick-bats, stones and clean rubble	-	0.60	11.10	-	0.03	100 m^2	**11.13**

E IN SITU CONCRETE/LARGE PRECAST CONCRETE

Item Excluding site overheads and profit	PC £	Labour hours	Labour £	Plant £	Material £	Unit	**Total rate £**
E10 MIXING/CASTING/CURING IN SITU CONCRETE							
General							
The concrete mixes used here are referred to as "Design", "Standard" and "Designated" mixes. The BS references on these are used to denote the concrete strength and mix volumes. Please refer to the definitions and the tables in the Concrete Work section in the Memoranda at the back of this book.							
Designed mix							
User specified performance of the concrete. Producer responsible for selecting appropriate mix. Strength testing is essential.							
Prescribed mix							
User specifies mix constituents and is responsible for ensuring that the concrete meets performance requirements. The mix proportion is essential.							
Standard mix							
Specified from the list in BS 5328 Pt 2 1991 s.4. Made with a restricted range of materials. Specification to include the proposed use of the material as well as; the standard mix reference, the type of cement, type and size of aggregate, slump (workability). Quality assurance required.							
Designated mix							
Mix specified in BS 5328 Pt 2: 1991 s.5. Producer to hold current product conformity certification and quality approval to BS 5750 Pt 1 (EN 29001). Quality assurance essential. The mix may not be modified.							
Note: The mixes below are all mixed in 113 litre mixers. Please see the Minor Works section of this book for mixes containing aggregates delivered in bulk bags.							
Concrete mixes; mixed on site; costs for producing concrete; prices for commonly used mixes for various types of work; based on bulk load 20 tonne rates for aggregates							
Roughest type mass concrete such as footings, road haunchings 300 thick							
1:3:6	58.24	1.20	22.14	-	58.24	m^3	**80.38**
1:3:6 sulphate resisting	62.11	1.20	22.14	-	62.11	m^3	**84.25**
As above but aggregates delivered in 10 tonne loads							
1:3:6	62.74	1.20	22.14	-	62.74	m^3	**84.88**
1:3:6 sulphate resisting	66.61	1.20	22.14	-	66.61	m^3	**88.75**
As above but aggregates delivered in 900 kg bulk bags							
1:3:6	85.94	1.20	22.14	-	85.94	m^3	**108.08**
1:3:6 sulphate resisting	89.81	1.20	22.14	-	89.81	m^3	**111.95**
Most ordinary use of concrete such as mass walls above ground, road slabs etc. and general reinforced concrete work							
1:2:4	68.56	1.20	22.14	-	68.56	m^3	**90.71**
1:2:4 sulphate resisting	74.04	1.20	22.14	-	74.04	m^3	**96.18**

E IN SITU CONCRETE/LARGE PRECAST CONCRETE

Item Excluding site overheads and profit	PC £	Labour hours	Labour £	Plant £	Material £	Unit	**Total rate £**
E10 MIXING/CASTING/CURING IN SITU CONCRETE - cont'd							
Concrete mixes; mixed on site - cont'd							
As above but aggregates delivered in 10 tonne loads							
1:2:4	73.06	1.20	22.14	-	73.06	m^3	**95.21**
1:2:4 sulphate resisting	78.54	1.20	22.14	-	78.54	m^3	**100.68**
As above but aggregates delivered in 900 kg bulk bags							
1:2:4	96.26	1.20	22.14	-	96.26	m^3	**118.41**
1:2:4 sulphate resisting	101.74	1.20	22.14	-	101.74	m^3	**123.88**
Watertight floors, pavements and walls, tanks pits steps paths surface of two course roads, reinforced concrete where extra strength is required							
1:1.5:3	78.42	1.20	22.14	-	78.42	m^3	**100.57**
As above but aggregates delivered in 10 tonne loads							
1:1.5:3	82.92	1.20	22.14	-	82.92	m^3	**105.07**
As above but aggregates delivered in 900 kg bulk bags							
1:1.5:3	106.12	1.20	22.14	-	106.12	m^3	**128.27**
Plain in situ concrete; site mixed; 10 N/mm^2 - 40 aggregate (1:3:6); aggregate delivery indicated							
Foundations							
ordinary Portland cement; 10 tonne ballast loads	80.38	1.00	18.50	-	82.39	m^3	**100.89**
ordinary Portland cement; 900 kg bulk bags	108.08	1.00	18.50	-	110.78	m^3	**129.28**
sulphate resistant cement; 10 tonne ballast loads	103.84	1.00	18.50	-	103.84	m^3	**122.34**
sulphate resistant cement; 900 kg bulk bags	111.95	1.00	18.50	-	114.75	m^3	**133.25**
Foundations; poured on or against earth or unblinded hardcore							
ordinary Portland cement; 10 tonne ballast loads	80.38	1.00	18.50	-	84.40	m^3	**102.90**
ordinary Portland cement; 900 kg bulk bags	108.08	1.00	18.50	-	113.48	m^3	**131.98**
sulphate resistant cement; 10 tonne ballast loads	103.84	1.00	18.50	-	109.03	m^3	**127.53**
sulphate resistant cement; 900 kg bulk bags	111.95	1.00	18.50	-	117.55	m^3	**136.05**
Isolated foundations							
ordinary Portland cement; 10 tonne ballast loads	80.38	1.10	20.35	-	82.39	m^3	**102.74**
ordinary Portland cement; 900 kg bulk bags	108.08	1.10	20.35	-	110.78	m^3	**131.13**
sulphate resistant cement; 10 tonne ballast loads	103.84	1.10	20.35	-	103.84	m^3	**124.19**
sulphate resistant cement; 900 kg bulk bags	111.95	1.10	20.35	-	114.75	m^3	**135.10**
Plain in situ concrete; site mixed; 21 N/mm^2 - 20 aggregate (1:2:4)							
Foundations							
ordinary Portland cement; 10 tonne ballast loads	98.71	1.00	18.50	-	101.18	m^3	**119.68**
ordinary Portland cement; 900 kg bulk bags	139.36	1.00	18.50	-	142.84	m^3	**161.34**
sulphate resistant cement; 10 tonne ballast loads	103.84	1.00	18.50	-	103.84	m^3	**122.34**
sulphate resistant cement; 900 kg bulk bags	111.95	1.00	18.50	-	114.75	m^3	**133.25**
Foundations; poured on or against earth or unblinded hardcore							
ordinary Portland cement; 900 kg bulk bags	139.36	1.65	30.52	-	142.84	m^3	**173.37**
sulphate resistant cement; 900 kg bulk bags	146.66	1.65	30.52	-	150.33	m^3	**180.85**
Isolated foundations							
ordinary Portland cement	139.36	2.20	40.70	-	142.84	m^3	**183.54**
sulphate resistant cement	146.66	2.20	40.70	-	150.33	m^3	**191.03**
Ready mix concrete; concrete mixed on site; Euromix							
Ready mix concrete mixed on site							
1:3:6 placed by barrow not more than 25 m distance (20-25 barrows/m^3)	79.50	1.00	18.50	-	79.50	m^3	**98.00**
1:2:4 placed by barrow not more than 25 m distance (20-25 barrows/m^3)	80.00	1.00	18.50	-	80.00	m^3	**98.50**

E IN SITU CONCRETE/LARGE PRECAST CONCRETE

Item Excluding site overheads and profit	PC £	Labour hours	Labour £	Plant £	Material £	Unit	**Total rate £**
Reinforced in situ concrete; site mixed; 21 N/mm² - 20 aggregate (1:2:4); aggregates delivered in 10 tonne loads							
Foundations							
ordinary Portland cement	98.71	2.20	40.70	-	101.18	m^3	**141.88**
sulphate resistant cement	103.84	2.20	40.70	-	106.44	m^3	**147.14**
Foundations; poured on or against earth or unblinded hardcore							
ordinary Portland cement	98.71	2.20	40.70	-	101.18	m^3	**141.88**
sulphate resistant cement	146.66	2.20	40.70	-	150.33	m^3	**191.03**
Isolated foundations							
ordinary Portland cement	98.71	2.75	50.88	-	101.18	m^3	**152.05**
sulphate resistant cement	146.66	2.75	50.88	-	150.33	m^3	**201.20**
Plain in situ concrete; ready mixed; Tarmac Southern, 10 N/mm mixes; suitable for mass concrete fill and blinding							
Foundations							
GEN1; Designated mix	70.40	1.50	27.75	-	70.40	m^3	**98.15**
ST2; Standard mix	75.59	1.50	27.75	-	79.37	m^3	**107.12**
Foundations; poured on or against earth or unblinded hardcore							
GEN1; Designated mix	70.40	1.57	29.14	-	70.40	m^3	**99.54**
ST2; Standard mix	75.59	1.57	29.14	-	81.26	m^3	**110.40**
Isolated foundations							
GEN1; Designated mix	70.40	2.00	37.00	-	70.40	m^3	**107.40**
ST2; Standard mix	75.59	2.00	37.00	-	79.37	m^3	**116.37**
Plain in situ concrete; ready mixed; Tarmac Southern, 15 N/mm mixes; suitable for oversite below suspended slabs and strip footings in non-aggressive soils							
Foundations							
GEN 2; Designated mix	72.63	1.50	27.75	-	76.26	m^3	**104.01**
ST3; Standard mix	75.90	1.50	27.75	-	79.69	m^3	**107.44**
Foundations; poured on or against earth or unblinded hardcore							
GEN2; Designated mix	72.63	1.57	29.14	-	72.63	m^3	**101.77**
ST3; Standard mix	75.90	1.57	29.14	-	75.90	m^3	**105.04**
Isolated foundations							
GEN2; Designated mix	72.63	2.00	37.00	-	72.63	m^3	**109.63**
ST3; Standard mix	75.90	2.00	37.00	-	75.90	m^3	**112.90**
Plain in situ concrete; ready mixed; Tarmac Southern; air entrained mixes suitable for paving							
Beds or slabs; house drives parking and external paving							
PAV 1; 35 N/mm²; Designated mix	74.80	1.50	27.75	-	78.54	m^3	**106.29**
Beds or slabs; heavy duty external paving							
PAV 2; 40 N/mm²; Designated mix	75.50	1.50	27.75	-	79.28	m^3	**107.03**
Reinforced in situ concrete; ready mixed; Tarmac Southern; 35 N/mm² mix; suitable for foundations in class 2 sulphate conditions							
Foundations							
RC 35; Designated mix	74.00	2.00	37.00	-	77.70	m^3	**114.70**
Foundations; poured on or against earth or unblinded hardcore							
RC 35; Designated mix	74.00	2.10	38.85	-	79.55	m^3	**118.40**
Isolated foundations							
RC 35; Designated mix	74.00	2.00	37.00	-	74.00	m^3	**111.00**

E IN SITU CONCRETE/LARGE PRECAST CONCRETE

Item Excluding site overheads and profit	PC £	Labour hours	Labour £	Plant £	Material £	Unit	**Total rate £**
E20 FORMWORK FOR IN SITU CONCRETE							
Plain vertical formwork; basic finish							
Sides of foundations							
height exceeding 1.00 m	-	2.00	37.00	-	4.35	m^2	**41.35**
height not exceeding 250 mm	-	1.00	18.50	-	1.22	m	**19.72**
height 250 - 500 mm	-	1.00	18.50	-	2.17	m	**20.67**
height 500 mm - 1.00 m	-	1.50	27.75	-	4.35	m	**32.10**
Sides of foundations; left in							
height over 1.00 m	-	2.00	37.00	-	9.10	m^2	**46.10**
height not exceeding 250 mm	-	1.00	18.50	-	4.72	m	**23.22**
height 250 - 500 mm	-	1.00	18.50	-	9.18	m	**27.68**
height 500 mm - 1.00 m	-	1.50	27.75	-	18.36	m	**46.11**
E30 REINFORCEMENT FOR IN SITU CONCRETE							
The rates for reinforcement shown for steel bar below are based on prices which would be supplied on a typical landscape contract. The steel prices shown have been priced on a selection of steel delivered to site where the total order quantity is in the region of 2 tonnes. The assumption is that should larger quantities be required, the work would fall outside the scope of the typical landscape contract defined in the front of this book. Keener rates can be obtained for larger orders.							
Reinforcement bars; BS 4449; hot rolled plain round mild steel; straight or bent							
Bars							
8 mm nominal size	550.00	27.00	499.50	-	550.00	tonne	**1049.50**
10 mm nominal size	550.00	26.00	481.00	-	550.00	tonne	**1031.00**
12 mm nominal size	550.00	25.00	462.50	-	550.00	tonne	**1012.50**
16 mm nominal size	550.00	24.00	444.00	-	550.00	tonne	**994.00**
20 mm nominal size	550.00	23.00	425.50	-	550.00	tonne	**975.50**
Reinforcement bar to concrete formwork							
8 mm bar							
100 ccs	2.15	0.33	6.11	-	2.15	m^2	**8.25**
200 ccs	1.04	0.25	4.63	-	1.04	m^2	**5.67**
300 ccs	0.71	0.15	2.77	-	0.71	m^2	**3.49**
10 mm bar							
100 ccs	3.35	0.33	6.11	-	3.35	m^2	**9.46**
200 ccs	1.65	0.25	4.63	-	1.65	m^2	**6.28**
300 ccs	1.10	0.15	2.77	-	1.10	m^2	**3.88**
12 mm bar							
100 ccs	4.84	0.33	6.11	-	4.84	m^2	**10.95**
200 ccs	2.42	0.25	4.63	-	2.42	m^2	**7.04**
300 ccs	1.59	0.15	2.77	-	1.59	m^2	**4.37**
16 mm bar							
100 ccs	8.63	0.40	7.40	-	8.63	m^2	**16.04**
200 ccs	4.34	0.33	6.11	-	4.34	m^2	**10.45**
300 ccs	2.92	0.25	4.63	-	2.92	m^2	**7.54**
25 mm bar							
100 ccs	21.18	0.40	7.40	-	21.18	m^2	**28.58**
200 ccs	10.56	0.33	6.11	-	10.56	m^2	**16.66**
300 ccs	7.04	0.25	4.63	-	7.04	m^2	**11.66**
32 mm bar							
100 ccs	34.70	0.40	7.40	-	34.70	m^2	**42.10**
200 ccs	17.32	0.33	6.11	-	17.32	m^2	**23.43**
300 ccs	11.55	0.25	4.63	-	11.55	m^2	**16.18**

E IN SITU CONCRETE/LARGE PRECAST CONCRETE

Item Excluding site overheads and profit	PC £	Labour hours	Labour £	Plant £	Material £	Unit	**Total rate £**
Reinforcement fabric; BS 4483; lapped; in beds or suspended slabs							
Fabric							
ref A98 (1.54 kg/m^2)	1.20	0.22	4.07	-	1.32	m^2	**5.39**
ref A142 (2.22 kg/m^2)	1.72	0.22	4.07	-	1.89	m^2	**5.96**
ref A193 (3.02 kg/m^2)	3.16	0.22	4.07	-	3.48	m^2	**7.55**
ref A252 (3.95 kg/m^2)	4.10	0.24	4.44	-	4.51	m^2	**8.95**
ref A393 (6.16 kg/m^2)	5.60	0.28	5.18	-	6.16	m^2	**11.34**

F MASONRY

Item Excluding site overheads and profit	PC £	Labour hours	Labour £	Plant £	Material £	Unit	Total rate £
F MARKET PRICES OF MATERIALS							
Cement; Builder Centre							
Portland cement	-	-	-	-	2.90	25 kg	-
Sulphate resistant cement	-	-	-	-	3.35	25 kg	-
White cement	-	-	-	-	7.30	25 kg	-
Sand; Builder Centre							
Building sand							
loose	-	-	-	-	30.50	m^3	-
900 kg bulk bags	-	-	-	-	31.00	nr	-
Sharp sand							
loose	-	-	-	-	30.50	m^3	-
900 kg bulk bags	-	-	-	-	30.50	nr	-
Sand; Yeoman Aggregates Ltd							
Sharp sand	-	-	-	-	18.50	tonne	-
Bricks; E.T. Clay Products							
Ibstock; facing bricks; 215 x 102.5 x 65 mm							
Leicester Red Stock	-	-	-	-	420.42	1000	-
Leicester Yellow Stock	-	-	-	-	354.90	1000	-
Himley Mixed Russet	-	-	-	-	366.36	1000	-
Himley Worcs. Mixture	-	-	-	-	354.90	1000	-
Roughdales Red Multi Rustic	-	-	-	-	390.40	1000	-
Ashdown Cottage Mixture	-	-	-	-	432.43	1000	-
Ashdown Crowborough Multi	-	-	-	-	502.10	1000	-
Ashdown Pevensey Multi	-	-	-	-	450.46	1000	-
Chailey Stock	-	-	-	-	490.31	1000	-
Dorking Multi Coloured	-	-	-	-	372.37	1000	-
Holbrook Smooth Red	-	-	-	-	462.47	1000	-
Stourbridge Kenilworth Multi	-	-	-	-	354.36	1000	-
Stourbridge Pennine Pastone	-	-	-	-	322.14	1000	-
Stratford Red Rustic	-	-	-	-	336.34	1000	-
Swanage Restoration Red	-	-	-	-	702.70	1000	-
Laybrook Sevenoaks Yellow	-	-	-	-	349.44	1000	-
Laybrook Arundel Yellow	-	-	-	-	371.28	1000	-
Laybrook Thakeham Red	-	-	-	-	316.68	1000	-
Funton Second Hard Stock	-	-	-	-	425.88	1000	-
Hanson Brick Ltd, London Brand; facing bricks; 215 x 102.5 x 65 mm							
Capel Multi Stock	-	-	-	-	301.60	1000	-
Rusper Stock	-	-	-	-	306.80	1000	-
Delph Autumn	-	-	-	-	343.20	1000	-
Regency	-	-	-	-	343.20	1000	-
Sandfaced	-	-	-	-	343.20	1000	-
Saxon Gold	-	-	-	-	343.20	1000	-
Tudor	-	-	-	-	343.20	1000	-
Windsor	-	-	-	-	343.20	1000	-
Autumn Leaf	-	-	-	-	343.20	1000	-
Claydon Red Multi	-	-	-	-	343.20	1000	-
Other facing bricks; 215 x 102.5 x 65 mm							
Soft Reds - Milton Hall	-	-	-	-	390.00	1000	-
Staffs Blues - Blue Smooth	-	-	-	-	483.60	1000	-
Staffs Blues - Blue Brindle	-	-	-	-	442.00	1000	-
Reclaimed (second hand) bricks							
Yellows	-	-	-	-	980.00	1000	-
Yellow Multi	-	-	-	-	980.00	1000	-
Mixed London Stock	-	-	-	-	700.00	1000	-
Red Multi	-	-	-	-	650.00	1000	-
Gaults	-	-	-	-	540.00	1000	-
Red Rubbers	-	-	-	-	780.00	1000	-
Common bricks	-	-	-	-	192.40	1000	-
Engineering bricks	-	-	-	-	291.20	1000	-
Kempston facing bricks; 215 x 102.5 x 65 mm							
Melford Yellow	-	-	-	-	343.20	1000	-

F MASONRY

Item Excluding site overheads and profit	PC £	Labour hours	Labour £	Plant £	Material £	Unit	**Total rate £**
F10 BRICK/BLOCK WALLING							
Note: Batching quantities for these mortar mixes may be found in the Memoranda at the back of this book.							
Mortar mixes; common mixes for various types of work; mortar mixed on site; prices based on builders merchant rates for cement							
Aggregates delivered in 900 kg bulk bags; mechanically mixed							
1:3	-	0.75	13.88	-	130.84	m^3	**144.72**
1:4	-	0.75	13.88	-	104.87	m^3	**118.74**
1:1:6	-	0.75	13.88	-	109.06	m^3	**122.93**
1:1:6 sulphate resisting	-	0.75	13.88	-	113.92	m^3	**127.79**
Mortar mixes; common mixes for various types of work; mortar mixed on site; prices based on builders merchant rates for cement							
Aggregates delivered in 10 tonne loads; mechanically mixed							
1:3	-	0.75	13.88	-	100.26	m^3	**114.13**
1:4	-	0.75	13.88	-	77.86	m^3	**91.74**
1:1:6	-	0.75	13.88	-	83.48	m^3	**97.36**
1:1:6 sulphate resisting	-	0.75	13.88	-	88.34	m^3	**102.21**
Variation in brick prices							
Add or subtract the following amounts for every £1.00/1000 difference in the PC price of the measured items below							
half brick thick	-	-	-	-	0.06	m^2	-
one brick thick	-	-	-	-	0.13	m^2	-
one and a half brick thick	-	-	-	-	0.19	m^2	-
two brick thick	-	-	-	-	0.25	m^2	-
Mortar (1:3) required per m^2 of brickwork; brick size 215 x 102.5 x 65 mm							
Half brick wall (103 mm)							
no frog	-	-	-	-	2.64	m^2	-
single frog	-	-	-	-	3.24	m^2	-
double frog	-	-	-	-	3.83	m^2	-
2 x half brick cavity wall (270 mm)							
no frog	-	-	-	-	4.55	m^2	-
single frog	-	-	-	-	5.40	m^2	-
double frog	-	-	-	-	6.60	m^2	-
One brick wall (215 mm)							
no frog	-	-	-	-	5.51	m^2	-
single frog	-	-	-	-	6.60	m^2	-
double frog	-	-	-	-	7.67	m^2	-
One and a half brick wall (328 mm)							
no frog	-	-	-	-	7.56	m^2	-
single frog	-	-	-	-	8.87	m^2	-
double frog	-	-	-	-	10.55	m^2	-
Mortar (1:3) required per m^2 of blockwork; blocks 440 x 215 mm							
Block thickness							
100 mm	-	-	-	-	0.84	m^2	-
140 mm	-	-	-	-	1.08	m^2	-
hollow blocks 440 x 215 mm	-	-	-	-	0.27	m^2	-
Movement of materials							
Loading to wheelbarrows and transporting to location; per 215 mm thick walls; maximum distance 25 m	-	0.42	7.71	-	-	m^2	**7.71**

F MASONRY

Item Excluding site overheads and profit	PC £	Labour hours	Labour £	Plant £	Material £	Unit	**Total rate £**
F10 BRICK/BLOCK WALLING - cont'd							
Class B engineering bricks; PC £260.00/1000; double Flemish bond in cement mortar (1:3)							
Mechanically offloading; maximum 25 m distance; loading to wheelbarrows; transporting to location; per 215 mm thick walls	-	0.42	7.71	-	-	m^2	**7.71**
Walls							
half brick thick	-	1.80	33.30	-	18.09	m^2	**51.39**
one brick thick	-	3.60	66.59	-	36.18	m^2	**102.78**
one and a half brick thick	-	5.40	99.89	-	62.55	m^2	**162.45**
two brick thick	-	7.20	133.19	-	72.37	m^2	**205.56**
Walls; curved; mean radius 6 m							
half brick thick	-	2.70	50.00	-	18.63	m^2	**68.63**
one brick thick	-	5.41	100.00	-	36.18	m^2	**136.18**
Walls; curved; mean radius 1.50 m							
half brick thick	-	3.60	66.59	-	18.66	m^2	**85.26**
one brick thick	-	7.20	133.19	-	36.18	m^2	**169.37**
Walls; tapering; one face battering; average							
one and a half brick thick	-	6.65	123.00	-	54.28	m^2	**177.28**
two brick thick	-	8.87	164.01	-	72.37	m^2	**236.37**
Walls; battering (retaining)							
one and a half brick thick	-	6.65	123.00	-	54.28	m^2	**177.28**
two brick thick	-	8.87	164.01	-	72.37	m^2	**236.37**
Isolated piers							
one brick thick	-	7.00	129.50	-	39.61	m^2	**169.11**
one and a half brick thick	-	9.00	166.50	-	59.41	m^2	**225.91**
two brick thick	-	10.00	185.00	-	80.36	m^2	**265.36**
three brick thick	-	12.40	229.40	-	118.82	m^2	**348.22**
Projections; vertical							
one brick x half brick	-	0.70	12.95	-	4.04	m	**16.99**
one brick x one brick	-	1.40	25.90	-	8.08	m	**33.98**
one and a half brick x one brick	-	2.10	38.85	-	12.12	m	**50.97**
two brick x one brick	-	2.30	42.55	-	16.16	m	**58.71**
Walls; half brick thick							
in honeycomb bond	-	1.80	33.30	-	12.94	m^2	**46.24**
in quarter bond	-	1.67	30.83	-	17.70	m^2	**48.53**
Facing bricks; PC £300.00/1000; English garden wall bond; in gauged mortar (1:1:6); facework one side							
Mechanically offloading; maximum 25 m distance; loading to wheelbarrows and transporting to location; per 215 mm thick walls	-	0.42	7.71	-	-	m^2	**7.71**
Walls							
half brick thick	-	1.80	33.30	-	20.40	m^2	**53.70**
half brick thick half brick thick (using site cut snap headers to form bond)	-	2.41	44.68	-	20.40	m^2	**65.08**
one brick thick	-	3.60	66.59	-	40.80	m^2	**107.40**
one and a half brick thick	-	5.40	99.89	-	61.20	m^2	**161.10**
two brick thick	-	7.20	133.19	-	81.61	m^2	**214.79**
Walls; curved; mean radius 6 m							
half brick thick	-	2.70	50.00	-	21.80	m^2	**71.80**
one brick thick	-	5.41	100.00	-	42.60	m^2	**142.60**
Walls; curved; mean radius 1.50 m							
half brick thick	-	3.60	66.59	-	21.35	m^2	**87.95**
one brick thick	-	7.20	133.19	-	41.70	m^2	**174.89**
Walls; tapering; one face battering; average							
one and a half brick thick	-	6.65	123.00	-	63.90	m^2	**186.91**
two brick thick	-	8.87	164.01	-	85.21	m^2	**249.21**
Walls; battering (retaining)							
one and a half brick thick	-	5.98	110.70	-	63.90	m^2	**174.61**
two brick thick	-	7.98	147.61	-	85.21	m^2	**232.81**

F MASONRY

Item Excluding site overheads and profit	PC £	Labour hours	Labour £	Plant £	Material £	Unit	**Total rate £**
Isolated piers; English bond; facework all round							
one brick thick	-	7.00	129.50	-	44.65	m^2	**174.15**
one and a half brick thick	-	9.00	166.50	-	66.97	m^2	**233.47**
two brick thick	-	10.00	185.00	-	90.44	m^2	**275.44**
three brick thick	-	12.40	229.40	-	133.94	m^2	**363.34**
Projections; vertical							
one brick x half brick	-	0.70	12.95	-	4.55	m	**17.50**
one brick x one brick	-	1.40	25.90	-	9.10	m	**35.00**
one and a half brick x one brick	-	2.10	38.85	-	13.65	m	**52.50**
two brick x one brick	-	2.30	42.55	-	18.20	m	**60.75**
Brickwork fair faced both sides; facing bricks in gauged mortar (1:1:6)							
Extra for fair face both sides; flush, struck, weathered, or bucket-handle pointing	-	0.67	12.33	-	-	m^2	**12.33**
Extra for cement mortar (1:3) in lieu of gauged mortar							
half brick thick	-	-	-	-	0.21	m^2	-
one brick thick	-	-	-	-	0.42	m^2	-
one and a half brick thick	-	-	-	-	0.63	m^2	-
two brick thick	-	-	-	-	0.84	m^2	-
Reclaimed bricks; PC £800.00/1000; English garden wall bond; in gauged mortar (1:1:6)							
Walls							
half brick thick (stretcher bond)	-	1.80	33.30	-	51.90	m^2	**85.20**
half brick thick (using site cut snap headers to form bond)	-	2.41	44.68	-	51.90	m^2	**96.58**
one brick thick	-	3.60	66.59	-	101.88	m^2	**168.48**
one and a half brick thick	-	5.40	99.89	-	152.82	m^2	**252.72**
two brick thick	-	7.20	133.19	-	203.77	m^2	**336.95**
Walls; curved; mean radius 6 m							
half brick thick	-	2.70	50.00	-	54.80	m^2	**104.80**
one brick thick	-	7.20	133.19	-	108.60	m^2	**241.79**
Walls; curved; mean radius 1.50 m							
half brick thick	-	1.60	29.60	-	57.20	m^2	**86.80**
one brick thick	-	3.20	59.20	-	113.40	m^2	**172.60**
Extra for cement mortar (1:3) in lieu of gauged mortar							
half brick thick	-	-	-	-	0.21	m^2	-
one brick thick	-	-	-	-	0.42	m^2	-
one and a half brick thick	-	-	-	-	0.63	m^2	-
two brick thick	-	-	-	-	0.84	m^2	-
Walls; stretcher bond; wall ties at 450 centres vertically and horizontally							
one brick thick	0.46	3.27	60.56	-	102.35	m^2	**162.91**
two brick thick	1.38	4.91	90.82	-	205.18	m^2	**296.00**
Brickwork fair-faced both sides; facing bricks in gauged mortar (1:1:6)							
Extra for fair-face both sides; flush, struck, weathered, or bucket-handle pointing	-	0.67	12.33	-	-	m^2	**12.33**
Brick copings							
Copings; all brick headers-on-edge; to BS 4729; two angles rounded 53 mm radius; flush pointing top and both sides as work proceeds; one brick wide; horizontal							
machine-made specials	25.65	0.49	9.11	-	27.10	m	**36.21**
hand-made specials	25.65	0.49	9.11	-	27.11	m	**36.22**
Extra over copings for two courses machine-made tile creasings, projecting 25 mm each side; 260 mm wide copings; horizontal	5.54	0.50	9.25	-	6.16	m	**15.41**
Copings; all brick headers-on-edge; flush pointing top and both sides as work proceeds; one brick wide; horizontal							
facing bricks PC £300.00/1000	1.00	0.49	9.11	-	4.30	m	**13.41**
engineering bricks PC £260.00/1000	3.47	0.49	9.11	-	3.64	m	**12.75**

F MASONRY

Item Excluding site overheads and profit	PC £	Labour hours	Labour £	Plant £	Material £	Unit	**Total rate £**
F10 BRICK/BLOCK WALLING - cont'd							
Dense aggregate concrete blocks; "Tarmac Topblock" or other equal and approved; in gauged mortar (1:2:9)							
Walls							
Solid blocks 7 N/mm^2							
440 x 215 x 100 mm thick	7.50	1.20	22.20	-	8.41	m^2	**30.61**
440 x 215 x 140 mm thick	11.90	1.30	24.05	-	12.93	m^2	**36.98**
Solid blocks 7 N/mm^2 laid flat							
440 x 100 x 215 mm thick	15.30	3.22	59.57	-	17.75	m^2	**77.32**
Hollow concrete blocks							
440 x 215 x 215 mm thick	16.50	1.30	24.05	-	17.53	m^2	**41.58**
Filling of hollow concrete blocks with concrete as work proceeds; tamping and compacting							
440 x 215 x 215 mm thick	18.04	0.20	3.70	-	18.04	m^2	**21.74**
F20 NATURAL STONE RUBBLE WALLING							
Granite walls							
Granite random rubble walls; laid dry							
200 mm thick; single faced	40.00	6.66	123.21	-	40.00	m^2	**163.21**
Granite walls; one face battering to 50 degrees; pointing faces							
450 mm (average) thick	90.00	5.00	92.50	-	97.47	m^2	**189.97**
Dry stone walling - General							
Preamble: In rural areas where natural stone is a traditional material, it may be possible to use dry stone walling or dyking as an alternative to fences or brick walls. Many local authorities are willing to meet the extra cost of stone walling in areas of high landscape value, and they may hold lists of available craftsmen. DSWA Office, Westmorland County Showground, Lane Farm, Crooklands, Milnthorpe, Cumbria, LA7 7NH; Tel: 01539 567953; E-mail: information@dswa.org.uk							
Note: Traditional walls are not built on concrete foundations.							
Dry stone wall; wall on concrete foundation (not included) dry stone coursed wall inclusive of locking stones and filling to wall with broken stone or rubble; walls up to 1.20 m high; battered; 2 sides fair faced							
Yorkstone	-	6.50	120.25	-	54.00	m^2	**174.25**
Cotswold stone	-	6.50	120.25	-	66.00	m^2	**186.25**
Purbeck	-	6.50	120.25	-	72.00	m^2	**192.25**
Rockery stone - General							
Preamble: Rockery stone prices vary considerably with source, carriage, distance and load. Typical PC prices are in the range of £60 - £80 per tonne collected.							

F MASONRY

Item Excluding site overheads and profit	PC £	Labour hours	Labour £	Plant £	Material £	Unit	**Total rate £**
Rockery stone; CED Ltd							
Boulders; maximum distance 25 m; by machine							
750 mm diameter	65.00	0.90	16.65	7.22	65.00	nr	**88.87**
1 m diameter	215.00	2.00	37.00	14.44	215.00	nr	**266.44**
1.5 m diameter	715.00	2.00	37.00	47.50	715.00	nr	**799.50**
2 m diameter	1735.00	2.00	37.00	47.50	1735.00	nr	**1819.50**
Boulders; maximum distance 25 m; by hand							
750 mm diameter	65.00	0.75	13.88	-	65.00	nr	**78.88**
1 m diameter	215.00	1.89	34.88	-	215.00	nr	**249.88**
F22 CAST STONE ASHLAR WALLING/DRESSINGS							
Haddonstone Ltd; cast stone piers; ornamental masonry in Portland Bath or terracotta finished cast stone							
Gate Pier S120; to foundations and underground work measured separately; concrete infill							
S120G base unit to pier; 699 x 699 x 172 mm	120.00	0.75	13.88	-	124.82	nr	**138.69**
S120F/F shaft base unit; 533 x 533 x 280 mm	96.00	1.00	18.50	-	100.94	nr	**119.44**
S120E/E main shaft unit; 533 x 533 x 280 mm; nr of units required dependent on height of pier	96.00	1.00	18.50	-	100.94	nr	**119.44**
S120D/D top shaft unit; 33 x 533 x 280 mm	96.00	1.00	18.50	-	100.94	nr	**119.44**
S120C pier cap unit; 737 x 737 x 114 mm	114.00	0.50	9.25	-	118.94	nr	**128.19**
S120B pier block unit; base for finial 533 x 533 x 64 mm	49.00	0.33	6.11	-	49.12	nr	**55.23**
Pier blocks; flat to receive gate finial							
S100B; 440 x 440 x 63 mm	31.00	0.33	6.17	-	31.28	nr	**37.45**
S120B; 546 x 546 x 64 mm	49.00	0.33	6.17	-	49.28	nr	**55.45**
S150B; 330 x 330 x 51 mm	17.00	0.33	6.17	-	17.28	nr	**23.45**
Pier caps; part weathered							
S100C; 915 x 915 x 150 mm	271.00	0.50	9.25	-	271.56	nr	**280.81**
S120C; 737 x 737 x 114 mm	114.00	0.50	9.25	-	114.28	nr	**123.53**
S150C; 584 x 584 x 120 mm	83.00	0.50	9.25	-	83.28	nr	**92.53**
Pier caps; weathered							
S230C; 1029 x 1029 x 175 mm	356.00	0.50	9.25	-	356.28	nr	**365.53**
S215C; 687 x 687 x 175 mm	155.00	0.50	9.25	-	155.28	nr	**164.53**
S210C; 584 x 584 x 175 mm	98.00	0.50	9.25	-	98.28	nr	**107.53**
Pier strings							
S100S; 800 x 800 x 55 mm	92.00	0.50	9.25	-	92.17	nr	**101.42**
S120S; 555 x 555 x 44 mm	42.00	0.50	9.25	-	42.17	nr	**51.42**
S150S; 457 x 457 x 48 mm	26.00	0.50	9.25	-	26.17	nr	**35.42**
Balls and bases							
E150A ball 535 mm and E150C collared base	249.36	0.50	9.25	-	249.53	nr	**258.78**
E120A ball 330 mm and E120C collared base	87.66	0.50	9.25	-	87.83	nr	**97.08**
E110A ball 230 mm and E110C collared base	63.83	0.50	9.25	-	64.00	nr	**73.25**
E100A ball 170 mm and E100B plain base	40.85	0.50	9.25	-	41.02	nr	**50.27**
Haddonstone Ltd; cast stone copings; ornamental masonry in Portland Bath or terracotta finished cast stone							
Copings for walls; bedded, jointed and pointed in approved coloured cement-lime mortar 1:2:9							
T100 weathered coping 102 mm high 178 mm wide x 914 mm	29.43	0.33	6.16	-	29.67	m	**35.83**
T140 weathered coping 102 mm high 337 mm wide x 914 mm	49.05	0.33	6.17	-	49.29	m	**55.46**
T200 weathered coping 127 mm high 508 mm wide x 750 mm	75.21	0.33	6.17	-	75.45	m	**81.62**
T170 weathered coping 108 mm high 483 mm wide x 914	79.57	0.00	6.17		79.81	m	**85.98**

F MASONRY

Item Excluding site overheads and profit	PC £	Labour hours	Labour £	Plant £	Material £	Unit	**Total rate £**
F22 CAST STONE ASHLAR WALLING/DRESSINGS - cont'd							
Haddonstone Ltd; cast stone copings - cont'd							
T340 raked coping 100-75 mm high 290 wide x 900 mm	45.78	0.33	6.17	-	46.02	m	**52.19**
T310 raked coping 89-76 mm high 381 wide x 914 mm	53.41	0.33	6.17	-	53.65	m	**59.82**
Bordeaux walling; Forticrete Ltd							
Dry stacked random sized units 150-400 mm long x 150 high cast stone wall mechanically interlocked with fibreglass pins; constructed to levelling pad of coarse compacted granular material backfilled behind the elevation with 300 wide granular drainage material; walls to 5.00 m high (retaining walls over heights shown below require individual design)							
Gravity wall - near vertical wall 250 mm thick	70.00	1.00	18.50	-	70.00	m^2	**88.50**
Gravity wall 9.5 deg battered	70.00	2.00	37.00	-	70.00	m^2	**107.00**
Copings to Bordeaux wall 70 thick random lengths	8.00	0.25	4.63	-	8.00	m	**12.63**
Retaining wall; as above but reinforced with Tensar geogrid 40 RE laid between every two courses horizontally into the face of the excavation (excavation not included)							
Near vertical wall 250 mm thick; 1.50 m of geogrid length							
up to 1.20 m high; 2 layers of geogrid	70.00	1.50	27.75	-	77.80	m^2	**105.55**
1.20 m - 1.50 m high; 3 layers of geogrid	70.00	2.00	37.00	-	81.70	m^2	**118.70**
1.50 - 1.8 m high; 4 courses of geogrid	70.00	2.25	41.63	-	85.60	m^2	**127.22**
Battered walls max 1:3 slope							
up to 1.20 m high; 3 layers of geogrid	70.00	2.50	46.25	-	81.70	m^2	**127.95**
1.20 m - 1.50 m high; 4 layers of geogrid	70.00	2.50	46.25	-	85.60	m^2	**131.85**
1.50 - 1.8 m high; 5 layers of geogrid	70.00	3.00	55.50	-	88.85	m^2	**144.35**
F30 ACCESSORIES/SUNDRY ITEMS FOR BRICK/BLOCK/STONE WALLING							
Damp proof courses; pitch polymer; 150 mm laps							
Horizontal							
width not exceeding 225 mm	4.10	1.14	21.09	-	5.51	m^2	**26.60**
width exceeding 225 mm	4.10	0.58	10.73	-	6.02	m^2	**16.75**
Vertical							
width not exceeding 225 mm	4.10	1.72	31.82	-	5.51	m^2	**37.33**
Two courses slates in cement mortar (1:3)							
Horizontal							
width exceeding 225 mm	8.86	3.46	64.01	-	12.28	m^2	**76.29**
Vertical							
width exceeding 225 mm	8.86	5.18	95.83	-	12.28	m^2	**108.11**

F MASONRY

Item Excluding site overheads and profit	PC £	Labour hours	Labour £	Plant £	Material £	Unit	**Total rate £**
F31 PRECAST CONCRETE SILLS/LINTELS/COPINGS/FEATURES							
Mix 21.00 N/mm² - 20 aggregate (1:2:4)							
Copings; once weathered; twice grooved							
152 x 75 mm	4.90	0.40	7.40	-	5.20	m	**12.60**
178 x 65 mm	7.10	0.40	7.40	-	7.39	m	**14.79**
305 x 75 mm	9.90	0.50	9.25	-	10.36	m	**19.61**
Pier caps; four sides weathered							
305 x 305 mm	5.90	1.00	18.50	-	6.04	nr	**24.54**
381 x 381 mm	6.20	1.00	18.50	-	6.34	nr	**24.84**
533 x 533 mm	6.25	1.20	22.20	-	6.39	nr	**28.59**

G STRUCTURAL/CARCASSING METAL/TIMBER

Item Excluding site overheads and profit	PC £	Labour hours	Labour £	Plant £	Material £	Unit	Total rate £
G31 PREFABRICATED TIMBER UNIT DECKING							
Timber decking							
Supports for timber decking; softwood joists to receive decking boards; joists at 400 mm centres; Southern Yellow pine							
38 x 88 mm	15.75	1.00	18.50	-	19.18	m^2	**37.67**
50 x 150 mm	10.85	1.00	18.50	-	13.79	m^2	**32.29**
50 x 125 mm	9.05	1.00	18.50	-	11.80	m^2	**30.30**
Hardwood decking Yellow Balau; grooved or smooth; 6 mm joints							
deck boards; 90 mm wide x 19 mm thick	19.05	1.00	18.50	-	21.16	m^2	**39.66**
deck boards; 145 mm wide x 21 mm thick	21.25	1.00	18.50	-	25.49	m^2	**43.99**
deck boards; 145 mm wide x 28 mm thick	28.73	1.00	18.50	-	33.72	m^2	**52.22**
Hardwood decking Ipe; smooth; 6 mm joints							
deck boards; 90 mm wide x 19 mm thick	40.60	1.00	18.50	-	42.71	m^2	**61.21**
deck boards; 145 mm wide x 19 mm thick	29.79	1.00	18.50	-	31.90	m^2	**50.40**
Western Red Cedar; 6 mm joints							
prime deck grade; 90 mm wide x 40 mm thick	32.27	1.00	18.50	-	37.61	m^2	**56.11**
prime deck grade; 142 mm wide x 40 mm thick	34.56	1.00	18.50	-	40.13	m^2	**58.63**
Handrails and base rail; fixed to posts at 2.00 m centres							
posts 100 x 100 x 1370 high	6.70	1.00	18.50	-	7.75	m	**26.25**
posts turned 1220 high	14.25	1.00	18.50	-	15.30	m	**33.80**
Handrails; balusters							
square balusters at 100 mm centres	38.00	0.50	9.25	-	38.50	m	**47.75**
square balusters at 300 mm centres	12.65	0.33	6.16	-	13.05	m	**19.21**
turned balusters at 100 mm centres	60.00	0.50	9.25	-	60.50	m	**69.75**
turned balusters at 300 mm centres	19.98	0.33	6.11	-	20.38	m	**26.48**

H CLADDING/COVERING

Item Excluding site overheads and profit	PC £	Labour hours	Labour £	Plant £	Material £	Unit	**Total rate £**
H51 NATURAL STONE SLAB CLADDING/FEATURES							
Sawn yorkstone cladding; Johnsons Wellfield Quarries Ltd Six sides sawn stone; rubbed face; sawn and jointed edges; fixed to blockwork (not included) with stainless steel fixings "Ancon Ltd" grade 304 stainless steel frame cramp and dowel 7 mm; cladding units drilled 4 x to receive dowels							
440 x 200 x 50 mm thick	88.95	1.87	34.59	-	92.20	m^2	**126.79**
H52 CAST STONE SLAB CLADDING/FEATURES							
Cast stone cladding; Haddonstone Ltd Reconstituted stone in Portland Bath or terracotta; fixed to blockwork or concrete (not included) with stainless steel fixings "Ancon Ltd" grade 304 stainless steel frame cramp and dowel M6 mm; cladding units drilled 4 x to receive dowels							
440 x 200 x 50 mm thick	94.65	1.87	34.59	-	97.90	m^2	**132.49**

J WATERPROOFING

Item Excluding site overheads and profit	PC £	Labour hours	Labour £	Plant £	Material £	Unit	Total rate £
J10 SPECIALIST WATERPROOF RENDERING							
"Sika" waterproof rendering; steel trowelled							
Walls; 20 thick; three coats; to concrete base							
width exceeding 300 mm	-	-	-	-	-	m^2	**48.84**
width not exceeding 300 mm	-	-	-	-	-	m^2	**76.53**
Walls; 25 thick; three coats; to concrete base							
width exceeding 300 mm	-	-	-	-	-	m^2	**55.37**
width not exceeding 300 mm	-	-	-	-	-	m^2	**87.93**
J20 MASTIC ASPHALT TANKING/DAMP PROOFING							
Tanking and damp proofing; mastic asphalt; to BS 6925; type T 1097; Coverite Ltd							
13 mm thick; one coat covering; to concrete base; flat; work subsequently covered							
width exceeding 300 mm	-	-	-	-	-	m^2	**11.24**
20 mm thick; two coat coverings; to concrete base; flat; work subsequently covered							
width exceeding 300 mm	-	-	-	-	-	m^2	**14.09**
30 mm thick; three coat coverings; to concrete base; flat; work subsequently covered							
width exceeding 300 mm	-	-	-	-	-	m^2	**19.13**
13 mm thick; two coat coverings; to brickwork base; vertical; work subsequently covered							
width exceeding 300 mm	-	-	-	-	-	m^2	**37.34**
20 mm thick; three coat coverings; to brickwork base; vertical; work subsequently covered							
width exceeding 300 mm	-	-	-	-	-	m^2	**50.98**
Internal angle fillets; work subsequently covered	-	-	-	-	-	m	**3.97**
Turning asphalt nibs into grooves; 20 mm deep	-	-	-	-	-	m	**2.52**
J30 LIQUID APPLIED TANKING/DAMP PROOFING							
Tanking and damp proofing; Ruberoid Building Products, "Synthaprufe" cold applied bituminous emulsion waterproof coating							
"Synthaprufe"; to smooth finished concrete or screeded slabs; flat; blinding with sand							
two coats	1.56	0.22	4.11	-	1.87	m^2	**5.98**
three coats	2.34	0.31	5.74	-	2.72	m^2	**8.46**
"Synthaprufe"; to fair-faced brickwork with flush joints, rendered brickwork, or smooth finished concrete walls; vertical							
two coats	1.76	0.29	5.29	-	2.08	m^2	**7.37**
three coats	2.58	0.40	7.40	-	2.98	m^2	**10.38**
Tanking and damp proofing; RIW Ltd							
Liquid asphaltic composition; to smooth finished concrete screeded slabs or screeded slabs; flat							
two coats	5.03	0.33	6.17	-	5.03	m^2	**11.20**
Liquid asphaltic composition; fair-faced brickwork with flush joints, rendered brickwork, or smooth finished concrete walls; vertical							
two coats	5.03	0.50	9.25	-	5.03	m^2	**14.28**
"Heviseal"; to smooth finished concrete or screeded slabs; to surfaces of ponds, tanks, planters; flat							
two coats	5.76	0.33	6.17	-	6.34	m^2	**12.50**

J WATERPROOFING

Item Excluding site overheads and profit	PC £	Labour hours	Labour £	Plant £	Material £	Unit	**Total rate £**
"Heviseal"; to fair-faced brickwork with flush joints, rendered brickwork, or smooth finished concrete walls; to surfaces of retaining walls, ponds, tanks, planters; vertical							
two coats	5.76	0.50	9.25	-	6.34	m^2	**15.59**
J40 FLEXIBLE SHEET TANKING/DAMP PROOFING							
Tanking and damp proofing; Grace Construction Products							
"Bitu-thene 2000"; 1.00 mm thick; overlapping and bonding; including sealing all edges							
to concrete slabs; flat	4.91	0.25	4.63	-	5.40	m^2	**10.03**
to brick/concrete walls; vertical	4.91	0.40	7.40	-	5.73	m^2	**13.13**
J50 GREEN ROOF SYSTEMS							
Preamble: A variety of systems are available which address all the varied requirements for a successful Green Roof. For installation by approved contractors only the prices shown are for budgeting purposes only as each installation is site specific and may incorporate some or all of the resources shown. Specifiers should verify that the systems specified include for design liability and inspections by the suppliers. The systems below assume commercial insulation levels are required to the space below the proposed Green Roof. Extensive Green Roofs are those of generally lightweight construction with low maintenance planting and shallow soil designed for aesthetics only; Intensive Green roofs are designed to allow use for recreation and trafficking. They require more maintenance and allow a greater variety of surfaces and plant types.							
Intensive Green Roof System; Bauder Ltd; soil based systems able to provide a variety of hard and soft landscaping; laid to the surface of an unprepared roof deck							
Vapour barrier laid to prevent intersticial condensation from spaces below the roof applied by torching to the roof deck							
VB4-Expal aluminium lined	-	-	-	-	-	m^2	**11.18**
Insulation laid and hot bitumen bonded to vapour barrier							
PIR insulation 100 mm	-	-	-	-	-	m^2	**24.35**
Underlayer to receive rootbarrier partially bonded to insulation by torching							
G4E	-	-	-	-	-	m^2	**11.03**
Root barrier							
"Plant E"; chemically treated root resistant capping sheet fully bonded to G4E underlayer by torching	-	-	-	-	-	m^2	**15.06**
Slip layers to absorb differential movement							
PE foil 2 layers laid to root barriers	-	-	-	-	-	m^2	**2.89**
Optional protection layer to prevent mechanical damage							
"Protection mat" 6 mm thick rubber matting loose laid	-	-	-	-	-	m^2	**9.14**

J WATERPROOFING

Item Excluding site overheads and profit	PC £	Labour hours	Labour £	Plant £	Material £	Unit	Total rate £
J50 GREEN ROOF SYSTEMS - cont'd							
Intensive Green Roof System - cont'd							
Drainage medium laid to root barrier							
"Drainage board" free draining EPS 50 mm thick	-	-	-	-	-	m^2	**10.14**
"Reservoir board" up to 21.5 litre water storage capacity EPS 75 mm thick	-	-	-	-	-	m^2	**12.64**
Filtration to prevent soil migration to drainage system							
"Filter fleece" 3 mm thick polyester geotextile loose laid over drainage/reservoir layer	-	-	-	-	-	m^2	**2.99**
For hard landscaped areas incorporate rigid drainage board laid to the protection mat							
PLT 60 drainage board	-	-	-	-	-	m^2	**16.64**
Extensive Green Roof System; Bauder Ltd; low maintenance soil free system incorporating single layer growing and planting medium							
Vapour barrier laid to prevent intersticial condensation applied by torching to the roof deck							
VB4-Expal aluminium lined	-	-	-	-	-	m^2	**11.18**
Insulation laid and hot bitumen bonded to vapour barrier							
PIR insulation 100 mm	-	-	-	-	-	m^2	**24.35**
Underlayer to receive rootbarrier partially bonded to insulation by torching							
G4E	-	-	-	-	-	m^2	**11.03**
Root barrier							
"Plant E"; chemically treated root resistant capping sheet fully bonded to G4E underlayer by torching	-	-	-	-	-	m^2	**15.06**
Landscape options							
Hydroplanting system; Bauder Ltd; to Extensive Green Roof as detailed above							
Ecomat 6 mm thick geotextile loose laid Hydroplanting with sedum and succulent coagulant directly onto plant substrate to 60 mm deep	-	-	-	-	-	m^2	**13.15**
Xeroflor vegetation blanket; Bauder Ltd; to Extensive Green Roof as detailed above							
Xeroflor Xf 301 pre cultivated sedum blanket incorporating 800 gram recycled fibre water retention layer laid loose	-	-	-	-	-	m^2	**34.75**
Waterproofing to upstands; Bauder Ltd							
Bauder Vapour Barrier; Bauder G4E and Bauder Plant E							
up to 200 mm high	-	-	-	-	-	m	**17.89**
up to 400 mm high	-	-	-	-	-	m	**25.17**
up to 600 mm high	-	-	-	-	-	m	**33.53**
Inverted waterproofing systems; Alumasc Exterior Building Products Ltd; to roof surfaces to receive Green Roof systems							
"Hydrotech 6125"; monolithic hot melt rubberised bitumen; applied in two 3 mm layers incorporating a polyester reinforcing sheet with 4 mm thick protection sheet and chemically impregnated root barrier; fully bonded into the Hydrotech; applied to plywood or suitably prepared wood float finish and primed concrete deck or screeds							
10 mm thick	-	-	-	-	-	m^2	**30.00**

J WATERPROOFING

Item Excluding site overheads and profit	PC £	Labour hours	Labour £	Plant £	Material £	Unit	**Total rate £**
"Alumasc Roofmate"; extruded polystyrene insulation; optional system; thickness to suit required U value; calculated at design stage; indicative thicknesses; laid to Hydrotech 6125							
0.25 U value; average requirement 120 mm	-	-	-	-	-	m^2	**20.80**
Warm Roof waterproofing systems; Alumasc Exterior Building Products Ltd (Euroroof); to roof surfaces to receive Green Roof systems							
"Derbigum" system							
"Nilperm" aluminium lined vapour barrier; 2 mm thick bonded in hot bitumen to the roof deck	-	-	-	-	-	m^2	**9.40**
"Korklite" insulation bonded to the vapour barrier in hot bitumen; U value dependent; 80 mm thick	-	-	-	-	-	m^2	**17.80**
"Hi-Ten Universal" 2 mm thick underlayer; fully bonded to the insulation	-	-	-	-	-	m^2	**7.65**
"Derbigum Anti-Root" cap sheet impregnated with root resisting chemical bonded to the underlayer	-	-	-	-	-	m^2	**15.75**
Intensive Green Roof Systems; Alumasc Exterior Building Products Ltd; components laid to the insulation over the Hydrotech or Derbigum waterproofing above							
Optional inclusion; moisture retention layer							
SSM-45; moisture mat	-	-	-	-	-	m^2	**5.45**
Drainage layer; "Floradrain" recycled polypropylene; providing water reservoir, multi-directional drainage and mechanical damage protection							
FD25; 25 mm deep; rolls inclusive of filter sheet SF	-	-	-	-	-	m^2	**18.50**
FD40; 40 mm deep; rolls inclusive of filter sheet SF	-	-	-	-	-	m^2	**24.10**
FD25; 25 mm deep; sheets excluding filter sheet	-	-	-	-	-	m^2	**15.00**
FD40; 40 mm deep; sheets excluding filter sheet	-	-	-	-	-	m^2	**18.50**
FD60; 60 mm deep; sheets excluding filter sheet	-	-	-	-	-	m^2	**29.20**
Drainage layer; "Elastodrain" recycled rubber mat; providing multi-directional drainage and mechanical damage protection							
EL200; 20 mm deep	-	-	-	-	-	m^2	**24.00**
"Zincolit" recycled crushed brick; optional drainage infil to Floradrain layers							
FD40; 17 litres per m^2	-	0.03	0.63	-	6.00	m^2	**6.63**
FD60; 27 litres per m^2	-	0.05	1.00	-	7.10	m^2	**8.10**
Filter sheet; rolled out onto drainage layer							
"Filter sheet TG" for Elastodrain range	-	-	-	-	-	m^2	**4.70**
"Filter sheet SF" for Floradrain range	-	-	-	-	-	m^2	**4.00**
Intensive substrate; lightweight growing medium laid to filter sheet	-	-	-	-	-	m^2	**21.00**
Extensive substrate; sedum carpet laid to filter sheet	-	-	-	-	-	m^2	**14.90**
Extensive substrate; rockery type plants laid to filter sheet	-	-	-	-	-	m^2	**19.50**
Semi-intensive substrate; heather with lavender laid to filter sheet	-	-	-	-	-	m^2	**26.80**

Item Excluding site overheads and profit	PC £	Labour hours	Labour £	Plant £	Material £	Unit	**Total rate £**
J50 GREEN ROOF SYSTEMS - cont'd							
Extensive Green Roof Systems; Alumasc Exterior Building Products Ltd; components laid to the Hydrotech or insulation layers above							
Moisture retention layer							
SSM-45; moisture mat	-	-	-	-	-	m^2	**5.45**
Drainage layer for flat roofs; "Floradrain"; recycled polypropylene; providing water reservoir, multi-directional drainage and mechanical damage protection; supplied in sheet or roll form							
FD25; 25 mm deep; rolls inclusive of filter sheet SF	-	-	-	-	-	m^2	**18.50**
FD40; 40 mm deep; rolls inclusive of filter sheet SF	-	-	-	-	-	m^2	**24.10**
FD25; 25 mm deep; sheets excluding filter sheet	-	-	-	-	-	m^2	**15.00**
FD40; 40 mm deep; sheets excluding filter sheet	-	-	-	-	-	m^2	**18.50**
FD60; 60 mm deep; sheets excluding filter sheet	-	-	-	-	-	m^2	**29.20**
Drainage layer for pitched roofs; "Floratec"; recycled polystyrene; providing water reservoir and multi-directional drainage; laid to moisture mat or to the waterproofing							
FS50	-	-	-	-	-	m^2	**14.30**
FS75	-	-	-	-	-	m^2	**16.80**
Landscape options							
Sedum Mat vegetation layer; Alumasc Exterior Building Products Ltd; to Extensive Green Roof as detailed above	-	-	-	-	-	m^2	**35.00**
Green Roof components; Alumasc Exterior Building Products Ltd							
Outlet inspection chambers							
"KS 15" 150 mm deep	-	-	-	-	-	nr	**118.00**
height extension piece 100 mm	-	-	-	-	-	nr	**42.70**
height extension piece 200 mm	-	-	-	-	-	nr	**46.20**
Outlet and irrigation control chambers							
"B32" 300 x 300 x 300 mm high	-	-	-	-	-	nr	**355.00**
"B52" 400 x 500 x 500 mm high	-	-	-	-	-	nr	**450.00**
Outlet damming piece for water retention	-	-	-	-	-	nr	**94.00**
Linear drainage channel; collects surface water from adjacent hard surfaces or down pipes for distribution to the drainage layer	-	-	-	-	-	m	**93.00**

M SURFACE FINISHES

Item Excluding site overheads and profit	PC £	Labour hours	Labour £	Plant £	Material £	Unit	**Total rate £**
M20 PLASTERED/RENDERED/ROUGHCAST COATINGS							
Cement:lime:sand (1:1:6); 19 mm thick; two coats; wood floated finish							
Walls							
width exceeding 300 mm; to brickwork or blockwork base	-	-	-	-	-	m^2	**41.80**
Extra over cement:sand:lime (1:1:6) coatings for decorative texture finish with water repellent cement							
combed or floated finish	-	-	-	-	-	m^2	**3.80**
M40 STONE/CONCRETE/QUARRY/CERAMIC TILING/MOSAIC							
Ceramic tiles; unglazed slip resistant; various colours and textures; jointing							
Floors							
level or to falls only not exceeding 15 degrees from horizontal; 150 x 150 x 8 mm thick	17.26	1.00	18.50	-	20.41	m^2	**38.91**
level or to falls only not exceeding 15 degrees from horizontal; 150 x 150 x 12 mm thick	21.03	1.00	18.50	-	24.36	m^2	**42.86**
Clay tiles - General							
Preamble: Typical specification - clay tiles should be to BS 6431 and shall be reasonably true to shape, flat, free from flaws, frost resistant and true to sample approved by the Landscape Architect prior to laying. Quarry tiles (or semi-vitrified tiles) shall be of external quality, either heather brown or blue, to size specified, laid on 1:2:4 concrete, with 20 maximum aggregate 100 thick, on 100 hardcore. The hardened concrete should be well wetted and the surplus water taken off. Clay tiles shall be thoroughly wetted immediately before laying and then drained and shall be bedded to 19 thick cement:sand (1:3) screed. Joints should be approximately 4 mm (or 3 mm for vitrified tiles) grouted in cement:sand (1:2) and cleaned off immediately.							
Quarry tiles; external quality; including bedding; jointing							
Floors							
level or to falls only not exceeding 15 degrees from horizontal; 150 x 150 x 12.5 mm thick; heather brown	21.57	0.80	14.80	-	24.93	m^2	**39.73**
level or to falls only not exceeding 15 degrees from horizontal; 225 x 225 x 29 mm thick; heather brown	21.57	0.67	12.33	-	24.93	m^2	**37.26**
level or to falls only not exceeding 15 degrees from horizontal; 150 x 150 x 12.5 mm thick; blue/black	13.70	1.00	18.50	-	16.67	m^2	**35.17**
level or to falls only not exceeding 15 degrees from horizontal; 194 x 194 x 12.5 mm thick; heather brown	12.62	0.80	14.80	-	15.53	m^2	**30.33**

M SURFACE FINISHES

Item Excluding site overheads and profit	PC £	Labour hours	Labour £	Plant £	Material £	Unit	Total rate £
M60 PAINTING/CLEAR FINISHING							
Prepare; touch up primer; two undercoats and one finishing coat of gloss oil paint; on metal surfaces							
General surfaces							
girth exceeding 300 mm	0.79	0.33	6.17	-	0.79	m^2	**6.96**
isolated surfaces; girth not exceeding 300 mm	0.22	0.13	2.47	-	0.22	m	**2.69**
isolated areas not exceeding 0.50 m^2 irrespective of girth	0.42	0.13	2.47	-	0.42	nr	**2.89**
Ornamental railings; each side measured overall							
girth exceeding 300 mm	0.79	0.75	13.88	-	0.79	m^2	**14.66**
Prepare; one coat primer; two undercoats and one finishing coat of gloss oil paint; on wood surfaces							
General surfaces							
girth exceeding 300 mm	1.06	0.40	7.40	-	1.06	m^2	**8.46**
isolated areas not exceeding 0.50 m^2 irrespective of girth	0.41	0.36	6.73	-	0.41	nr	**7.14**
isolated surfaces; girth not exceeding 300 mm	0.35	0.20	3.70	-	0.35	m	**4.05**
Prepare; two coats of creosote; on wood surfaces							
General surfaces							
girth exceeding 300 mm	0.18	0.21	3.84	-	0.18	m^2	**4.02**
isolated surfaces; girth not exceeding 300 mm	0.06	0.12	2.13	-	0.06	m	**2.19**
Prepare, proprietary solution primer; two coats of dark stain; on wood surfaces							
General surfaces							
girth exceeding 300 mm	0.12	0.10	1.85	-	0.13	m^2	**1.98**
isolated surfaces; girth not exceeding 300 mm	0.02	0.05	0.93	-	0.02	m	**0.94**
Three coats "Dimex Shield"; to clean, dry surfaces; in accordance with manufacturer's instructions							
Brick or block walls							
girth exceeding 300 mm	0.96	0.28	5.18	-	0.96	m^2	**6.14**
Cement render or concrete walls							
girth exceeding 300 mm	0.76	0.25	4.63	-	0.76	m^2	**5.38**
Two coats resin based paint; "Sandtex Matt"; in accordance with manufacturer's instructions							
Brick or block walls							
girth exceeding 300 mm	6.47	0.20	3.70	-	6.47	m^2	**10.17**
Cement render or concrete walls							
girth exceeding 300 mm	4.40	0.17	3.08	-	4.40	m^2	**7.48**

P BUILDING FABRIC SUNDRIES

Item Excluding site overheads and profit	PC £	Labour hours	Labour £	Plant £	Material £	Unit	**Total rate £**
P30 TRENCHES/PIPEWAYS/PITS FOR BURIED ENGINEERING SERVICES							
Excavating trenches; using 3 tonne tracked excavator; to receive pipes; grading bottoms; earthwork support; filling with excavated material to within 150 mm of finished surfaces and compacting; completing fill with topsoil; disposal of surplus soil							
Services not exceeding 200 mm nominal size							
average depth of run not exceeding 0.50 m	0.83	0.12	2.22	1.00	1.00	m	**4.22**
average depth of run not exceeding 0.75 m	0.83	0.16	3.01	1.38	1.00	m	**5.40**
average depth of run not exceeding 1.00 m	0.83	0.28	5.24	2.41	1.00	m	**8.65**
average depth of run not exceeding 1.25 m	0.83	0.38	7.09	3.24	0.83	m	**11.17**
Excavating trenches; using 3 tonne tracked excavator; to receive pipes; grading bottoms; earthwork support; filling with imported granular material and compacting; disposal of surplus soil							
Services not exceeding 200 mm nominal size							
average depth of run not exceeding 0.50 m	4.86	0.09	1.60	0.70	6.66	m	**8.96**
average depth of run not exceeding 0.75 m	7.29	0.11	2.00	0.87	9.99	m	**12.87**
average depth of run not exceeding 1.00 m	9.72	0.14	2.58	1.14	13.32	m	**17.04**
average depth of run not exceeding 1.25 m	12.15	0.23	4.23	1.94	16.65	m	**22.82**
Excavating trenches; using 3 tonne tracked excavator; to receive pipes; grading bottoms; earthwork support; filling with lean mix concrete; disposal of surplus soil							
Services not exceeding 200 mm nominal size							
average depth of run not exceeding 0.50 m	11.34	0.11	1.97	0.36	13.14	m	**15.47**
average depth of run not exceeding 0.75 m	17.01	0.13	2.41	0.45	19.71	m	**22.56**
average depth of run not exceeding 1.00 m	22.68	0.17	3.08	0.60	26.28	m	**29.96**
average depth of run not exceeding 1.25 m	28.35	0.23	4.16	0.90	32.85	m	**37.91**
Earthwork support; providing support to opposing faces of excavation; moving along as work proceeds; A Plant Acrow							
Maximum depth not exceeding 2.00 m							
distance between opposing faces not exceeding 2.00 m	-	0.80	14.80	17.70	-	m	**32.50**

Item Excluding site overheads and profit	PC £	Labour hours	Labour £	Plant £	Material £	Unit	**Total rate £**
Q10 KERBS/EDGINGS/CHANNELS/PAVING ACCESSORIES							
Foundations to kerbs							
Excavating trenches; width 300 mm; 3 tonne excavator; disposal off site							
depth 300 mm	-	0.10	1.85	0.45	1.48	m	**3.78**
depth 400 mm	-	0.11	2.06	0.50	1.98	m	**4.54**
By hand	-	0.60	11.14	-	2.46	m	**13.61**
Excavating trenches; width 450 mm; 3 tonne excavator; disposal off site							
depth 300 mm	-	0.13	2.31	0.56	2.22	m	**5.10**
depth 400 mm	-	0.14	2.64	0.64	2.97	m	**6.25**
Foundations to kerbs, edgings, or channels; in situ concrete; 21 N/mm^2 - 20 aggregate ((1:2:4) site mixed); one side against earth face, other against formwork (not included); site mixed concrete							
Site mixed concrete							
150 wide x 100 mm deep	-	0.13	2.47	-	1.48	m	**3.95**
150 wide x 150 mm deep	-	0.17	3.08	-	2.22	m	**5.30**
200 wide x 150 mm deep	-	0.20	3.70	-	2.96	m	**6.66**
300 wide x 150 mm deep	-	0.23	4.32	-	4.44	m	**8.76**
600 wide x 200 mm deep	-	0.29	5.29	-	11.85	m	**17.13**
Ready mixed concrete							
150 wide x 100 mm deep	-	0.13	2.47	-	1.11	m	**3.58**
150 wide x 150 mm deep	-	0.17	3.08	-	1.66	m	**4.74**
200 wide x 150 mm deep	-	0.20	3.70	-	2.22	m	**5.92**
300 wide x 150 mm deep	-	0.23	4.32	-	3.32	m	**7.64**
600 wide x 200 mm deep	-	0.29	5.29	-	8.86	m	**14.15**
Formwork; sides of foundations (this will usually be required to one side of each kerb foundation adjacent to road sub-bases)							
100 mm deep	-	0.04	0.77	-	0.15	m	**0.92**
150 mm deep	-	0.04	0.77	-	0.22	m	**0.99**
Precast concrete kerbs, channels, edgings etc.; to BS 340; Marshalls Mono; bedding, jointing and pointing in cement mortar (1:3); including haunching with in situ concrete; 11.50 N/mm^2 - 40 aggregate one side							
Kerbs; straight							
150 x 305 mm; ref HB1	6.48	0.50	9.25	-	8.89	m	**18.14**
125 x 255 mm; ref HB2; SP	3.11	0.44	8.22	-	5.52	m	**13.74**
125 x 150 mm; ref BN	2.10	0.40	7.40	-	4.51	m	**11.91**
Dropper kerbs; left and right handed							
125 x 255 - 150 mm; ref DL1 or DR1	4.69	0.50	9.25	-	7.10	m	**16.35**
125 x 255 - 150 mm; ref DL2 or DR2	4.69	0.50	9.25	-	7.10	m	**16.35**
Quadrant kerbs							
305 mm radius	7.88	0.50	9.25	-	9.49	nr	**18.74**
455 mm radius	8.47	0.50	9.25	-	10.88	nr	**20.13**
Straight kerbs or channels; to radius; 125 x 255 mm							
0.90 m radius (2 units per quarter circle)	5.93	0.80	14.80	-	8.41	m	**23.21**
1.80 m radius (4 units per quarter circle)	5.93	0.73	13.45	-	8.41	m	**21.86**
2.40 m radius (5 units per quarter circle)	5.93	0.68	12.54	-	8.41	m	**20.95**
3.00 m radius (5 units per quarter circle)	5.93	0.67	12.33	-	8.41	m	**20.74**
4.50 m radius (2 units per quarter circle)	5.93	0.60	11.11	-	8.41	m	**19.52**
6.10 m radius (11 units per quarter circle)	5.93	0.58	10.73	-	8.41	m	**19.14**
7.60 m radius (14 units per quarter circle)	5.93	0.57	10.57	-	8.41	m	**18.98**
9.15 m radius (17 units per quarter circle)	5.93	0.56	10.28	-	8.41	m	**18.69**
10.70 m radius (20 units per quarter circle)	5.93	0.56	10.28	-	8.41	m	**18.69**
12.20 m radius (22 units per quarter circle)	5.93	0.53	9.74	-	8.41	m	**18.15**

Q PAVING/PLANTING/FENCING/SITE FURNITURE

Item Excluding site overheads and profit	PC £	Labour hours	Labour £	Plant £	Material £	Unit	Total rate £
Kerbs; "Conservation Kerb" units; to simulate natural granite kerbs							
255 x 150 x 914 mm; laid flat	17.42	0.57	10.57	-	21.75	m	**32.32**
150 x 255 x 914 mm; laid vertical	17.42	0.57	10.57	-	21.04	m	**31.61**
145 x 255 mm; radius internal 3.25 m	22.05	0.83	15.42	-	26.22	m	**41.63**
150 x 255 mm; radius external 3.40 m	21.06	0.83	15.42	-	25.20	m	**40.62**
145 x 255 mm; radius internal 6.50 m	20.19	0.67	12.33	-	23.81	m	**36.14**
145 x 255 mm; radius external 6.70 m	19.59	0.67	12.33	-	23.69	m	**36.03**
150 x 255 mm; radius internal 9.80 m	19.48	0.67	12.33	-	23.58	m	**35.91**
150 x 255 mm; radius external 10.00 m	18.93	0.67	12.33	-	22.54	m	**34.88**
305 x 305 x 255 mm; solid quadrants	24.81	0.57	10.57	-	28.43	nr	**39.00**
Channels; square							
125 x 225 x 915 mm long; ref CS1	2.96	0.40	7.40	-	11.03	m	**18.43**
125 x 150 x 915 mm long; ref CS2	2.39	0.40	7.40	-	8.91	m	**16.31**
Channels; dished							
305 x 150 x 915 mm long; ref CD	8.35	0.40	7.40	-	17.16	m	**24.56**
150 x 100 x 915 mm	4.91	0.40	7.40	-	9.84	m	**17.24**
Precast concrete edging units; including haunching with in situ concrete; 11.50 N/mm² - 40 aggregate both sides							
Edgings; rectangular, bullnosed, or chamfered							
50 x 150 mm	1.28	0.33	6.17	-	5.30	m	**11.46**
125 x 150 mm bullnosed	1.32	0.33	6.17	-	5.34	m	**11.51**
50 x 200 mm	1.92	0.33	6.17	-	5.94	m	**12.10**
50 x 250 mm	2.22	0.33	6.17	-	6.24	m	**12.41**
50 x 250 mm flat top	2.61	0.33	6.17	-	6.63	m	**12.80**
Marshalls Mono; small element precast concrete kerb system; Keykerb Large (KL) upstand of 100 - 125 mm; on 150 mm concrete foundation including haunching with in situ concrete 1:3:6 1 side							
Bullnosed or half battered; 100 x 127 x 200							
laid straight	9.43	0.80	14.80	-	10.27	m	**25.07**
radial blocks laid to curve; 8 blocks/1/4 circle - 500 mm radius	7.25	1.00	18.50	-	8.10	m	**26.60**
radial blocks laid to curve; 8 radial blocks, alternating 8 standard blocks/1/4 circle - 1000 mm radius	11.97	1.25	23.13	-	12.81	m	**35.94**
radial blocks, alternating 16 standard blocks/1/4 circle - 1500 mm radius	11.01	1.50	27.75	-	11.86	m	**39.61**
internal angle 90 degree	3.98	0.20	3.70	-	3.98	nr	**7.68**
external angle	3.98	0.20	3.70	-	3.98	nr	**7.68**
drop crossing kerbs; KL half battered to KL splay; LH and RH	16.66	1.00	18.50	-	17.50	pair	**36.00**
drop crossing kerbs; KL half battered to KS bullnosed	16.20	1.00	18.50	-	17.04	pair	**35.54**
Marshalls Mono; small element precast concrete kerb system; Keykerb Small (KS) upstand of 25 - 50 mm; on 150 mm concrete foundation including haunching with in situ concrete 1:3:6: 1 side							
Half battered							
laid straight	6.62	0.80	14.80	-	7.46	m	**22.26**
radial blocks laid to curve; 8 blocks/1/4 circle - 500 mm radius	5.44	1.00	18.50	-	6.28	m	**24.78**
radial blocks laid to curve; 8 radial blocks, alternating 8 standard blocks/1/4 circle - 1000 mm radius	8.75	1.25	23.13	-	9.59	m	**32.72**
radial blocks, alternating 16 standard blocks/1/4 circle - 1500 mm radius	9.37	1.50	27.75	-	10.21	m	**37.96**
internal angle 90 degree	3.98	0.20	3.70	-	3.98	nr	**7.68**
external angle	3.98	0.20	3.70	-	3.98	nr	**7.68**

Q PAVING/PLANTING/FENCING/SITE FURNITURE

Item Excluding site overheads and profit	PC £	Labour hours	Labour £	Plant £	Material £	Unit	Total rate £
Q10 KERBS/EDGINGS/CHANNELS/PAVING ACCESSORIES - cont'd							
Dressed natural stone kerbs - General Preamble: BS 435 includes the following conditions for dressed natural stone kerbs. The kerbs are to be good, sound and uniform in texture and free from defects; worked straight or to radius, square and out of wind, with the top front and back edges parallel or concentric to the dimensions specified. All drill and pick holes shall be removed from dressed faces. Standard dressings shall be in accordance with one of three illustrations in BS 435; and designated as either fine picked, single axed or nidged, or rough punched.							
Dressed natural stone kerbs; to BS435; CED Ltd; on concrete foundations (not included); including haunching with in situ concrete; 11.50 N/mm^2 - 40 aggregate one side							
Granite kerbs; 125 x 250 mm							
special quality, straight, random lengths	14.50	0.80	14.80	-	17.59	m	**32.39**
Granite kerbs; 125 x 250 mm; curved to mean radius 3 m							
special quality, random lengths	16.20	0.91	16.84	-	19.29	m	**36.12**
Second-hand granite setts; 100 x 100 mm; bedding in cement mortar (1:4); on 150 mm deep concrete foundations; including haunching with in situ concrete; 11.50 N/mm^2 - 40 aggregate one side							
Edgings							
300 mm wide	13.54	1.33	24.67	-	17.14	m	**41.81**
Brick or block stretchers; bedding in cement mortar (1:4); on 150 mm deep concrete foundations, including haunching with in situ concrete; 11.50 N/mm^2 - 40 aggregate one side							
Single course							
concrete paving blocks; PC £7.32/m^2; 200 x 100 x 60 mm	1.46	0.31	5.69	-	3.95	m	**9.64**
engineering bricks; PC £260.00/1000; 215 x 102.5 x 65 mm	1.16	0.40	7.40	-	3.70	m	**11.10**
paving bricks; PC £450.00/1000; 215 x 102.5 x 65 mm	2.00	0.40	7.40	-	4.49	m	**11.89**
Two courses							
concrete paving blocks; PC £7.32/m^2; 200 x 100 x 60 mm	2.93	0.40	7.40	-	5.83	m	**13.23**
engineering bricks; PC £260.00/1000; 215 x 102.5 x 65 mm	2.31	0.57	10.57	-	5.33	m	**15.90**
paving bricks; PC £450.00/1000; 215 x 102.5 x 65 mm	4.00	0.57	10.57	-	6.90	m	**17.47**
Three courses							
concrete paving blocks; PC £7.32/m^2; 200 x 100 x 60 mm	4.39	0.44	8.22	-	7.29	m	**15.51**
engineering bricks; PC £260.00/1000; 215 x 102.5 x 65 mm	3.47	0.67	12.33	-	6.54	m	**18.87**
paving bricks; PC £450.00/1000; 215 x 102.5 x 65 mm	6.00	0.67	12.33	-	8.90	m	**21.23**

Q PAVING/PLANTING/FENCING/SITE FURNITURE

Item Excluding site overheads and profit	PC £	Labour hours	Labour £	Plant £	Material £	Unit	**Total rate £**
Bricks on edge; bedding in cement mortar (1:4); on 150 mm deep concrete foundations; including haunching with in situ concrete; 11.50 N/mm^2 - 40 aggregate one side							
One brick wide							
engineering bricks; 215 x 102.5 x 65 mm	3.47	0.57	10.57	-	6.64	m	**17.21**
paving bricks; 215 x 102.5 x 65 mm	6.00	0.57	10.57	-	9.00	m	**19.57**
Two courses; stretchers laid on edge; 225 mm wide							
engineering bricks; 215 x 102.5 x 65 mm	6.93	1.20	22.20	-	11.20	m	**33.40**
paving bricks; 215 x 102.5 x 65 mm	12.00	1.20	22.20	-	15.92	m	**38.12**
Extra over bricks on edge for standard kerbs to one side; haunching in concrete							
125 x 255 mm; ref HB2; SP	3.11	0.44	8.22	-	5.52	m	**13.74**
Channels; bedding in cement mortar (1:3); joints pointed flush; on concrete foundations (not included)							
Three courses stretchers; 350 mm wide; quarter bond to form dished channels							
engineering bricks; PC £260.00/1000; 215 x 102.5 x 65 mm	3.47	1.00	18.50	-	4.57	m	**23.07**
paving bricks; PC £450.00/1000; 215 x 102.5 x 65 mm	6.00	1.00	18.50	-	6.93	m	**25.43**
Three courses granite setts; 340 mm wide; to form dished channels							
340 mm wide	13.54	2.00	37.00	-	15.66	m	**52.66**
Timber edging boards; fixed with 50 x 50 x 750 mm timber pegs at 1000 mm centres (excavations and hardcore under edgings not included)							
Straight							
38 x 150 mm treated softwood edge boards	2.23	0.10	1.85	-	2.23	m	**4.08**
50 x 150 mm treated softwood edge boards	2.42	0.10	1.85	-	2.42	m	**4.27**
38 x 150 mm hardwood (iroko) edge boards	7.17	0.10	1.85	-	7.17	m	**9.02**
50 x 150 mm hardwood (iroko) edge boards	9.03	0.10	1.85	-	9.03	m	**10.88**
Curved							
38 x 150 mm treated softwood edge boards	2.23	0.20	3.70	-	2.23	m	**5.93**
50 x 150 mm treated softwood edge boards	2.42	0.25	4.63	-	2.42	m	**7.05**
38 x 150 mm hardwood (iroko) edge boards	7.17	0.20	3.70	-	7.17	m	**10.87**
50 x 150 mm hardwood (iroko) edge boards	9.03	0.25	4.63	-	9.03	m	**13.65**
Permaloc "AshphaltEdge"; RTS Ltd; extruded aluminium alloy L shaped edging with 5.33 mm exposed upper lip; edging fixed to roadway base and edge profile with 250 mm steel fixing spike; laid to straight or curvilinear road edge; subsequently filled with macadam (not included)							
Depth of macadam							
38 mm	7.72	0.02	0.31	-	7.91	m	**8.22**
51 mm	8.43	0.02	0.32	-	8.64	m	**8.96**
64 mm	9.09	0.02	0.34	-	9.32	m	**9.65**
76 mm	10.55	0.02	0.37	-	10.55	m	**10.92**
102 mm	11.92	0.02	0.38	-	12.22	m	**12.60**

Q PAVING/PLANTING/FENCING/SITE FURNITURE

Item Excluding site overheads and profit	PC £	Labour hours	Labour £	Plant £	Material £	Unit	Total rate £
Q10 KERBS/EDGINGS/CHANNELS/PAVING ACCESSORIES - cont'd							
Permaloc "Cleanline"; RTS Ltd; heavy duty straight profile edging; for edgings to soft landscape beds or turf areas; 3.2 mm x 102 high; 3.2 mm thick with 4.75 mm exposed upper lip; fixed to form straight or curvilinear edge with 305 mm fixing spike							
Milled aluminium							
100 deep	6.64	0.02	0.34	-	6.81	m	**7.14**
Black							
100 deep	7.31	0.02	0.34	-	7.49	m	**7.83**
Permaloc "Permastrip"; RTS Ltd; heavy duty L shaped profile maintenance strip; 3.2 mm x 89 mm high with 5.2 mm exposed top lip; for straight or gentle curves on paths or bed turf interfaces; fixed to form straight or curvilinear edge with standard 305 mm stake; other stake lengths available							
Milled aluminium							
89 deep	6.64	0.02	0.34	-	6.81	m	**7.14**
Black							
89 deep	7.31	0.02	0.34	-	7.49	m	**7.83**
Permaloc "Proline"; RTS Ltd; medium duty straight profiled maintenance strip; 3.2 mm x 102 mm high with 3.18 mm exposed top lip; for straight or gentle curves on paths or bed turf interfaces; fixed to form straight or curvilinear edge with standard 305 mm stake; other stake lengths available							
Milled aluminium							
89 deep	5.03	0.02	0.34	-	5.16	m	**5.49**
Black							
89 deep	5.80	0.02	0.34	-	5.95	m	**6.28**
Q20 GRANULAR SUB-BASES TO ROADS/PAVINGS							
Herbicides; Scotts Ltd							
"Casoron G" (residual) herbicide; treating substrate before laying base							
at 1 kg/125 m^2	3.62	0.33	6.17	-	3.98	100 m^2	**10.14**
Hardcore bases; obtained off site; PC £18.00/m^3							
By machine							
100 mm thick	1.80	0.05	0.93	0.95	1.80	m^2	**3.67**
100 mm thick	18.00	0.50	9.25	9.47	18.00	m^3	**36.72**
150 mm thick	2.70	0.07	1.23	1.22	2.70	m^2	**5.15**
200 mm thick	3.60	0.08	1.48	1.29	3.60	m^2	**6.37**
300 mm thick	5.40	0.07	1.23	1.36	5.40	m^2	**7.99**
exceeding 300 mm thick	18.00	0.17	3.08	3.15	21.60	m^3	**27.84**
By hand							
100 mm thick	1.80	0.20	3.70	0.12	1.80	m^2	**5.62**
150 mm thick	2.70	0.30	5.56	0.12	3.24	m^2	**8.92**
200 mm thick	3.60	0.40	7.40	0.20	4.32	m^2	**11.92**
300 mm thick	5.40	0.60	11.10	0.12	6.48	m^2	**17.70**
exceeding 300 mm thick	18.00	2.00	37.00	0.41	21.60	m^3	**59.01**

Q PAVING/PLANTING/FENCING/SITE FURNITURE

Item Excluding site overheads and profit	PC £	Labour hours	Labour £	Plant £	Material £	Unit	**Total rate £**
Hardcore; difference for each £1.00 increase/decrease in PC price per m³; price will vary with type and source of hardcore							
average 75 mm thick	-	-	-	-	0.08	m²	-
average 100 mm thick	-	-	-	-	0.10	m²	-
average 150 mm thick	-	-	-	-	0.15	m²	-
average 200 mm thick	-	-	-	-	0.20	m²	-
average 250 mm thick	-	-	-	-	0.25	m²	-
average 300 mm thick	-	-	-	-	0.30	m²	-
exceeding 300 mm thick	-	-	-	-	1.10	m³	-
Type 1 granular fill base; PC £18.50/tonne (£40.70/m³ compacted)							
By machine							
100 mm thick	4.07	0.03	0.52	0.31	4.07	m²	**4.90**
150 mm thick	6.11	0.03	0.46	0.47	6.11	m²	**7.04**
250 mm thick	10.18	0.02	0.38	0.79	10.18	m²	**11.35**
over 250 mm thick	40.70	0.20	3.70	2.85	40.70	m³	**47.25**
By hand (mechanical compaction)							
100 mm thick	4.07	0.17	3.08	0.04	4.07	m²	**7.19**
150 mm thick	6.11	0.25	4.63	0.06	6.11	m²	**10.79**
250 mm thick	10.18	0.42	7.71	0.10	10.18	m²	**17.99**
over 250 mm thick	40.70	0.47	8.64	0.10	40.70	m³	**49.45**
Surface treatments							
Sand blinding; to hardcore base (not included); 25 mm thick	0.76	0.03	0.62	-	0.76	m²	**1.38**
Sand blinding; to hardcore base (not included); 50 mm thick	1.53	0.05	0.93	-	1.53	m²	**2.45**
Filter fabrics; to hardcore base (not included)	0.41	0.01	0.19	-	0.43	m²	**0.62**
Q21 IN SITU CONCRETE ROADS/PAVINGS							
Unreinforced concrete; on prepared sub-base (not included)							
Roads; 21.00 N/mm² - 20 aggregate (1:2:4) mechanically mixed on site							
100 mm thick	9.87	0.13	2.31	-	10.36	m²	**12.68**
150 mm thick	14.81	0.17	3.08	-	15.55	m²	**18.63**
Reinforced in situ concrete; mechanically mixed on site; normal Portland cement; on hardcore base (not included); reinforcement (not included)							
Roads; 11.50 N/mm² - 40 aggregate (1:3:6)							
100 mm thick	8.04	0.40	7.40	-	8.24	m²	**15.64**
150 mm thick	12.06	0.60	11.10	-	12.36	m²	**23.46**
200 mm thick	16.08	0.80	14.80	-	16.88	m²	**31.68**
250 mm thick	20.09	1.00	18.50	0.20	20.59	m²	**39.29**
300 mm thick	24.11	1.20	22.20	0.20	24.72	m²	**47.11**
Roads; 21.00 N/mm² - 20 aggregate (1:2:4)							
100 mm thick	9.87	0.40	7.40	-	10.12	m²	**17.52**
150 mm thick	14.81	0.60	11.10	-	15.18	m²	**26.28**
200 mm thick	19.74	0.80	14.80	-	20.24	m²	**35.04**
250 mm thick	24.68	1.00	18.50	0.20	25.29	m²	**43.98**
300 mm thick	29.61	1.20	22.20	0.20	30.35	m²	**52.75**

Q PAVING/PLANTING/FENCING/SITE FURNITURE

Item Excluding site overheads and profit	PC £	Labour hours	Labour £	Plant £	Material £	Unit	Total rate £
Q21 IN SITU CONCRETE ROADS/PAVINGS - cont'd							
Reinforced in situ concrete - cont'd							
Roads; 25.00 N/mm² - 20 aggregate GEN 4 ready mixed							
100 mm thick	7.41	0.40	7.40	-	7.59	m²	**14.99**
150 mm thick	11.11	0.60	11.10	-	11.39	m²	**22.49**
200 mm thick	14.81	0.80	14.80	-	15.18	m²	**29.98**
250 mm thick	18.51	1.00	18.50	0.20	18.97	m²	**37.67**
300 mm thick	22.21	1.20	22.20	0.20	22.77	m²	**45.17**
Reinforced in situ concrete; ready mixed; discharged directly into location from supply lorry; normal Portland cement; on hardcore base (not included); reinforcement (not included)							
Roads; 11.50 N/mm² - 40 aggregate (1:3:6)							
100 mm thick	7.56	0.16	2.96	-	7.56	m²	**10.52**
150 mm thick	11.34	0.24	4.44	-	11.34	m²	**15.78**
200 mm thick	15.12	0.36	6.66	-	15.12	m²	**21.78**
250 mm thick	18.90	0.54	9.99	0.20	18.90	m²	**29.08**
300 mm thick	22.68	0.66	12.21	0.20	22.68	m²	**35.08**
Roads; 21.00 N/mm² - 20 aggregate (1:2:4)							
100 mm thick	7.61	0.16	2.96	-	7.61	m²	**10.57**
150 mm thick	11.41	0.24	4.44	-	11.41	m²	**15.86**
200 mm thick	15.22	0.36	6.66	-	15.22	m²	**21.88**
250 mm thick	19.02	0.54	9.99	0.20	19.02	m²	**29.21**
300 mm thick	22.83	0.66	12.21	0.20	22.83	m²	**35.23**
Roads; 26.00 N/mm² - 20 aggregate (1:1:5:3)							
100 mm thick	7.62	0.16	2.96	-	7.62	m²	**10.58**
150 mm thick	11.44	0.24	4.44	-	11.44	m²	**15.88**
200 mm thick	15.22	0.36	6.66	-	15.22	m²	**21.88**
250 mm thick	19.06	0.54	9.99	0.20	19.06	m²	**29.24**
300 mm thick	22.87	0.66	12.21	0.20	22.87	m²	**35.28**
Roads; PAV1 concrete - 35 N/mm²; designated mix							
100 mm thick	7.48	0.16	2.96	-	7.67	m²	**10.63**
150 mm thick	11.22	0.24	4.44	-	11.50	m²	**15.94**
200 mm thick	14.96	0.36	6.66	-	15.33	m²	**21.99**
250 mm thick	18.70	0.54	9.99	0.20	19.16	m²	**29.35**
300 mm thick	22.44	0.66	12.21	0.20	23.00	m²	**35.41**
Concrete sundries							
Treating surfaces of unset concrete; grading to cambers, tamping with 75 mm thick steel shod tamper or similar	-	0.13	2.47	-	-	m²	**2.47**
Expansion joints							
13 mm thick joint filler; formwork							
width or depth not exceeding 150 mm	1.49	0.20	3.70	-	2.07	m	**5.77**
width or depth 150 - 300 mm	1.08	0.25	4.63	-	2.24	m	**6.87**
width or depth 300 - 450 mm	0.49	0.30	5.55	-	2.23	m	**7.78**
25 mm thick joint filler; formwork							
width or depth not exceeding 150 mm	1.95	0.20	3.70	-	2.53	m	**6.23**
width or depth 150 - 300 mm	1.95	0.25	4.63	-	3.11	m	**7.74**
width or depth 300 - 450 mm	1.95	0.30	5.55	-	3.69	m	**9.24**
Sealants; sealing top 25 mm of joint with rubberized bituminous compound	1.04	0.25	4.63	-	1.04	m	**5.66**
Formwork for in situ concrete							
Sides of foundations							
height not exceeding 250 mm	0.37	0.03	0.62	-	0.47	m	**1.08**
height 250 - 500 mm	0.49	0.04	0.74	-	0.71	m	**1.45**
height 500 mm - 1.00 m	0.49	0.05	0.93	-	0.76	m	**1.68**
height exceeding 1.00 m	1.46	2.00	37.00	-	4.35	m²	**41.35**

Q PAVING/PLANTING/FENCING/SITE FURNITURE

Item Excluding site overheads and profit	PC £	Labour hours	Labour £	Plant £	Material £	Unit	**Total rate £**
Extra over formwork for curved work 6 m radius	-	0.25	4.63	-	-	m	**4.63**
Steel road forms; to edges of beds or faces of foundations							
150 mm wide	-	0.20	3.70	1.10	-	m	**4.80**
Reinforcement; fabric; BS 4483; side laps 150 mm; head laps 300 mm; mesh 200 x 200 mm; in roads, footpaths or pavings							
Fabric							
ref A142 (2.22 kg/m^2)	1.72	0.08	1.54	-	1.89	m^2	**3.43**
ref A193 (3.02 kg/m^2)	3.16	0.08	1.54	-	3.48	m^2	**5.01**
Q22 COATED MACADAM/ASPHALT ROADS/PAVINGS							
Coated macadam/asphalt roads/pavings - General							
Preamble: The prices for all in situ finishings to roads and footpaths include for work to falls, crossfalls or slopes not exceeding 15 degrees from horizontal; for laying on prepared bases (not included) and for rolling with an appropriate roller.							
Users should note the new terminology for the surfaces described below which is to European standard descriptions. The now redundant descriptions for each course are shown in brackets.							
Macadam surfacing; Spadeoak Construction Co Ltd; surface (wearing) course; 20 mm of 6 mm dense bitumen macadam to BS4987-1 2001 ref 7.5							
Machine lay; areas 1000 m^2 and over							
limestone aggregate	-	-	-	-	-	m^2	**5.26**
granite aggregate	-	-	-	-	-	m^2	**5.32**
red	-	-	-	-	-	m^2	**9.45**
Hand lay; areas 400 m^2 and over							
limestone aggregate	-	-	-	-	-	m^2	**6.69**
granite aggregate	-	-	-	-	-	m^2	**6.75**
red	-	-	-	-	-	m^2	**11.12**
Macadam surfacing; Spadeoak Construction Co Ltd; surface (wearing) course; 30 mm of 10 mm dense bitumen macadam to BS4987-1 2001 ref 7.4							
Machine lay; areas 1000 m^2 and over							
limestone aggregate	-	-	-	-	-	m^2	**6.35**
granite aggregate	-	-	-	-	-	m^2	**6.43**
red	-	-	-	-	-	m^2	**13.31**
Hand lay; areas 400 m^2 and over							
limestone aggregate	-	-	-	-	-	m^2	**7.84**
granite aggregate	-	-	-	-	-	m^2	**7.93**
red	-	-	-	-	-	m^2	**15.24**

Q PAVING/PLANTING/FENCING/SITE FURNITURE

Item Excluding site overheads and profit	PC £	Labour hours	Labour £	Plant £	Material £	Unit	Total rate £
Q22 COATED MACADAM/ASPHALT ROADS/PAVINGS - cont'd							
Macadam surfacing; Spadeoak Construction Co Ltd; surface (wearing) course; 40 mm of 10 mm dense bitumen macadam to BS4987-1 2001 ref 7.4							
Machine lay; areas 1000 m^2 and over							
limestone aggregate	-	-	-	-	-	m^2	8.08
granite aggregate	-	-	-	-	-	m^2	8.20
red	-	-	-	-	-	m^2	15.09
Hand lay; areas 400 m^2 and over							
limestone aggregate	-	-	-	-	-	m^2	9.69
granite aggregate	-	-	-	-	-	m^2	9.81
red	-	-	-	-	-	m^2	17.13
Macadam surfacing; Spadeoak Construction Co Ltd; binder (base) course; 50 mm of 20 mm dense bitumen macadam to BS4987-1 2001 ref 6.5							
Machine lay; areas 1000 m^2 and over							
limestone aggregate	-	-	-	-	-	m^2	8.10
granite aggregate	-	-	-	-	-	m^2	8.23
Hand lay; areas 400 m^2 and over							
limestone aggregate	-	-	-	-	-	m^2	9.71
granite aggregate	-	-	-	-	-	m^2	9.85
Macadam surfacing; Spadeoak Construction Co Ltd; binder (base) course; 60 mm of 20 mm dense bitumen macadam to BS4987-1 2001 ref 6.5							
Machine lay; areas 1000 m^2 and over							
limestone aggregate	-	-	-	-	-	m^2	8.91
granite aggregate	-	-	-	-	-	m^2	9.06
Hand lay; areas 400 m^2 and over							
limestone aggregate	-	-	-	-	-	m^2	10.57
granite aggregate	-	-	-	-	-	m^2	10.73
Macadam surfacing; Spadeoak Construction Co Ltd; base (roadbase) course; 75 mm of 28 mm dense bitumen macadam to BS4987-1 2001 ref 5.2							
Machine lay; areas 1000 m^2 and over							
limestone aggregate	-	-	-	-	-	m^2	10.93
granite aggregate	-	-	-	-	-	m^2	11.12
Hand lay; areas 400 m^2 and over							
limestone aggregate	-	-	-	-	-	m^2	12.73
granite aggregate	-	-	-	-	-	m^2	12.94
Macadam surfacing; Spadeoak Construction Co Ltd; base (roadbase) course; 100 mm of 28 mm dense bitumen macadam to BS4987-1 2001 ref 5.2							
Machine lay; areas 1000 m^2 and over							
limestone aggregate	-	-	-	-	-	m^2	13.54
granite aggregate	-	-	-	-	-	m^2	13.80
Hand lay; areas 400 m^2 and over							
limestone aggregate	-	-	-	-	-	m^2	15.51
granite aggregate	-	-	-	-	-	m^2	15.79

Q PAVING/PLANTING/FENCING/SITE FURNITURE

Item Excluding site overheads and profit	PC £	Labour hours	Labour £	Plant £	Material £	Unit	**Total rate £**
Macadam surfacing; Spadeoak Construction Co Ltd; base (roadbase) course; 150 mm of 28 mm dense bitumen macadam in two layers to BS4987-1 2001 ref 5.2							
Machine lay; areas 1000 m^2 and over							
limestone aggregate	-	-	-	-	-	m^2	**21.35**
granite aggregate	-	-	-	-	-	m^2	**21.74**
Hand lay; areas 400 m^2 and over							
limestone aggregate	-	-	-	-	-	m^2	**24.92**
granite aggregate	-	-	-	-	-	m^2	**25.33**
Base (roadbase) course; 200 mm of 28 mm dense bitumen macadam in two layers to BS4987-1 2001 ref 5.2							
Machine lay; areas 1000 m^2 and over							
limestone aggregate	-	-	-	-	-	m^2	**27.13**
granite aggregate	-	-	-	-	-	m^2	**27.65**
Hand lay; areas 400 m^2 and over							
limestone aggregate	-	-	-	-	-	m^2	**21.08**
granite aggregate	-	-	-	-	-	m^2	**31.63**
Resin bound macadam pavings; machine laid; Colourpave Ltd							
"Naturatex" clear resin bound macadam to pedestrian or vehicular hard landscape areas; laid to base course (not included)							
25 mm thick to pedestrian areas	-	-	-	-	-	m^2	**22.00**
30 mm thick to vehicular areas	-	-	-	-	-	m^2	**25.00**
"Colourtex" coloured resin bound macadam to pedestrian or vehicular hard landscape areas							
25 mm thick	-	-	-	-	-	m^2	**20.00**
Marking car parks							
Car parking space division strips; in accordance with BS 3262; laid hot at 115 degrees C; on bitumen macadam surfacing							
minimum daily rate	-	-	-	-	-	item	**500.00**
Stainless metal road studs							
100 x 100 mm	5.15	0.25	4.63	-	5.15	nr	**9.78**
Q23 GRAVEL/HOGGIN/WOODCHIP ROADS/ PAVINGS							
Excavation and path preparation							
Excavating; 300 mm deep; to width of path; depositing excavated material at sides of excavation							
width 1.00 m	-	-	-	2.20	-	m^2	**2.20**
width 1.50 m	-	-	-	1.84	-	m^2	**1.84**
width 2.00 m	-	-	-	1.57	-	m^2	**1.57**
width 3.00 m	-	-	-	1.32	-	m^2	**1.32**
Excavating trenches; in centre of pathways; 100 flexible drain pipes; filling with clean broken stone or gravel rejects							
300 x 450 mm deep	2.88	0.10	1.85	1.10	2.92	m	**5.87**
Hand trimming and compacting reduced surface of pathway; by machine							
width 1.00 m	-	0.05	0.93	0.12	-	m	**1.05**
width 1.50 m	-	0.04	0.82	0.11	-	m	**0.93**
width 2.00 m	-	0.04	0.74	0.10	-	m	**0.84**
width 3.00 m	-	0.04	0.74	0.10	-	m	**0.84**
Permeable membranes; to trimmed and compacted surface of pathway							
"Terram 1000"	0.41	0.02	0.37	-	0.43	m^2	**0.80**

Q PAVING/PLANTING/FENCING/SITE FURNITURE

Item Excluding site overheads and profit	PC £	Labour hours	Labour £	Plant £	Material £	Unit	Total rate £
Q23 GRAVEL/HOGGIN/WOODCHIP ROADS/ PAVINGS - cont'd							
Permaloc "AshphaltEdge"; RTS Ltd; extruded aluminium alloy L shaped edging with 5.33 mm exposed upper lip; edging fixed to roadway base and edge profile with 250 mm steel fixing spike; laid to straight or curvilinear road edge; subsequently filled with macadam (not included)							
Depth of macadam							
38 mm	7.72	0.02	0.31	-	7.91	m	**8.22**
51 mm	8.43	0.02	0.32	-	8.64	m	**8.96**
64 mm	9.09	0.02	0.34	-	9.32	m	**9.65**
76 mm	10.55	0.02	0.37	-	10.55	m	**10.92**
102 mm	11.92	0.02	0.38	-	12.22	m	**12.60**
Permaloc "Cleanline"; RTS Ltd; heavy duty straight profile edging; for edgings to soft landscape beds or turf areas; 3.2 mm x 102 high; 3.2 mm thick with 4.75 mm exposed upper lip; fixed to form straight or curvilinear edge with 305 mm fixing spike							
Milled aluminium							
100 deep	6.64	0.02	0.34	-	6.81	m	**7.14**
Black							
100 deep	7.31	0.02	0.34	-	7.49	m	**7.83**
Permaloc "Permastrip"; RTS Ltd; heavy duty L shaped profile maintenance strip; 3.2 mm x 89 mm high with 5.2 mm exposed top lip; for straight or gentle curves on paths or bed turf interfaces; fixed to form straight or curvilinear edge with standard 305 mm stake; other stake lengths available							
Milled aluminium							
89 deep	6.64	0.02	0.34	-	6.81	m	**7.14**
Black							
89 deep	7.31	0.02	0.34	-	7.49	m	**7.83**
Permaloc "Proline"; RTS Ltd; medium duty straight profiled maintenance strip; 3.2 mm x 102 mm high with 3.18 mm exposed top lip; for straight or gentle curves on paths or bed turf interfaces; fixed to form straight or curvilinear edge with standard 305 mm stake; other stake lengths available							
Milled aluminium							
89 deep	5.03	0.02	0.34	-	5.16	m	**5.49**
Black							
89 deep	5.80	0.02	0.34	-	5.95	m	**6.28**
Filling to make up levels							
Obtained off site; hardcore; PC £18.00/m^3							
150 mm thick	2.88	0.04	0.74	0.43	2.88	m^2	**4.05**
Obtained off site; granular fill type 1; PC £18.50/tonne (£40.70/m^3 compacted)							
100 mm thick	4.07	0.03	0.52	0.31	4.07	m^2	**4.90**
150 mm thick	6.11	0.03	0.46	0.47	6.11	m^2	**7.04**
Surface treatments							
Sand blinding; to hardcore (not included)							
50 mm thick	1.53	0.04	0.74	-	1.68	m^2	**2.42**
Filter fabric; to hardcore (not included)	0.41	0.01	0.19	-	0.43	m^2	**0.62**

Q PAVING/PLANTING/FENCING/SITE FURNITURE

Item Excluding site overheads and profit	PC £	Labour hours	Labour £	Plant £	Material £	Unit	**Total rate £**
Footpath gravels; porous self-binding gravel							
CED Ltd; Cedec gravel; self-binding; laid to inert (non-limestone) base measured separately; compacting							
Red, silver or gold; 50 mm thick	9.60	0.03	0.46	0.35	9.60	m^2	**10.41**
Grundon Ltd; Coxwell self-binding path gravels laid and compacted to excavation or base measured separately							
50 mm thick	2.87	0.03	0.46	0.35	2.87	m^2	**3.68**
Breedon Special Aggregates; "Golden Gravel" or equivalent; rolling wet; on hardcore base (not included); for pavements; to falls and crossfalls and to slopes not exceeding 15 degrees from horizontal; over 300 mm wide							
50 mm thick	8.65	0.03	0.46	0.28	8.65	m^2	**9.39**
75 mm thick	12.98	0.10	1.85	0.36	12.98	m^2	**15.19**
Breedon Special Aggregates; 'Wayfarer' specially formulated fine gravel for use on golf course pathways							
50 mm thick	7.70	0.03	0.46	0.28	7.70	m^2	**8.44**
75 mm thick	11.54	0.05	0.93	0.48	11.54	m^2	**12.95**
Hoggin (stabilized);PC £31.10/m^3 on hardcore base (not included); to falls and crossfalls and to slopes not exceeding 15 degrees from horizontal; over 300 mm wide							
100 mm thick	3.11	0.03	0.62	0.46	4.67	m^2	**5.74**
150 mm thick	4.67	0.05	0.93	0.69	7.00	m^2	**8.62**
Ballast; as dug; watering; rolling; on hardcore base (not included)							
100 mm thick	3.33	0.03	0.62	0.46	4.16	m^2	**5.24**
150 mm thick	5.00	0.05	0.93	0.69	6.24	m^2	**7.86**
Footpath gravels; porous loose gravels							
Breedon Special Aggregates; Breedon Buff decorative limestone chippings							
50 mm thick	3.93	0.01	0.19	0.09	4.13	m^2	**4.40**
75 mm thick	10.81	0.01	0.23	0.11	11.36	m^2	**11.71**
Breedon Special Aggregates; Breedon Buff decorative limestone chippings							
50 mm thick	3.93	0.01	0.19	0.09	4.13	m^2	**4.40**
75 mm thick	5.90	0.01	0.23	0.11	6.20	m^2	**6.54**
Breedon Special Aggregates; Brindle or Moorland Black chippings							
50 mm thick	5.19	0.01	0.19	0.09	5.45	m^2	**5.73**
75 mm thick	7.79	0.01	0.23	0.11	8.18	m^2	**8.53**
Breedon Special Aggregates; Slate Chippings; plum/blue/green							
50 mm thick	7.48	0.01	0.19	0.09	7.86	m^2	**8.13**
75 mm thick	11.22	0.01	0.23	0.11	11.79	m^2	**12.14**
Washed shingle; on prepared base (not included)							
25 - 50 size, 25 mm thick	0.83	0.02	0.33	0.08	0.88	m^2	**1.28**
25 - 50 size, 75 mm thick	2.50	0.05	0.97	0.24	2.63	m^2	**3.84**
50 - 75 size, 25 mm thick	0.83	0.02	0.37	0.09	0.88	m^2	**1.34**
50 - 75 size, 75 mm thick	2.50	0.07	1.23	0.30	2.63	m^2	**4.16**
Pea shingle; on prepared base (not included)							
10 - 15 size, 25 mm thick	0.76	0.02	0.33	0.08	0.80	m^2	**1.21**
5 - 10 size, 75 mm thick	2.29	0.05	0.97	0.24	2.41	m^2	**3.62**

Q PAVING/PLANTING/FENCING/SITE FURNITURE

Item Excluding site overheads and profit	PC £	Labour hours	Labour £	Plant £	Material £	Unit	Total rate £
Q23 GRAVEL/HOGGIN/WOODCHIP ROADS/ PAVINGS - cont'd							
Bark surfaces; Melcourt Industries Ltd (items labelled FSC are Forest Stewardship Council certified)							
Bark chips; to surface of pathways by machine; material delivered in 80 m^3 loads; levelling and spreading by hand (excavation and preparation not included)							
Walk Chips; 100 mm thick; FSC (25 m^3 loads)	3.15	0.03	0.62	0.27	3.30	m^2	**4.19**
Walk Chips; 100 mm thick; FSC (80 m^3 loads)	1.84	0.03	0.62	0.27	1.94	m^2	**2.83**
Woodfibre; 100 mm thick; FSC (80 m^3 loads)	1.70	0.03	0.62	0.27	1.78	m^2	**2.67**
Bound aggregates; Addagrip Surface Treatments UK Ltd; natural decorative resin bonded surface dressing laid to concrete, macadam or to plywood panels priced separately							
Primer coat to macadam or concrete base	-	-	-	-	-	m^2	**4.00**
Golden pea gravel 1-3 mm							
buff adhesive	-	-	-	-	-	m^2	**22.00**
red adhesive	-	-	-	-	-	m^2	**22.00**
green adhesive	-	-	-	-	-	m^2	**22.00**
Golden pea gravel 2-5 mm							
buff adhesive	-	-	-	-	-	m^2	**25.00**
Chinese bauxite 1-3 mm							
buff adhesive	-	-	-	-	-	m^2	**22.00**
Q24 INTERLOCKING BRICK/BLOCK ROADS/PAVINGS							
Precast concrete block edgings; PC £7.60/m^2; 200 x 100 x 60 mm; on prepared base (not included); haunching one side							
Edgings; butt joints							
stretcher course	0.67	0.17	3.08	-	2.92	m	**6.00**
header course	1.35	0.27	4.93	-	3.75	m	**8.68**
Precast concrete vehicular paving blocks; Marshalls Plc; on prepared base (not included); on 50 mm compacted sharp sand bed; blocks laid in 7 mm loose sand and vibrated; joints filled with sharp sand and vibrated; level and to falls only							
"Trafica" paving blocks; 450 x 450 x 70 mm							
"Perfecta" finish; colour natural	22.53	0.50	9.25	0.09	24.77	m^2	**34.11**
"Perfecta" finish; colour buff	26.08	0.50	9.25	0.09	28.41	m^2	**37.75**
"Saxon" finish; colour natural	19.82	0.50	9.25	0.09	21.99	m^2	**31.33**
"Saxon" finish; colour buff	22.78	0.50	9.25	0.09	25.02	m^2	**34.36**
Precast concrete vehicular paving blocks; "Keyblock" Marshalls Plc; on prepared base (not included); on 50 mm compacted sharp sand bed; blocks laid in 7 mm loose sand and vibrated; joints filled with sharp sand and vibrated; level and to falls only							
Herringbone bond							
200 x 100 x 60 mm; natural grey	6.74	1.50	27.75	0.09	8.79	m^2	**36.63**
200 x 100 x 60 mm; colours	7.32	1.50	27.75	0.09	9.40	m^2	**37.24**
200 x 100 x 80 mm; natural grey	7.50	1.50	27.75	0.09	9.59	m^2	**37.43**
200 x 100 x 80 mm; colours	8.46	1.50	27.75	0.09	10.60	m^2	**38.44**

Q PAVING/PLANTING/FENCING/SITE FURNITURE

Item Excluding site overheads and profit	PC £	Labour hours	Labour £	Plant £	Material £	Unit	Total rate £
Basketweave bond							
200 x 100 x 60 mm; natural grey	6.74	1.20	22.20	0.09	8.79	m^2	**31.08**
200 x 100 x 60 mm; colours	7.32	1.20	22.20	0.09	9.40	m^2	**31.69**
200 x 100 x 80 mm; natural grey	7.50	1.20	22.20	0.09	9.59	m^2	**31.88**
200 x 100 x 80 mm; colours	8.46	1.20	22.20	0.09	10.60	m^2	**32.89**
Precast concrete vehicular paving blocks; Charcon Hard Landscaping; on prepared base (not included); on 50 mm compacted sharp sand bed; blocks laid in 7 mm loose sand and vibrated; joints filled with sharp sand and vibrated; level and to falls only							
"Europa" concrete blocks							
200 x 100 x 60 mm; natural grey	7.65	1.50	27.75	0.09	9.75	m^2	**37.59**
200 x 100 x 60 mm; colours	8.35	1.50	27.75	0.09	10.48	m^2	**38.32**
"Parliament" concrete blocks							
200 x 100 x 65 mm; natural grey	21.13	1.50	27.75	0.09	23.37	m^2	**51.21**
200 x 100 x 65 mm; colours	21.13	1.50	27.75	0.09	23.37	m^2	**51.21**
Recycled polyethylene grassblocks; ADP Netlon Turf Systems; interlocking units laid to prepared base or rootzone (not included)							
"ADP Netpave 50"; load bearing 150 tonnes per m^2; 500 x 500 x 50 mm deep							
minimum area 50 m^2	13.00	0.20	3.70	0.18	15.33	m^2	**19.21**
200 - 699 m^2	12.40	0.20	3.70	0.18	14.73	m^2	**18.61**
700 - 1299 m^2	11.85	0.20	3.70	0.18	14.18	m^2	**18.06**
1300 m^2 or over	11.10	0.20	3.70	0.18	13.43	m^2	**17.31**
"ADP Netpave 25"; load bearing: light vehicles and pedestrians; 500 x 500 x 25 mm deep; laid onto established grass surface							
minimum area 100 m^2	11.70	0.10	1.85	-	11.70	m^2	**13.55**
400 - 1299 m^2	11.40	0.10	1.85	-	11.40	m^2	**13.25**
1300 - 2599 m^2	11.10	0.10	1.85	-	11.10	m^2	**12.95**
2600 m^2 or over	10.50	0.10	1.85	-	10.50	m^2	**12.35**
"ADP Turfguard"; extruded polyethethylene flexible mesh laid to existing grass surface or newly seeded areas to provide surface protection from traffic including vehicle or animal wear and tear							
Turfguard Standard 30 m x 2 m; up to 300 m^2	2.57	0.01	0.15	-	2.57	m^2	**2.73**
Turfguard Standard 30 m x 2 m; up to 600 m^2	2.29	0.01	0.15	-	2.29	m^2	**2.44**
Turfguard Standard 30 m x 2 m; up to 1440 m^2	2.08	0.01	0.15	-	2.08	m^2	**2.23**
Turfguard Standard 30 m x 2 m; over 1440 m^2	1.91	0.01	0.15	-	1.91	m^2	**2.06**
Turfguard Premium 30 m x 2 m; up to 300 m^2	2.74	0.01	0.15	-	2.74	m^2	**2.89**
Turfguard Premium 30 m x 2 m; up to 600 m^2	2.46	0.01	0.15	-	2.46	m^2	**2.61**
Turfguard Premium 30 m x 2 m; over 1440 m^2	2.24	0.01	0.15	-	2.24	m^2	**2.39**
Turfguard Premium 30 m x 2 m; over 1440 m^2	2.08	0.01	0.15	-	2.08	m^2	**2.23**
"Grassroad"; Cooper Clarke Group; heavy duty for car parking and fire paths verge hardening and shallow embankments							
Honeycomb cellular polyproylene interconnecting paviors with integral downstead anti-shear cleats including topsoil but excluding edge restraints; to granular sub-base (not included)							
635 x 330 x 42 overall laid to a module of 622 x 311 x 32	-	-	-	-	-	m^2	**24.26**
Extra over for green colour	-	-	-	-	-	m^2	**0.49**

Q PAVING/PLANTING/FENCING/SITE FURNITURE

Item Excluding site overheads and profit	PC £	Labour hours	Labour £	Plant £	Material £	Unit	**Total rate £**
Q25 SLAB/BRICK/SETT/COBBLE PAVINGS							
Bricks - General							
Preamble: BS 3921 includes the following specification for bricks for paving: bricks shall be hard, well burnt, non-dusting, resistant to frost and sulphate attack and true to shape, size and sample.							
Movement of materials							
Mechanically offloading bricks; loading wheelbarrows; transporting maximum 25 m distance	-	0.20	3.70	-	-	m^2	**3.70**
Edge restraints; to brick paving; on prepared base (not included); 65 mm thick bricks; PC £300.00/1000; haunching one side							
Header course							
200 x 100 mm; butt joints	3.00	0.27	4.93	-	5.40	m	**10.33**
210 x 105 mm; mortar joints	2.67	0.50	9.25	-	5.25	m	**14.50**
Stretcher course							
200 x 100 mm; butt joints	1.50	0.17	3.08	-	3.74	m	**6.82**
210 x 105 mm; mortar joints	1.36	0.33	6.17	-	3.69	m	**9.86**
Variation in brick prices; add or subtract the following amounts for every £1.00/1000 difference in the PC price							
Edgings							
100 wide	-	-	-	-	0.05	10 m	-
200 wide	-	-	-	-	0.10	10 m	-
102.5 wide	-	-	-	-	0.04	10 m	-
215 wide	-	-	-	-	0.09	10 m	-
Clay brick pavings; on prepared base (not included); bedding on 50 mm sharp sand; kiln dried sand joints							
Pavings; 200 x 100 x 65 mm wirecut chamfered paviors							
brick; PC £450.00/1000	22.50	1.44	26.71	0.23	24.95	m^2	**51.88**
Clay brick pavings; 200 x 100 x 50; laid to running stretcher, or stack bond only; on prepared base (not included); bedding on cement:sand (1:4) pointing mortar as work proceeds							
PC £600.00/1000							
laid on edge	47.62	4.76	88.10	-	54.94	m^2	**143.03**
laid on edge but pavior 65 mm thick	40.00	3.81	70.48	-	47.12	m^2	**117.60**
laid flat	25.97	2.20	40.70	-	30.28	m^2	**70.98**
PC £500.00/1000							
laid on edge	39.68	4.76	88.10	-	46.80	m^2	**134.90**
laid on edge but pavior 65 mm thick	33.33	3.81	70.48	-	40.29	m^2	**110.77**
laid flat	21.64	2.20	40.70	-	25.84	m^2	**66.54**
PC £400.00/1000							
laid flat	17.32	2.20	40.70	-	21.40	m^2	**62.10**
laid on edge	31.74	4.76	88.10	-	38.67	m^2	**126.76**
laid on edge but pavior 65 mm thick	26.66	3.81	70.48	-	33.46	m^2	**103.94**
PC £300.00/1000							
laid on edge	23.81	4.76	88.10	-	30.53	m^2	**118.63**
laid on edge but pavior 65 mm thick	20.00	3.81	70.48	-	26.63	m^2	**97.10**
laid flat	12.99	2.20	40.70	-	16.96	m^2	**57.66**

Q PAVING/PLANTING/FENCING/SITE FURNITURE

Item Excluding site overheads and profit	PC £	Labour hours	Labour £	Plant £	Material £	Unit	**Total rate £**
Clay brick pavings; 200 x 100 x 50; butt jointed laid herringbone or basketweave pattern only; on prepared base (not included); bedding on 50 mm sharp sand							
PC £600.00/1000							
laid flat	30.00	1.44	26.71	0.29	32.64	m^2	**59.63**
PC £500.00/1000							
laid flat	25.00	1.44	26.71	0.29	27.51	m^2	**54.50**
PC £400.00/1000							
laid flat	20.00	1.44	26.71	0.29	22.39	m^2	**49.38**
PC £300.00/1000							
laid flat	15.00	1.44	26.71	0.29	17.26	m^2	**44.25**
Clay brick pavings; 215 x 102.5 x 65 mm; on prepared base (not included); bedding on cement:sand (1:4) pointing mortar as work proceeds							
Paving bricks; PC £600.00/1000; herringbone bond							
laid on edge	35.55	3.55	65.77	-	41.33	m^2	**107.09**
laid flat	23.70	2.37	43.85	-	28.59	m^2	**72.44**
Paving bricks; PC £600.00/1000; basketweave bond							
laid on edge	35.55	2.37	43.85	-	41.33	m^2	**85.18**
laid flat	23.70	1.58	29.24	-	28.59	m^2	**57.83**
Paving bricks; PC £600.00/1000; running or stack bond							
laid on edge	35.55	1.90	35.09	-	41.33	m^2	**76.41**
laid flat	23.70	1.26	23.39	-	28.59	m^2	**51.98**
Paving bricks; PC £500.00/1000; herringbone bond							
laid on edge	29.63	3.55	65.77	-	33.17	m^2	**98.94**
laid flat	19.75	2.37	43.85	-	24.64	m^2	**68.49**
Paving bricks; PC £500.00/1000; basketweave bond							
laid on edge	29.63	2.37	43.85	-	33.17	m^2	**77.02**
laid flat	19.75	1.58	29.24	-	24.64	m^2	**53.88**
Paving bricks; PC £500.00/1000; running or stack bond							
laid on edge	29.63	1.90	35.09	-	33.17	m^2	**68.26**
laid flat	19.75	1.26	23.39	-	24.64	m^2	**48.03**
Paving bricks; PC £400.00/1000; herringbone bond							
laid on edge	23.70	3.55	65.77	-	27.10	m^2	**92.87**
laid flat	15.80	2.37	43.85	-	20.60	m^2	**64.45**
Paving bricks; PC £400.00/1000; basketweave bond							
laid on edge	23.70	2.37	43.85	-	27.10	m^2	**70.95**
laid flat	15.80	1.58	29.24	-	20.60	m^2	**49.83**
Paving bricks; PC £400.00/1000; running or stack bond							
laid on edge	23.70	1.90	35.09	-	27.10	m^2	**62.19**
laid flat	15.80	1.26	23.39	-	20.60	m^2	**43.99**
Paving bricks; PC £300.00/1000; herringbone bond							
laid on edge	17.77	3.55	65.77	-	20.58	m^2	**86.35**
laid flat	11.85	2.37	43.85	-	16.74	m^2	**60.59**
Paving bricks; PC £300.00/1000; basketweave bond							
laid on edge	17.77	2.37	43.85	-	20.58	m^2	**64.43**
laid flat	11.85	1.58	29.24	-	16.74	m^2	**45.98**
Paving bricks; PC £300.00/1000; running or stack bond							
laid on edge	17.77	1.90	35.08	-	20.58	m^2	**55.66**
laid flat	11.85	1.26	23.39	-	16.74	m^2	**40.13**
Cutting							
curved cutting	-	0.44	8.22	4.74	-	m	**12.96**
raking cutting	-	0.33	6.17	3.75	-	m	**9.91**

Item Excluding site overheads and profit	PC £	Labour hours	Labour £	Plant £	Material £	Unit	**Total rate £**
Q25 SLAB/BRICK/SETT/COBBLE PAVINGS - cont'd							
Add or subtract the following amounts for every £10.00/1000 difference in the prime cost of bricks							
Butt joints							
200 x 100	-	-	-	-	0.50	m²	-
215 x 102.5	-	-	-	-	0.45	m²	-
10 mm mortar joints							
200 x 100	-	-	-	-	0.43	m²	-
215 x 102.5	-	-	-	-	0.40	m²	-
Precast concrete pavings; Charcon Hard Landscaping; to BS 7263; on prepared sub-base (not included); bedding on 25 mm thick cement:sand mortar (1:4); butt joints; straight both ways; jointing in cement:sand (1:3) brushed in; on 50 mm thick sharp sand base							
Pavings; natural grey							
450 x 450 x 70 mm chamfered	13.09	0.44	8.22	-	16.91	m²	**25.13**
450 x 450 x 50 mm chamfered	10.22	0.44	8.22	-	14.04	m²	**22.26**
600 x 300 x 50 mm	8.89	0.44	8.22	-	12.71	m²	**20.93**
400 x 400 x 65 mm chamfered	17.75	0.40	7.40	-	21.57	m²	**28.97**
450 x 600 x 50 mm	8.52	0.44	8.22	-	12.34	m²	**20.56**
600 x 600 x 50 mm	6.97	0.40	7.40	-	10.79	m²	**18.19**
750 x 600 x 50 mm	6.67	0.40	7.40	-	10.49	m²	**17.89**
900 x 600 x 50 mm	5.70	0.40	7.40	-	9.52	m²	**16.92**
Pavings; coloured							
450 x 450 x 70 mm chamfered	17.43	0.44	8.22	-	21.25	m²	**29.47**
450 x 600 x 50 mm	10.52	0.44	8.22	-	14.34	m²	**22.56**
400 x 400 x 65 mm chamfered	17.75	0.40	7.40	-	21.57	m²	**28.97**
600 x 600 x 50 mm	8.47	0.40	7.40	-	12.29	m²	**19.69**
750 x 600 x 50 mm	7.84	0.40	7.40	-	11.66	m²	**19.06**
900 x 600 x 50 mm	6.73	0.40	7.40	-	10.55	m²	**17.95**
Precast concrete pavings; Charcon Hard Landscaping; to BS 7263; on prepared sub-base (not included); bedding on 25 mm thick cement:sand mortar (1:4); butt joints; straight both ways; jointing in cement:sand (1:3) brushed in; on 50 mm thick sharp sand base							
"Appalacian" rough textured exposed aggregate pebble paving							
600 mm x 600 mm x 65 mm	21.77	0.50	9.25	-	24.61	m²	**33.86**
Pavings; Marshalls Plc; spot bedding on 5 nr pads of cement:sand mortar (1:4); on sharp sand							
"Blister Tactile" pavings; specially textured slabs for guidance of blind pedestrians; red or buff							
400 x 400 x 50 mm	23.50	0.50	9.25	-	25.66	m²	**34.91**
450 x 450 x 50 mm	20.74	0.50	9.25	-	22.83	m²	**32.08**
"Metric Four Square" pavings							
496 x 496 x 50 mm; exposed river gravel aggregate	65.20	0.50	9.25	-	66.78	m²	**76.03**
"Metric Four Square" cycle blocks							
496 x 496 x 50 mm; exposed aggregate	65.20	0.25	4.63	-	67.31	m²	**71.93**

Q PAVING/PLANTING/FENCING/SITE FURNITURE

Item Excluding site overheads and profit	PC £	Labour hours	Labour £	Plant £	Material £	Unit	**Total rate £**
Precast concrete pavings; Marshalls Plc; "Heritage" imitation riven yorkstone paving; on prepared sub-base measured separately; bedding on 25 mm thick cement:sand mortar (1:4); pointed straight both ways cement:sand (1:3)							
Square and rectangular paving							
450 x 300 x 38 mm	31.30	1.00	18.50	-	34.55	m^2	**53.05**
450 x 450 x 38 mm	19.06	0.75	13.88	-	22.30	m^2	**36.18**
600 x 300 x 38 mm	20.83	0.80	14.80	-	24.07	m^2	**38.87**
600 x 450 x 38 mm	21.13	0.75	13.88	-	24.37	m^2	**38.24**
600 x 600 x 38 mm	20.41	0.50	9.25	-	23.71	m^2	**32.96**
Extra labours for laying the a selection of the above sizes to random rectangular pattern	-	0.33	6.17	-	-	m^2	**6.17**
Radial paving for circles							
circle with centre stone and first ring (8 slabs), 450 x 230/560 x 38 mm; diameter 1.54 m (total area 1.86 m^2)	55.14	1.50	27.75	-	59.38	nr	**87.13**
circle with second ring (16 slabs), 450 x 300/460 x 38 mm; diameter 2.48 m (total area 4.83 m^2)	55.14	4.00	74.00	-	152.91	nr	**226.91**
circle with third ring (16 slabs), 450 x 470/625 x 38 mm; diameter 3.42 m (total area 9.18 m^2)	55.14	8.00	148.00	-	279.01	nr	**427.01**
Stepping stones							
380 dia x 38 mm	3.92	0.20	3.70	-	8.51	nr	**12.21**
asymmetrical 560 x 420 x 38 mm	5.85	0.20	3.70	-	10.44	nr	**14.14**
Precast concrete pavings; Marshalls Plc; "Chancery" imitation reclaimed riven yorkstone paving; on prepared sub-base measured separately; bedding on 25 mm thick cement:sand mortar (1:4); pointed straight both ways cement:sand (1:3)							
Square and rectangular paving							
300 x 300 x 45 mm	21.84	1.00	18.50	-	25.08	m^2	**43.58**
450 x 300 x 45 mm	20.59	0.90	16.65	-	23.83	m^2	**40.48**
600 x 300 x 45 mm	19.98	0.80	14.80	-	23.22	m^2	**38.02**
600 x 450 x 45 mm	20.05	0.75	13.88	-	23.30	m^2	**37.17**
450 x 450 x 45 mm	18.13	0.75	13.88	-	21.37	m^2	**35.25**
600 x 600 x 45 mm	20.52	0.50	9.25	-	23.77	m^2	**33.02**
Extra labours for laying the a selection of the above sizes to random rectangular pattern	-	0.33	6.17	-	-	m^2	**6.17**
Radial paving for circles							
circle with centre stone and first ring (8 slabs), 450 x 230/560 x 38 mm; diameter 1.54 m (total area 1.86 m^2)	66.07	1.50	27.75	-	70.31	nr	**98.06**
circle with second ring (16 slabs), 450 x 300/460 x 38 mm; diameter 2.48 m (total area 4.83 m^2)	170.55	4.00	74.00	-	181.44	nr	**255.44**
circle with third ring (16 slabs), 450 x 470/625 x 38 mm; diameter 3.42 m (total area 9.18 m^2)	310.39	8.00	148.00	-	330.90	nr	**478.90**
Squaring off set for 2 ring circle							
16 slabs; 2.72 m^2	181.58	1.00	18.50	-	181.58	nr	**200.08**
Marshalls Plc; "La Linia Pavings"; fine textured exposed aggregate 80 mm thick pavings in various sizes to designed laying patterns; laid to 50 mm sharp sand bed on Type 1 base all priced separately							
Bonded laying patterns 300 x 300							
light granite/anthracite basalt	27.08	0.75	13.88	-	27.76	m^2	**41.63**
Indian granite/yellow	27.08	0.75	13.88	-	27.76	m^2	**41.63**
Random scatter pattern incorporating 100 x 200, 200 x 200 and 300 x 200 units							
light granite/anthracite basalt	27.08	1.00	18.50	-	27.08	m^2	**45.58**
Indian granite/yellow	27.08	1.00	18.50	-	27.08	m^2	**45.58**

Q PAVING/PLANTING/FENCING/SITE FURNITURE

Item Excluding site overheads and profit	PC £	Labour hours	Labour £	Plant £	Material £	Unit	Total rate £
Q25 SLAB/BRICK/SETT/COBBLE PAVINGS - cont'd							
Marshalls Plc; "La Linia Grande Paving"; as above but 140 mm thick							
Bonded laying patterns 300 x 300							
all colours	49.82	0.75	13.88	-	49.82	m^2	**63.70**
Random scatter pattern incorporating 100 x 200, 200 x 200 and 300 x 200 units							
light granite/anthracite basalt	49.82	1.00	18.50	-	49.82	m^2	**68.32**
Extra over to the above for incorporating inlay stones to patterns; blue, light granite or anthracite basalt							
to prescribed patterns	315.06	1.50	27.75	-	315.06	m^2	**342.81**
as individual units; triangular 200 x 200 x 282 mm	1.78	0.05	0.93	-	1.78	nr	**2.71**
Pedestrian deterrent pavings; Marshalls Plc; on prepared base (not included); bedding on 25 mm cement:sand (1:3); cement:sand (1:3) joints							
"Lambeth" pyramidal pavings							
600 x 600 x 75 mm	15.91	0.25	4.63	-	19.15	m^2	**23.78**
"Thaxted" pavings; granite sett appearance							
600 x 600 x 75 mm	18.45	0.33	6.17	-	21.69	m^2	**27.86**
Pedestrian deterrent pavings; Townscape Products Ltd; on prepared base (not included); bedding on 25 mm cement:sand (1:3); cement:sand (1:3) joints							
"Strata" striated textured slab pavings; giving bonded appearance; grey							
600 x 600 x 60 mm	27.92	0.50	9.25	-	31.16	m^2	**40.41**
"Geoset" raised chamfered studs pavings; grey							
600 x 600 x 60 mm	27.92	0.50	9.25	-	31.16	m^2	**40.41**
"Abbey" square cobble pattern pavings; reinforced							
600 x 600 x 65 mm	26.39	0.80	14.80	-	29.63	m^2	**44.43**
Edge restraints; to block paving; on prepared base (not included); 200 x 100 x 80 mm; PC £7.50/m^2; haunching one side							
Header course							
200 x 100 mm; butt joints	1.50	0.27	4.93	-	3.90	m	**8.83**
Stretcher course							
200 x 100 mm; butt joints	0.75	0.17	3.08	-	2.99	m	**6.07**
Concrete paviors; Marshalls Plc; on prepared base (not included); bedding on 50 mm sand; kiln dried sand joints swept in							
"Keyblock" paviors							
200 x 100 x 60 mm; grey	6.74	0.40	7.40	0.09	8.46	m^2	**15.95**
200 x 100 x 60 mm; colours	7.32	0.40	7.40	0.09	9.04	m^2	**16.53**
200 x 100 x 80 mm; grey	7.50	0.44	8.22	0.09	9.22	m^2	**17.53**
200 x 100 x 80 mm; colours	8.46	0.44	8.22	0.09	10.18	m^2	**18.49**

Q PAVING/PLANTING/FENCING/SITE FURNITURE

Item Excluding site overheads and profit	PC £	Labour hours	Labour £	Plant £	Material £	Unit	**Total rate £**
Concrete cobble paviors; Charcon Hard Landscaping; Concrete Products; on prepared base (not included); bedding on 50 mm sand; kiln dried sand joints swept in							
Paviors							
"Woburn" blocks; 100 - 201 x 134 x 80 mm; random sizes	22.20	0.67	12.33	0.09	24.47	m^2	**36.89**
"Woburn" blocks; 100 - 201 x 134 x 80 mm; single size	22.20	0.50	9.25	0.09	24.47	m^2	**33.81**
"Woburn" blocks; 100 - 201 x 134 x 60 mm; random sizes	18.30	0.67	12.33	0.09	20.47	m^2	**32.90**
"Woburn" blocks; 100 - 201 x 134 x 60 mm; single size	18.30	0.50	9.25	0.09	20.45	m^2	**29.79**
Concrete setts; on 25 mm sand; compacted; vibrated; joints filled with sand; natural or coloured; well rammed hardcore base (not included)							
Marshalls Plc; Tegula cobble paving							
60 mm thick; random sizes	18.12	0.57	10.57	0.09	20.89	m^2	**31.55**
60 mm thick; single size	18.12	0.45	8.41	0.09	20.89	m^2	**29.39**
80 mm thick; random sizes	20.87	0.57	10.57	0.09	23.78	m^2	**34.44**
80 mm thick; single size	20.87	0.45	8.41	0.09	23.78	m^2	**32.28**
cobbles 80 x 80 x 60 mm thick; traditional	31.64	0.56	10.28	0.09	34.30	m^2	**44.67**
Cobbles							
Charcon Hard Landscaping; Country setts							
100 mm thick; random sizes	30.33	1.00	18.50	0.09	32.20	m^2	**50.79**
100 mm thick; single size	30.33	0.67	12.33	0.09	32.96	m^2	**45.38**
Natural stone, slab or granite paving - General							
Preamble: Provide paving slabs of the specified thickness in random sizes but not less than 25 slabs per 10 m^2 of surface area, to be laid in parallel courses with joints alternately broken and laid to falls.							
Reconstituted Yorkstone aggregate pavings; Marshalls Plc; "Saxon" on prepared sub-base measured separately; bedding on 25 mm thick cement:sand mortar (1:4); on 50 mm thick sharp sand base							
Square and rectangular paving in buff; butt joints straight both ways							
300 x 300 x 35 mm	29.08	0.88	16.28	-	33.43	m^2	**49.71**
600 x 300 x 35 mm	18.65	0.71	13.23	-	23.00	m^2	**36.23**
450 x 450 x 50 mm	20.89	0.82	15.26	-	25.24	m^2	**40.50**
600 x 600 x 35 mm	14.31	0.55	10.18	-	18.66	m^2	**28.84**
600 x 600 x 50 mm	17.98	0.60	11.19	-	22.33	m^2	**33.52**
Square and rectangular paving in natural; butt joints straight both ways							
300 x 300 x 35 mm	24.31	0.88	16.28	-	28.66	m^2	**44.94**
450 x 450 x 50 mm	17.79	0.77	14.24	-	22.14	m^2	**36.38**
600 x 300 x 35 mm	16.80	0.82	15.26	-	21.15	m^2	**36.42**
600 x 600 x 35 mm	12.35	0.55	10.18	-	16.70	m^2	**26.88**
600 x 600 x 50 mm	14.98	0.66	12.21	-	19.33	m^2	**31.54**
Radial paving for circles; 20 mm joints							
circle with centre stone and first ring (8 slabs), 450 x 230/560 x 35 mm; diameter 1.54 m (total area 1.86 m^2)	56.94	1.50	27.75	-	61.42	nr	**89.17**
circle with second ring (16 slabs), 450 x 300/460 x 35 mm; diameter 2.48 m (total area 4.83 m^2)	138.70	4.00	74.00	-	150.26	nr	**224.26**
circle with third ring (24 slabs), 450 x 310/430 x 35 mm; diameter 3.42 m (total area 9.18 m^2)	261.34	8.00	148.00	-	283.45	nr	**431.45**

Q PAVING/PLANTING/FENCING/SITE FURNITURE

Item Excluding site overheads and profit	PC £	Labour hours	Labour £	Plant £	Material £	Unit	**Total rate £**
Q25 SLAB/BRICK/SETT/COBBLE PAVINGS - cont'd							
Granite setts; bedding on 25 mm cement:sand (1:3)							
Natural granite setts; 100 x 100 mm to 125 x 150 mm; x 150 to 250 mm length; riven surface; silver grey							
new; standard grade	18.89	2.00	37.00	-	25.73	m^2	**62.73**
new; high grade	24.44	2.00	37.00	-	31.29	m^2	**68.29**
reclaimed; cleaned	40.00	2.00	37.00	-	47.64	m^2	**84.64**
Natural stone, slate or granite flag pavings; CED Ltd; on prepared base (not included); bedding on 25 mm cement:sand (1:3); cement:sand (1:3) joints							
Yorkstone; riven laid random rectangular							
new slabs; 40 - 60 mm thick	57.00	1.71	31.71	-	65.70	m^2	**97.41**
reclaimed slabs, Cathedral grade; 50 - 75 mm thick	69.50	2.80	51.80	-	79.45	m^2	**131.25**
Donegal quartzite slabs; standard tiles							
200 x random lengths x 15-25 mm	56.50	3.50	64.75	-	59.50	m^2	**124.25**
250 x random lengths x 15-25 mm	56.50	3.30	61.05	-	59.50	m^2	**120.55**
300 x random lengths x 15-25 mm	56.50	3.10	57.35	-	59.50	m^2	**116.85**
350 x random lengths x 15-25 mm	56.50	2.80	51.80	-	59.50	m^2	**111.30**
400 x random lengths x 15-25 mm	56.50	2.50	46.25	-	59.50	m^2	**105.75**
450 x random lengths x 15-25 mm	56.50	2.33	43.17	-	59.50	m^2	**102.67**
Natural Yorkstone, pavings or edgings; Johnsons Wellfield Quarries; sawn 6 sides; 50 mm thick; on prepared base measured separately; bedding on 25 mm cement:sand (1:3); cement:sand (1:3) joints							
Paving							
laid to random rectangular pattern	56.25	1.71	31.71	-	62.48	m^2	**94.19**
laid to coursed laying pattern; 3 sizes	61.25	1.72	31.90	-	67.17	m^2	**99.06**
Paving; single size							
600 x 600 mm	66.25	0.85	15.72	-	72.42	m^2	**88.14**
600 x 400 mm	66.25	1.00	18.50	-	72.42	m^2	**90.92**
300 x 200 mm	75.00	2.00	37.00	-	81.60	m^2	**118.60**
215 x 102.5 mm	77.50	2.50	46.25	-	84.23	m^2	**130.48**
Paving; cut to template off site; 600 x 600; radius							
1.00 m	170.00	3.33	61.67	-	172.85	m^2	**234.52**
2.50 m	170.00	2.00	37.00	-	172.85	m^2	**209.85**
5.00 m	170.00	2.00	37.00	-	172.85	m^2	**209.85**
Edgings							
100 mm wide x random lengths	7.00	0.50	9.25	-	7.64	m	**16.89**
100 mm x 100 mm	7.75	0.50	9.25	-	10.99	m	**20.24**
100 mm x 200 mm	15.50	0.50	9.25	-	19.13	m	**28.38**
250 mm wide x random lengths	15.31	0.40	7.40	-	18.93	m	**26.33**
500 mm wide x random lengths	30.63	0.33	6.17	-	35.01	m	**41.18**
Yorkstone edgings; 600 mm long x 250 mm wide; cut to radius							
1.00 m to 3.00	48.75	0.50	9.25	-	51.47	m	**60.72**
3.00 m to 5.00 m	48.75	0.44	8.22	-	51.47	m	**59.69**
exceeding 5.00 m	48.75	0.40	7.40	-	51.47	m	**58.87**

Q PAVING/PLANTING/FENCING/SITE FURNITURE

Item Excluding site overheads and profit	PC £	Labour hours	Labour £	Plant £	Material £	Unit	**Total rate £**
Natural Yorkstone, pavings or edgings; Johnsons Wellfield Quarries; sawn 6 sides; 75 mm thick; on prepared base measured separately; bedding on 25 mm cement:sand (1:3); cement:sand (1:3) joints							
Paving							
laid to random rectangular pattern	68.15	0.95	17.57	-	74.41	m^2	**91.99**
laid to coursed laying pattern; 3 sizes	73.15	0.95	17.57	-	79.66	m^2	**97.24**
Paving; single size							
600 x 600 mm	78.15	0.95	17.57	-	84.91	m^2	**102.49**
600 x 400 mm	78.15	0.95	17.57	-	84.91	m^2	**102.49**
300 x 200 mm	88.00	0.75	13.88	-	95.25	m^2	**109.13**
215 x 102.5 mm	90.00	2.50	46.25	-	97.35	m^2	**143.60**
Paving; cut to template off site; 600 x 600; radius							
1.00 m	200.00	4.00	74.00	-	202.85	m^2	**276.85**
2.50 m	200.00	2.50	46.25	-	202.85	m^2	**249.10**
5.00 m	200.00	2.50	46.25	-	202.85	m^2	**249.10**
Edgings							
100 mm wide x random lengths	8.75	0.60	11.10	-	9.47	m	**20.57**
100 mm x 100 mm	0.90	0.60	11.10	-	3.80	m	**14.90**
100 mm x 200 mm	1.80	0.50	9.25	-	4.74	m	**13.99**
250 mm wide x random lengths	18.29	0.50	9.25	-	22.06	m	**31.31**
500 mm wide x random lengths	36.58	0.40	7.40	-	41.26	m	**48.66**
Edgings; 600 mm long x 250 mm wide; cut to radius							
1.00 m to 3.00	56.25	0.60	11.10	-	59.35	m	**70.45**
3.00 m to 5.00 m	56.25	0.50	9.25	-	59.35	m	**68.60**
exceeding 5.00 m	56.25	0.44	8.22	-	59.35	m	**67.57**
CED Ltd; Indian sandstone, riven pavings or edgings; 25 - 35 mm thick; on prepared base measured separately; bedding on 25 mm cement:sand (1:3); cement:sand (1:3) joints							
Paving							
laid to random rectangular pattern	20.80	2.40	44.41	-	24.69	m^2	**69.10**
laid to coursed laying pattern; 3 sizes	20.80	2.00	37.00	-	24.69	m^2	**61.69**
Paving; single size							
600 x 600 mm	20.80	1.00	18.50	-	24.69	m^2	**43.19**
600 x 400 mm	20.80	1.25	23.13	-	24.69	m^2	**47.82**
400 x 400 mm	20.80	1.67	30.83	-	24.69	m^2	**55.53**
Natural stone, slate or granite flag pavings; CED Ltd; on prepared base (not included); bedding on 25 mm cement:sand (1:3); cement:sand (1:3) joints							
Granite paving; sawn 6 sides; textured top							
new slabs; silver grey; 50 mm thick	29.00	1.71	31.71	-	33.45	m^2	**65.16**
new slabs; blue grey; 50 mm thick	34.00	1.71	31.71	-	38.70	m^2	**70.41**
new slabs; yellow grey; 50 mm thick	36.00	1.71	31.71	-	40.80	m^2	**72.51**
new slabs; black; 50 mm thick	53.00	1.71	31.71	-	58.65	m^2	**90.36**
Edgings; silver grey							
100 mm wide x random lengths	4.10	1.00	18.50	-	6.03	m	**24.53**
100 mm x 100 mm	4.10	1.50	27.75	-	6.85	m	**34.60**
100 mm long x 200 mm wide	8.20	0.60	11.10	-	10.95	m	**22.05**
250 mm wide x random lengths	10.25	0.75	13.88	-	14.13	m	**28.00**
300 mm wide x random lengths	12.30	0.75	13.88	-	15.15	m	**29.03**

Item Excluding site overheads and profit	PC £	Labour hours	Labour £	Plant £	Material £	Unit	**Total rate £**
Q25 SLAB/BRICK/SETT/COBBLE PAVINGS - cont'd							
Cobble pavings - General Cobbles should be embedded by hand, tight-butted, endwise to a depth of 60% of their length. A dry grout of rapid-hardening cement:sand (1:2) shall be brushed over the cobbles until the interstices are filled to the level of the adjoining paving. Surplus grout shall then be brushed off and a light, fine spray of water applied over the area.							
Cobble pavings Cobbles; to present a uniform colour in panels; or varied in colour as required							
Scottish beach cobbles; 200 - 100 mm	16.75	2.00	37.00	-	23.27	m^2	**60.27**
Scottish beach cobbles; 100 - 75 mm	11.85	2.50	46.25	-	18.37	m^2	**64.62**
Scottish beach cobbles; 75 - 50 mm	6.90	3.33	61.67	-	13.42	m^2	**75.08**
Concrete cycle blocks; Marshalls Plc; bedding in cement:sand (1:4) "Metric 4 Square" cycle stand blocks, smooth grey concrete							
496 x 496 x 100 mm	29.87	0.25	4.63	-	30.30	nr	**34.93**
Concrete cycle blocks; Townscape Products Ltd; on 100 mm concrete (1:2:4); on 150 mm hardcore; bedding in cement:sand (1:4) Cycle blocks							
"Cycle Bloc"; in white concrete	20.40	0.25	4.63	-	20.98	nr	**25.60**
"Mountain Cycle Bloc"; in white concrete	28.23	0.25	4.63	-	28.81	nr	**33.43**
Grass concrete - General Preamble: Grass seed should be a perennial ryegrass mixture, with the proportion depending on expected traffic. Hardwearing winter sportsground mixtures are suitable for public areas. Loose gravel, shingle or sand is liable to be kicked out of the blocks; rammed hoggin or other stabilized material should be specified.							
Grass concrete; Grass Concrete Ltd; on blinded Granular Type 1 sub-base (not included) "Grasscrete" in situ concrete continuously reinforced surfacing; including expansion joints at 10 m centres; soiling; seeding							
ref GC2; 150 mm thick; traffic up to 40.00 tonnes	-	-	-	-	-	m^2	**36.18**
ref GC1; 100 mm thick; traffic up to 13.30 tonnes	-	-	-	-	-	m^2	**28.12**
ref GC3; 76 mm thick; traffic up to 4.30 tonnes	-	-	-	-	-	m^2	**24.73**
Grass concrete; Grass Concrete Ltd; 406 x 406 mm blocks; on 20 mm sharp sand; on blinded MOT type 1 sub-base (not included); level and to falls only; including filling with topsoil; seeding with dwarf rye grass at £3.00/kg Pavings							
ref GB103; 103 mm thick	13.09	0.40	7.40	0.18	15.01	m^2	**22.59**
ref GB83; 83 mm thick	11.79	0.36	6.73	0.18	13.71	m^2	**20.62**
Grass concrete; Charcon Hard Landscaping; on 25 mm sharp sand; including filling with topsoil; seeding with dwarf rye grass at £3.00/kg "Grassgrid" grass/concrete paving blocks							
366 x 274 x 100 mm thick	14.87	0.38	6.94	0.18	16.79	m^2	**23.91**

Q PAVING/PLANTING/FENCING/SITE FURNITURE

Item Excluding site overheads and profit	PC £	Labour hours	Labour £	Plant £	Material £	Unit	**Total rate £**
Grass concrete; Marshalls Plc; on 25 mm sharp sand; including filling with topsoil; seeding with dwarf rye grass at £3.00/kg							
Concrete grass pavings							
"Grassguard 130"; for light duty applications; (80 mm prepared base not included)	15.66	0.38	6.94	0.18	18.14	m^2	**25.26**
"Grassguard 160"; for medium duty applications; (80 - 150 mm prepared base not included)	18.46	0.46	8.54	0.18	21.01	m^2	**29.73**
"Grassguard 180"; for heavy duty applications; (150 mm prepared base not included)	20.66	0.60	11.10	0.18	23.26	m^2	**34.54**
Full mortar bedding							
Extra over pavings for bedding on 25 mm cement:sand (1:4); in lieu of spot bedding on sharp sand	-	0.03	0.46	-	0.86	m^2	**1.32**
Q26 SPECIAL SURFACINGS/PAVINGS FOR SPORT/GENERAL AMENITY							
Market prices of surfacing materials							
Surfacings; Melcourt Industries Ltd (items labelled FSC are Forest Stewardship Council certified)							
Playbark 10/50®; per 25 m^3 load	-	-	-	-	51.60	m^3	-
Playbark 10/50®; per 80 m^3 load	-	-	-	-	38.60	m^3	-
Playbark 8/25®; per 25 m^3 load	-	-	-	-	50.10	m^3	-
Playbark 8/25®; per 80 m^3 load	-	-	-	-	37.10	m^3	-
Playchips®; per 25 m^3 load; FSC	-	-	-	-	34.50	m^3	-
Playchips®; per 80 m^3 load; FSC	-	-	-	-	21.50	m^3	-
Kushyfall; per 25 m^3 load; FSC	-	-	-	-	32.70	m^3	-
Kushyfall; per 80 m^3 load; FSC	-	-	-	-	19.70	m^3	-
Softfall; per 25 m^3 load	-	-	-	-	20.40	m^3	-
Softfall; per 80 m^3 load	-	-	-	-	13.40	m^3	-
Playsand; per 10 t load	-	-	-	-	81.36	m^3	-
Playsand; per 20 t load	-	-	-	-	69.12	m^3	-
Walk Chips; per 25 m^3 load; FSC	-	-	-	-	31.45	m^3	-
Walk Chips; per 80 m^3 load; FSC	-	-	-	-	18.45	m^3	-
Woodfibre; per 25 m^3 load; FSC	-	-	-	-	29.95	m^3	-
Woodfibre; per 80 m^3 load; FSC	-	-	-	-	16.95	m^3	-
Natural sports pitches; White Horse Contractors Ltd; sports pitches; gravel raft construction							
Professional standard;							
football pitch 6500 m^2	-	-	-	-	-	nr	**150, 000**
cricket wicket - county standard	-	-	-	-	-	nr	**40, 000**
Playing fields; Sport England compliant							
football pitch 6500 m^2	-	-	-	-	-	nr	**80,000**
cricket wicket; playing field standard	-	-	-	-	-	nr	**25,000**
track and infield	-	-	-	-	-	nr	**125,000**
Artificial surfaces and finishes - General							
Preamble: Advice should also be sought from the Technical Unit for Sport, the appropriate regional office of the Sports Council or the National Playing Fields Association. Some of the following prices include base work whereas others are for a specialist surface only on to a base prepared and costed separately.							

Q PAVING/PLANTING/FENCING/SITE FURNITURE

Item Excluding site overheads and profit	PC £	Labour hours	Labour £	Plant £	Material £	Unit	Total rate £
Q26 SPECIAL SURFACINGS/PAVINGS FOR SPORT/GENERAL AMENITY - cont'd							
Artificial sports pitches; White Horse Contractors Ltd; 3rd generation rubber crumb							
Sports pitches to respective National governing body standards; inclusive of all excavation, drainage; lighting, fencing etc.							
football 106 m x 71 m	-	-	-	-	-	nr	**450,000**
rugby union multi-use pitch 120 x75	-	-	-	-	-	nr	**550,000**
hockey; water based; 101.4 m x 63 m	-	-	-	-	-	nr	**460,000**
athletic track; 8 lane; International Amateur Athletic Federation	-	-	-	-	-	nr	**750,000**
Multi-sport pitches; inclusive of all excavation, drainage; lighting, fencing etc.							
sand dressed; 101.4 m x 63 m	-	-	-	-	-	nr	**365,000**
sand filled; 101.4 m x 63 m	-	-	-	-	-	nr	**340,000**
polymeric (synthetic bound rubber)	-	-	-	-	-	m^2	**70.00**
macadam	-	-	-	-	-	m^2	**42.00**
Sports areas; Baylis Landscape Contractors							
Sports tracks; polyurethane rubber surfacing; on bitumen-macadam (not included); prices for 5500 m^2 minimum							
"International"	-	-	-	-	-	m^2	**46.58**
"Club Grade"	-	-	-	-	-	m^2	**27.73**
Sports areas; multi-component polyurethane rubber surfacing; on bitumen-macadam; finished with polyurethane or acrylic coat; green or red							
"Permaprene"	-	-	-	-	-	m^2	**65.22**
Sports areas; Exclusive Leisure Ltd							
Cricketweave artificial wicket system; including proprietary sub-base (all by specialist sub-contractor)							
28 x 2.74 m; hard porous base	-	-	-	-	-	nr	**5544.00**
28 x 2.74 m; tarmac base	-	-	-	-	-	nr	**6699.00**
Cricketweave practice batting end							
11 x 2 .74 m; hard porous base	-	-	-	-	-	each	**2850.00**
11 x 2 .74 m; tarmac base	-	-	-	-	-	each	**3600.00**
11 x 3.65 m; hard porous base	-	-	-	-	-	each	**3800.00**
11 x 3.65 m; tarmac base	-	-	-	-	-	each	**4800.00**
Cricketweave practice bowling end							
8 x 2.74 m; hard porous base	-	-	-	-	-	each	**2090.00**
8 x 2.74 m; tarmac base	-	-	-	-	-	each	**2640.00**
Supply and erection of single bay cricket cage							
18.3 x 3.65 x 3.2 m high	-	-	-	-	-	each	**1932.00**
Tennis courts; Duracourt (Spadeoak) Ltd							
Hard playing surfaces to SAPCA Code of Practice minimum requirements; laid on 65 mm thick macadam base on 150 mm thick stone foundation, to include lines, nets, posts, brick edging, 2.75 m high fence and perimeter drainage (excluding excavation, levelling or additional foundation); based on court size 36.6 m x 18.3 m - 670 m^2							
"Durapore"; 'All Weather' porous macadam and acrylic colour coating	-	-	-	-	-	m^2	**37.48**
"Tiger Turf TT"; sand filled artificial grass	-	-	-	-	-	m^2	**51.69**
"Tiger Turf Grand Prix"; short pile sand filled artificial grass	-	-	-	-	-	m^2	**54.44**
"DecoColour"; impervious acrylic hardcourt (200 mm foundation)	-	-	-	-	-	m^2	**49.97**

Item Excluding site overheads and profit	PC £	Labour hours	Labour £	Plant £	Material £	Unit	**Total rate £**
"DecoTurf"; cushioned impervious acrylic tournament surface (200 mm foundation)	-	-	-	-	-	m^2	**58.57**
"Porous Kushion Kourt"; porous cushioned acrylic surface	-	-	-	-	-	m^2	**59.15**
"EasiClay"; synthetic clay system	-	-	-	-	-	m^2	**61.20**
"Canada Tenn"; American green clay, fast dry surface (no macadam but including irrigation)	-	-	-	-	-	m^2	**61.20**
Playgrounds; Baylis Landscape Contractors; "Rubaflex" in situ playground surfacing, porous; on prepared stone/granular base Type 1 (not included) and macadam base course (not included)							
Black							
15 mm thick (0.50 m critical fall height)	-	-	-	-	-	m^2	**27.15**
35 mm thick (1.00 m critical fall height)	-	-	-	-	-	m^2	**41.85**
60 mm thick (1.50 m critical fall height)	-	-	-	-	-	m^2	**55.45**
Playgrounds; Baylis Landscape Contractors; "Rubaflex" in situ playground surfacing, porous; on prepared stone/granular base Type 1 (not included) and macadam base course (not included)							
Coloured							
15 mm thick (0.50 m critical fall height)	-	-	-	-	-	m^2	**64.48**
35 mm thick (1.00 m critical fall height)	-	-	-	-	-	m^2	**70.15**
60 mm thick (1.50 m critical fall height)	-	-	-	-	-	m^2	**79.19**
Playgrounds; Wicksteed Leisure Ltd							
Safety tiles; on prepared base (not included)							
1000 x 1000 x 60 mm; red or green	47.00	0.13	2.31	-	47.00	m^2	**49.31**
1000 x 1000 x 60 mm; black	44.00	0.13	2.31	-	44.00	m^2	**46.31**
1000 x 1000 x 43 mm; red or green	44.00	0.13	2.31	-	44.00	m^2	**46.31**
1000 x 1000 x 43 mm; black	41.00	0.13	2.31	-	41.00	m^2	**43.31**
Playgrounds; SMP Playgrounds Ltd							
Tiles; on prepared base (not included)							
Premier 25; 1000 x 1000 x 25 mm; black; for general use	38.00	0.20	3.70	-	44.16	m^2	**47.86**
Premier 70; 1000 x 1000 x 70 mm; black; for higher equipment	64.00	0.20	3.70	-	70.16	m^2	**73.86**
Playgrounds; Melcourt Industries Ltd							
"Playbark"; on drainage layer (not included); to BS5696; minimum 300 mm settled depth							
"Playbark 8/25", 8 - 25 mm particles; red/brown	11.13	0.35	6.49	-	12.24	m^2	**18.73**
"Playbark 10/50"; 10 - 50 mm particles; red/brown	12.87	0.35	6.49	-	14.15	m^2	**20.64**
Playgrounds; timber edgings							
Timber edging boards; 50 x 50 x 750 mm timber pegs at 1000 mm centres; excavations and hardcore under edgings (not included)							
50 x 150 mm; hardwood (iroko) edge boards	9.03	0.10	1.85	-	9.03	m	**10.88**
38 x 150 mm; hardwood (iroko) edge boards	7.17	0.10	1.85	-	7.17	m	**9.02**
50 x 150 mm; treated softwood edge boards	2.42	0.10	1.85	-	2.42	m	**4.27**
38 x 150 mm; treated softwood edge boards	2.23	0.10	1.85	-	2.23	m	**4.08**

Q PAVING/PLANTING/FENCING/SITE FURNITURE

Item Excluding site overheads and profit	PC £	Labour hours	Labour £	Plant £	Material £	Unit	Total rate £
Q30 SEEDING/TURFING							
Seeding/turfing - General Preamble: The following market prices generally reflect the manufacturer's recommended retail prices. Trade and bulk discounts are often available on the prices shown. The manufacturer's of these products generally recommend application rates. Note the following rates reflect the average rate for each product.							
Market prices of pre-seeding materials							
Rigby Taylor Ltd							
turf fertilizer; "Mascot Outfield 6-9-6"	-	-	-	-	1.79	100 m^2	-
Boughton Loam							
screened topsoil; 100 mm	-	-	-	-	48.60	m^3	-
screened Kettering loam; 3 mm	-	-	-	-	81.63	m^3	-
screened Kettering loam - sterilised; 3 mm	-	-	-	-	90.63	m^3	-
topdressing; sand soil mixtures; 90/10 to 50/50	-	-	-	-	72.00	m^3	-
Market prices of landscape chemicals at suggested application rates							
Residual herbicides							
Embargo G; Rigby Taylor Ltd; residual pre and post emergent herbicide							
80 - 125 kg/ha; annual meadow-grass, black grass, charlock, common chickweed, common mouse-ear, common orache, common poppy, corn marigold, corn spurrey, fat-hen, groundsel, hedge mustard, scentless mayweed, small nettle, sow-thistle, stinking chamomile, wild-oat	-	-	-	-	0.48	100 m^2	-
Premiere; Scotts Horticulture; granular weedkiller approved for use in new plantings of trees and shrubs							
Application rate 1 kg/100 m^2 (100 kg/ha)	-	-	-	-	30.38	100 m^2	-
Casoron G; Scotts Horticulture; control of broadleaved weeds; existing weeds and germinating weeds; a wide range of annual and perennial weeds							
Selective; 560 g/100 m^2: control of germinating annual and perennial weeds and light to moderate infestation of established annuals	-	-	-	-	2.53	100 m^2	-
Selective; 1 kg/100 m^2: control more persistent weeds in the above category	-	-	-	-	2.53	100 m^2	-
Selective; 1.25 kg/100 m^2: control established weed	-	-	-	-	5.65	100 m^2	-
Selective; 1 kg/280 m^2	-	-	-	-	1.46	100 m^2	-
Contact herbicides; Scotts Horticulture							
Dextrone X (Diquat Paraquat)							
8.5 L/ha	-	-	-	-	0.77	100 m^2	-
3.0 L/ha	-	-	-	-	0.28	100 m^2	-
Speedway 2 (Diquat Paraquat)							
1220 m^2/4 kg	-	-	-	-	2.25	100 m^2	-
3333 m^2/4 kg	-	-	-	-	0.82	100 m^2	-
Selective herbicides							
Rigby Taylor; junction for broadleaf weeds in turf grasses							
Application rate; 1.2 L/ha	-	-	-	-	0.45	100 m^2	-

Q PAVING/PLANTING/FENCING/SITE FURNITURE

Item Excluding site overheads and profit	PC £	Labour hours	Labour £	Plant £	Material £	Unit	**Total rate £**
Rigby Taylor; Greenor systemic selective treatment of weeds in turfgrass							
Application rate; 4 L/ha	-	-	-	-	0.88	100 m^2	-
Rigby Taylor Bastion -T; control of weeds in both established and newly seeded turf							
Application rate	-	-	-	-	0.88	100 m^2	-
Scotts; Intrepid 2; for broadleaf weed control in grass							
Application rate; 7700 m^2/5 L	-	-	-	-	0.52	100 m^2	-
Scotts; Re-Act; selective herbicide; established managed amenity turf, and newly seeded grass; controls many annual and perennial weeds; will not vaporise in hot conditions							
Application rate; 9090 - 14,285 m^2/5 L	-	-	-	-	0.45	100 m^2	-
Bayer Environmental Science; Asulox; post-emergence translocated herbicide for the control of bracken and docks							
Application rate; docks; 17857 m^2/5 L	-	-	-	-	0.42	100 m^2	-
Application rate; bracken; 4545 m^2/5 L	-	-	-	-	1.65	100 m^2	-
Bayer Environmental Science; Clovotox; selective herbicide for the control of clover and other difficult broadleaved weeds in turf							
Application rate; 4545 m^2/5 L	-	-	-	-	0.77	100 m^2	-
Dicotox Extra; Bayer Environmental Science; showerproof herbicide for the control of certain broadleaved weeds on turf and for the control of heather and woody weeds in forestry situations; also controls ragwort							
Application rate; 2.8 L/ha; plantains; hawksweed buttercups, etc.	-	-	-	-	0.22	100 m^2	-
Application rate; 5.6 L/ha; birdsfoot clovers, mouse-ear etc. (1 application)	-	-	-	-	0.43	100 m^2	-
Application rate; 5.6 L/ha; pearlwort, yarrow (2 applications)	-	-	-	-	0.87	100 m^2	-
Application rate; 13 L/ha; woody weeds and heather in established trees and shrub areas	-	-	-	-	1.01	100 m^2	-
Dormone; Bayer Environmental Science; selective herbicide for the control of broadleaved weeds in amenity grassland situations; may also be used in situations in or near water							
Application rate; 2.8 L/ha	-	-	-	-	0.18	100 m^2	-
Application rate; 5.6 L/ha	-	-	-	-	0.36	100 m^2	-
Near water; application rate; 4.5 L/ha	-	-	-	-	0.29	100 m^2	-
Near water; application rate; 9 L/ha	-	-	-	-	0.58	100 m^2	-
Spearhead; Bayer Environmental Science; selective herbicide for the control of broadleaved weeds in established turf							
Application rate; 4.5 L/ha	-	-	-	-	0.84	100 m^2	-
Total herbicides; Note: these application rates will vary dependent on season							
Rigby Taylor Ltd; Gallup Biograde amenity; Glyphosate 360 g/l formulation							
General use; woody weeds; ash beech bracken bramble; 3 L/ha	-	-	-	-	0.27	100 m^2	-
Annual and perennial grasses; heather (peat soils); 4 L/ha	-	-	-	-	0.36	100 m^2	-
Pre-planting; general clearance; 5 L/ha	-	-	-	-	0.45	100 m^2	-
Heather; mineral soils; 6 L/ha	-	-	-	-	0.54	100 m^2	-
Rhododendron; 10 L/ha	-	-	-	-	0.90	100 m^2	-

Item Excluding site overheads and profit	PC £	Labour hours	Labour £	Plant £	Material £	Unit	Total rate £
Q30 SEEDING/TURFING - cont'd							
Total herbicides - cont'd							
Gallup Hi-aktiv Amenity; 490 g/L formulation							
General use; woody weeds; ash beech bracken bramble; 2.2 L/ha	-	-	-	-	0.23	100 m^2	-
Annual and perennial grasses; heather (peat soils); 2.9 L/ha	-	-	-	-	0.31	100 m^2	-
Pre-planting; general clearance; 3.7 L/ha	-	-	-	-	0.39	100 m^2	-
Heather; mineral soils; 4.4 L/ha	-	-	-	-	0.47	100 m^2	-
Rhododendron; 7.3 L/ha	-	-	-	-	0.77	100 m^2	-
Roundup Pro Biactive; Scotts							
Application rate; 5 L/ha	-	-	-	-	0.53	100 m^2	-
Casoron G; Scotts							
Total; 2.25 kg/100 m^2	-	-	-	-	10.17	100 m^2	-
Kerb							
"Kerb Flowable (Propyzamide)"	-	-	-	-	2.16	100 m^2	-
"Kerb Granules" (2 x 2000 tree pack)	-	-	-	-	0.06	tree	-
Woody weed herbicides; Rigby Taylor Ltd							
Timbrel; summer applied; water based; Bayer Environmental Science; selective scrub and brushwood herbicide							
2.0 L/ha; bramble, briar, broom, gorse, nettle	-	-	-	-	0.98	100 m^2	-
4.0 L/ha; alder, birch, blackthorn, dogwood, elder, poplar, rosebay willowherb, sycamore	-	-	-	-	1.97	100 m^2	-
6.0 L/ha; beech, box, buckthorn, elm, hazel, hornbeam, horse chestnut, lime, maple, privet, rowan, Spanish chestnut, willow, wild pear	-	-	-	-	2.95	100 m^2	-
8.0 L/ha; ash, oak, rhododendron	-	-	-	-	3.93	100 m^2	-
Timbrel; winter applied; paraffin or diesel based; Bayer Environmental Science; selective scrub and brushwood herbicide							
3.0 L/ha; bramble, briar, broom, gorse, nettle	-	-	-	-	1.47	100 m^2	-
6.0 L/ha; alder, ash, beech, birch, blackthorn, box, buckthorn, dogwood, elder, elm, hazel, hornbeam, horse chestnut, lime, maple, oak, poplar, privet, rowan, Spanish chestnut, sycamore, willow, wild pear	-	-	-	-	2.95	100 m^2	-
10.0 L/ha; hawthorn, laurel, rhododendron	-	-	-	-	4.92	100 m^2	-
Aquatic herbicides							
Casoron G; control of broadleaved weeds; existing weeds and germinating weeds; a wide range of annual and perennial weeds							
Aquatic: 1,665 m^2/25 kg	-	-	-	-	6.78	100 m^2	-
Aquatic: 5,555 m^2/25 kg	-	-	-	-	2.03	100 m^2	-
Fungicides							
Rigby Taylor; Masalon; systemic							
Application rate; 8 L/ha	-	-	-	-	7.18	100 m^2	-
Scotts UK Daconil turf							
Application rate; 333 m^2/L	-	-	-	-	5.13	100 m^2	-
Moss control							
Enforcer; Scotts; surface biocide and mosskiller for the control of mosses in turf or on external hard surfaces							
Turf areas; application rate 1 L/200 m^2	-	-	-	-	6.13	100 m^2	-
Hard surfaces; moss and fungi control; application rate 1 L/588 m^2	-	-	-	-	2.11	100 m^2	-

Q PAVING/PLANTING/FENCING/SITE FURNITURE

Item Excluding site overheads and profit	PC £	Labour hours	Labour £	Plant £	Material £	Unit	Total rate £
Insect control							
Merit Turf; insecticide for the control of chafer grubs and leatherjackets; Bayer Environmental Science							
Application rate; 10 kg/3333 m^2	-	-	-	-	5.20	100 m^2	-
Mildothane Turf Liquid; casting worm control and fungicide; Bayer Environmental Science							
Application rate; 5 L/6667 m^2	-	-	-	-	0.07	100 m^2	-
Spray mark indicator							
Rigby Taylor; Trail Blazer; mixed with chemicals; price per 1000 L applied	-	-	-	-	19.75	1000 L	-
Market prices of turf fertilizers; recommended "average" application rates							
Scotts UK; turf fertilizers							
grass fertilizer; "Longlife Fine Turf - Spring & Summer"	-	-	-	-	4.83	100 m^2	-
grass fertilizer; "Greenmaster - Invigorator"	-	-	-	-	2.63	100 m^2	-
grass fertilizer; slow release; "Sierraform", 18+24+5	-	-	-	-	4.98	100 m^2	-
grass fertilizer; slow release; "Sierraform", 18+9+18+Fe+Mn	-	-	-	-	4.15	100 m^2	-
grass fertilizer; slow release; "Sierraform", 15+0+26	-	-	-	-	4.98	100 m^2	-
grass fertilizer; slow release; "Sierraform", 22+5+10	-	-	-	-	3.32	100 m^2	-
grass fertilizer; slow release; "Sierraform", 16+0+15+Fe+Mn	-	-	-	-	4.15	100 m^2	-
grass fertilizer; water soluble; "Sierrasol", 28+5+18+TE	-	-	-	-	7.43	100 m^2	-
grass fertilizer; water soluble; "Sierrasol", 20+5+30+TE	-	-	-	-	5.78	100 m^2	-
grass fertilizer; controlled release; "Sierrablen", 28+5+5+Fe	-	-	-	-	4.91	100 m^2	-
grass fertilizer; controlled release; "Sierrablen", 27+5+5+Fe	-	-	-	-	5.14	100 m^2	-
grass fertilizer; controlled release; "Sierrablen", 15+0+22+Fe	-	-	-	-	4.91	100 m^2	-
grass fertilizer; controlled release; "Sierrablen Fine", 38+0+0	-	-	-	-	1.83	100 m^2	-
grass fertilizer; controlled release; "Sierrablen Fine", 25+5+12	-	-	-	-	3.06	100 m^2	-
grass fertilizer; controlled release; "Sierrablen Fine", 21+0+20	-	-	-	-	3.06	100 m^2	-
grass fertilizer; controlled release; "Sierrablen Fine", 15+0+29	-	-	-	-	3.06	100 m^2	-
grass fertilizer; controlled release; "Sierrablen Mini", 0+0+37	-	-	-	-	4.42	100 m^2	-
grass fertilizer; "Greenmaster Turf Tonic"	-	-	-	-	2.44	100 m^2	-
grass fertilizer; "Greenmaster Spring & Summer"	-	-	-	-	3.49	100 m^2	-
grass fertilizer; "Greenmaster Zero Phosphate"	-	-	-	-	3.49	100 m^2	-
grass fertilizer; "Greenmaster Mosskiller"	-	-	-	-	3.89	100 m^2	-
grass fertilizer; "Greenmaster Extra"	-	-	-	-	4.43	100 m^2	-
grass fertilizer; "Greenmaster Autumn"	-	-	-	-	3.76	100 m^2	-
grass fertilizer; "Greenmaster Double K"	-	-	-	-	3.76	100 m^2	-
grass fertilizer; "Greenmaster NK"	-	-	-	-	3.76	100 m^2	-
outfield turf fertilizer; "Sportsmaster PS3"	-	-	-	-	1.59	100 m^2	-
outfield turf fertilizer; "Sportsmaster PS4"	-	-	-	-	1.59	100 m^2	-
outfield turf fertilizer; "Sportsmaster PS5"	-	-	-	-	2.04	100 m^2	-
outfield turf fertilizer; "Sportsmaster Fairway"	-	-	-	-	3.60	100 m^2	-
outfield turf fertilizer; "Sportsmaster Municipality"	-	-	-	-	4.93	100 m^2	-
"TPMC"	-	-	-	-	4.35	80 lt	-

Q PAVING/PLANTING/FENCING/SITE FURNITURE

Item Excluding site overheads and profit	PC £	Labour hours	Labour £	Plant £	Material £	Unit	Total rate £
Q30 SEEDING/TURFING - cont'd							
Market prices of turf fertilizers - cont'd							
Rigby Taylor Ltd; 35 g/m^2							
grass fertilizer; "Mascot Microfine 20-0-15" + 2% Mg	-	-	-	-	8.28	100 m^2	-
grass fertilizer; "Mascot Microfine 14-4-14"	-	-	-	-	5.08	100 m^2	-
grass fertilizer; "Mascot Microfine 18-0-0" + 4% Fe	-	-	-	-	7.40	100 m^2	-
grass fertilizer; "Mascot Microfine 12-0-10" 2% Mg + 2% Fe	-	-	-	-	5.13	100 m^2	-
grass fertilizer; "Mascot Microfine OC1 8-0-0 + 2% Fe"	-	-	-	-	3.91	100 m^2	-
grass fertilizer; "Mascot Microfine 5-0-20" + 6% Fe + 2% Mg	-	-	-	-	5.65	100 m^2	-
grass fertilizer; "Mascot Fine Turf 11-5-5"	-	-	-	-	2.98	100 m^2	-
grass fertilizer/weedkiller; "Fine Turf Weed & Feed"	-	-	-	-	1.71	100 m^2	-
grass fertilizer/mosskiller; "Fine Turf Lawn Sand"	-	-	-	-	6.93	100 m^2	-
outfield fertilizer; "Mascot Outfield 9-7-7"	-	-	-	-	2.52	100 m^2	-
outfield fertilizer; "Mascot Outfield 7-7-7"	-	-	-	-	2.16	100 m^2	-
outfield fertilizer; "Mascot Outfield 20-10-10"	-	-	-	-	2.83	100 m^2	-
outfield fertilizer; "Mascot Outfield 3-12-12"	-	-	-	-	2.54	100 m^2	-
liquid fertilizer; "Vitax 50/50 Fine Turf"; 56 ml/100 m^2	-	-	-	-	15.40	100 m^2	-
liquid fertilizer; "Vitax 50/50 Fine Turf Special"; 56 ml/100 m^2	-	-	-	-	12.53	100 m^2	-
liquid fertilizer; "Vitax 50/50 Autumn & Winter"; 56 ml/100 m^2	-	-	-	-	13.34	100 m^2	-
Cultivation							
Treating soil with "Paraquat-Diquat" weedkiller at rate of 5 litre/ha; PC £9.13 per litre; in accordance with manufacturer's instructions; including all safety precautions							
by machine	-	-	-	0.25	0.46	100 m^2	**0.71**
by hand	-	0.40	7.40	-	0.46	100 m^2	**7.86**
Ripping up subsoil; using approved subsoiling machine; minimum depth 250 mm below topsoil; at 1.20 m centres; in							
gravel or sandy clay	-	-	-	2.40	-	100 m^2	**2.40**
soil compacted by machines	-	-	-	2.80	-	100 m^2	**2.80**
clay	-	-	-	3.00	-	100 m^2	**3.00**
chalk or other soft rock	-	-	-	6.01	-	100 m^2	**6.01**
Extra for subsoiling at 1 m centres	-	-	-	0.60	-	100 m^2	**0.60**
Breaking up existing ground; using pedestrian operated tine cultivator or rotavator							
100 mm deep	-	0.22	4.07	1.90	-	100 m^2	**5.97**
150 mm deep	-	0.28	5.09	2.37	-	100 m^2	**7.46**
200 mm deep	-	0.37	6.78	3.16	-	100 m^2	**9.95**
As above but in heavy clay or wet soils							
100 mm deep	-	0.44	8.14	3.80	-	100 m^2	**11.94**
150 mm deep	-	0.66	12.21	5.70	-	100 m^2	**17.91**
200 mm deep	-	0.82	15.26	7.12	-	100 m^2	**22.38**
Breaking up existing ground; using tractor drawn tine cultivator or rotavator							
single pass							
100 mm deep	-	-	-	0.27	-	100 m^2	**0.27**
150 mm deep	-	-	-	0.34	-	100 m^2	**0.34**
200 mm deep	-	-	-	0.45	-	100 m^2	**0.45**
600 mm deep	-	-	-	1.36	-	100 m^2	**1.36**
Cultivating ploughed ground; using disc, drag, or chain harrow							
4 passes	-	-	-	1.63	-	100 m^2	**1.63**

Q PAVING/PLANTING/FENCING/SITE FURNITURE

Item Excluding site overheads and profit	PC £	Labour hours	Labour £	Plant £	Material £	Unit	**Total rate £**
Rolling cultivated ground lightly; using self-propelled agricultural roller	-	0.06	1.03	0.60	-	100 m^2	**1.63**
Importing and storing selected and approved topsoil; to BS 3882; from source not exceeding 13 km from site; inclusive of settlement							
small quantities, less than 15 m^3	32.20	-	-	-	38.64	m^3	-
over 15 m^3	18.50	-	-	-	22.20	m^3	-
Spreading and lightly consolidating approved topsoil (imported or from spoil heaps); in layers not exceeding 150 mm; travel distance from spoil heaps not exceeding 100 m; by machine (imported topsoil not included)							
minimum depth 100 mm	-	1.55	28.68	37.20	-	100 m^2	**65.87**
minimum depth 150 mm	-	2.33	43.17	55.98	-	100 m^2	**99.15**
minimum depth 300 mm	-	4.67	86.33	111.96	-	100 m^2	**198.29**
minimum depth 450 mm	-	6.99	129.31	167.76	-	100 m^2	**297.08**
Spreading and lightly consolidating approved topsoil (imported or from spoil heaps); in layers not exceeding 150 mm; travel distance from spoil heaps not exceeding 100 m; by hand (imported topsoil not included)							
minimum depth 100 mm	-	20.00	370.07	-	-	100 m^2	**370.07**
minimum depth 150 mm	-	30.01	555.11	-	-	100 m^2	**555.11**
minimum depth 300 mm	-	60.01	1110.22	-	-	100 m^2	**1110.22**
minimum depth 450 mm	-	90.02	1665.33	-	-	100 m^2	**1665.33**
Extra over for spreading topsoil to slopes 15 - 30 degrees by machine or hand	-	-	-	-	-	10%	-
Extra over for spreading topsoil to slopes over 30 degrees by machine or hand	-	-	-	-	-	25%	-
Extra over for spreading topsoil from spoil heaps travel exceeding 100 m; by machine							
100 - 150 m	-	0.01	0.23	0.07	-	m^3	**0.30**
150 - 200 m	-	0.02	0.34	0.11	-	m^3	**0.45**
200 - 300 m	-	0.03	0.51	0.17	-	m^3	**0.68**
Extra over spreading topsoil for travel exceeding 100 m; by hand							
100 m	-	0.83	15.42	-	-	m^3	**15.42**
200 m	-	1.67	30.84	-	-	m^3	**30.84**
300 m	-	2.50	46.25	-	-	m^3	**46.25**
Evenly grading; to general surfaces to bring to finished levels							
by machine (tractor mounted rotavator)	-	-	-	0.01	-	m^2	**0.01**
by pedestrian operated rotavator	-	-	0.07	0.04	-	m^2	**0.11**
by hand	-	0.01	0.19	-	-	m^2	**0.19**
Extra over grading for slopes 15 - 30 degrees by machine or hand	-	-	-	-	-	10%	-
Extra over grading for slopes over 30 degrees by machine or hand	-	-	-	-	-	25%	-
Apply screened topdressing to grass surfaces; spread using Tru-Lute							
Sand soil mixes 90/10 to 50/50	-	-	0.04	0.03	0.11	m^2	**0.18**
Spread only existing cultivated soil to final levels using Tru-Lute							
Cultivated soil	-	-	0.04	0.03	-	m^2	**0.07**
Clearing stones; disposing off site; to distance not exceeding 13 km							
by hand; stones not exceeding 50 mm in any direction; loading to skip 4.6 m^3	-	0.01	0.19	0.03	-	m^2	**0.22**
by mechanical stone rake; stones not exceeding 50 mm in any direction; loading to 15 m^3 truck by mechanical loader	-	-	0.04	0.07	-	m^2	**0.11**
Lightly cultivating; weeding; to fallow areas; disposing debris off site; to distance not exceeding 13 km							
by hand	-	0.01	0.26		0.07	m^2	**0.34**

Q PAVING/PLANTING/FENCING/SITE FURNITURE

Item Excluding site overheads and profit	PC £	Labour hours	Labour £	Plant £	Material £	Unit	Total rate £
Q30 SEEDING/TURFING - cont'd							
Surface applications; soil additives; pre-seeding; material delivered to a maximum of 25 m from area of application; applied; by machine							
Soil conditioners; to cultivated ground; ground limestone; PC £24.40/tonne; including turning in							
0.25 kg/m^2 = 2.50 tonnes/ha	0.61	-	-	2.86	0.61	100 m^2	**3.47**
0.50 kg/m^2 = 5.00 tonnes/ha	1.22	-	-	2.86	1.22	100 m^2	**4.08**
0.75 kg/m^2 = 7.50 tonnes/ha	1.83	-	-	2.86	1.83	100 m^2	**4.69**
1.00 kg/m^2 = 10.00 tonnes/ha	2.44	-	-	2.86	2.44	100 m^2	**5.30**
Soil conditioners; to cultivated ground; medium bark; based on deliveries of 25 m^3 loads; PC £32.65/m^3; including turning in							
1 m^3 per 40 m^2 = 25 mm thick	0.82	-	-	0.09	0.82	m^2	**0.90**
1 m^3 per 20 m^2 = 50 mm thick	1.63	-	-	0.14	1.63	m^2	**1.77**
1 m^3 per 13.33 m^2 = 75 mm thick	2.45	-	-	0.19	2.45	m^2	**2.64**
1 m^3 per 10 m^2 = 100 mm thick	3.27	-	-	0.25	3.27	m^2	**3.52**
Soil conditioners; to cultivated ground; mushroom compost; delivered in 25 m^3 loads; PC £19.20/m^3; including turning in							
1 m^3 per 40 m^2 = 25 mm thick	0.48	0.02	0.31	0.09	0.48	m^2	**0.87**
1 m^3 per 20 m^2 = 50 mm thick	0.96	0.03	0.57	0.14	0.96	m^2	**1.67**
1 m^3 per 13.33 m^2 = 75 mm thick	1.44	0.04	0.74	0.19	1.44	m^2	**2.37**
1 m^3 per 10 m^2 = 100 mm thick	1.92	0.05	0.93	0.25	1.92	m^2	**3.10**
Soil conditioners; to cultivated ground; mushroom compost; delivered in 35 m^3 loads; PC £10.86/m^3; including turning in							
1 m^3 per 40 m^2 = 25 mm thick	0.27	0.02	0.31	0.09	0.27	m^2	**0.66**
1 m^3 per 20 m^2 = 50 mm thick	0.54	0.03	0.57	0.14	0.54	m^2	**1.25**
1 m^3 per 13.33 m^2 = 75 mm thick	0.81	0.04	0.74	0.19	0.81	m^2	**1.75**
1 m^3 per 10 m^2 = 100 mm thick	1.09	0.05	0.93	0.25	1.09	m^2	**2.26**
Surface applications and soil additives; pre-seeding; material delivered to a maximum of 25 m from area of application; applied; by hand							
Soil conditioners; to cultivated ground; ground limestone; PC £24.40/tonne; including turning in							
0.25 kg/m^2 = 2.50 tonnes/ha	0.61	1.20	22.20	-	0.61	100 m^2	**22.81**
0.50 kg/m^2 = 5.00 tonnes/ha	1.22	1.33	24.66	-	1.22	100 m^2	**25.88**
0.75 kg/m^2 = 7.50 tonnes/ha	1.83	1.50	27.75	-	1.83	100 m^2	**29.58**
1.00 kg/m^2 = 10.00 tonnes/ha	2.44	1.71	31.71	-	2.44	100 m^2	**34.15**
Soil conditioners; to cultivated ground; medium bark; based on deliveries of 25 m^3 loads; PC £32.65/m^3; including turning in							
1 m^3 per 40 m^2 = 25 mm thick	0.82	0.02	0.41	-	0.82	m^2	**1.23**
1 m^3 per 20 m^2 = 50 mm thick	1.63	0.04	0.82	-	1.63	m^2	**2.45**
1 m^3 per 13.33 m^2 = 75 mm thick	2.45	0.07	1.23	-	2.45	m^2	**3.68**
1 m^3 per 10 m^2 = 100 mm thick	3.27	0.08	1.48	-	3.27	m^2	**4.75**
Soil conditioners; to cultivated ground; mushroom compost; delivered in 25 m^3 loads; PC £19.20/m^3; including turning in							
1 m^3 per 40 m^2 = 25 mm thick	0.48	0.02	0.41	-	0.48	m^2	**0.89**
1 m^3 per 20 m^2 = 50 mm thick	0.96	0.04	0.82	-	0.96	m^2	**1.78**
1 m^3 per 13.33 m^2 = 75 mm thick	1.44	0.07	1.23	-	1.44	m^2	**2.67**
1 m^3 per 10 m^2 = 100 mm thick	1.92	0.08	1.48	-	1.92	m^2	**3.40**
Soil conditioners; to cultivated ground; mushroom compost; delivered in 55 m^3 loads; PC £7.50/m^3; including turning in							
1 m^3 per 40 m^2 = 25 mm thick	0.19	0.02	0.41	-	0.19	m^2	**0.60**
1 m^3 per 20 m^2 = 50 mm thick	0.33	0.04	0.82	-	0.33	m^2	**1.15**
1 m^3 per 13.33 m^2 = 75 mm thick	0.56	0.07	1.23	-	0.56	m^2	**1.79**
1 m^3 per 10 m^2 = 100 mm thick	0.75	0.08	1.48	-	0.75	m^2	**2.23**

Q PAVING/PLANTING/FENCING/SITE FURNITURE

Item Excluding site overheads and profit	PC £	Labour hours	Labour £	Plant £	Material £	Unit	**Total rate £**
Preparation of seedbeds - General							
Preamble: For preliminary operations see "Cultivation" section.							
Preparation of seedbeds; soil preparation							
Lifting selected and approved topsoil from spoil heaps; passing through 6 mm screen; removing debris	-	0.08	1.54	4.50	0.01	m^3	**6.05**
Topsoil; supply only; PC £18.50/m^3; allowing for 20% settlement							
25 mm	-	-	-	-	0.56	m^2	-
50 mm	-	-	-	-	1.11	m^2	-
100 mm	-	-	-	-	2.22	m^2	-
150 mm	-	-	-	-	3.33	m^2	-
200 mm	-	-	-	-	4.44	m^2	-
250 mm	-	-	-	-	5.55	m^2	-
300 mm	-	-	-	-	6.66	m^2	-
400 mm	-	-	-	-	8.88	m^2	-
450 mm	-	-	-	-	9.99	m^2	-
Spreading topsoil to form seedbeds (topsoil not included); by machine							
25 mm deep	-	-	0.05	0.10	-	m^2	**0.15**
50 mm deep	-	-	0.06	0.13	-	m^2	**0.19**
75 mm deep	-	-	0.07	0.15	-	m^2	**0.22**
100 mm deep	-	0.01	0.09	0.20	-	m^2	**0.29**
150 mm deep	-	0.01	0.14	0.29	-	m^2	**0.43**
Spreading only topsoil to form seedbeds (topsoil not included); by hand							
25 mm deep	-	0.03	0.46	-	-	m^2	**0.46**
50 mm deep	-	0.03	0.62	-	-	m^2	**0.62**
75 mm deep	-	0.04	0.79	-	-	m^2	**0.79**
100 mm deep	-	0.05	0.93	-	-	m^2	**0.93**
150 mm deep	-	0.08	1.39	-	-	m^2	**1.39**
Bringing existing topsoil to a fine tilth for seeding; by raking or harrowing; stones not to exceed 6 mm; by machine	-	-	0.07	0.04	-	m^2	**0.12**
Bringing existing topsoil to a fine tilth for seeding; by raking or harrowing; stones not to exceed 6 mm; by hand	-	0.01	0.16	-	-	m^2	**0.16**
Preparation of seedbeds; soil treatments							
For the following topsoil improvement and seeding operations add or subtract the following amounts for every £0.10 difference in the material cost price							
35 g/m^2	-	-	-	-	0.35	100 m^2	-
50 g/m^2	-	-	-	-	0.50	100 m^2	-
70 g/m^2	-	-	-	-	0.70	100 m^2	-
100 g/m^2	-	-	-	-	1.00	100 m^2	-
125 kg/ha	-	-	-	-	12.50	ha	-
150 kg/ha	-	-	-	-	15.00	ha	-
175 kg/ha	-	-	-	-	17.50	ha	-
200 kg/ha	-	-	-	-	20.00	ha	-
225 kg/ha	-	-	-	-	22.50	ha	-
250 kg/ha	-	-	-	-	25.00	ha	-
300 kg/ha	-	-	-	-	30.00	ha	-
350 kg/ha	-	-	-	-	35.00	ha	-
400 kg/ha	-	-	-	-	40.00	ha	-
500 kg/ha	-	-	-	-	50.00	ha	-
700 kg/ha	-	-	-	-	70.00	ha	-
1000 kg/ha	-	-	-	-	100.00	ha	-
1250 kg/ha	-	-	-	-	125.00	ha	-

Q PAVING/PLANTING/FENCING/SITE FURNITURE

Item Excluding site overheads and profit	PC £	Labour hours	Labour £	Plant £	Material £	Unit	**Total rate £**
Q30 SEEDING/TURFING - cont'd							
Preparation of seedbeds - cont'd							
Pre-seeding fertilizers (6:9:6); PC £0.39/kg; to seedbeds; by machine							
35 g/m²	1.36	-	-	0.19	1.36	100 m²	**1.55**
50 g/m²	1.94	-	-	0.19	1.94	100 m²	**2.13**
70 g/m²	2.72	-	-	0.19	2.72	100 m²	**2.91**
100 g/m²	3.89	-	-	0.19	3.89	100 m²	**4.08**
125 g/m²	4.86	-	-	0.19	4.86	100 m²	**5.05**
125 kg/ha	48.60	-	-	18.76	48.60	ha	**67.36**
250 kg/ha	97.20	-	-	18.76	97.20	ha	**115.96**
300 kg/ha	116.64	-	-	18.76	116.64	ha	**135.40**
350 kg/ha	136.08	-	-	18.76	136.08	ha	**154.84**
400 kg/ha	155.52	-	-	18.76	155.52	ha	**174.28**
500 kg/ha	194.40	-	-	30.01	194.40	ha	**224.41**
700 kg/ha	272.16	-	-	30.01	272.16	ha	**302.17**
1250 kg/ha	486.00	-	-	50.02	486.00	ha	**536.02**
Pre-seeding fertilizers (6:9:6); PC £0.39/kg; to seedbeds; by hand							
35 g/m²	1.36	0.17	3.08	-	1.36	100 m²	**4.44**
50 g/m²	1.94	0.17	3.08	-	1.94	100 m²	**5.03**
70 g/m²	2.72	0.17	3.08	-	2.72	100 m²	**5.81**
100 g/m²	3.89	0.20	3.70	-	3.89	100 m²	**7.59**
125 g/m²	4.86	0.20	3.70	-	4.86	100 m²	**8.56**
Pre-emergent granular applied weedkillers; Rigby Taylor; Embargo in accordance with manufacturer's instructions; including all safety precautions; by hand							
0.8 kg	3.87	0.33	6.17	-	3.87	100 m²	**10.03**
1 kg	4.83	0.33	6.17	-	4.83	100 m²	**11.00**
1.25 kg	6.04	0.33	6.17	-	6.04	100 m²	**12.21**
1.70 kg	8.22	0.40	7.40	-	8.22	100 m²	**15.62**
2.25 kg	10.88	0.50	9.25	-	10.87	100 m²	**20.12**
Seeding grass areas - General							
Preamble: The British Standard recommendations for seed and seeding of grass areas are contained in BS 4428: 1989. The prices given in this section are based on compliance with the standard.							
Market prices of grass seed							
Preamble: The prices shown are for supply only at one number 20 kg or 25 kg bag purchase price unless otherwise stated. Rates shown are based on the manufacturer's maximum recommendation for each seed type. Trade and bulk discounts are often available on the prices shown for quantities of more than one bag.							
Bowling greens; fine lawns; ornamental turf; croquet lawns							
"British Seed Houses"; ref A1 Greens; 35 g/m²	-	-	-	-	17.32	100 m²	-
"DLF Trifolium"; ref J1 Golf & Bowling Greens; 34 - 50 g/m²	-	-	-	-	33.10	100 m²	-
"DLF Trifolium"; ref J2 Greens High Performance; 34 - 50 g/m²	-	-	-	-	56.85	100 m²	-
"DLF Trifolium"; ref Pro 20 Fineturf; 35 - 50 g/m²	-	-	-	-	14.05	100 m²	-
"DLF Trifolium"; ref Pro 10 Traditional Green; 35 - 50 g/m²	-	-	-	-	23.70	100 m²	-

Q PAVING/PLANTING/FENCING/SITE FURNITURE

Item Excluding site overheads and profit	PC £	Labour hours	Labour £	Plant £	Material £	Unit	**Total rate £**
Tennis courts; cricket squares							
"British Seed Houses"; ref A2 Lawns & Tennis; 35 g/m²	-	-	-	-	11.72	100 m²	-
"British Seed Houses"; ref A5 Cricket Square; 35 g/m²	-	-	-	-	18.94	100 m²	-
"DLF Trifolium"; ref J2 Greens High Performance; 34 - 50 g/m²	-	-	-	-	56.85	100 m²	-
"DLF Trifolium"; ref Taskmaster; 18 - 25 g/m²	-	-	-	-	9.88	100 m²	-
"DLF Trifolium"; ref Pro 35 Universal; 35 g/m²	-	-	-	-	10.85	100 m²	-
Amenity grassed areas; general purpose lawns							
"British Seed Houses"; ref A3 Landscape; 25 - 50 g/m²	-	-	-	-	13.38	100 m²	-
"DLF Trifolium"; ref J5 Tees Fairways & Cricket; 18 - 25 g/m²	-	-	-	-	13.00	100 m²	-
"DLF Trifolium"; ref Taskmaster; 18 - 25 g/m²	-	-	-	-	9.88	100 m²	-
"DLF Trifolium"; ref Pro 50 Quality Lawn; 25 - 35 g/m²	-	-	-	-	10.08	100 m²	-
"DLF Trifolium"; ref Pro 120 Slowgrowth; 25 - 35 g/m²	-	-	-	-	12.36	100 m²	-
"Rigby Taylor"; ref Mascot R15 General Landscape; 35 g/m²	-	-	-	-	13.98	100 m²	-
Conservation; country parks; slopes and banks							
"British Seed Houses"; ref A4 Low-maintenance; 17 - 35 g/m²	-	-	-	-	11.68	100 m²	-
"British Seed Houses"; ref A16 Country Park; 8 - 19 g/m²	-	-	-	-	6.90	100 m²	-
"British Seed Houses"; ref A17 Legume and Clover; 2 g/m²	-	-	-	-	12.82	100 m²	-
Shaded areas							
"British Seed Houses"; ref A6 Supra Shade; 50 g/m²	-	-	-	-	23.19	100 m²	-
"DLF Trifolium"; ref J1 Golf & Bowling Greens; 34 - 50 g/m²	-	-	-	-	33.10	100 m²	-
"DLF Trifolium"; ref Pro 60 Greenshade; 35 - 50 g/m²	-	-	-	-	17.65	100 m²	-
"Rigby Taylor"; ref Mascot R18 Shade/Drought Tolerance; 35 g/m²	-	-	-	-	21.82	100 m²	-
Sports pitches; rugby; soccer pitches							
"British Seed Houses"; ref A7 Sportsground; 20 g/m²	-	-	-	-	7.42	100 m²	-
"DLF Trifolium"; ref Sportsmaster; 18 - 30g/m²	-	-	-	-	9.60	100 m²	-
"DLF Trifolium"; ref Pro 70 Recreation; 15 - 35 g/m²	-	-	-	-	8.61	100 m²	-
"DLF Trifolium"; ref Pro 75 Stadia; 15 - 35g/m²	-	-	-	-	12.60	100 m²	-
"DLF Trifolium"; ref Pro 80 Renovator; 17 - 35 g/m²	-	-	-	-	9.10	100 m²	-
"Rigby Taylor"; ref Mascot R11 Football & Rugby; 35 g/m²	-	-	-	-	13.50	100 m²	-
"Rigby Taylor"; ref Mascot R12 General Playing Fields; 35 g/m²	-	-	-	-	13.50	100 m²	-
Outfields							
"British Seed Houses"; ref A7 Sportsground; 20 g/m²	-	-	-	-	7.42	100 m²	-
"British Seed Houses"; ref A9 Outfield; 17 - 35 g/m²	-	-	-	-	9.63	100 m²	-
"DLF Trifolium"; ref J4 Fairways & Cricket Outfields; 12 - 25 g/m²	-	-	-	-	9.68	100 m²	-
"DLF Trifolium"; ref Pro 40 Tee & Fairway; 35 g/m²	-	-	-	-	11.10	100 m²	-
"DLF Trifolium"; ref Pro 70 Recreation; 15 - 35 g/m²	-	-	-	-	8.61	100 m²	-
"Rigby Taylor"; ref Mascot R4 Cricket Outfields; 35 g/m²	-	-	-	-	19.65	100 m²	-

Item Excluding site overheads and profit	PC £	Labour hours	Labour £	Plant £	Material £	Unit	**Total rate £**
Q30 SEEDING/TURFING - cont'd							
Market prices of grass seed - cont'd							
Hockey pitches							
"DLF Trifolium"; ref J4 Fairways & Cricket Outfields; 12 - 25 g/m^2	-	-	-	-	9.68	100 m^2	-
"DLF Trifolium"; ref Pro 70 Recreation; 15 - 35 g/m^2	-	-	-	-	8.61	100 m^2	-
"Rigby Taylor"; ref Mascot R10 Cricket & Hockey Outfield; 35 g/m^2	-	-	-	-	13.88	100 m^2	-
Parks							
"British Seed Houses"; ref A7 Sportsground 20 g/m^2	-	-	-	-	7.42	100 m^2	-
"British Seed Houses"; ref A9 Outfield; 17 - 35 g/m^2	-	-	-	-	9.63	100 m^2	-
"DLF Trifolium"; ref Pro 120 Slowgrowth; 25 - 35 g/m^2	-	-	-	-	12.36	100 m^2	-
"Rigby Taylor"; ref Mascot R16 Landscape & Ornamental; 35 g/m^2	-	-	-	-	17.67	100 m^2	-
Informal playing fields							
"DLF Trifolium"; ref J5 Tees Fairways & Cricket; 18 - 25 g/m^2	-	-	-	-	13.00	100 m^2	-
"DLF Trifolium"; ref Pro 45 Tee & Fairway Plus; 35 g/m^2	-	-	-	-	10.85	100 m^2	-
Caravan sites							
"British Seed Houses"; ref A9 Outfield; 17 - 35 g/m^2	-	-	-	-	9.63	100 m^2	-
Sports pitch re-seeding and repair							
"British Seed Houses"; ref A8 Pitch Renovator; 20 - 35 g/m^2	-	-	-	-	8.75	100 m^2	-
"British Seed Houses"; ref A20 Ryesport; 20 - 35 g/m^2	-	-	-	-	10.63	100 m^2	-
"DLF Trifolium"; ref Sportsmaster; 18 - 30 g/m^2	-	-	-	-	9.60	100 m^2	-
"DLF Trifolium"; ref Pro 80 Renovator; 17 - 35 g/m^2	-	-	-	-	9.10	100 m^2	-
"DLF Trifolium"; ref Pro 81 Premier Renovation; 17 - 35 g/m^2	-	-	-	-	9.84	100 m^2	-
"Rigby Taylor"; ref Mascot R14 Premier Winter Games Renovation; 35 g/m^2	-	-	-	-	16.44	100 m^2	-
Racecourses; gallops; polo grounds; horse rides							
"British Seed Houses"; ref A14 Racecourse; 25 - 30 g/m^2	-	-	-	-	9.22	100 m^2	-
"DLF Trifolium"; ref Taskmaster; 18 - 25 g/m^2	-	-	-	-	9.88	100 m^2	-
"DLF Trifolium"; ref Pro 65 Gallop; 17 - 35 g/m^2	-	-	-	-	13.65	100 m^2	-
Motorway and road verges							
"British Seed Houses"; ref A18 Road Verge; 6 - 15 g/m^2	-	-	-	-	4.95	100 m^2	-
"DLF Trifolium"; ref Pro 120 Slowgrowth; 25 - 35 g/m^2	-	-	-	-	12.36	100 m^2	-
"DLF Trifolium"; ref Pro 85 DOT; 10 g/m^2	-	-	-	-	3.07	100 m^2	-
Golf courses; tees							
"British Seed Houses"; ref A10 Golf Tee, 35 - 50 g/m^2	-	-	-	-	19.75	100 m^2	-
"DLF Trifolium"; ref J3 Golf Tees & Fairways; 18 - 30 g/m^2	-	-	-	-	16.35	100 m^2	-
"DLF Trifolium"; ref Taskmaster; 18 - 25 g/m^2	-	-	-	-	9.88	100 m^2	-
"DLF Trifolium"; ref Pro 40 Tee & Fairway; 35 g/m^2	-	-	-	-	11.10	100 m^2	-
"DLF Trifolium"; ref Pro 45 Tee & Fairway Plus; 35 g/m^2	-	-	-	-	10.85	100 m^2	-
"Rigby Taylor"; ref Mascot R4 Golf Tees & Fairways; 35 g/m^2	-	-	-	-	19.66	100 m^2	-

Q PAVING/PLANTING/FENCING/SITE FURNITURE

Item Excluding site overheads and profit	PC £	Labour hours	Labour £	Plant £	Material £	Unit	**Total rate £**
Golf courses; greens							
"British Seed Houses"; ref A11 Golf Green; 35 g/m^2	-	-	-	-	18.99	100 m^2	-
"British Seed Houses"; ref A13 Golf Roughs; 8 g/m^2	-	-	-	-	2.49	100 m^2	-
"DLF Trifolium"; ref J1 Golf & Bowling Greens; 34 - 50 g/m^2	-	-	-	-	33.10	100 m^2	-
"DLF Trifolium"; ref Greenmaster; 8 g/m^2	-	-	-	-	12.42	100 m^2	-
"DLF Trifolium"; ref Pro 5 Economy Green; 35 - 50 g/m^2	-	-	-	-	16.20	100 m^2	-
"Rigby Taylor"; ref Mascot R1 Greenkeeper; 35 g/m^2	-	-	-	-	24.86	100 m^2	-
Golf courses; fairways							
"British Seed Houses"; ref A12 Golf Fairway; 15 - 25 g/m^2	-	-	-	-	8.72	100 m^2	-
"DLF Trifolium"; ref J3 Golf Tees & Fairways; 18 - 30 g/m^2	-	-	-	-	16.35	100 m^2	-
"DLF Trifolium"; ref J5 Tees Fairways & Cricket; 18 - 30 g/m^2	-	-	-	-	15.60	100 m^2	-
"DLF Trifolium"; ref Pro 40 Tee & Fairway; 35 g/m^2	-	-	-	-	11.10	100 m^2	-
"DLF Trifolium"; ref Pro 45 Tee & Fairway Plus; 35 g/m^2	-	-	-	-	10.85	100 m^2	-
"Rigby Taylor"; ref Mascot R6 Golf Fairways; 35 g/m^2	-	-	-	-	13.92	100 m^2	-
Golf courses; roughs							
"DLF Trifolium"; ref Pro 25 Grow-Slow; 17 - 35 g/m^2	-	-	-	-	12.85	100 m^2	-
"Rigby Taylor"; ref Mascot R127 Golf Links & Rough; 35 g/m^2	-	-	-	-	2.19	100 m^2	-
Waste land; spoil heaps; quarries							
"British Seed Houses"; ref A15 Reclamation; 15 - 20 g/m^2	-	-	-	-	6.42	100 m^2	-
"DLF Trifolium"; ref Pro 95 Land Reclamation; 12 - 35 g/m^2	-	-	-	-	14.11	100 m^2	-
"DLF Trifolium"; ref Pro 105 Fertility; 5 g/m^2	-	-	-	-	2.98	100 m^2	-
Low maintenance; housing estates; amenity grassed areas							
"British Seed Houses"; ref A19 Housing Estate; 25 - 35 g/m^2	-	-	-	-	10.02	100 m^2	-
"British Seed Houses"; ref A22 Low Maintenance; 25 - 35 g/m^2	-	-	-	-	14.04	100 m^2	-
"DLF Trifolium"; ref Pro 120 Slowgrowth; 25 - 35 g/m^2	-	-	-	-	12.36	100 m^2	-
Saline coastal; roadside areas							
"British Seed Houses"; ref A21 Coastal/Saline Restoration; 15 - 20 g/m^2	-	-	-	-	7.20	100 m^2	-
"DLF Trifolium"; ref Pro 90 Coastal; 15 - 35 g/m^2	-	-	-	-	12.36	100 m^2	-
Turf production							
"British Seed Houses"; ref A25 Ley Mixture; 160 kg/ha	-	-	-	-	536.00	ha	-
"British Seed Houses"; ref A24 Wear & Tear; 185 kg/ha	-	-	-	-	607.73	ha	-
Market prices of wild flora seed mixtures							
Acid soils							
"British Seed Houses"; ref WF1 (Annual Flowering); 1.00 - 2.00 g/m^2	-	-	-	-	19.50	100 m^2	-
Neutral soils							
"British Seed Houses"; ref WF3 (Neutral Soils); 0.50 - 1.00 g/m^2	-	-	-	-	9.75	100 m^2	-

Q PAVING/PLANTING/FENCING/SITE FURNITURE

Item Excluding site overheads and profit	PC £	Labour hours	Labour £	Plant £	Material £	Unit	**Total rate £**
Q30 SEEDING/TURFING - cont'd							
Market prices of wild flora and grass seed mixtures							
General purpose							
"DLF Trifolium"; ref Pro Flora 8 Old English Country Meadow Mix; 5 g/m^2	-	-	-	-	12.50	100 m^2	-
"DLF Trifolium"; ref Pro Flora 9 General purpose; 5 g/m^2	-	-	-	-	10.00	100 m^2	-
Acid soils							
"British Seed Houses"; ref WFG2 (Annual Meadow); 5 g/m^2	-	-	-	-	22.25	100 m^2	-
"DLF Trifolium"; ref Pro Flora 2 Acidic soils; 5 g/m^2	-	-	-	-	8.75	100 m^2	-
Neutral soils							
"British Seed Houses"; ref WFG4 (Neutral Meadow); 5 g/m^2	-	-	-	-	24.45	100 m^2	-
"British Seed Houses"; ref WFG13 (Scotland); 5 g/m^2	-	-	-	-	21.55	100 m^2	-
"DLF Trifolium"; ref Pro Flora 3 Damp loamy soils; 5 g/m^2	-	-	-	-	12.25	100 m^2	-
Calcareous soils							
"British Seed Houses"; ref WFG5 (Calcareous Soils); 5 g/m^2	-	-	-	-	22.80	100 m^2	-
"DLF Trifolium"; ref Pro Flora 4 Calcareous soils; 5 g/m^2	-	-	-	-	10.50	100 m^2	-
Heavy clay soils							
"British Seed Houses"; ref WFG6 (Clay Soils); 5 g/m^2	-	-	-	-	27.50	100 m^2	-
"British Seed Houses"; ref WFG12 (Ireland); 5 g/m^2	-	-	-	-	21.63	100 m^2	-
"DLF Trifolium"; ref Pro Flora 5 Wet loamy soils; 5 g/m^2	-	-	-	-	33.00	100 m^2	-
Sandy soils							
"British Seed Houses"; ref WFG7 (Free Draining Soils); 5 g/m^2	-	-	-	-	32.25	100 m^2	-
"British Seed Houses"; ref WFG11 (Ireland); 5 g/m^2	-	-	-	-	22.75	100 m^2	-
"British Seed Houses"; ref WFG14 (Scotland); 5 g/m^2	-	-	-	-	25.00	100 m^2	-
"DLF Trifolium"; ref Pro Flora 6 Dry free draining loamy soils; 5 g/m^2	-	-	-	-	29.25	100 m^2	-
Shaded areas							
"British Seed Houses"; ref WFG8 (Woodland and Hedgerow); 5 g/m^2	-	-	-	-	24.00	100 m^2	-
"DLF Trifolium"; ref Pro Flora 7 Hedgerow and light shade; 5 g/m^2	-	-	-	-	33.75	100 m^2	-
Educational							
"British Seed Houses"; ref WFG15 (Schools and Colleges); 5 g/m^2	-	-	-	-	33.75	100 m^2	-
Wetlands							
"British Seed Houses"; ref WFG9 (Wetlands and Ponds); 5 g/m^2	-	-	-	-	29.75	100 m^2	-
"DLF Trifolium"; ref Pro Flora 5 Wet loamy soils; 5 g/m^2	-	-	-	-	33.00	100 m^2	-
Scrub and moorland							
"British Seed Houses"; ref WFG10 (Cornfield Annuals); 5 g/m^2	-	-	-	-	32.80	100 m^2	-
Hedgerow							
"DLF Trifolium"; ref Pro Flora 7 Hedgerow and light shade; 5 g/m^2	-	-	-	-	33.75	100 m^2	-
Vacant sites							
"DLF Trifolium"; ref Pro Flora 1 Cornfield annuals; 5 g/m^2	-	-	-	-	30.00	100 m^2	-

Q PAVING/PLANTING/FENCING/SITE FURNITURE

Item Excluding site overheads and profit	PC £	Labour hours	Labour £	Plant £	Material £	Unit	**Total rate £**
Regional Environmental mixes							
"British Seed Houses"; ref RE1 (Traditional Hay); 5 g/m^2	-	-	-	-	25.95	100 m^2	-
"British Seed Houses"; ref RE2 (Lowland Meadow); 5 g/m^2	-	-	-	-	33.75	100 m^2	-
"British Seed Houses"; ref RE3 (Riverflood Plain/Water Meadow); 5 g/m^2	-	-	-	-	33.25	100 m^2	-
"British Seed Houses"; ref RE4 (Lowland Limestone); 5 g/m^2	-	-	-	-	33.25	100 m^2	-
"British Seed Houses"; ref RE5 (Calcareous Sub-mountain Restoration); 5 g/m^2	-	-	-	-	37.10	100 m^2	-
"British Seed Houses"; ref RE6 (Upland Limestone); 5g/m^2	-	-	-	-	47.45	100 m^2	-
"British Seed Houses"; ref RE7 (Acid Sub-mountain Restoration); 5 g/m^2	-	-	-	-	32.25	100 m^2	-
"British Seed Houses"; ref RE8 (Coastal Reclamation); 5 g/m^2	-	-	-	-	41.45	100 m^2	-
"British Seed Houses"; ref RE9 (Farmland Mixture); 5 g/m^2	-	-	-	-	28.85	100 m^2	-
"British Seed Houses"; ref RE10 (Marginal Land); 5 g/m^2	-	-	-	-	42.75	100 m^2	-
"British Seed Houses"; ref RE11 (Heath Scrubland); 5 g/m^2	-	-	-	-	12.30	100 m^2	-
"British Seed Houses"; ref RE12 (Drought Land); 5 g/m^2	-	-	-	-	25.95	100 m^2	-
Seeding							
Seeding labours only in two operations; by machine (for seed prices see above)							
35 g/m^2	-	-	-	0.51	-	100 m^2	**0.51**
Grass seed; spreading in two operations; PC £3.00/kg; (for changes in material prices please refer to table above); by machine							
35 g/m^2	-	-	-	0.51	10.50	100 m^2	**11.01**
50 g/m^2	-	-	-	0.51	15.00	100 m^2	**15.51**
70 g/m^2	-	-	-	0.51	21.00	100 m^2	**21.51**
100 g/m^2	-	-	-	0.51	30.00	100 m^2	**30.51**
125 kg/ha	-	-	-	50.95	375.00	ha	**425.95**
150 kg/ha	-	-	-	50.95	450.00	ha	**500.95**
200 kg/ha	-	-	-	50.95	600.00	ha	**650.95**
250 kg/ha	-	-	-	50.95	750.00	ha	**800.95**
300 kg/ha	-	-	-	50.95	900.00	ha	**950.95**
350 kg/ha	-	-	-	50.95	1050.00	ha	**1100.95**
400 kg/ha	-	-	-	50.95	1200.00	ha	**1250.95**
500 kg/ha	-	-	-	50.95	1500.00	ha	**1550.95**
700 kg/ha	-	-	-	50.95	2100.00	ha	**2150.95**
1400 kg/ha	-	-	-	50.95	4200.00	ha	**4250.95**
Extra over seeding by machine for slopes over 30 degrees (allowing for the actual area but measured in plan)							
35 g/m^2	-	-	-	0.08	1.57	100 m^2	**1.65**
50 g/m^2	-	-	-	0.08	2.25	100 m^2	**2.33**
70 g/m^2	-	-	-	0.08	3.15	100 m^2	**3.23**
100 g/m^2	-	-	-	0.08	4.50	100 m^2	**4.58**
125 kg/ha	-	-	-	7.64	56.25	ha	**63.89**
150 kg/ha	-	-	-	7.64	67.50	ha	**75.14**
200 kg/ha	-	-	-	7.64	90.00	ha	**97.64**
250 kg/ha	-	-	-	7.64	112.50	ha	**120.14**
300 kg/ha	-	-	-	7.64	135.00	ha	**142.64**
350 kg/ha	-	-	-	7.64	157.50	ha	**165.14**
400 kg/ha	-	-	-	7.64	180.00	ha	**187.64**
500 kg/ha	-	-	-	7.64	225.00	ha	**232.64**
700 kg/ha	-	-	-	7.64	315.00	ha	**322.64**
1400 kg/ha	-	-	-	7.64	630.00	ha	**637.64**

Q PAVING/PLANTING/FENCING/SITE FURNITURE

Item Excluding site overheads and profit	PC £	Labour hours	Labour £	Plant £	Material £	Unit	Total rate £
Q30 SEEDING/TURFING - cont'd							
Seeding labours only in two operations; by machine (for seed prices see above)							
35 g/m^2	-	0.17	3.08	-	-	100 m^2	**3.08**
Grass seed; spreading in two operations; PC £3.00/kg; (for changes in material prices please refer to table above); by hand							
35 g/m^2	-	0.17	3.08	-	10.50	100 m^2	**13.58**
50 g/m^2	-	0.17	3.08	-	15.00	100 m^2	**18.08**
70 g/m^2	-	0.17	3.08	-	21.00	100 m^2	**24.08**
100 g/m^2	-	0.20	3.70	-	30.00	100 m^2	**33.70**
125 g/m^2	-	0.20	3.70	-	37.50	100 m^2	**41.20**
Extra over seeding by hand for slopes over 30 degrees (allowing for the actual area but measured in plan)							
35 g/m^2	1.56	-	0.07	-	1.56	100 m^2	**1.63**
50 g/m^2	2.25	-	0.07	-	2.25	100 m^2	**2.32**
70 g/m^2	3.15	-	0.07	-	3.15	100 m^2	**3.22**
100 g/m^2	4.50	-	0.08	-	4.50	100 m^2	**4.58**
125 g/m^2	5.61	-	0.08	-	5.61	100 m^2	**5.69**
Harrowing seeded areas; light chain harrow	-	-	-	0.09	-	100 m^2	**0.09**
Raking over seeded areas							
by mechanical stone rake	-	-	-	2.18	-	100 m^2	**2.18**
by hand	-	0.80	14.80	-	-	100 m^2	**14.80**
Rolling seeded areas; light roller							
by tractor drawn roller	-	-	-	0.54	-	100 m^2	**0.54**
by pedestrian operated mechanical roller	-	0.08	1.54	0.41	-	100 m^2	**1.95**
by hand drawn roller	-	0.17	3.08	-	-	100 m^2	**3.08**
Extra over harrowing, raking or rolling seeded areas for slopes over 30 degrees; by machine or hand	-	-	-	-	-	25%	**-**
Turf edging; to seeded areas; 300 mm wide	-	0.05	0.88	-	2.20	m^2	**3.08**
Liquid sod; Turf Management							
Spray on grass system of grass plantlets fertilizer, bio-degradable mulch carrier, root enhancer and water							
to prepared ground	-	-	-	-	-	m^2	**1.80**
Preparation of turf beds							
Rolling turf to be lifted; lifting by hand or mechanical turf stripper; stacks to be not more than 1 m high							
cutting only preparing to lift; pedestrian turf cutter	-	0.75	13.88	9.48	-	100 m^2	**23.36**
lifting and stacking; by hand	-	8.33	154.17	-	-	100 m^2	**154.17**
Rolling up; moving to stacks							
distance not exceeding 100 m	-	2.50	46.25	-	-	100 m^2	**46.25**
extra over rolling and moving turf to stacks to transport per additional 100 m	-	0.83	15.42	-	-	100 m^2	**15.42**
Lifting selected and approved topsoil from spoil heaps							
passing through 6 mm screen; removing debris	-	0.17	3.08	9.00	-	m^3	**12.09**
Extra over lifting topsoil and passing through screen for imported topsoil; plus 20% allowance for settlement	18.50	-	-	-	22.20	m^3	**-**

Q PAVING/PLANTING/FENCING/SITE FURNITURE

Item Excluding site overheads and profit	PC £	Labour hours	Labour £	Plant £	Material £	Unit	**Total rate £**
Topsoil; PC £18.50/m^3; plus 20% allowance for settlement							
25 mm deep	-	-	-	-	0.56	m^2	-
50 mm deep	-	-	-	-	1.11	m^2	-
100 mm deep	-	-	-	-	2.22	m^2	-
150 mm deep	-	-	-	-	3.33	m^2	-
200 mm deep	-	-	-	-	4.44	m^2	-
250 mm deep	-	-	-	-	5.55	m^2	-
300 mm deep	-	-	-	-	6.66	m^2	-
400 mm deep	-	-	-	-	8.88	m^2	-
450 mm deep	-	-	-	-	9.99	m^2	-
Spreading topsoil to form turfbeds (topsoil not included); by machine							
25 mm deep	-	-	0.05	0.10	-	m^2	**0.15**
50 mm deep	-	-	0.06	0.13	-	m^2	**0.19**
75 mm deep	-	-	0.07	0.15	-	m^2	**0.22**
100 mm deep	-	0.01	0.09	0.20	-	m^2	**0.29**
150 mm deep	-	0.01	0.14	0.29	-	m^2	**0.43**
Spreading topsoil to form turfbeds (topsoil not included); by hand							
25 mm deep	-	0.03	0.46	-	-	m^2	**0.46**
50 mm deep	-	0.03	0.62	-	-	m^2	**0.62**
75 mm deep	-	0.04	0.79	-	-	m^2	**0.79**
100 mm deep	-	0.05	0.93	-	-	m^2	**0.93**
150 mm deep	-	0.08	1.39	-	-	m^2	**1.39**
Bringing existing topsoil to a fine tilth for turfing by raking or harrowing; stones not to exceed 6 mm; by machine	-	-	0.07	0.04	-	m^2	**0.12**
Bringing existing topsoil to a fine tilth for turfing by raking or harrowing; stones not to exceed 6 mm; by hand	-	0.01	0.17	-	-	m^2	**0.17**
Turfing							
Turfing; laying only; to stretcher bond; butt joints; including providing and working from barrow plank runs where necessary to surfaces not exceeding 30 degrees from horizontal; average size of turves 0.50 yd2 (0.418 m^2); 2.39 turves per m^2							
specially selected lawn turves from previously lifted stockpile	-	0.08	1.39	-	-	m^2	**1.39**
cultivated lawn turves; to large open areas	-	0.06	1.08	-	-	m^2	**1.08**
cultivated lawn turves; to domestic or garden areas	-	0.08	1.44	-	-	m^2	**1.44**
road verge quality turf	-	0.04	0.74	-	-	m^2	**0.74**
Industrially grown turf; PC prices listed represent the general range of industrial turf prices for sportsfields and amenity purposes; prices will vary with quantity and site location							
"Rolawn"							
ref RB Medallion; sports fields, domestic lawns, general landscape; full loads 1720 m^2	1.60	0.07	1.29	-	1.60	m^2	**2.89**
ref RB Medallion; sports fields, domestic lawns, general landscape; part loads;	2.20	0.07	1.29	-	2.20	m^2	**3.49**
Tensar Ltd							
"Tensar Turf Mat" reinforced turf for embankments; laid to embankments 3.3 m^2 turves	3.50	0.11	2.04	-	3.50	m^2	**5.54**

Q PAVING/PLANTING/FENCING/SITE FURNITURE

Item Excluding site overheads and profit	PC £	Labour hours	Labour £	Plant £	Material £	Unit	Total rate £
Q30 SEEDING/TURFING - cont'd							
Industrially grown turf - cont'd							
"Inturf"							
ref Inturf 1; fine lawns, golf greens, bowling greens	2.35	0.10	1.76	-	2.35	m^2	**4.11**
ref Inturf 2; football grounds, parks, hardwearing areas	1.35	0.05	0.97	-	1.35	m^2	**2.32**
ref Inturf 2 Bargold; fine turf ; football grounds, parks, hardwearing areas	1.60	0.05	0.97	-	1.60	m^2	**2.57**
ref Inturf 3; hockey grounds, polo, medium wearing areas	1.60	0.05	1.00	-	1.60	m^2	**2.60**
Custom Grown Turf; specific seed mixtures to suit soil or site conditions	8.10	0.08	1.48	-	8.10	m^2	**9.58**
Reinforced turf; ADP Netlon Advanced Turf; ADP Netlon Turf Systems; blended mesh fibre elements incorporated into rootzone; rootzone spread and levelled over, cultivated, prepared and reduced and levelled ground (not included); compacted with vibratory roller							
"ADP ATS 400" with selected turf and fertilizer 100 mm thick							
100 - 500 m^2	18.00	0.03	0.62	0.82	18.00	m^2	**19.43**
over 500 m^2	16.50	0.03	0.62	0.82	16.50	m^2	**17.93**
"ADP ATS 400" with selected turf and fertilizer 150 mm thick							
100 - 500 m^2	24.00	0.04	0.74	0.92	24.00	m^2	**25.66**
over 500 m^2	22.00	0.04	0.74	0.92	22.00	m^2	**23.66**
"ADP ATS 400" with selected turf and fertilizer 200 mm thick							
100 - 500 m^2	26.00	0.05	0.93	1.07	26.00	m^2	**28.00**
over 500 m^2	24.00	0.05	0.93	1.07	24.00	m^2	**26.00**
Firming turves with wooden beater	-	0.01	0.19	-	-	m^2	**0.19**
Rolling turfed areas; light roller							
by tractor with turf tyres and roller	-	-	-	0.54	-	100 m^2	**0.54**
by pedestrian operated mechanical roller	-	0.08	1.54	0.41	-	100 m^2	**1.95**
by hand drawn roller	-	0.17	3.08	-	-	100 m^2	**3.08**
Dressing with finely sifted topsoil; brushing into joints	0.02	0.05	0.93	-	0.02	m^2	**0.95**
Turfing; laying only							
to slopes over 30 degrees; to diagonal bond (measured as plan area - add 15% to these rates for the incline area of 30 degree slopes)	-	0.12	2.22	-	-	m^2	**2.22**
Extra over laying turfing for pegging down turves							
wooden or galvanized wire pegs; 200 mm long; 2 pegs per 0.50 yd^2	1.44	0.01	0.25	-	1.44	m^2	**1.68**
Artificial grass; Artificial Lawn Company; laid to sharp sand bed priced separately 15 kg kiln sand brushed in per m^2							
"Leisure"; 24 mm thick artificial sports turf; sand filled	-	-	-	-	-	m^2	**26.17**
"Greenfield"; for general use; budget surface; sand filled	-	-	-	-	-	m^2	**20.94**
"Garden"; patios conservatories and pool surrounds; sand filled	-	-	-	-	-	m^2	**21.90**
"Premier"; lawns and patios	-	-	-	-	-	m^2	**28.08**
"Elite"; grass/sand and rubber filled	-	-	-	-	-	m^2	**31.45**
"Grassflex"; safety surfacing for play areas	-	-	-	-	-	m^2	**44.97**

Q PAVING/PLANTING/FENCING/SITE FURNITURE

Item Excluding site overheads and profit	PC £	Labour hours	Labour £	Plant £	Material £	Unit	**Total rate £**
Maintenance operations (Note: the following rates apply to aftercare maintenance executed as part of a landscaping contract only)							
Initial cutting; to turfed areas							
20 mm high; using pedestrian guided power driven cylinder mower; including boxing off cuttings (stone picking and rolling not included)	-	0.18	3.33	0.28	-	100 m^2	**3.61**
Repairing damaged grass areas							
scraping out; removing slurry; from ruts and holes; average 100 mm deep	-	0.13	2.47	-	-	m^2	**2.47**
100 mm topsoil	-	0.13	2.47	-	2.22	m^2	**4.69**
Repairing damaged grass areas; sowing grass seed to match existing or as specified; to individually prepared worn patches							
35 g/m^2	0.11	0.01	0.19	-	0.12	m^2	**0.31**
50 g/m^2	0.15	0.01	0.19	-	0.17	m^2	**0.36**
Sweeping leaves; disposing off site; motorized vacuum sweeper or rotary brush sweeper							
areas of maximum 2500 m^2 with occasional large tree and established boundary planting; 4.6 m^3 (1 skip of material to be removed)	-	0.40	7.40	3.19	-	100 m^2	**10.59**
Leaf clearance; clearing grassed area of leaves and other extraneous debris							
Using equipment towed by tractor							
large grassed areas with perimeters of mature trees such as sports fields and amenity areas	-	0.01	0.23	0.04	-	100 m^2	**0.27**
large grassed areas containing ornamental trees and shrub beds	-	0.03	0.46	0.06	-	100 m^2	**0.52**
Using pedestrian operated mechanical equipment and blowers							
grassed areas with perimeters of mature trees such as sports fields and amenity areas	-	0.04	0.74	0.05	-	100 m^2	**0.79**
grassed areas containing ornamental trees and shrub beds	-	0.10	1.85	0.13	-	100 m^2	**1.98**
verges	-	0.07	1.23	0.08	-	100 m^2	**1.32**
By hand							
grassed areas with perimeters of mature trees such as sports fields and amenity areas	-	0.05	0.93	0.10	-	100 m^2	**1.02**
grassed areas containing ornamental trees and shrub beds	-	0.08	1.54	0.17	-	100 m^2	**1.71**
verges	-	1.00	18.50	1.99	-	100 m^2	**20.49**
Removal of arisings							
areas with perimeters of mature trees	-	0.01	0.10	0.09	0.80	100 m^2	**0.99**
areas containing ornamental trees and shrub beds	-	0.02	0.31	0.34	2.00	100 m^2	**2.65**
Cutting grass to specified height; per cut							
multi unit gang mower	-	0.59	10.88	17.43	-	ha	**28.31**
ride-on triple cylinder mower	-	0.01	0.26	0.13	-	100 m^2	**0.39**
ride-on triple rotary mower	-	0.01	0.26	-	-	100 m^2	**0.26**
pedestrian mower	-	0.18	3.33	0.66	-	100 m^2	**3.99**
Cutting grass to banks; per cut							
side arm cutter bar mower	-	0.02	0.43	0.32	-	100 m^2	**0.75**
Cutting rough grass; per cut							
power flail or scythe cutter	-	0.04	0.65	-	-	100 m^2	**0.65**
Extra over cutting grass for slopes not exceeding 30 degrees	-	-	-	-	-	10%	**-**
Extra over cutting grass for slopes exceeding 30 degrees	-	-	-	-	-	40%	**-**
Cutting fine sward							
pedestrian operated seven-blade cylinder lawn mower	-	0.14	2.59	0.21	-	100 m^2	**2.80**
Extra over cutting fine sward for boxing off cuttings							
pedestrian mower	-	0.03	0.52	0.04	-	100 m^2	**0.56**

Q PAVING/PLANTING/FENCING/SITE FURNITURE

Item Excluding site overheads and profit	PC £	Labour hours	Labour £	Plant £	Material £	Unit	Total rate £
Q30 SEEDING/TURFING - cont'd							
Maintenance operations - cont'd							
Cutting areas of rough grass							
scythe	-	1.00	18.50	-	-	100 m^2	**18.50**
sickle	-	2.00	37.00	-	-	100 m^2	**37.00**
petrol operated strimmer	-	0.30	5.56	0.42	-	100 m^2	**5.98**
Cutting areas of rough grass which contain trees or whips							
petrol operated strimmer	-	0.40	7.40	0.56	-	100 m^2	**7.96**
Extra over cutting rough grass for on site raking up and dumping	-	0.33	6.17	-	-	100 m^2	**6.17**
Trimming edge of grass areas; edging tool							
with petrol powered strimmer	-	0.13	2.47	0.19	-	100 m	**2.65**
by hand	-	0.67	12.33	-	-	100 m	**12.33**
Marking out pitches using approved line marking compound; including initial setting out and marking							
discus, hammer, javelin or shot putt area	3.15	2.00	37.00	-	3.15	nr	**40.15**
cricket square	2.10	2.00	37.00	-	2.10	nr	**39.10**
cricket boundary	7.35	8.00	148.00	-	7.35	nr	**155.35**
grass tennis court	3.15	4.00	74.00	-	3.15	nr	**77.15**
hockey pitch	10.50	8.00	148.00	-	10.50	nr	**158.50**
football pitch	10.50	8.00	148.00	-	10.50	nr	**158.50**
rugby pitch	10.50	8.00	148.00	-	10.50	nr	**158.50**
eight lane running track; 400 m	21.00	16.00	296.00	-	21.00	nr	**317.00**
Re-marking out pitches using approved line marking compound							
discus, hammer, javelin or shot putt area	2.10	0.50	9.25	-	2.10	nr	**11.35**
cricket square	2.10	0.50	9.25	-	2.10	nr	**11.35**
grass tennis court	7.35	1.00	18.50	-	7.35	nr	**25.85**
hockey pitch	7.35	1.00	18.50	-	7.35	nr	**25.85**
football pitch	7.35	1.00	18.50	-	7.35	nr	**25.85**
rugby pitch	7.35	1.00	18.50	-	7.35	nr	**25.85**
eight lane running track; 400 m	21.00	2.50	46.25	-	21.00	nr	**67.25**
Rolling grass areas; light roller							
by tractor drawn roller	-	-	-	0.54	-	100 m^2	**0.54**
by pedestrian operated mechanical roller	-	0.08	1.54	0.41	-	100 m^2	**1.95**
by hand drawn roller	-	0.17	3.08	-	-	100 m^2	**3.08**
Aerating grass areas; to a depth of 100 mm							
using tractor-drawn aerator	-	0.06	1.08	0.96	-	100 m^2	**2.04**
using pedestrian-guided motor powered solid or slitting tine turf aerator	-	0.18	3.24	2.82	-	100 m^2	**6.06**
using hollow tine aerator; including sweeping up and dumping corings	-	0.50	9.25	5.64	-	100 m^2	**14.89**
using hand aerator or fork	-	1.67	30.83	-	-	100 m^2	**30.83**
Extra over aerating grass areas for on site sweeping up and dumping corings	-	0.17	3.08	-	-	100 m^2	**3.08**
Switching off dew; from fine turf areas	-	0.20	3.70	-	-	100 m^2	**3.70**
Scarifying grass areas to break up thatch; removing dead grass							
using tractor-drawn scarifier	-	0.07	1.29	0.30	-	100 m^2	**1.59**
using self-propelled scarifier; including removing and disposing of grass on site	-	0.33	6.17	0.14	-	100 m^2	**6.31**
Harrowing grass areas							
using drag mat	-	0.03	0.52	0.30	-	100 m^2	**0.82**
using chain harrow	-	0.04	0.65	0.38	-	100 m^2	**1.03**
using drag mat	-	2.80	51.81	30.25	-	ha	**82.05**
using chain harrow	-	3.50	64.75	37.81	-	ha	**102.56**
Extra for scarifying and harrowing grass areas for disposing excavated material off site; to tip not exceeding 13 km; loading by machine							
slightly contaminated	-	-	-	1.65	15.00	m^3	**16.65**
rubbish	-	-	-	1.65	15.00	m^3	**16.65**
inert material	-	-	-	1.10	10.00	m^3	**11.10**

Q PAVING/PLANTING/FENCING/SITE FURNITURE

Item Excluding site overheads and profit	PC £	Labour hours	Labour £	Plant £	Material £	Unit	**Total rate £**
For the following topsoil improvement and seeding operations add or subtract the following amounts for every £0.10 difference in the material cost price							
35 g/m²	-	-	-	-	0.35	100 m²	-
50 g/m²	-	-	-	-	0.50	100 m²	-
70 g/m²	-	-	-	-	0.70	100 m²	-
100 g/m²	-	-	-	-	1.00	100 m²	-
125 kg/ha	-	-	-	-	12.50	ha	-
150 kg/ha	-	-	-	-	15.00	ha	-
175 kg/ha	-	-	-	-	17.50	ha	-
200 kg/ha	-	-	-	-	20.00	ha	-
225 kg/ha	-	-	-	-	22.50	ha	-
250 kg/ha	-	-	-	-	25.00	ha	-
300 kg/ha	-	-	-	-	30.00	ha	-
350 kg/ha	-	-	-	-	35.00	ha	-
400 kg/ha	-	-	-	-	40.00	ha	-
500 kg/ha	-	-	-	-	50.00	ha	-
700 kg/ha	-	-	-	-	70.00	ha	-
1000 kg/ha	-	-	-	-	100.00	ha	-
1250 kg/ha	-	-	-	-	125.00	ha	-
Topdressing fertilizers (7:7:7); PC £0.61/kg; to seedbeds; by machine							
35 g/m²	2.16	-	-	0.19	2.16	100 m²	**2.35**
50 g/m²	3.09	-	-	0.19	3.09	100 m²	**3.28**
300 kg/ha	185.25	-	-	18.76	185.25	ha	**204.01**
350 kg/ha	216.13	-	-	18.76	216.13	ha	**234.88**
400 kg/ha	247.00	-	-	18.76	247.00	ha	**265.76**
500 kg/ha	308.75	-	-	30.01	308.75	ha	**338.76**
Topdressing fertilizers (7:7:7); PC £0.61/kg; to seedbeds; by hand							
35 g/m²	2.16	0.17	3.08	-	2.16	100 m²	**5.25**
50 g/m²	3.09	0.17	3.08	-	3.09	100 m²	**6.17**
70 g/m²	4.32	0.17	3.08	-	4.32	100 m²	**7.41**
Watering turf; evenly; at a rate of 5 litre/m²							
using movable spray lines powering 3 nr sprinkler heads with a radius of 15 m and allowing for 60% overlap (irrigation machinery costs not included)	-	0.02	0.29	-	-	100 m²	**0.29**
using sprinkler equipment and with sufficient water pressure to run 1 nr 15 m radius sprinkler	-	0.02	0.37	-	-	100 m²	**0.37**
using hand-held watering equipment	-	0.25	4.63	-	-	100 m²	**4.63**
Q31 PLANTING							
Note: For market prices of landscape chemicals, please see section Q30 or Q35							
Market prices of mulching materials							
Melcourt Industries Ltd; 25 m³ loads (Items labelled FSC are Forest Stewardship Council certified)							
"Ornamental Bark Mulch"	-	-	-	-	47.40	m³	-
"Bark Nuggets"®	-	-	-	-	42.90	m³	-
"Graded Bark Flakes"	-	-	-	-	47.30	m³	-
"Amenity Bark Mulch"; FSC	-	-	-	-	32.65	m³	-
"Contract Bark Mulch"; FSC	-	-	-	-	30.40	m³	-
"Spruce Ornamental"; FSC	-	-	-	-	33.40	m³	-
"Decorative Biomulch"®	-	-	-	-	31.95	m³	-
"Rustic Biomulch"®	-	-	-	-	35.70	m³	-
"Mulch 2000"	-	-	-	-	25.90	m³	-
"Forest BioMulch"®	-	-	-	-	29.45	m³	

Q PAVING/PLANTING/FENCING/SITE FURNITURE

Item Excluding site overheads and profit	PC £	Labour hours	Labour £	Plant £	Material £	Unit	**Total rate £**
Q31 PLANTING - cont'd							
Market prices of mulching materials - cont'd							
Melcourt Industries Ltd; 50 m^3 loads							
"Mulch 2000"	-	-	-	-	15.20	m^3	-
Melcourt Industries Ltd; 70 m^3 loads							
"Contract Bark Mulch"; FSC	-	-	-	-	18.40	m^3	-
Melcourt Industries Ltd; 80 m^3 loads							
"Ornamental Bark Mulch"	-	-	-	-	34.40	m^3	-
"Bark Nuggets"®	-	-	-	-	29.90	m^3	-
"Graded Bark Flakes"	-	-	-	-	34.30	m^3	-
"Amenity Bark Mulch"; FSC	-	-	-	-	19.65	m^3	-
"Spruce Ornamental"; FSC	-	-	-	-	20.40	m^3	-
"Decorative Biomulch"®	-	-	-	-	18.95	m^3	-
"Rustic Biomulch"®	-	-	-	-	22.90	m^3	-
"Forest BioMulch"®	-	-	-	-	16.45	m^3	-
Market prices of planting materials (Note: the rates shown generally reflect the manufacturer's recommended retail prices; trade and bulk discounts are often available on the prices shown)							
Topsoil							
general purpose screened	-	-	-	-	18.50	m^3	-
50/50; as dug/10 mm screened	-	-	-	-	25.95	m^3	-
Mulch; Melcourt Industries Ltd; 25 m^3 loads							
"Composted Fine Bark"; FSC	-	-	-	-	29.40	m^3	-
"Humus 2000"	-	-	-	-	24.35	m^3	-
"Spent Mushroom Compost"	-	-	-	-	19.20	m^3	-
"Topgrow"	-	-	-	-	28.30	m^3	-
Mulch; Melcourt Industries Ltd; 50 m^3 loads							
"Humus 2000"	-	-	-	-	13.85	m^3	-
Mulch; Melcourt Industries Ltd; 60 m^3 loads							
"Spent Mushroom Compost"	-	-	-	-	7.50	m^3	-
Mulch; Melcourt Industries Ltd; 65 m^3 loads							
"Composted Fine Bark"	-	-	-	-	17.40	m^3	-
"Super Humus"	-	-	-	-	15.10	m^3	-
"Topgrow"	-	-	-	-	17.30	m^3	-
Fertilizers; Scotts UK							
"Enmag"; 70 g/m^2	-	-	-	-	12.59	100 m^2	-
fertilizer; controlled release; "Osmocote Exact", 15+10+12+2MgO+TE, tablet; 5 gr each	-	-	-	-	3.21	100 nr	-
Fertilizers; Scotts UK; granular "Sierrablen Flora", 15+9+9+3 MgO; controlled release fertilizer; costs for recommended application rates							
transplant	-	-	-	-	0.10	nr	-
whip	-	-	-	-	0.15	nr	-
feathered	-	-	-	-	0.20	nr	-
light standard	-	-	-	-	0.20	nr	-
standard	-	-	-	-	0.25	nr	-
selected standard	-	-	-	-	0.35	nr	-
heavy standard	-	-	-	-	0.41	nr	-
extra heavy standard	-	-	-	-	0.51	nr	-
16 -18 cm girth	-	-	-	-	0.56	nr	-
18 - 20 cm girth	-	-	-	-	0.61	nr	-
20 - 22 cm girth	-	-	-	-	0.71	nr	-
22 - 24 cm girth	-	-	-	-	0.76	nr	-
24 - 26 cm girth	-	-	-	-	0.81	nr	-
Fertilizers; Farmura Environmental Ltd; Seanure Root Dip							
to transplants	-	-	-	-	0.23	10 nr	-
medium whips	-	-	-	-	0.06	nr	-
standard trees	-	-	-	-	0.39	nr	-

Q PAVING/PLANTING/FENCING/SITE FURNITURE

Item Excluding site overheads and profit	PC £	Labour hours	Labour £	Plant £	Material £	Unit	**Total rate £**
Fertilizers; Farmura Environmental Ltd; Seanure Soilbuilder							
soil amelioration; 70 g/m^2	-	-	-	-	9.02	100 m^2	-
to plant pits; 300 x 300 x 300	-	-	-	-	0.04	nr	-
to plant pits; 600 x 600 x 600	-	-	-	-	0.42	nr	-
to tree pits; 1.00 x 1.00 x 1.00	-	-	-	-	1.93	nr	-
Fertilizers; Scotts UK							
"TPMC" tree and shrub planting compost	-	-	-	-	4.35	bag	-
Fertilizers; Rigby Taylor Ltd; fertilizer application rates 35 g/m^2 unless otherwise shown							
straight fertilizer; "Bone Meal"; at 70 g/m^2	-	-	-	-	6.46	100 m^2	-
straight fertilizer; "Sulphate of Ammonia"	-	-	-	-	1.97	100 m^2	-
straight fertilizer; "Sulphate of Iron"	-	-	-	-	2.52	100 m^2	-
straight fertilizer; "Sulphate of Potash"	-	-	-	-	2.89	100 m^2	-
straight fertilizer; "Super Phosphate Powder"	-	-	-	-	0.61	100 m^2	-
liquid fertilizer; "Vitax 50/50 Soluble Iron"; 56 ml/100 m^2	-	-	-	-	8.87	100 m^2	-
liquid fertilizer; "Vitax 50/50 Standard"; 56 ml/100 m^2	-	-	-	-	11.97	100 m^2	-
liquid fertilizer; "Vitax 50/50 Extra"; 56 ml/100 m^2	-	-	-	-	14.04	100 m^2	-
Wetting Agents; Rigby Taylor Ltd							
wetting agent; "Breaker Advance Liquid"; 10 L	-	-	-	-	1.48	100 m^2	-
wetting agent; "Breaker Advance Granules"; 20 kg	-	-	-	-	11.56	100 m^2	-
Market prices of trees, shrubs and plants; Notcutts Nurseries Ltd							
Shrubs							
Acer palmatum Dissectum Atropurpureum; 30-45 3 L	-	-	-	-	17.64	nr	-
Arbutus unedo rubra; 30-45 3 L	-	-	-	-	7.97	nr	-
Aucuba japonica Variegata; 30-45 3 L	-	-	-	-	3.36	nr	-
Berberis candidula; 20-30 2 L	-	-	-	-	2.11	nr	-
Berberis thunbergii; 30-45 3 L	-	-	-	-	1.59	nr	-
Berberis thunbergii Rose Glow; 30-45 2 L	-	-	-	-	3.18	nr	-
Buddleja davidii Black Knight; 45-60 3 L	-	-	-	-	2.05	nr	-
Buddleja globosa; 45-60 3 L	-	-	-	-	3.31	nr	-
Buxus sempervirens; 30-45 3 L	-	-	-	-	3.31	nr	-
Caryopteris clandonensis; 30-45 3 L	-	-	-	-	2.79	nr	-
Ceanothus dentatus; 60-90 3 L	-	-	-	-	3.65	nr	-
Ceanothus thyrsiflorus Repens; 30-45 2 L	-	-	-	-	2.51	nr	-
Chaenomeles superba Jet Trail; 45-60 3 L	-	-	-	-	2.51	nr	-
Choisya ternata; 30-45 3 L	-	-	-	-	2.79	nr	-
Cornus alba; 60-90 3 L	-	-	-	-	1.59	nr	-
Cornus alba Elegantissima; 60-90 3 L	-	-	-	-	2.74	nr	-
Cornus stolonifera Flayiramea; 60-90 3 L	-	-	-	-	2.05	nr	-
Corylus avellana; 45-60 3 L	-	-	-	-	1.59	nr	-
Cotinus coggygria Royal Purple; 30-45 3 L	-	-	-	-	3.75	nr	-
Cotoneaster salicifolius Exburyensis; 45-50 3 L	-	-	-	-	2.74	nr	-
Cotoneaster suecicus watereri; 60-90 3 L	-	-	-	-	2.05	nr	-
Elaeagnus pungens maculata; 45-60 3 L	-	-	-	-	3.65	nr	-
Elaeagnus x ebbingei; 45-60 3 L	-	-	-	-	2.51	nr	-
Elaeagnus x ebbingei Limelight; 45-60 3 L	-	-	-	-	3.65	nr	-
Escallonia Edinensis; 45-60 3 L	-	-	-	-	2.39	nr	-
Euonymus fortunei Blondy; 20-30 2 L	-	-	-	-	3.18	nr	-
Euonymus fortunei Darts Blanket; 30-45 2 L	-	-	-	-	1.59	nr	-
Forsythia intermedia Lynwood; 60-90 3 L	-	-	-	-	2.05	nr	-
Genista hispanica; 20-30 2 L	-	-	-	-	2.05	nr	-
Griselinia littoralis; 45-60 3 L	-	-	-	-	3.36	nr	-
Hebe albicans; 20-30 2 L	-	-	-	-	2.05	nr	-
Hebe Great Orme; 20-30 3 L	-	-	-	-	2.96	nr	-
Hebe pinguifolia Pagei; 15-20 2 L	-	-	-	-	1.72	nr	-
Hebe rakaiensis; 20-30 2 L	-	-	-	-	1.72	nr	-
Hedera helix Glacier; 45-60 1.5 L	-	-	-	-	2.39	nr	-

Q PAVING/PLANTING/FENCING/SITE FURNITURE

Item Excluding site overheads and profit	PC £	Labour hours	Labour £	Plant £	Material £	Unit	Total rate £
Q31 PLANTING - cont'd							
Market prices of trees, shrubs and plants; Notcutts Nurseries Ltd - cont'd							
Hippophae rhamnoides; 45-60 3 L	-	-	-	-	1.82	nr	-
Hydrangea macrophylla Blue Wave; 45-60 4 L	-	-	-	-	4.79	nr	-
Hypericum Hidcote; 30-45 3 L	-	-	-	-	2.05	nr	-
Ilex aquifolium Pyramidalis; 60-90 3 L	-	-	-	-	4.45	nr	-
Lavandula angustifolia Hidcote; 15-20 1.5 L	-	-	-	-	1.59	nr	-
Lavandula angustifolia Hidcote; 20-30 3 L	-	-	-	-	2.51	nr	-
Lavatera Barnsley; 45-60 3 L	-	-	-	-	2.79	nr	-
Leucothoe walteri Rainbow; 30-45 3 L	-	-	-	-	5.12	nr	-
Leycesteria formosa; 30-45 3 L	-	-	-	-	3.18	nr	-
Ligustrum ovalifolium; 45-60 2 L	-	-	-	-	1.59	nr	-
Ligustrum ovalifolium Aureum; 30-45 2 L	-	-	-	-	2.05	nr	-
Ligustrum nitida Baggesens Gold; 30-45 3 L	-	-	-	-	2.17	nr	-
Mahonia japonica; 45-60 3 L	-	-	-	-	2.96	nr	-
Olearia haastii; 30-45 3 L	-	-	-	-	2.51	nr	-
Osmanthus delavayi; 30-45 3 L	-	-	-	-	3.65	nr	-
Pachysandra terminalis; 15-20 1.5 L	-	-	-	-	1.82	nr	-
Pernettya mucronata Male; 20-30 3 L	-	-	-	-	2.96	nr	-
Philadelphus Belle Etoile; 45-60 3 L	-	-	-	-	2.05	nr	-
Philadelphus coronarius Aureus; 30-45 3 L	-	-	-	-	3.18	nr	-
Philadelphus Virginal; 45-60 3 L	-	-	-	-	2.74	nr	-
Phormium tenax; 30-45 3 L	-	-	-	-	3.98	nr	-
Photinia x fraseri Red Robin; 30-45 3 L	-	-	-	-	4.09	nr	-
Potentilla arbuscula; 20-30 3 L	-	-	-	-	1.82	nr	-
Potentilla fruticosa Pretty Polly; 20-30 3 L	-	-	-	-	2.79	nr	-
Prunus lusitanica; 45-60 3 L	-	-	-	-	2.05	nr	-
Pyracantha Orange Glow; 60-90 3 L	-	-	-	-	2.05	nr	-
Pyracantha Teton; 30-45 3 L	-	-	-	-	2.51	nr	-
Ribes sanguineum King Edward VII; 45-60 3 L	-	-	-	-	2.39	nr	-
Rosmarinus officinalis; 20-30 3 L	-	-	-	-	2.51	nr	-
Rubus Betty Ashburner; 45-60 2 L	-	-	-	-	1.48	nr	-
Rubus Tridel Benenden; 45-60 3 L	-	-	-	-	2.74	nr	-
Salix lanata; 45-60 3 L	-	-	-	-	2.96	nr	-
Salvia officinalis Icterina; 20-30 3 L	-	-	-	-	2.96	nr	-
Sambucus nigra Aurea; 45-60 3 L	-	-	-	-	1.59	nr	-
Sambucus racemosa Plumosa Aurea; 45-60 3 L	-	-	-	-	4.50	nr	-
Santolina chamaecyparissus; 20-30 3 L	-	-	-	-	1.82	nr	-
Sarcococca confusa; 20-30 2 L	-	-	-	-	3.18	nr	-
Senecio monroi; 20-25 2 L	-	-	-	-	2.96	nr	-
Senecio Sunshine; 30-45 3 L	-	-	-	-	1.82	nr	-
Skimmia japonica; 30-45 3 L	-	-	-	-	2.96	nr	-
Spiraea x arguta; 45-60 3 L	-	-	-	-	2.05	nr	-
Spiraea bumalda x Goldflame; 20-30 2 L	-	-	-	-	1.82	nr	-
Stephanandra incisa Crispa; 30-45 2 L	-	-	-	-	1.82	nr	-
Symphoricarpos albus; 45-60 3 L	-	-	-	-	1.82	nr	-
Symphoricarpos x chenaultii Hancock; 45-60 2 L	-	-	-	-	1.38	nr	-
Tamarix tetrandra; 45-60 3 L	-	-	-	-	2.74	nr	-
Viburnum x bodnantense Dawn; 45-60 3 L	-	-	-	-	3.88	nr	-
Viburnum opulus; 60-90 3 L	-	-	-	-	1.82	nr	-
Viburnum opulus Sterile; 45-60 3 L	-	-	-	-	2.74	nr	-
Viburnum plicatum Pink Beauty; 45-60 3 L	-	-	-	-	4.89	nr	-
Viburnum tinus; 30-45 3 L	-	-	-	-	2.22	nr	-
Viburnum tinus Variegatum; 20-30 3 L	-	-	-	-	3.31	nr	-
Container grown climbers							
Clematis Jackmanii; 60-90 2 L	-	-	-	-	3.67	nr	-
Clematis montana Elizabeth; 60-90 2 L	-	-	-	-	3.67	nr	-
Hydrangea petiolaris; 45-60 3 L	-	-	-	-	4.01	nr	-
Jasminum nudiflorum; 45-60 2 L	-	-	-	-	2.94	nr	-
Lonicera japonica Halliana; 60-90 3 L	-	-	-	-	2.94	nr	-
Parthenocissus quinquefolia; 60-90 2 L	-	-	-	-	2.70	nr	-
Passiflora caerulea; 60-90 2 L	-	-	-	-	3.94	nr	-
Vitis coignetiae; 60-90 2 L	-	-	-	-	6.20	nr	-
Wisteria sinensis; 60-90 3 L	-	-	-	-	7.33	nr	-

Q PAVING/PLANTING/FENCING/SITE FURNITURE

Item Excluding site overheads and profit	PC £	Labour hours	Labour £	Plant £	Material £	Unit	**Total rate £**
Container grown specimen shrubs							
Amelanchier lamarckii; 90-120 10 L	-	-	-	-	9.69	nr	-
Aralia elata; 120-150 10 L	-	-	-	-	22.77	nr	-
Aucuba japonica; 45-60 10 L	-	-	-	-	10.83	nr	-
Berberis darwinii; 45-60 10 L	-	-	-	-	10.83	nr	-
Choisya ternata; 45-60 10 L	-	-	-	-	10.83	nr	-
Cotoneaster cornubia; 90-120 10 L	-	-	-	-	9.69	nr	-
Cotoneaster Exburiensis; 90-120 10 L	-	-	-	-	10.83	nr	-
Cotoneaster watereri; 90-120 10 L	-	-	-	-	9.69	nr	-
Elaeagnus pungens Maculata; 45-60 10 L	-	-	-	-	14.23	nr	-
Elaeagnus ebbingei; 60-90 10 L	-	-	-	-	9.69	nr	-
Elaeagnus ebbingei Limelight; 60-90 10 L	-	-	-	-	14.23	nr	-
Escallonia rubra Crimson Spire; 60-90 10 L	-	-	-	-	9.69	nr	-
Euonymus fortunei Emerald N Gold; 30-45 7.3 L	-	-	-	-	9.11	nr	-
Fatsia japonica; 90-120 15 L	-	-	-	-	14.23	nr	-
Ilex aquifolium; 90-120 10 L	-	-	-	-	10.83	nr	-
Ilex aquifolium Golden King; 90-120 10 L	-	-	-	-	17.08	nr	-
Ilex aquifolium Argentea Marginata; 90-120 10 L	-	-	-	-	17.08	nr	-
Mahonia japonica; 60-90 10 L	-	-	-	-	14.23	nr	-
Phormium tenax; 60-90 10 L	-	-	-	-	15.94	nr	-
Phormium tenax Purpureum; 60-90 10 L	-	-	-	-	15.94	nr	-
Phormium Yellow Wave; 60-90 10 L	-	-	-	-	20.49	nr	-
Photinia x Fraseri Red Robin; 60-90 10 L	-	-	-	-	14.23	nr	-
Prunus x cistena Crimson Dwarf; 45-60 10 L	-	-	-	-	14.23	nr	-
Prunus laurocerasus; 90-120 10 L	-	-	-	-	9.69	nr	-
Prunus laurocerasus Otto Luyken; 45-60 10 L	-	-	-	-	9.69	nr	-
Prunus laurocerasus Zabeliana; 45-60 10 L	-	-	-	-	9.69	nr	-
Prunus lusitanica; 45-60 10 L	-	-	-	-	10.83	nr	-
Pyracantha Mohave; 45-60 10 L	-	-	-	-	9.69	nr	-
Pyracantha Orange Glow; 60-90 10 L	-	-	-	-	9.69	nr	-
Pyracantha Red Column; 60-90 10 L	-	-	-	-	9.69	nr	-
Pyracantha Soleil d'Or; 60-90 10 L	-	-	-	-	9.69	nr	-
Pyracantha Teton; 45-60 10 L	-	-	-	-	9.69	nr	-
Rhus typhina; 90-120 10 L	-	-	-	-	9.69	nr	-
Rosmarinus officinalis; 30-45 10 L	-	-	-	-	9.69	nr	-
Rubus cockburnianus Golden Vale; 90-120 10 L	-	-	-	-	11.38	nr	-
Skimmia Rubella; 30-45 7.5 L	-	-	-	-	11.95	nr	-
Spiraea japonica Little Princess; 30-45 7.5 L	-	-	-	-	10.83	nr	-
Spiraea Snowmound; 45-60 10 L	-	-	-	-	9.69	nr	-
Spiraea thunbergii; 45-60 10 L	-	-	-	-	9.69	nr	-
Bamboos							
Arundinaria auricoma; 40-50 7.5 L	-	-	-	-	20.25	nr	-
Container grown conifers							
X Cupressocyparis leylandii; 120-150 10 L	-	-	-	-	8.45	nr	-
X Cupressocyparis leylandii Castlewellan; 120-150 10 L	-	-	-	-	8.45	nr	-
Juniperus communis Repanda; 30-45 10 L	-	-	-	-	9.59	nr	-
Juniperus x media pfitzeriana Aurea; 45-60 10L	-	-	-	-	9.59	nr	-
Juniperus sabina Tamariscifolia; 45-60 10 L	-	-	-	-	9.59	nr	-
Pinus mugo; 45-60 10 L	-	-	-	-	14.10	nr	-
Pinus sylvestris; 90-120 10 L	-	-	-	-	9.59	nr	-
Pinus sylvestris; 150-180 50 L	-	-	-	-	67.65	nr	-
Pinus sylvestris; 180-210 80 L	-	-	-	-	118.40	nr	-
Chamaecyparis lawsoniana Ellwoodii; 30-45 2 L	-	-	-	-	2.49	nr	-
Chamaecyparis lawsoniana Ellwoods Gold; 30-45 2 L	-	-	-	-	3.28	nr	-
Chamaecyparis lawsoniana Ellwoods Pillar; 30-45 2 L	-	-	-	-	3.28	nr	-
X Cupressocyparis leylandii; 60-90 3 L	-	-	-	-	2.03	nr	-
X Cupressocyparis leylandii Castlewellan; 60-90 3 L	-	-	-	-	2.03	nr	-
Juniperus communis Compressa; 30-45 2 L	-	-	-	-	2.94	nr	-
Juniperus communis Repanda; 30-45 3 L	-	-	-	-	2.04	nr	-

Q PAVING/PLANTING/FENCING/SITE FURNITURE

Item Excluding site overheads and profit	PC £	Labour hours	Labour £	Plant £	Material £	Unit	Total rate £
Q31 PLANTING - cont'd							
Market prices of trees, shrubs and plants; Notcutts Nurseries Ltd - cont'd							
Juniperus conferta; 30-45 3 L	-	-	-	-	3.15	nr	-
Juniperus media Mint Julep; 30-45 3 L	-	-	-	-	3.28	nr	-
Juniperus media pfitzeriana; 30-45 3 L	-	-	-	-	3.04	nr	-
Juniperus media pfitzeriana Aurea; 30-45 3 L	-	-	-	-	3.28	nr	-
Juniperus sabina Tamariscifolia; 30-45 3 L	-	-	-	-	3.04	nr	-
Juniperus squamata Blue Star; 30-45 2 L	-	-	-	-	3.62	nr	-
Larix decidua; 60-90 3 L	-	-	-	-	1.81	nr	-
Picea albertiana Conica; 20-30 3 L	-	-	-	-	3.15	nr	-
Pinus mugo; 20-30 2 L	-	-	-	-	2.49	nr	-
Pinus nigra; 45-60 3 L	-	-	-	-	2.15	nr	-
Pinus nigra austriaca; 45-60 3 L	-	-	-	-	1.81	nr	-
Pinus nigra maritima; 45-60 3 L	-	-	-	-	1.81	nr	-
Pinus sylvestris; 45-60 3 L	-	-	-	-	1.81	nr	-
Transplants 45-60							
Acer campestre	-	-	-	-	0.22	nr	-
Acer platanoides	-	-	-	-	0.22	nr	-
Acer pseudoplatanus	-	-	-	-	0.24	nr	-
Alnus cordata	-	-	-	-	0.24	nr	-
Alnus glutinosa	-	-	-	-	0.24	nr	-
Alnus incana	-	-	-	-	0.24	nr	-
Betula pendula	-	-	-	-	0.24	nr	-
Carpinus betulus	-	-	-	-	0.27	nr	-
Cornus sanguinea	-	-	-	-	0.27	nr	-
Corylus avellana	-	-	-	-	0.27	nr	-
Crataegus monogyna	-	-	-	-	0.17	nr	-
Fagus sylvatica	-	-	-	-	0.29	nr	-
Fraxinus excelsior	-	-	-	-	0.27	nr	-
Hippophae rhamnoides	-	-	-	-	0.31	nr	-
Prunus avium	-	-	-	-	0.27	nr	-
Quercus robur	-	-	-	-	0.31	nr	-
Rosa rugosa	-	-	-	-	0.29	nr	-
Rosa rugosa Alba	-	-	-	-	0.29	nr	-
Sambucus nigra	-	-	-	-	0.27	nr	-
Sorbus aria	-	-	-	-	0.33	nr	-
Sorbus intermedia	-	-	-	-	0.35	nr	-
Viburnum opulus	-	-	-	-	0.35	nr	-
Transplants 60-90							
Acer campestre	-	-	-	-	0.24	nr	-
Acer platanoides	-	-	-	-	0.27	nr	-
Alnus cordata	-	-	-	-	0.27	nr	-
Amelanchier canadensis	-	-	-	-	0.31	nr	-
Cornus alba	-	-	-	-	0.27	nr	-
Crataegus monogyna	-	-	-	-	0.27	nr	-
Fagus sylvatica	-	-	-	-	0.35	nr	-
Fraxinus excelsior	-	-	-	-	0.29	nr	-
Populus alba	-	-	-	-	0.29	nr	-
Populus nigra italica	-	-	-	-	0.31	nr	-
Populus trichocarpa	-	-	-	-	0.31	nr	-
Prunus avium	-	-	-	-	0.31	nr	-
Salix alba	-	-	-	-	0.24	nr	-
Salix caprea	-	-	-	-	0.29	nr	-
Salix fragilis	-	-	-	-	0.27	nr	-
Salix rosmarinifolia	-	-	-	-	0.27	nr	-
Salix viminalis	-	-	-	-	0.27	nr	-
Salix vitellina	-	-	-	-	0.27	nr	-
Sorbus aucuparia	-	-	-	-	0.29	nr	-
Herbaceous							
Achillea fil. Cloth of Gold; 10-15 2 L	-	-	-	-	2.04	nr	-
Alchemilla mollis; 10-15 2 L	-	-	-	-	2.04	nr	-
Anemone hybrida Honorine Jobert; 10-15 3 L	-	-	-	-	2.50	nr	-

Q PAVING/PLANTING/FENCING/SITE FURNITURE

Item Excluding site overheads and profit	PC £	Labour hours	Labour £	Plant £	Material £	Unit	**Total rate £**
Artemisia absin. Lambrook Silver; 10-15 3 L	-	-	-	-	2.78	nr	-
Artemisia Powis Castle; 10-15 3 L	-	-	-	-	2.78	nr	-
Bergenia Bressingham Salmon; 20-30 2 L	-	-	-	-	2.55	nr	-
Convallaria majalis; 10-15 2 L	-	-	-	-	2.04	nr	-
Euphorbia amygdaloides Purpurea; 15-20 2 L	-	-	-	-	2.50	nr	-
Euphorbia wulfenii; 10-15 2 L	-	-	-	-	2.95	nr	-
Geranium oxonianum Wargrave Pink; 10-15 3 L	-	-	-	-	2.04	nr	-
Geranium Johnson Blue; 10-15 3 L	-	-	-	-	2.04	nr	-
Hosta August Moon; 10-15 2 L	-	-	-	-	2.83	nr	-
Hosta sieboldii; 10-15 2 L	-	-	-	-	2.83	nr	-
Iris foetidissima; 10-15 2 L	-	-	-	-	2.50	nr	-
Papaver orientale; 10-15 3 L	-	-	-	-	2.04	nr	-
Penstemon Garnet; 10-15 3 L	-	-	-	-	2.55	nr	-
Pulmonaria angustifolia; 10-15 2 L	-	-	-	-	2.04	nr	-
Salvia East Friesland; 10-15 3 L	-	-	-	-	2.04	nr	-
Sedum spectabilis Autumn Joy; 10-15 3 L	-	-	-	-	2.04	nr	-
Tiarella cordifolia; 10-15 2 L	-	-	-	-	2.04	nr	-
Waldsteinia ternata; 10-15 2 L	-	-	-	-	2.04	nr	-
Bare root trees 1.50 - 1.80 feathered							
Acer campestre	-	-	-	-	2.27	nr	-
Acer platanoides	-	-	-	-	1.72	nr	-
Acer pseudoplatanus	-	-	-	-	1.72	nr	-
Alnus cordata	-	-	-	-	1.72	nr	-
Alnus glutinosa	-	-	-	-	1.72	nr	-
Alnus incana	-	-	-	-	3.41	nr	-
Amelanchier canadensis	-	-	-	-	6.83	nr	-
Betula pendula	-	-	-	-	1.72	nr	-
Carpinus betulus	-	-	-	-	2.85	nr	-
Fagus sylvatica	-	-	-	-	3.98	nr	-
Fagus sylvatica purpurea	-	-	-	-	6.26	nr	-
Fraxinus excelsior	-	-	-	-	2.85	nr	-
Populus alba	-	-	-	-	1.72	nr	-
Populus nigra italica	-	-	-	-	1.72	nr	-
Populus tremula	-	-	-	-	6.26	nr	-
Prunus avium	-	-	-	-	2.85	nr	-
Prunus avium 'Plena'	-	-	-	-	6.83	nr	-
Quercus robur	-	-	-	-	10.83	nr	-
Quercus rubra	-	-	-	-	5.12	nr	-
Sorbus lutescens	-	-	-	-	3.41	nr	-
Sorbus aucuparia	-	-	-	-	2.27	nr	-
Bare root trees 1.80 - 2.10 feathered							
Acer campestre	-	-	-	-	2.27	nr	-
Alnus cordata	-	-	-	-	3.41	nr	-
Acer platanoides	-	-	-	-	3.41	nr	-
Acer pseudoplatanus	-	-	-	-	3.41	nr	-
Alnus glutinosa	-	-	-	-	3.41	nr	-
Alnus incana	-	-	-	-	2.85	nr	-
Amelanchier canadensis	-	-	-	-	13.66	nr	-
Betula pendula	-	-	-	-	3.41	nr	-
Carpinus betulus	-	-	-	-	5.12	nr	-
Fagus sylvatica	-	-	-	-	6.26	nr	-
Fagus sylvatica purpurea	-	-	-	-	10.83	nr	-
Fraxinus excelsior	-	-	-	-	4.55	nr	-
Populus alba	-	-	-	-	3.98	nr	-
Populus nigra italica	-	-	-	-	3.98	nr	-
Populus tremula	-	-	-	-	3.98	nr	-
Prunus avium	-	-	-	-	4.55	nr	-
Prunus avium 'Plena'	-	-	-	-	9.11	nr	-
Quercus robur	-	-	-	-	10.83	nr	-
Quercus rubra	-	-	-	-	11.95	nr	-
Sorbus aucuparia	-	-	-	-	5.12	nr	-
Sorbus lutescens	-	-	-	-	10.25	nr	-

Q PAVING/PLANTING/FENCING/SITE FURNITURE

Item Excluding site overheads and profit	PC £	Labour hours	Labour £	Plant £	Material £	Unit	Total rate £
Q31 PLANTING - cont'd							
Market prices of trees, shrubs and plants; Notcutts Nurseries Ltd - cont'd							
Bare root trees 8-10 cm							
Acer campestre	-	-	-	-	12.52	nr	-
Acer platanoides	-	-	-	-	10.25	nr	-
Acer pseudoplatanus	-	-	-	-	10.25	nr	-
Alnus cordata	-	-	-	-	10.25	nr	-
Alnus glutinosa	-	-	-	-	10.25	nr	-
Alnus incana	-	-	-	-	10.25	nr	-
Amelanchier canadensis	-	-	-	-	28.47	nr	-
Betula pendula	-	-	-	-	9.11	nr	-
Carpinus betulus	-	-	-	-	13.66	nr	-
Fagus sylvatica	-	-	-	-	13.66	nr	-
Fraxinus excelsior	-	-	-	-	9.00	nr	-
Fagus sylvatica Purpurea	-	-	-	-	22.77	nr	-
Populus tremula	-	-	-	-	12.52	nr	-
Populus alba	-	-	-	-	10.83	nr	-
Populus nigra Italica	-	-	-	-	10.83	nr	-
Prunus avium	-	-	-	-	11.38	nr	-
Prunus avium Plena	-	-	-	-	14.81	nr	-
Quercus robor	-	-	-	-	13.11	nr	-
Quercus rubra	-	-	-	-	19.36	nr	-
Sorbus aucuparia	-	-	-	-	12.52	nr	-
Bare root trees 10-12 cm sel. standard							
Acer campestre	-	-	-	-	15.94	nr	-
Alnus cordata	-	-	-	-	11.38	nr	-
Acer platanoides	-	-	-	-	11.38	nr	-
Acer pseudoplatanus	-	-	-	-	11.38	nr	-
Amelanchier canadensis	-	-	-	-	45.55	nr	-
Betula pendula	-	-	-	-	11.38	nr	-
Carpinus betulus	-	-	-	-	26.70	nr	-
Fagus sylvatica	-	-	-	-	20.49	nr	-
Fagus sylvatica purpurea	-	-	-	-	34.16	nr	-
Fraxinus excelsior	-	-	-	-	13.66	nr	-
Populus alba	-	-	-	-	14.23	nr	-
Populus nigra italica	-	-	-	-	14.23	nr	-
Populus tremula	-	-	-	-	18.22	nr	-
Prunus avium	-	-	-	-	13.66	nr	-
Prunus avium 'Plena'	-	-	-	-	18.22	nr	-
Quercus robur	-	-	-	-	23.92	nr	-
Quercus rubra	-	-	-	-	23.92	nr	-
Sorbus aucuparia	-	-	-	-	17.08	nr	-
Sorbus lutescens	-	-	-	-	20.49	nr	-
Bare root trees 12-14 cm heavy standard							
Acer campestre	-	-	-	-	27.33	nr	-
Alnus cordata	-	-	-	-	22.77	nr	-
Acer platanoides	-	-	-	-	22.77	nr	-
Acer pseudoplatanus	-	-	-	-	22.77	nr	-
Alnus glutinosa	-	-	-	-	22.77	nr	-
Alnus incana	-	-	-	-	22.77	nr	-
Amelanchier canadensis	-	-	-	-	79.70	nr	-
Betula pendula	-	-	-	-	22.77	nr	-
Carpinus betulus	-	-	-	-	31.88	nr	-
Fagus sylvatica	-	-	-	-	31.88	nr	-
Fagus sylvatica Purpurea	-	-	-	-	51.23	nr	-
Fraxinus excelsior	-	-	-	-	27.33	nr	-
Populus alba	-	-	-	-	25.05	nr	-
Populus nigra italica	-	-	-	-	14.23	nr	-
Populus tremula	-	-	-	-	18.22	nr	-
Prunus avium	-	-	-	-	27.33	nr	-
Prunus avium 'Plena'	-	-	-	-	31.88	nr	-

Q PAVING/PLANTING/FENCING/SITE FURNITURE

Item Excluding site overheads and profit	PC £	Labour hours	Labour £	Plant £	Material £	Unit	**Total rate £**
Quercus robur	-	-	-	-	36.44	nr	-
Quercus rubra	-	-	-	-	36.44	nr	-
Sorbus aucuparia	-	-	-	-	25.05	nr	-
Sorbus lutescens	-	-	-	-	34.16	nr	-
Market prices of trees; James Coles & Sons (Nurseries) Ltd							
Bare root trees 14-16 cm extra heavy standard							
Acer campestre	-	-	-	-	72.00	nr	-
Acer platanoides	-	-	-	-	16.50	nr	-
Acer pseudoplatanus	-	-	-	-	33.50	nr	-
Alnus glutinosa	-	-	-	-	37.00	nr	-
Alnus incana	-	-	-	-	37.00	nr	-
Betula pendula	-	-	-	-	40.00	nr	-
Carpinus betulus	-	-	-	-	49.00	nr	-
Fraxinus excelsior	-	-	-	-	33.50	nr	-
Prunus avium	-	-	-	-	36.50	nr	-
Prunus avium 'Plena'	-	-	-	-	50.40	nr	-
Quercus robur	-	-	-	-	44.80	nr	-
Quercus rubra	-	-	-	-	44.80	nr	-
Sorbus aucuparia	-	-	-	-	49.00	nr	-
Sorbus aria	-	-	-	-	72.80	nr	-
Rootball trees; extra over for rootballing and wrapping in hessian	-	-	-	-	15.00	nr	-
Market prices of aquatic plants in prepared growing medium of approved hessian bag filled with soil and fertilizer; tied, slit and weighted and placed in position in pool to receive three submerged floating leafed marginal or swamp plants; Anglo Aquarium Plant Co Ltd							
Water lilies							
Nymphaea in variety - white	-	-	-	-	8.15	nr	-
Nymphaea in variety - deep red	-	-	-	-	8.15	nr	-
Deep water aquatics							
Aponogeton distachyos	-	-	-	-	2.80	nr	-
Nymphoides peltata	-	-	-	-	2.80	nr	-
Marginal plants							
Acorus calamus	-	-	-	-	1.96	nr	-
Butomus umbellatus	-	-	-	-	1.96	nr	-
Caltha palustris	-	-	-	-	1.96	nr	-
Carex acuta	-	-	-	-	1.96	nr	-
Iris pseudacorus	-	-	-	-	1.96	nr	-
Juncus effusus	-	-	-	-	1.96	nr	-
Mentha aquatica	-	-	-	-	1.96	nr	-
Mimulus luteus	-	-	-	-	1.96	nr	-
Phragmites (australis)	-	-	-	-	1.96	nr	-
Ranunculus lingua	-	-	-	-	1.96	nr	-
Scirpus lacustris	-	-	-	-	1.96	nr	-
Sagittaria sagittifolia var. leucopetala	-	-	-	-	1.96	nr	-
Typha latifolia	-	-	-	-	1.96	nr	-
Submerged oxygenating plants							
Ceratophyllum demersum; containerised	-	-	-	-	36.00	100	-
Lagarosiphon major (Elodea crispa); containerised	-	-	-	-	36.00	100	-
Ranunculus aquatilis; containerised	-	-	-	-	36.00	100	-

Q PAVING/PLANTING/FENCING/SITE FURNITURE

Item Excluding site overheads and profit	PC £	Labour hours	Labour £	Plant £	Material £	Unit	Total rate £
Q31 PLANTING - cont'd							
Market prices of bulbs; prices per each; purchased in the quantities shown; autumn bulbs							
Dutch Hyacinths (prepared for early flowering); 17-18 cm							
"Anne Marie", rose-pink to salmon-pink	-	-	-	-	32.00	100	-
"Delft Blue", porcelain blue	-	-	-	-	32.00	100	-
"Ostara", deep navy blue	-	-	-	-	34.00	100	-
"White Pearl", pure white	-	-	-	-	35.00	100	-
Dutch Hyacinths (not prepared); 16-17 cm							
"Anne Marie", rose-pink to salmon-pink, early	-	-	-	-	26.00	100	-
"City of Haarlem", primrose yellow, late	-	-	-	-	24.00	100	-
"Delft Blue", porcelain blue, medium	-	-	-	-	25.00	100	-
"Ostara", deep navy blue, early	-	-	-	-	25.00	100	-
"White Pearl", pure white, early	-	-	-	-	28.00	100	-
Tulips, Single Early, flowering mid-April							
"Apricot Beauty", soft salmon pink	-	-	-	-	145.00	1000	-
"Purple Prince", mallow purple	-	-	-	-	110.00	1000	-
Tulips, Double Early, flowering late April							
"Electra", deep cherry red	-	-	-	-	170.00	1000	-
"Mr van der Hoef", canary yellow	-	-	-	-	190.00	1000	-
"Mixture"	-	-	-	-	170.00	1000	-
Tulips, Triumph, flowering late April to May							
"Abba", glowing tomato red	-	-	-	-	95.00	1000	-
"Arabian Mystery", rich violet-purple, edged white	-	-	-	-	160.00	1000	-
Tulips, Darwin Hybrid, flowering late April to May							
"Golden Apeldoorn", golden yellow, black and bronze-green centre	-	-	-	-	80.00	1000	-
"Pink Impression", rose pink	-	-	-	-	110.00	1000	-
"Darwin Hybrid Mixture"	-	-	-	-	95.00	1000	-
Tulips, Single Late, flowering early May							
"Blushing Girl", creamy white, deep maroon red edge	-	-	-	-	170.00	1000	-
"Maureen", marble white	-	-	-	-	120.00	1000	-
Tulips, Lily Flowered, flowering early May							
"Aladdin", scarlet, yellow edge	-	-	-	-	120.00	1000	-
"Ballerina", soft orange and lemon yellow	-	-	-	-	85.00	1000	-
Tulips, Parrot, flowering mid-May							
"Apricot Parrot", pale apricot	-	-	-	-	150.00	1000	-
"Rococo", velvet red, fiery red edges, green crests	-	-	-	-	140.00	1000	-
Tulips, Double Late, flowering late April to May							
"Angelique", soft pink	-	-	-	-	155.00	1000	-
"Creme Upstar", primrose yellow, magenta-rose margins	-	-	-	-	110.00	1000	-
Tulips, Greigii, flowering March to early April							
"Cape Cod", apricot, yellow edges	-	-	-	-	95.00	1000	-
"Mary Ann", carmine red, white edges	-	-	-	-	120.00	1000	-
Narcissi, Daffodils							
"Dutch Master", golden yellow perianth and trumpet, March	-	-	-	-	85.00	1000	-
"Magnet", creamy white perianth, bright yellow trumpet, April	-	-	-	-	100.00	1000	-
"Mount Hood", pure white perianth, ivory white trumpet, April	-	-	-	-	95.00	1000	-
"Yellow River", large flowered, pure yellow	-	-	-	-	190.00	1000	-
Narcissi, Daffodils, dwarf							
"Little Beauty", white perianth, yellow trumpet, March to April	-	-	-	-	175.00	1000	-
"Little Gem", yellow perianth and trumpet, March to April	-	-	-	-	100.00	1000	-
"Topolino", creamy white perianth, pale lemon trumpet, March to April	-	-	-	-	100.00	1000	-

Q PAVING/PLANTING/FENCING/SITE FURNITURE

Item Excluding site overheads and profit	PC £	Labour hours	Labour £	Plant £	Material £	Unit	**Total rate £**
Narcissi, Large Cupped							
"Big Gun", large flowered, white perianth, golden yellow cup, April	-	-	-	-	280.00	1000	-
"Fragrant Rose", creamy white perianth, deep pink cup, April	-	-	-	-	385.00	1000	-
"St. Patricks Day", lemon yellow, March to April	-	-	-	-	85.00	1000	-
Narcissi, Small Cupped							
"Audubon", white perianth, pale primrose cup with dark pink edges	-	-	-	-	450.00	1000	-
"Verger", white perianth, lemon yellow cup with orange rim, April	-	-	-	-	80.00	1000	-
Narcissi, Double							
"Pencrebar", dwarf, deep golden yellow, April	-	-	-	-	200.00	1000	-
"White Lion", white with soft yellow central petals, April	-	-	-	-	85.00	1000	-
Narcissi, Triandrus							
"Ice Wings", multi-headed, pure white, April	-	-	-	-	230.00	1000	-
"Thalia", 3-4 pure white flowers per stem, April	-	-	-	-	180.00	1000	-
Narcissi, Cyclamineus							
"Jack Snipe", creamy white perianth, primrose cup, long-lasting, March to April	-	-	-	-	160.00	1000	-
Narcissi, Jonquilla							
"Fruit Cup", white with soft yellow cup, April	-	-	-	-	125.00	1000	-
"Sweetness", deep golden yellow, bowl shaped cups, April	-	-	-	-	70.00	1000	-
Narcissi, Tazetta							
"Martinette", yellow with red cups, scented, March to April	-	-	-	-	110.00	1000	-
"Minnow", 2-4 flowers per stem, creamy white with flat yellow cups	-	-	-	-	45.00	1000	-
Crocus, Autumn Flowering							
"Niveus", pure white with scarlet stigma, 6-9 cm	-	-	-	-	400.00	1000	-
"Speciosus", rich violet-blue, 12-15 cm	-	-	-	-	40.00	1000	-
Crocus, Dutch Large Spring Flowering							
"Golden Yellow", 8-9 cm	-	-	-	-	37.00	1000	-
"Pickwick", silvery white, striped with purple, 8-9 cm	-	-	-	-	50.00	1000	-
"Remembrance", purple-blue, 8-9 cm	-	-	-	-	55.00	1000	-
"Mixed", 9-10 cm	-	-	-	-	55.00	1000	-
Crocus, Winter and Spring Flowering							
"Ancyrensis Golden Bunch", bright orange-yellow, 4-10cm	-	-	-	-	26.00	1000	-
"Sieberi Violet Queen", amethyst violet with deep yellow centre, 4-10 cm	-	-	-	-	45.00	1000	-
Crocus, Crysanthus, Spring Flowering							
"Advance", yellow and violet	-	-	-	-	40.00	1000	-
"Mixed"	-	-	-	-	23.00	1000	-
Iris, Dutch							
"Bronze Beauty", bronzy orange	-	-	-	-	85.00	1000	-
"Gipsy Beauty", purple blue standards, veined yellow falls	-	-	-	-	70.00	1000	-
"Silvery Beauty", deep blue standards, white falls with orange blotch	-	-	-	-	70.00	1000	-
Iris, Species							
"Danfordiae", bright canary yellow, scented	-	-	-	-	45.00	1000	-
"Reticulata", violet with yellow blotch	-	-	-	-	50.00	1000	-
"Reticulata Harmony", blue standards, blue and yellow falls with white-edged blotch	-	-	-	-	50.00	1000	-
"Reticulata Mixture"	-	-	-	-	42.00	1000	-

Item Excluding site overheads and profit	PC £	Labour hours	Labour £	Plant £	Material £	Unit	**Total rate £**
Q31 PLANTING - cont'd							
Market prices of bulbs; prices per each; purchased in the quantities shown; spring bulbs							
Allium							
"Giganteum", giant head of pale violet, height 120-150 cm	-	-	-	-	200.00	100	-
"Neapolitanum", pure white, height 50-60 cm	-	-	-	-	15.00	1000	-
Amaryllis							
"Belladonna", pale rosy-pink, 50-75 cm	-	-	-	-	11.00	10	-
"Parkeri Alba", pure white, 40-60 cm	-	-	-	-	21.00	10	-
Anemones							
"Blanda Pink Star", large bright pink flowers, 9-12 cm	-	-	-	-	110.00	1000	-
"De Caen", single flower, mixed colours, 25-30 cm	-	-	-	-	50.00	1000	-
Freesia							
Single Flowering Mixture	-	-	-	-	28.00	1000	-
Double Flowering Mixture	-	-	-	-	30.00	1000	-
Galanthus (Snowdrop)							
"Nivalis", single	-	-	-	-	55.00	1000	-
"Nivalis Flore Pleno", double	-	-	-	-	180.00	1000	-
Muscari (Grape Hyacinth)							
"Armeniacum", bright blue	-	-	-	-	30.00	1000	-
"Botryoides Album", white	-	-	-	-	85.00	1000	-
Ranunculus							
"Asiaticus Mixture", paeony flowered	-	-	-	-	100.00	1000	-
Scilla							
"Campanulata Blue", The Spanish Bluebell (Hyacinthoides hispanica blue)	-	-	-	-	110.00	1000	-
Sparaxis							
"Tricolor Mixture"	-	-	-	-	25.00	1000	-
Planting - General							
Preamble: The British Standard recommendations for Nursery Stock are in BS 3936: pts 1-5. Planting of trees and shrubs as well as forestry is covered by the appropriate sections of BS 4428. Transplanting of semi-mature trees is covered by BS 4043. Prices for all planting work are deemed to include carrying out planting in accordance with BS 4428 and good horticultural practice.							
Site protection; temporary protective fencing							
Cleft chestnut rolled fencing; to 100 mm diameter chestnut posts; driving into firm ground at 3 m centres; pales at 50 mm centres							
900 mm high	3.32	0.11	1.97	-	6.14	m	**8.11**
1100 mm high	3.90	0.11	1.97	-	6.56	m	**8.53**
1500 mm high	6.18	0.11	1.97	-	8.84	m	**10.81**
Extra over temporary protective fencing for removing and making good (no allowance for re-use of material)	-	0.07	1.23	0.20	-	m	**1.44**
Cultivation							
Treating soil with "Paraquat-Diquat" weedkiller at rate of 5 litre/ha; PC £9.13 per litre; in accordance with manufacturer's instructions; including all safety precautions							
by machine	-	-	-	0.25	0.46	100 m^2	**0.71**
by hand	-	0.40	7.40	-	0.46	100 m^2	**7.86**

Q PAVING/PLANTING/FENCING/SITE FURNITURE

Item Excluding site overheads and profit	PC £	Labour hours	Labour £	Plant £	Material £	Unit	**Total rate £**
Ripping up subsoil; using approved subsoiling machine; minimum depth 250 mm below topsoil; at 1.20 m centres; in							
gravel or sandy clay	-	-	-	2.40	-	100 m^2	**2.40**
soil compacted by machines	-	-	-	2.80	-	100 m^2	**2.80**
clay	-	-	-	3.00	-	100 m^2	**3.00**
chalk or other soft rock	-	-	-	6.01	-	100 m^2	**6.01**
Extra for subsoiling at 1 m centres	-	-	-	0.60	-	100 m^2	**0.60**
Breaking up existing ground; using pedestrian operated tine cultivator or rotavator							
100 mm deep	-	0.22	4.07	1.90	-	100 m^2	**5.97**
150 mm deep	-	0.28	5.09	2.37	-	100 m^2	**7.46**
200 mm deep	-	0.37	6.78	3.16	-	100 m^2	**9.95**
As above but in heavy clay or wet soils							
100 mm deep	-	0.44	8.14	3.80	-	100 m^2	**11.94**
150 mm deep	-	0.66	12.21	5.70	-	100 m^2	**17.91**
200 mm deep	-	0.82	15.26	7.12	-	100 m^2	**22.38**
Breaking up existing ground; using tractor drawn tine cultivator or rotavator							
100 mm deep	-	-	-	0.27	-	100 m^2	**0.27**
150 mm deep	-	-	-	0.34	-	100 m^2	**0.34**
200 mm deep	-	-	-	0.45	-	100 m^2	**0.45**
600 mm deep	-	-	-	1.36	-	100 m^2	**1.36**
Cultivating ploughed ground; using disc, drag, or chain harrow							
4 passes	-	-	-	1.63	-	100 m^2	**1.63**
Rolling cultivated ground lightly; using self-propelled agricultural roller	-	0.06	1.03	0.60	-	100 m^2	**1.63**
Importing only selected and approved topsoil; from source not exceeding 13 km from site							
1 - 14 m^3	18.50	-	-	-	55.50	m^3	-
over 15 m^3	18.50	-	-	-	18.50	m^3	-
Spreading and lightly consolidating approved topsoil (imported or from spoil heaps); in layers not exceeding 150 mm; travel distance from spoil heaps not exceeding 100 m; by machine (imported topsoil not included)							
minimum depth 100 mm	-	1.55	28.68	37.20	-	100 m^2	**65.87**
minimum depth 150 mm	-	2.33	43.17	55.98	-	100 m^2	**99.15**
minimum depth 300 mm	-	4.67	86.33	111.96	-	100 m^2	**198.29**
minimum depth 450 mm	-	6.99	129.31	167.76	-	100 m^2	**297.08**
Spreading and lightly consolidating approved topsoil (imported or from spoil heaps); in layers not exceeding 150 mm; travel distance from spoil heaps not exceeding 100 m; by hand (imported topsoil not included)							
minimum depth 100 mm	-	20.00	370.07	-	-	100 m^2	**370.07**
minimum depth 150 mm	-	30.01	555.11	-	-	100 m^2	**555.11**
minimum depth 300 mm	-	60.01	1110.22	-	-	100 m^2	**1110.22**
minimum depth 450 mm	-	90.02	1665.33	-	-	100 m^2	**1665.33**
Extra over for spreading topsoil to slopes 15 - 30 degrees by machine or hand	-	-	-	-	-	10%	-
Extra over for spreading topsoil to slopes over 30 degrees by machine or hand	-	-	-	-	-	25%	-
Extra over spreading topsoil for travel exceeding 100 m; by machine							
100 - 150 m	-	0.02	0.37	0.12	-	m^3	**0.49**
150 - 200 m	-	0.03	0.49	0.16	-	m^3	**0.65**
200 - 300 m	-	0.04	0.66	0.21	-	m^3	**0.87**
Extra over spreading topsoil for travel exceeding 100 m; by hand							
100 m	-	2.50	46.25	-	-	m^3	**46.25**
200 m	-	3.50	64.75	-	-	m^3	**64.75**
300 m	-	4.50	83.25	-	-	m^3	**83.25**

Q PAVING/PLANTING/FENCING/SITE FURNITURE

Item Excluding site overheads and profit	PC £	Labour hours	Labour £	Plant £	Material £	Unit	Total rate £
Q31 PLANTING - cont'd							
Cultivation - cont'd							
Evenly grading; to general surfaces to bring to finished levels							
by machine (tractor mounted rotavator)	-	-	-	0.01	-	m^2	**0.01**
by pedestrian operated rotavator	-	-	0.07	0.04	-	m^2	**0.11**
by hand	-	0.01	0.19	-	-	m^2	**0.19**
Extra over grading for slopes 15 - 30 degrees by machine or hand	-	-	-	-	-	10%	-
Extra over grading for slopes over 30 degrees by machine or hand	-	-	-	-	-	25%	-
Clearing stones; disposing off site; to distance not exceeding 13 km							
by hand; stones not exceeding 50 mm in any direction; loading to skip 4.6 m^3	-	0.01	0.19	0.03	-	m^2	**0.22**
by mechanical stone rake; stones not exceeding 50 mm in any direction; loading to 15 m^3 truck by mechanical loader	-	-	0.04	0.07	-	m^2	**0.11**
Lightly cultivating; weeding; to fallow areas; disposing debris off site; to distance not exceeding 13 km							
by hand	-	0.01	0.26	-	0.07	m^2	**0.34**
Preparation of planting operations							
For the following topsoil improvement and planting operations add or subtract the following amounts for every £0.10 difference in the material cost price							
35 g/m^2	-	-	-	-	0.35	100 m^2	-
50 g/m^2	-	-	-	-	0.50	100 m^2	-
70 g/m^2	-	-	-	-	0.70	100 m^2	-
100 g/m^2	-	-	-	-	1.00	100 m^2	-
150 kg/ha	-	-	-	-	15.00	ha	-
200 kg/ha	-	-	-	-	20.00	ha	-
250 kg/ha	-	-	-	-	25.00	ha	-
300 kg/ha	-	-	-	-	30.00	ha	-
400 kg/ha	-	-	-	-	40.00	ha	-
500 kg/ha	-	-	-	-	50.00	ha	-
700 kg/ha	-	-	-	-	70.00	ha	-
1000 kg/ha	-	-	-	-	100.00	ha	-
1250 kg/ha	-	-	-	-	125.00	ha	-
Selective herbicides; in accordance with manufacturer's instructions; PC £29.23 per litre; by machine; application rate							
30 ml/100 m^2	-	-	-	0.30	0.88	100 m^2	**1.18**
35 ml/100 m^2	-	-	-	0.30	1.02	100 m^2	**1.32**
40 ml/100 m^2	-	-	-	0.30	1.17	100 m^2	**1.47**
50 ml/100 m^2	-	-	-	0.30	1.46	100 m^2	**1.76**
3.00 l/ha	-	-	-	30.01	87.67	ha	**117.69**
3.50 l/ha	-	-	-	30.01	102.29	ha	**132.30**
4.00 l/ha	-	-	-	30.01	116.90	ha	**146.91**
Selective herbicides; in accordance with manufacturer's instructions; PC £29.23 per litre; by hand; application rate							
30 ml/m^2	-	0.17	3.08	-	0.88	100 m^2	**3.96**
35 ml/m^2	-	0.17	3.08	-	1.02	100 m^2	**4.11**
40 ml/m^2	-	0.17	3.08	-	1.17	100 m^2	**4.25**
50 ml/m^2	-	0.17	3.08	-	1.46	100 m^2	**4.55**
3.00 l/ha	-	16.67	308.33	-	87.67	ha	**396.01**
3.50 l/ha	-	16.67	308.33	-	102.29	ha	**410.62**
4.00 l/ha	-	16.67	308.34	-	116.90	ha	**425.24**

Q PAVING/PLANTING/FENCING/SITE FURNITURE

Item Excluding site overheads and profit	PC £	Labour hours	Labour £	Plant £	Material £	Unit	**Total rate £**
General herbicides; in accordance with manufacturer's instructions; "Knapsack" spray application; see "market rates of chemicals" for costs of specific applications							
knapsack sprayer; selective spraying around bases of plants	-	0.35	6.49	-	-	100 m^2	**6.49**
knapsack sprayer; spot spraying	-	0.20	3.70	-	-	100 m^2	**3.70**
knapsack sprayer; mass spraying	-	0.20	3.70	-	-	100 m^2	**3.70**
granular distribution by hand or hand applicator	-	0.40	7.40	-	-	100 m^2	**7.40**
Fertilizers; in top 150 mm of topsoil; at 35 g/m^2							
fertilizer (18:0:0+Mg+Fe)	7.40	0.12	2.27	-	7.40	100 m^2	**9.67**
"Enmag"	6.30	0.12	2.27	-	6.61	100 m^2	**8.88**
fertilizer (7:7:7)	2.16	0.12	2.27	-	2.16	100 m^2	**4.43**
"Super Phosphate Powder"	2.13	0.12	2.27	-	2.23	100 m^2	**4.50**
fertilizer (20:10:10)	2.83	0.12	2.27	-	2.83	100 m^2	**5.09**
"Bone meal"	3.23	0.12	2.27	-	3.39	100 m^2	**5.66**
Fertilizers; in top 150 mm of topsoil at 70 g/m^2							
fertilizer (18:0:0+Mg+Fe)	14.80	0.12	2.27	-	14.80	100 m^2	**17.07**
"Enmag"	12.59	0.12	2.27	-	13.22	100 m^2	**15.49**
fertilizer (7:7:7)	4.32	0.12	2.27	-	4.32	100 m^2	**6.59**
"Super Phosphate Powder"	4.25	0.12	2.27	-	4.47	100 m^2	**6.73**
fertilizer (20:10:10)	5.65	0.12	2.27	-	5.65	100 m^2	**7.92**
"Bone meal"	6.46	0.12	2.27	-	6.78	100 m^2	**9.05**
Spreading topsoil; from dump not exceeding 100 m distance; by machine (topsoil not included)							
at 1 m^3 per 13 m^2; 75 mm thick	-	0.56	10.28	28.09	-	100 m^2	**38.37**
at 1 m^3 per 10 m^2; 100 mm thick	-	0.89	16.44	28.89	-	100 m^2	**45.33**
at 1 m^3 per 6.50 m^2; 150 mm thick	-	1.11	20.54	50.62	-	100 m^2	**71.15**
at 1 m^3 per 5 m^2; 200 mm thick	-	1.48	27.38	67.29	-	100 m^2	**94.67**
Spreading topsoil; from dump not exceeding 100 m distance; by hand (topsoil not included)							
at 1 m^3 per 13 m^2; 75 mm thick	-	15.00	277.56	-	-	100 m^2	**277.56**
at 1 m^3 per 10 m^2; 100 mm thick	-	20.00	370.07	-	-	100 m^2	**370.07**
at 1 m^3 per 6.50 m^2; 150 mm thick	-	30.01	555.11	-	-	100 m^2	**555.11**
at 1 m^3 per 5 m^2; 200 mm thick	-	36.67	678.47	-	-	100 m^2	**678.47**
Imported topsoil; tipped 100 m from area of application; by machine							
at 1 m^3 per 13 m^2; 75 mm thick	138.75	0.56	10.28	28.09	166.50	100 m^2	**204.87**
at 1 m^3 per 10 m^2; 100 mm thick	185.00	0.89	16.44	28.89	222.00	100 m^2	**267.33**
at 1 m^3 per 6.50 m^2; 150 mm thick	277.50	1.11	20.54	50.62	333.00	100 m^2	**404.15**
at 1 m^3 per 5 m^2; 200 mm thick	370.00	1.48	27.38	67.29	444.00	100 m^2	**538.67**
Imported topsoil; tipped 100 m from area of application; by hand							
at 1 m^3 per 13 m^2; 75 mm thick	138.75	15.00	277.56	-	166.50	100 m^2	**444.06**
at 1 m^3 per 10 m^2; 100 mm thick	185.00	20.00	370.07	-	222.00	100 m^2	**592.07**
at 1 m^3 per 6.50 m^2; 150 mm thick	277.50	30.01	555.11	-	333.00	100 m^2	**888.11**
at 1 m^3 per 5 m^2; 200 mm thick	370.00	36.67	678.47	-	444.00	100 m^2	**1122.47**
Composted bark and manure soil conditioner (20 m^3 loads); from not further than 25 m from location; cultivating into topsoil by machine							
50 mm thick	136.50	2.86	52.86	1.58	143.32	100 m^2	**197.77**
100 mm thick	273.00	6.05	111.91	1.58	286.65	100 m^2	**400.14**
150 mm thick	409.50	8.90	164.74	1.58	429.98	100 m^2	**596.30**
200 mm thick	546.00	12.90	238.74	1.58	573.30	100 m^2	**813.62**
Composted bark soil conditioner (20 m^3 loads); placing on beds by mechanical loader; spreading and rotavating into topsoil by machine							
50 mm thick	136.50	-	-	6.11	136.50	100 m^2	**142.61**
100 mm thick	273.00	-	-	10.36	286.65	100 m^2	**297.01**
150 mm thick	409.50	-	-	14.78	429.98	100 m^2	**444.75**
200 mm thick	546.00	-	-	19.11	573.30	100 m^2	**592.41**

Item Excluding site overheads and profit	PC £	Labour hours	Labour £	Plant £	Material £	Unit	Total rate £
Q31 PLANTING - cont'd							
Preparation of planting operations - cont'd							
Mushroom compost (20 m^3 loads); from not further than 25 m from location; cultivating into topsoil by machine							
50 mm thick	96.00	2.86	52.86	1.58	96.00	100 m^2	**150.44**
100 mm thick	192.00	6.05	111.91	1.58	192.00	100 m^2	**305.49**
150 mm thick	288.00	8.90	164.74	1.58	288.00	100 m^2	**454.32**
200 mm thick	384.00	12.90	238.74	1.58	384.00	100 m^2	**624.32**
Mushroom compost (20 m^3 loads); placing on beds by mechanical loader; spreading and rotavating into topsoil by machine							
50 mm thick	96.00	-	-	6.11	96.00	100 m^2	**102.11**
100 mm thick	192.00	-	-	10.36	192.00	100 m^2	**202.36**
150 mm thick	288.00	-	-	14.78	288.00	100 m^2	**302.78**
200 mm thick	384.00	-	-	19.11	384.00	100 m^2	**403.11**
Manure (60 m^3 loads); from not further than 25 m from location; cultivating into topsoil by machine							
50 mm thick	106.25	2.86	52.86	1.58	106.25	100 m^2	**160.69**
100 mm thick	212.50	6.05	111.91	1.58	223.13	100 m^2	**336.61**
150 mm thick	318.75	8.90	164.74	1.58	334.69	100 m^2	**501.01**
200 mm thick	425.00	12.90	238.74	1.58	446.25	100 m^2	**686.57**
Surface applications and soil additives; pre-planting; from not further than 25 m from location; by machine							
Ground limestone soil conditioner; including turning in to cultivated ground							
0.25 kg/m^2 = 2.50 tonnes/ha	0.61	-	-	2.86	0.61	100 m^2	**3.47**
0.50 kg/m^2 = 5.00 tonnes/ha	1.22	-	-	2.86	1.22	100 m^2	**4.08**
0.75 kg/m^2 = 7.50 tonnes/ha	1.83	-	-	2.86	1.83	100 m^2	**4.69**
1.00 kg/m^2 = 10.00 tonnes/ha	2.44	-	-	2.86	2.44	100 m^2	**5.30**
Medium bark soil conditioner; A.H.S. Ltd; including turning in to cultivated ground; delivered in 15 m^3 loads							
1 m^3 per 40 m^2 = 25 mm thick	0.82	-	-	0.09	0.82	m^2	**0.90**
1 m^3 per 20 m^2 = 50 mm thick	1.63	-	-	0.14	1.63	m^2	**1.77**
1 m^3 per 13.33 m^2 = 75 mm thick	2.45	-	-	0.19	2.45	m^2	**2.64**
1 m^3 per 10 m^2 = 100 mm thick	3.27	-	-	0.25	3.27	m^2	**3.52**
Mushroom compost soil conditioner; A.H.S. Ltd; including turning in to cultivated ground; delivered in 20 m^3 loads							
1 m^3 per 40 m^2 = 25 mm thick	0.48	0.02	0.31	-	0.48	m^2	**0.79**
1 m^3 per 20 m^2 = 50 mm thick	0.96	0.03	0.57	-	0.96	m^2	**1.53**
1 m^3 per 13.33 m^2 = 75 mm thick	1.44	0.04	0.74	-	1.44	m^2	**2.18**
1 m^3 per 10 m^2 = 100 mm thick	1.92	0.05	0.93	-	1.92	m^2	**2.84**
Mushroom compost soil conditioner; A.H.S. Ltd; including turning in to cultivated ground; delivered in 35 m^3 loads							
1 m^3 per 40 m^2 = 25 mm thick	0.27	0.02	0.31	-	0.27	m^2	**0.58**
1 m^3 per 20 m^2 = 50 mm thick	0.54	0.03	0.57	-	0.54	m^2	**1.11**
1 m^3 per 13.33 m^2 = 75 mm thick	0.81	0.04	0.74	-	0.81	m^2	**1.55**
1 m^3 per 10 m^2 = 100 mm thick	1.09	0.05	0.93	-	1.09	m^2	**2.01**
Surface applications and soil additives; pre-planting; from not further than 25 m from location; by hand							
Ground limestone soil conditioner; including turning in to cultivated ground							
0.25 kg/m^2 = 2.50 tonnes/ha	0.61	1.20	22.20	-	0.61	100 m^2	**22.81**
0.50 kg/m^2 = 5.00 tonnes/ha	1.22	1.33	24.66	-	1.22	100 m^2	**25.88**
0.75 kg/m^2 = 7.50 tonnes/ha	1.83	1.50	27.75	-	1.83	100 m^2	**29.58**
1.00 kg/m^2 = 10.00 tonnes/ha	2.44	1.71	31.71	-	2.44	100 m^2	**34.15**

Q PAVING/PLANTING/FENCING/SITE FURNITURE

Item Excluding site overheads and profit	PC £	Labour hours	Labour £	Plant £	Material £	Unit	**Total rate £**
Medium bark soil conditioner; A.H.S. Ltd; including turning in to cultivated ground; delivered in 15 m^3 loads							
1 m^3 per 40 m^2 = 25 mm thick	0.82	0.02	0.41	-	0.82	m^2	**1.23**
1 m^3 per 20 m^2 = 50 mm thick	1.63	0.04	0.82	-	1.63	m^2	**2.45**
1 m^3 per 13.33 m^2 = 75 mm thick	2.45	0.07	1.23	-	2.45	m^2	**3.68**
1 m^3 per 10 m^2 = 100 mm thick	3.27	0.08	1.48	-	3.27	m^2	**4.75**
Mushroom compost soil conditioner; Melcourt Industries Ltd; including turning in to cultivated ground; delivered in 25 m^3 loads							
1 m^3 per 40 m^2 = 25 mm thick	0.48	0.02	0.41	-	0.48	m^2	**0.89**
1 m^3 per 20 m^2 = 50 mm thick	0.96	0.04	0.82	-	0.96	m^2	**1.78**
1 m^3 per 13.33 m^2 = 75 mm thick	1.44	0.07	1.23	-	1.44	m^2	**2.67**
1 m^3 per 10 m^2 = 100 mm thick	1.92	0.08	1.48	-	1.92	m^2	**3.40**
Mushroom compost soil conditioner; A.H.S. Ltd; including turning in to cultivated ground; delivered in 35 m^3 loads							
1 m^3 per 40 m^2 = 25 mm thick	0.27	0.02	0.41	-	0.27	m^2	**0.68**
1 m^3 per 20 m^2 = 50 mm thick	0.54	0.04	0.82	-	0.54	m^2	**1.36**
1 m^3 per 13.33 m^2 = 75 mm thick	0.81	0.07	1.23	-	0.81	m^2	**2.05**
1 m^3 per 10 m^2 = 100 mm thick	1.09	0.08	1.48	-	1.09	m^2	**2.57**
"Super Humus" mixed bark and manure conditioner; Melcourt Industries Ltd; delivered in 25 m^3 loads							
1 m^3 per 40 m^2 = 25 mm thick	0.68	0.02	0.41	-	0.72	m^2	**1.13**
1 m^3 per 20 m^2 = 50 mm thick	1.36	0.04	0.82	-	1.43	m^2	**2.25**
1 m^3 per 13.33 m^2 = 75 mm thick	2.05	0.07	1.23	-	2.15	m^2	**3.38**
1 m^3 per 10 m^2 = 100 mm thick	2.73	0.08	1.48	-	2.87	m^2	**4.35**
Tree planting; pre-planting operations							
Excavating tree pits; depositing soil alongside pits; by machine							
600 mm x 600 mm x 600 mm deep	-	0.15	2.72	0.66	-	nr	**3.39**
900 mm x 900 mm x 600 mm deep	-	0.33	6.11	1.49	-	nr	**7.59**
1.00 m x 1.00 m x 600 mm deep	-	0.61	11.35	1.84	-	nr	**13.19**
1.25 m x 1.25 m x 600 mm deep	-	0.96	17.77	2.88	-	nr	**20.65**
1.00 m x 1.00 m x 1.00 m deep	-	1.02	18.92	3.07	-	nr	**21.99**
1.50 m x 1.50 m x 750 mm deep	-	1.73	31.93	5.18	-	nr	**37.11**
1.50 m x 1.50 m x 1.00 m deep	-	2.30	42.46	6.89	-	nr	**49.35**
1.75 m x 1.75 m x 1.00 mm deep	-	3.13	57.95	9.40	-	nr	**67.34**
2.00 m x 2.00 m x 1.00 mm deep	-	4.09	75.68	12.27	-	nr	**87.95**
Excavating tree pits; depositing soil alongside pits; by hand							
600 mm x 600 mm x 600 mm deep	-	0.44	8.14	-	-	nr	**8.14**
900 mm x 900 mm x 600 mm deep	-	1.00	18.50	-	-	nr	**18.50**
1.00 m x 1.00 m x 600 mm deep	-	1.13	20.81	-	-	nr	**20.81**
1.25 m x 1.25 m x 600 mm deep	-	1.93	35.70	-	-	nr	**35.70**
1.00 m x 1.00 m x 1.00 m deep	-	2.06	38.11	-	-	nr	**38.11**
1.50 m x 1.50 m x 750 mm deep	-	3.47	64.19	-	-	nr	**64.19**
1.75 m x 1.50 m x 750 mm deep	-	4.05	74.92	-	-	nr	**74.92**
1.50 m x 1.50 m x 1.00 m deep	-	4.63	85.66	-	-	nr	**85.66**
2.00 m x 2.00 m x 750 mm deep	-	6.17	114.14	-	-	nr	**114.14**
2.00 m x 2.00 m x 1.00 m deep	-	8.23	152.26	-	-	nr	**152.26**
Breaking up subsoil in tree pits; to a depth of 200 mm	-	0.03	0.62	-	-	m^2	**0.62**
Spreading and lightly consolidating approved topsoil (imported or from spoil heaps); in layers not exceeding 150 mm; distance from spoil heaps not exceeding 100 m (imported topsoil not included); by machine							
minimum depth 100 mm	-	1.55	28.68	37.20	-	100 m^2	**65.87**
minimum depth 150 mm	-	2.33	43.17	55.98	-	100 m^2	**99.15**
minimum depth 300 mm	-	4.67	86.33	111.96	-	100 m^2	**198.29**
minimum depth 450 mm	-	6.00	120.01	107.70	-	100 m^2	**297.08**

Q PAVING/PLANTING/FENCING/SITE FURNITURE

Item Excluding site overheads and profit	PC £	Labour hours	Labour £	Plant £	Material £	Unit	Total rate £
Q31 PLANTING - cont'd							
Tree planting; pre-planting operations - cont'd							
Spreading and lightly consolidating approved topsoil (imported or from spoil heaps); in layers not exceeding 150 mm; distance from spoil heaps not exceeding 100 m (imported topsoil not included); by hand							
minimum depth 100 mm	-	20.00	370.07	-	-	100 m^2	**370.07**
minimum depth 150 mm	-	30.01	555.11	-	-	100 m^2	**555.11**
minimum depth 300 mm	-	60.01	1110.22	-	-	100 m^2	**1110.22**
minimum depth 450 mm	-	90.02	1665.33	-	-	100 m^2	**1665.33**
Extra for filling tree pits with imported topsoil; PC £18.50/m^3; plus allowance for 20% settlement							
depth 100 mm	-	-	-	-	2.22	m^2	-
depth 150 mm	-	-	-	-	3.33	m^2	-
depth 200 mm	-	-	-	-	4.44	m^2	-
depth 300 mm	-	-	-	-	6.66	m^2	-
depth 400 mm	-	-	-	-	8.88	m^2	-
depth 450 mm	-	-	-	-	9.99	m^2	-
depth 500 mm	-	-	-	-	11.10	m^2	-
depth 600 mm	-	-	-	-	13.32	m^2	-
Add or deduct the following amounts for every £0.50 change in the material price of topsoil							
depth 100 mm	-	-	-	-	0.06	m^2	-
depth 150 mm	-	-	-	-	0.09	m^2	-
depth 200 mm	-	-	-	-	0.12	m^2	-
depth 300 mm	-	-	-	-	0.18	m^2	-
depth 400 mm	-	-	-	-	0.24	m^2	-
depth 450 mm	-	-	-	-	0.27	m^2	-
depth 500 mm	-	-	-	-	0.30	m^2	-
depth 600 mm	-	-	-	-	0.36	m^2	-
Structural soils (Amsterdam tree sand); Heicom; non compressive soil mixture for tree planting in areas to receive compressive surface treatments							
Backfilling and lightly compacting in layers; excavation, disposal, moving of material from delivery position and surface treatments not included; by machine							
individual treepits	59.15	0.50	9.25	2.25	59.15	m^3	**70.65**
in trenches	59.15	0.42	7.71	1.88	59.15	m^3	**68.74**
Backfilling and lightly compacting in layers; excavation, disposal, moving of material from delivery position and surface treatments not included; by hand							
individual treepits	59.15	1.60	29.60	-	59.15	m^3	**88.75**
in trenches	59.15	1.33	24.61	-	59.15	m^3	**83.76**
Tree staking							
J Toms Ltd; extra over trees for tree stake(s); driving 500 mm into firm ground; trimming to approved height; including two tree ties to approved pattern							
one stake; 75 mm diameter x 2.40 m long	4.01	0.20	3.70	-	4.01	nr	**7.71**
two stakes; 60 mm diameter x 1.65 m long	5.56	0.30	5.55	-	5.56	nr	**11.11**
two stakes; 75 mm diameter x 2.40 m long	7.26	0.30	5.55	-	7.26	nr	**12.81**
three stakes; 75 mm diameter x 2.40 m long	10.88	0.36	6.66	-	10.88	nr	**17.54**

Q PAVING/PLANTING/FENCING/SITE FURNITURE

Item Excluding site overheads and profit	PC £	Labour hours	Labour £	Plant £	Material £	Unit	Total rate £
Tree anchors							
Platipus Anchors Ltd; extra over trees for tree anchors							
ref RF1 rootball kit; for 75 to 220 mm girth, 2 to 4.5 m high inclusive of "Plati-Mat" PM1	27.91	1.00	18.50	-	27.91	nr	**46.41**
ref RF2; rootball kit; for 220 to 450 mm girth, 4.5 to 7.5 m high inclusive of "Plati-Mat" Pm2	45.72	1.33	24.61	-	45.72	nr	**70.33**
ref RF3; rootball kit; for 450 to 750 mm girth, 7.5 to 12 m high "Plati-Mat" PM3	96.61	1.50	27.75	-	96.61	nr	**124.36**
ref CG1; guy fixing kit; 75to 220 mm girth, 2 to 4.5 m high	18.17	1.67	30.83	-	18.17	nr	**49.00**
ref CG2; guy fixing kit; 220 to 450 mm girth, 4.5 to 7.5 m high	31.96	2.00	37.00	-	31.96	nr	**68.96**
installation tools; Drive Rod for RF1/CG1 Kits	-	-	-	-	58.91	nr	-
installation tools; Drive Rod for RF2/CG2 Kits	-	-	-	-	88.76	nr	-
Extra over trees for land drain to tree pits; 100 mm diameter perforated flexible agricultural drain; including excavating drain trench; laying pipe; backfilling	0.97	1.00	18.50	-	1.07	m	**19.57**
Tree planting; tree pit additives							
Melcourt Industries Ltd; "Topgrow"; incorporating into topsoil at 1 part "Topgrow" to 3 parts excavated topsoil; supplied in 75 L bags; pit size							
600 mm x 600 mm x 600 mm	1.22	0.02	0.37	-	1.22	nr	**1.59**
900 mm x 900 mm x 900 mm	4.13	0.06	1.11	-	4.13	nr	**5.24**
1.00 m x 1.00 m x 1.00 m	5.66	0.24	4.44	-	5.66	nr	**10.10**
1.25 m x 1.25 m x 1.25 m	11.07	0.40	7.40	-	11.07	nr	**18.47**
1.50 m x 1.50 m x 1.50 m	19.13	0.90	16.65	-	19.13	nr	**35.77**
Melcourt Industries Ltd; "Topgrow"; incorporating into topsoil at 1 part "Topgrow" to 3 parts excavated topsoil; supplied in 60 m^3 loose loads; pit size							
600 mm x 600 mm x 600 mm	0.93	0.02	0.31	-	0.93	nr	**1.24**
900 mm x 900 mm x 900 mm	3.15	0.05	0.93	-	3.15	nr	**4.08**
1.00 m x 1.00 m x 1.00 m	4.33	0.20	3.70	-	4.33	nr	**8.03**
1.25 m x 1.25 m x 1.25 m	8.45	0.33	6.17	-	8.45	nr	**14.61**
1.50 m x 1.50 m x 1.50 m	14.60	0.75	13.88	-	14.60	nr	**28.47**
Alginure Products; "Alginure Root Dip"; to bare rooted plants at 1 part "Alginure Root Dip" to 3 parts water							
transplants; at 3000/15 kg bucket	1.17	0.07	1.23	-	1.17	100 nr	**2.40**
standard trees; at 112/15 kg bucket of dip	0.31	0.03	0.62	-	0.31	nr	**0.93**
medium shrubs; at 600/15 kg bucket	0.06	0.02	0.31	-	0.06	nr	**0.37**
Tree planting - root barriers							
English Woodlands; "Root Director", one-piece root control planters for installation at time of planting, to divert root growth down away from pavements and out for anchorage; excavation measured separately							
RD 1050; 1050 x 1050 mm to 1300 x 1300 mm at base	76.07	0.25	4.63	-	76.07	nr	**80.69**
RD 640; 640 x 640 mm to 870 x 870 mm at base	49.40	0.25	4.63	-	49.40	nr	**54.02**
Greenleaf Horticulture; linear root deflection barriers; installed to trench measured separately							
Re-Root 2000, 1.5 mm thick	5.20	0.05	0.93	-	5.20	m	**6.13**
Re-Root 2000, 1.0 mm thick	4.40	0.05	0.93	-	4.40	m	**5.33**
Re-Root 600, 1.0 mm thick	6.29	0.05	0.93	-	6.29	m	**7.21**

Q PAVING/PLANTING/FENCING/SITE FURNITURE

Item Excluding site overheads and profit	PC £	Labour hours	Labour £	Plant £	Material £	Unit	Total rate £
Q31 PLANTING - cont'd							
Greenleaf Horticulture - irrigation systems; "Root Rain" tree pit irrigation systems							
"Metro" - "Small"; 35 mm pipe diameter 1.25 m long for specimen shrubs and standard trees							
plastic	5.82	0.25	4.63	-	5.82	nr	**10.45**
plastic with chain	7.62	0.25	4.63	-	7.62	nr	**12.25**
metal with chain	9.44	0.25	4.63	-	9.44	nr	**14.06**
"Medium" 35 mm pipe diameter 1.75 m long for standard and selected standard trees							
plastic	6.04	0.29	5.29	-	6.04	nr	**11.33**
plastic with chain	8.00	0.29	5.29	-	8.00	nr	**13.29**
metal with chain	9.79	0.29	5.29	-	9.79	nr	**15.08**
"Large" 35 mm pipe diameter 2.50 m long for selected standards and extra heavy standards							
plastic	7.15	0.33	6.17	-	7.15	nr	**13.32**
plastic with chain	8.95	0.33	6.17	-	8.95	nr	**15.12**
metal with chain	11.96	0.33	6.17	-	11.96	nr	**18.13**
"Large" 35 mm pipe diameter 2.50 m long for selected standards and extra heavy standards							
plastic	7.15	0.33	6.17	-	7.15	nr	**13.32**
plastic with chain	8.95	0.33	6.17	-	8.95	nr	**15.12**
metal with chain	11.96	0.33	6.17	-	11.96	nr	**18.13**
"Urban" irrigation and aeration systems; for large capacity general purpose irrigation to parkland and street verge planting							
RRUrb1; 3.0 m pipe	-	0.29	5.29	-	14.07	nr	**19.36**
RRUrb2; 5.0 m pipe	16.30	0.33	6.17	-	16.30	nr	**22.47**
RRUrb3; 8.0 m pipe	17.76	0.40	7.40	-	17.76	nr	**25.16**
"Precinct" irrigation and aeration systems; heavy cast aluminium inlet for heavily trafficked locations							
"5 m pipe"	30.25	0.33	6.17	-	30.25	nr	**36.42**
"8 m pipe"	32.70	0.40	7.40	-	32.70	nr	**40.10**
Mulching of tree pits; Melcourt Industries Ltd (items labelled FSC are Forest Stewardship Council certified)							
Spreading mulch; to individual trees; maximum distance 25 m (mulch not included)							
50 mm thick	-	0.05	0.90	-	-	m^2	**0.90**
75 mm thick	-	0.07	1.35	-	-	m^2	**1.35**
100 mm thick	-	0.10	1.80	-	-	m^2	**1.80**
Mulch; "Bark Nuggets®"; to individual trees; delivered in 80 m^3 loads; maximum distance 25 m							
50 mm thick	1.50	0.05	0.90	-	1.57	m^2	**2.47**
75 mm thick	2.24	0.07	1.35	-	2.36	m^2	**3.70**
100 mm thick	2.99	0.10	1.80	-	3.14	m^2	**4.94**
Mulch; "Bark Nuggets®"; to individual trees; delivered in 25 m^3 loads; maximum distance 25 m							
50 mm thick	2.15	0.05	0.90	-	2.25	m^2	**3.15**
75 mm thick	3.22	0.05	0.93	-	3.38	m^2	**4.31**
100 mm thick	4.29	0.07	1.23	-	4.50	m^2	**5.74**
Mulch; "Amenity Bark Mulch"; FSC; to individual trees; delivered in 80 m^3 loads; maximum distance 25 m							
50 mm thick	0.98	0.05	0.90	-	1.03	m^2	**1.93**
75 mm thick	1.47	0.07	1.35	-	1.55	m^2	**2.90**
100 mm thick	1.97	0.10	1.80	-	2.06	m^2	**3.86**
Mulch; "Amenity Bark Mulch"; FSC; to individual trees; delivered in 25 m^3 loads; maximum distance 25 m							
50 mm thick	1.63	0.05	0.90	-	1.71	m^2	**2.61**
75 mm thick	2.45	0.07	1.35	-	2.57	m^2	**3.92**
100 mm thick	3.27	0.07	1.23	-	3.43	m^2	**4.66**

Q PAVING/PLANTING/FENCING/SITE FURNITURE

Item Excluding site overheads and profit	PC £	Labour hours	Labour £	Plant £	Material £	Unit	Total rate £
Trees; planting labours only							
Bare root trees; including backfilling with previously excavated material (all other operations and materials not included)							
light standard; 6 - 8 cm girth	-	0.35	6.47	-	-	nr	**6.47**
standard; 8 - 10 cm girth	-	0.40	7.40	-	-	nr	**7.40**
selected standard; 10 - 12 cm girth	-	0.58	10.73	-	-	nr	**10.73**
heavy standard; 12 - 14 cm girth	-	0.83	15.42	-	-	nr	**15.42**
extra heavy standard; 14 - 16 cm girth	-	1.00	18.50	-	-	nr	**18.50**
Rootballed trees; including backfilling with previously excavated material (all other operations and materials not included)							
standard; 8 - 10 cm girth	-	0.50	9.25	-	-	nr	**9.25**
selected standard; 10 - 12 cm girth	-	0.60	11.10	-	-	nr	**11.10**
heavy standard; 12 - 14 cm girth	-	0.80	14.80	-	-	nr	**14.80**
extra heavy standard;14 - 16 cm girth	-	1.50	27.75	-	-	nr	**27.75**
16 - 18 cm girth	-	1.30	23.98	21.38	-	nr	**45.36**
18 - 20 cm girth	-	1.60	29.60	26.40	-	nr	**56.00**
20 - 25 cm girth	-	4.50	83.25	86.67	-	nr	**169.92**
25 - 30 cm girth	-	6.00	111.00	113.90	-	nr	**224.90**
30 - 35 cm girth	-	11.00	203.50	200.13	-	nr	**403.63**
Tree planting; rootballed trees; advanced nursery stock and semi-mature - General							
Preamble: The cost of planting semi-mature trees will depend on the size and species, and on the access to the site for tree handling machines. Prices should be obtained for individual trees and planting.							
Tree planting; bare root trees; nursery stock; James Coles & Sons (Nurseries) Ltd							
"Acer platanoides"; including backfillling with excavated material (other operations not included)							
light standard; 6 - 8 cm girth	8.40	0.35	6.47	-	8.40	nr	**14.88**
standard; 8 - 10 cm girth	10.75	0.40	7.40	-	10.75	nr	**18.15**
selected standard; 10 - 12 cm girth	16.90	0.58	10.73	-	16.90	nr	**27.63**
heavy standard; 12 - 14 cm girth	28.00	0.83	15.42	-	28.00	nr	**43.42**
extra heavy standard; 14 - 16 cm girth	16.50	1.00	18.50	-	16.50	nr	**35.00**
"Carpinus betulus"; including backfillling with excavated material (other operations not included)							
light standard; 6 - 8 cm girth	11.75	0.35	6.47	-	11.75	nr	**18.23**
standard; 8 - 10 cm girth	16.10	0.40	7.40	-	16.10	nr	**23.50**
selected standard; 10 - 12 cm girth	26.70	0.58	10.73	-	26.70	nr	**37.43**
heavy standard; 12 - 14 cm girth	26.50	0.83	15.42	-	26.50	nr	**41.92**
extra heavy standard; 14 - 16 cm girth	40.00	1.00	18.50	-	40.00	nr	**58.50**
"Fraxinus excelsior"; including backfillling with excavated material (other operations not included)							
light standard; 6 - 8 cm girth	7.00	0.35	6.47	-	7.00	nr	**13.47**
standard; 8 - 10 cm girth	9.00	0.40	7.40	-	9.00	nr	**16.40**
selected standard; 10 - 12 cm girth	14.75	0.58	10.73	-	14.75	nr	**25.48**
heavy standard; 12 - 14 cm girth	26.50	0.83	15.36	-	26.50	nr	**41.86**
extra heavy standard; 14 - 16 cm girth	33.50	1.00	18.50	-	33.50	nr	**52.00**
"Prunus avium Plena"; including backfillling with excavated material (other operations not included)							
light standard; 6 - 8 cm girth	7.00	0.36	6.73	-	7.00	nr	**13.73**
standard; 8 - 10 cm girth	9.00	0.40	7.40	-	9.00	nr	**16.40**
selected standard; 10 - 12 cm girth	25.20	0.58	10.73	-	25.20	nr	**35.93**
heavy standard; 12 - 14 cm girth	42.00	0.83	15.42	-	42.00	nr	**57.42**
extra heavy standard; 14 - 16 cm girth	50.40	1.00	18.50	-	50.40	nr	**68.90**

Q PAVING/PLANTING/FENCING/SITE FURNITURE

Item Excluding site overheads and profit	PC £	Labour hours	Labour £	Plant £	Material £	Unit	Total rate £
Q31 PLANTING - cont'd							
Tree planting; bare root trees - cont'd							
"Quercus robur"; including backfillling with excavated material (other operations not included)							
light standard; 6 - 8 cm girth	11.20	0.35	6.47	-	11.20	nr	**17.67**
standard; 8 - 10 cm girth	17.50	0.40	7.40	-	17.50	nr	**24.90**
selected standard; 10 - 12 cm girth	24.00	0.58	10.73	-	24.00	nr	**34.73**
heavy standard; 12 - 14 cm girth	39.20	0.83	15.42	-	39.20	nr	**54.62**
extra heavy standard; 14 - 16 cm girth	44.80	1.00	18.50	-	44.80	nr	**63.30**
"Robinia pseudoacacia Frisia"; including backfillling with excavated material (other operations not included)							
light standard; 6 - 8 cm girth	19.00	0.35	6.47	-	19.00	nr	**25.48**
standard; 8 - 10 cm girth	25.75	0.40	7.40	-	25.75	nr	**33.15**
selected standard; 10 - 12 cm girth	37.50	0.58	10.73	-	37.50	nr	**48.23**
Tree planting; rootballed trees; nursery stock; James Coles & Sons (Nurseries) Ltd							
"Acer platanoides"; including backfillling with excavated material (other operations not included)							
standard; 8 - 10 cm girth	25.75	0.48	8.88	-	25.75	nr	**34.63**
selected standard; 10 - 12 cm girth	31.90	0.56	10.37	-	31.90	nr	**42.27**
heavy standard; 12 - 14 cm girth	43.00	0.76	14.14	-	43.00	nr	**57.14**
extra heavy standard; 14 - 16 cm girth	51.50	1.20	22.20	-	51.50	nr	**73.70**
"Carpinus betulus"; including backfillling with excavated material (other operations not included)							
standard; 8 - 10 cm girth	31.10	0.48	8.88	-	31.10	nr	**39.98**
selected standard; 10 - 12 cm girth	41.70	0.56	10.37	-	41.70	nr	**52.07**
heavy standard; 12 - 14 cm girth	54.25	0.76	14.14	-	54.25	nr	**68.39**
extra heavy standard; 14 - 16 cm girth	64.00	1.20	22.20	-	64.00	nr	**86.20**
"Fraxinus excelsior"; including backfillling with excavated material (other operations not included)							
standard; 8 - 10 cm girth	24.00	0.48	8.88	-	24.00	nr	**32.88**
selected standard; 10 - 12 cm girth	29.75	0.56	10.37	-	29.75	nr	**40.12**
heavy standard; 12 - 14 cm girth	41.50	0.76	14.14	-	41.50	nr	**55.64**
extra heavy standard; 14 - 16 cm girth	48.50	1.20	22.20	-	48.50	nr	**70.70**
"Prunus avium Plena"; including backfillling with excavated material (other operations not included)							
standard; 8 - 10 cm girth	32.50	0.40	7.40	-	32.50	nr	**39.90**
selected standard; 10 - 12 cm girth	32.50	0.56	10.37	-	32.50	nr	**42.87**
heavy standard; 12 - 14 cm girth	57.00	0.76	14.14	-	57.00	nr	**71.14**
extra heavy standard; 14 - 16 cm girth	65.40	1.20	22.20	-	65.40	nr	**87.60**
"Quercus robur"; including backfillling with excavated material (other operations not included)							
standard; 8 - 10 cm girth	32.50	0.48	8.88	-	32.50	nr	**41.38**
selected standard; 10 - 12 cm girth	39.00	0.56	10.37	-	39.00	nr	**49.37**
heavy standard; 12 - 14 cm girth	54.20	0.76	14.14	-	54.20	nr	**68.34**
extra heavy standard; 14 - 16 cm girth	59.80	1.20	22.20	-	59.80	nr	**82.00**
"Robinia pseudoacacia Frisia"; including backfillling with excavated material (other operations not included)							
standard; 8 - 10 cm girth	40.75	0.48	8.88	-	40.75	nr	**49.63**
selected standard; 10 - 12 cm girth	52.50	0.56	10.37	-	52.50	nr	**62.87**
heavy standard; 12 - 14 cm girth	85.00	0.76	14.14	-	85.00	nr	**99.14**

Q PAVING/PLANTING/FENCING/SITE FURNITURE

Item Excluding site overheads and profit	PC £	Labour hours	Labour £	Plant £	Material £	Unit	**Total rate £**
Tree planting; "Airpot" container grown trees; advanced nursery stock and semi-mature; Deepdale Trees Ltd							
"Acer platanoides Emerald Queen"; including backfilling with excavated material (other operations not included)							
16 - 18 cm girth	95.00	1.98	36.63	40.14	95.00	nr	**171.77**
18 - 20 cm girth	130.00	2.18	40.29	42.91	130.00	nr	**213.20**
20 - 25 cm girth	190.00	2.38	43.96	48.16	190.00	nr	**282.12**
25 - 30 cm girth	250.00	2.97	54.95	72.37	250.00	nr	**377.32**
30 - 35 cm girth	450.00	3.96	73.26	80.27	450.00	nr	**603.53**
"Aesculus briotti"; including backfilling with excavated material (other operations not included)							
16 - 18 cm girth	90.00	1.98	36.63	40.14	90.00	nr	**166.77**
18 - 20 cm girth	130.00	1.60	29.60	42.91	130.00	nr	**202.51**
20 - 25 cm girth	180.00	2.38	43.96	48.16	180.00	nr	**272.12**
25 - 30 cm girth	240.00	2.97	54.95	72.37	240.00	nr	**367.32**
30 - 35 cm girth	330.00	3.96	73.26	80.27	330.00	nr	**483.53**
"Prunus avium Flora Plena"; including backfilling with excavated material (other operations not included)							
16 - 18 cm girth	95.00	1.98	36.63	40.14	95.00	nr	**171.77**
18 - 20 cm girth	130.00	1.60	29.60	42.91	130.00	nr	**202.51**
20 - 25 cm girth	190.00	2.38	43.96	48.16	190.00	nr	**282.12**
25 - 30 cm girth	250.00	2.97	54.95	72.37	250.00	nr	**377.32**
30 - 35 cm girth	350.00	3.96	73.26	80.27	350.00	nr	**503.53**
"Quercus palustris - Pin Oak"; including backfilling with excavated material (other operations not included)							
16 - 18 cm girth	95.00	1.98	36.63	40.14	95.00	nr	**171.77**
18 - 20 cm girth	130.00	1.60	29.60	42.91	130.00	nr	**202.51**
20 - 25 cm girth	190.00	2.38	43.96	48.16	190.00	nr	**282.12**
25 - 30 cm girth	250.00	2.97	54.95	72.37	250.00	nr	**377.32**
30 - 35 cm girth	375.00	3.96	73.26	80.27	375.00	nr	**528.53**
"Betula pendula multistem"; including backfilling with excavated material (other operations not included)							
3.0 - 3.5 m high	140.00	1.98	36.63	40.14	140.00	nr	**216.77**
3.5 - 4.0 m high	175.00	1.60	29.60	42.91	175.00	nr	**247.51**
4.0 - 4.5 m high	225.00	2.38	43.96	48.16	225.00	nr	**317.12**
4.5 - 5.0 m high	275.00	2.97	54.95	72.37	275.00	nr	**402.32**
5.0 - 6.0 m high	375.00	3.96	73.26	80.27	375.00	nr	**528.53**
6.0 - 7.0 m high	495.00	4.50	83.25	96.05	495.00	nr	**674.30**
"Pinus sylvestris"; including backfilling with excavated material (other operations not included)							
3.0 - 3.5 m high	300.00	1.98	36.63	40.14	300.00	nr	**376.77**
3.5 - 4.0 m high	375.00	1.60	29.60	42.91	375.00	nr	**447.51**
4.0 - 4.5 m high	450.00	2.38	43.96	48.16	450.00	nr	**542.12**
4.5 - 5.0 m high	500.00	2.97	54.95	72.37	500.00	nr	**627.32**
5.0 - 6.0 m high	650.00	3.96	73.26	80.27	650.00	nr	**803.53**
6.0 - 7.0 m high	950.00	4.50	83.25	98.85	950.00	nr	**1132.10**
Tree planting; "Airpot" container grown trees; semi-mature and mature trees; Deepdale Trees Ltd; planting and backfilling; planted by telehandler or by crane; delivery included; all other operations priced separately							
Semi-mature trees indicative prices							
40 - 45 cm girth	550.00	4.00	74.00	48.00	550.00	nr	**672.00**
45 - 50 cm girth	750.00	4.00	74.00	48.00	750.00	nr	**872.00**
55 - 60 cm girth	1350.00	6.00	111.00	48.00	1350.00	nr	**1509.00**
60 - 70 cm girth	2500.00	7.00	129.50	65.00	2500.00	nr	**2694.50**
70 - 80 cm girth	3500.00	7.50	138.75	82.00	3500.00	nr	**3720.75**
[illegible] cm girth	4500.00	8.00	148.00	96.00	4500.00	nr	**4744.00**

Q PAVING/PLANTING/FENCING/SITE FURNITURE

Item Excluding site overheads and profit	PC £	Labour hours	Labour £	Plant £	Material £	Unit	Total rate £
Q31 PLANTING - cont'd							
Tree planting; rootballed trees; advanced nursery stock and semi-mature; Lorenz von Ehren							
"Acer platanoides Emerald Queen"; including backfilling with excavated material (other operations not included)							
16 - 18 cm girth	80.00	1.30	23.98	3.75	80.00	nr	**107.73**
18 - 20 cm girth	93.00	1.60	29.60	3.75	93.00	nr	**126.35**
20 - 25 cm girth	115.00	4.50	83.25	18.04	115.00	nr	**216.29**
25 - 30 cm girth	210.00	6.00	111.00	22.40	210.00	nr	**343.40**
30 - 35 cm girth	340.00	11.00	203.50	29.88	340.00	nr	**573.38**
"Aesculus briotti"; including backfilling with excavated material (other operations not included)							
16 - 18 cm girth	125.00	1.30	23.98	3.75	125.00	nr	**152.73**
18 - 20 cm girth	150.00	1.60	29.60	3.75	150.00	nr	**183.35**
20 - 25 cm girth	210.00	4.50	83.25	18.04	210.00	nr	**311.29**
25 - 30 cm girth	270.00	6.00	111.00	22.40	270.00	nr	**403.40**
30 - 35 cm girth	380.00	11.00	203.50	29.88	380.00	nr	**613.38**
"Prunus avium Flora Plena"; including backfilling with excavated material (other operations not included)							
16 - 18 cm girth	71.00	1.30	23.98	3.75	71.00	nr	**98.73**
18 - 20 cm girth	100.00	1.60	29.60	3.75	100.00	nr	**133.35**
20 - 25 cm girth	110.00	4.50	83.25	18.04	110.00	nr	**211.29**
25 - 30 cm girth	150.00	6.00	111.00	22.40	150.00	nr	**283.40**
30 - 35 cm girth	315.00	11.00	203.50	26.15	315.00	nr	**544.65**
"Quercus palustris - Pin Oak"; including backfilling with excavated material (other operations not included)							
16 - 18 cm girth	82.00	1.30	23.98	3.75	82.00	nr	**109.73**
18 - 20 cm girth	110.00	1.60	29.60	3.75	110.00	nr	**143.35**
20 - 25 cm girth	150.00	4.50	83.25	18.04	150.00	nr	**251.29**
25 - 30 cm girth	220.00	6.00	111.00	22.40	220.00	nr	**353.40**
30 - 35 cm girth	320.00	11.00	203.50	29.88	320.00	nr	**553.38**
"Tilia cordata Green Spire"; including backfilling with excavated material (other operations not included)							
16 - 18 cm girth	68.00	1.30	23.98	3.75	68.00	nr	**95.73**
18 - 20 cm girth	86.00	1.60	29.60	3.75	86.00	nr	**119.35**
20 - 25 cm girth	104.00	4.50	83.25	18.04	104.00	nr	**205.29**
25 - 30 cm girth; 5 x transplanted 4.0 - 5.0 m tall	160.00	6.00	111.00	22.40	160.00	nr	**293.40**
30 - 35 cm girth; 5 x transplanted 5.0 - 7.0 m tall	260.00	11.00	203.50	29.88	260.00	nr	**493.38**
Tree planting; rootballed trees; semi-mature and mature trees; Lorenz von Ehren; planting and backfilling; planted by tele handler or by crane; delivery included; all other operations priced separately							
Semi-mature trees							
40 - 45 cm girth	680.00	8.00	148.00	70.50	680.00	nr	**898.50**
45 - 50 cm girth	970.00	8.00	148.00	95.85	970.00	nr	**1213.85**
50 - 60 cm girth	1390.00	10.00	185.00	115.13	1390.00	nr	**1690.13**
60 - 70 cm girth	2200.00	15.00	277.50	194.00	2200.00	nr	**2671.50**
70 - 80 cm girth	3100.00	18.00	333.00	186.00	3100.00	nr	**3619.00**
80 - 90 cm girth	4600.00	18.00	333.00	186.00	4600.00	nr	**5119.00**
90 - 100 cm girth	5400.00	18.00	333.00	186.00	5400.00	nr	**5919.00**
Tree planting; tree protection - General							
Preamble: Care must be taken to ensure that tree grids and guards are removed when trees grow beyond the specified diameter of guard.							

Q PAVING/PLANTING/FENCING/SITE FURNITURE

Item Excluding site overheads and profit	PC £	Labour hours	Labour £	Plant £	Material £	Unit	**Total rate £**
Tree planting; tree protection							
Crowders Nurseries; "Crowders Tree Tube"; olive green							
1200 mm high x 80 x 80 mm	0.75	0.07	1.23	-	0.75	nr	**1.98**
stakes; 1500 mm high for "Crowders Tree Tube"; driving into ground	0.50	0.05	0.93	-	0.50	nr	**1.43**
Crowders Nurseries; expandable plastic tree guards; including 25 mm softwood stakes							
500 mm high	0.55	0.17	3.08	-	0.55	nr	**3.63**
1.00 m high	0.61	0.17	3.08	-	0.61	nr	**3.69**
English Woodlands; Weldmesh tree guards; nailing to tree stakes (tree stakes not included)							
1800 mm high x 200 mm diameter	13.00	0.33	6.16	-	13.00	nr	**19.16**
1800 mm high x 225 mm diameter	14.00	0.33	6.16	-	14.00	nr	**20.16**
1800 mm high x 250 mm diameter	14.75	0.25	4.63	-	14.75	nr	**19.38**
1800 mm high x 300 mm diameter	15.50	0.33	6.16	-	15.50	nr	**21.66**
J. Toms Ltd; "Spiral Guard" perforated PVC guards; for trees 10 to 40 mm diameter; white, black or grey							
450 mm	0.18	0.03	0.62	-	0.18	nr	**0.80**
600 mm	0.20	0.03	0.62	-	0.20	nr	**0.82**
750 mm	0.27	0.03	0.62	-	0.27	nr	**0.89**
English Woodlands; plastic mesh tree guards supplied in 50 m rolls							
13 mm x 13 mm small mesh; roll width 60 cm; black	0.95	0.04	0.66	-	1.32	nr	**1.98**
13 mm x 13 mm small mesh; roll width 120 cm; black	1.65	0.06	1.09	-	2.19	nr	**3.28**
English Woodlands; plastic mesh tree guard in pre-cut pieces, 60 cm x 150 mm, supplied flat packed	0.64	0.06	1.09	-	1.01	nr	**2.10**
Tree guards of 3 nr 2.40 m x 100 mm stakes; driving 600 mm into firm ground; bracing with timber braces at top and bottom; including 3 strands barbed wire	0.29	1.00	18.50	-	0.29	nr	**18.79**
English Woodlands; strimmer guard in heavy duty black plastic, 225 mm high	2.40	0.07	1.23	-	2.40	nr	**3.63**
Tubex Ltd; "Standard Treeshelter" inclusive of 25 mm stake; prices shown for quantities of 500 nr							
0.6 m high	0.52	0.05	0.93	-	0.70	each	**1.63**
0.75 m high	0.64	0.05	0.93	-	0.97	each	**1.90**
1.2 m high	2.17	0.05	0.93	-	2.47	each	**3.40**
1.5 m high	1.25	0.05	0.93	-	1.61	each	**2.53**
Tubex Ltd; "Shrubshelter" inclusive of 25 mm stake; prices shown for quantities of 500 nr							
"Ecostart" shelter for forestry transplants and seedlings,	0.60	0.07	1.23	-	0.60	nr	**1.83**
0.6 m high	1.15	0.07	1.23	-	1.33	nr	**2.56**
0.75 m high	1.60	0.07	1.23	-	1.82	nr	**3.05**
Tree guards; Netlon							
Extra over trees for spraying with antidessicant spray; "Wiltpruf"							
selected standards; standards; light standards	2.62	0.20	3.70	-	2.62	nr	**6.32**
standards; heavy standards	4.36	0.25	4.63	-	4.36	nr	**8.99**
Hedges							
Excavating trench for hedges; depositing soil alongside trench; by machine							
300 mm deep x 300 mm wide	-	0.03	0.56	0.27	-	m	**0.83**
300 mm deep x 450 mm wide	-	0.05	0.83	0.41	-	m	**1.24**
Excavating trench for hedges; depositing soil alongside trench; by hand							
300 mm deep x 300 mm wide	-	0.12	2.22	-	-	m	**2.22**
300 mm deep x 450 mm wide	-	0.23	4.17	-	-	m	**4.17**

Q PAVING/PLANTING/FENCING/SITE FURNITURE

Item Excluding site overheads and profit	PC £	Labour hours	Labour £	Plant £	Material £	Unit	Total rate £
Q31 PLANTING - cont'd							
Hedges - cont'd							
Setting out; notching out; excavating trench; breaking up subsoil to minimum depth 300 mm							
minimum 400 mm deep	-	0.25	4.63	-	-	m	**4.63**
Hedge planting; including backfill with excavated topsoil; PC £0.32/nr							
single row; 200 mm centres	1.65	0.06	1.16	-	1.65	m	**2.81**
single row; 300 mm centres	1.10	0.06	1.03	-	1.10	m	**2.13**
single row; 400 mm centres	0.82	0.04	0.77	-	0.82	m	**1.60**
single row; 500 mm centres	0.66	0.03	0.62	-	0.66	m	**1.28**
double row; 200 mm centres	3.30	0.17	3.08	-	3.30	m	**6.38**
double row; 300 mm centres	2.20	0.13	2.47	-	2.20	m	**4.67**
double row; 400 mm centres	1.65	0.08	1.54	-	1.65	m	**3.19**
double row; 500 mm centres	1.32	0.07	1.23	-	1.32	m	**2.55**
Extra over hedges for incorporating manure; at 1 m^3 per 30 m	0.71	0.03	0.46	-	0.71	m	**1.17**
Topiary; clipped topiary; Lorenz von Ehren; German field grown clipped and transplanted as detailed; planted to plantpit; including backfilling with excavated material and TPMC							
Buxus sempervirens (box); balls							
300 diameter, 3 x transplanted; container grown or rootballed	11.50	0.25	4.63	-	19.54	nr	**24.16**
500 diameter, 4 x transplanted; wire rootballed	35.00	1.50	27.75	5.47	94.95	nr	**128.17**
900 diameter, 5 x transplanted; wire rootballed	185.00	2.75	50.88	5.47	284.24	nr	**340.58**
1300 diameter, 6 x transplanted; wire rootballed	610.00	3.10	57.35	6.56	707.06	nr	**770.98**
Buxus sempervirens (box); pyramids							
500 high, 3 x transplanted; container grown or rootballed	27.00	1.50	27.75	5.47	86.95	nr	**120.17**
900 high, 4 x transplanted; wire rootballed	128.00	2.75	50.88	5.47	227.24	nr	**283.58**
1300 high, 5 x transplanted; wire rootballed	357.00	3.10	57.35	6.56	454.06	nr	**517.98**
Buxus sempervirens (box); truncated pyramids							
500 high, 3 x transplanted; container grown or rootballed	71.00	1.50	27.75	5.47	130.95	nr	**164.17**
900 high, 4 x transplanted; wire rootballed	275.00	2.75	50.88	5.47	374.24	nr	**430.58**
1300 high, 5 x transplanted, wire rootballed	850.00	3.10	57.35	6.56	947.06	nr	**1010.98**
Buxus sempervirens (box); cubes							
500 square, 4 x transplanted; rootballed	75.00	1.50	27.75	5.47	134.95	nr	**168.17**
900 square, 5 x transplanted; wire rootballed	335.00	2.75	50.88	5.47	434.24	nr	**490.58**
Taxus baccata (yew); balls							
500 diameter, 4 x transplanted; rootballed	50.00	1.50	27.75	5.47	109.95	nr	**143.17**
900 diameter, 5 x transplanted; wire rootballed	157.00	2.75	50.88	5.47	256.24	nr	**312.58**
1300 diameter, 7 x transplanted, wire rootballed	535.00	3.10	57.35	6.56	632.06	nr	**695.98**
Taxus baccata (yew); cones							
800 high, 4 x transplanted; wire rootballed	50.00	1.50	27.75	5.47	109.95	nr	**143.17**
1500 high, 5 x transplanted; wire rootballed	132.00	2.00	37.00	5.47	205.85	nr	**248.32**
2500 high, 7 x transplanted; wire rootballed	430.00	3.10	57.35	6.56	527.06	nr	**590.98**
Taxus baccata (yew); cubes							
500 square, 4 x transplanted; rootballed	60.00	1.50	27.75	5.47	119.95	nr	**153.17**
900 square, 5 x transplanted; wire rootballed	280.00	2.75	50.88	5.47	379.24	nr	**435.58**
Taxus baccata (yew); pyramids							
900 high, 4 x transplanted; wire rootballed	115.00	1.50	27.75	5.47	174.95	nr	**208.17**
1500 high, 5 x transplanted; wire rootballed	280.00	3.10	57.35	6.56	377.06	nr	**440.98**
2500 high, 7 x transplanted; wire rootballed	780.00	4.00	74.00	10.94	904.49	nr	**989.42**
Carpinus betulus (common hornbeam); columns, round base							
800 wide, 2000 high, 4 x transplanted, wire rootballed	143.00	3.10	57.35	6.56	240.06	nr	**303.98**
800 wide, 2750 high, 4 x transplanted, wire rootballed	285.00	4.00	74.00	10.94	409.49	nr	**494.42**

Q PAVING/PLANTING/FENCING/SITE FURNITURE

Item Excluding site overheads and profit	PC £	Labour hours	Labour £	Plant £	Material £	Unit	**Total rate £**
Carpinus betulus (common hornbeam); columns, square base							
800 wide, 2000 high, 4 x transplanted, wire rootballed	178.00	3.10	57.35	6.56	275.06	nr	**338.98**
800 wide, 2750 high, 4 x transplanted, wire rootballed	320.00	4.00	74.00	10.94	444.49	nr	**529.42**
Carpinus betulus (common hornbeam); cones							
3 m high, 5 x transplanted; wire rootballed	225.00	4.00	74.00	10.94	349.49	nr	**434.42**
4 m high, 6 x transplanted; wire rootballed	500.00	5.00	92.50	10.94	647.70	nr	**751.14**
Carpinus betulus 'Fastigiata'; pyramids							
4 m high, 6 x transplanted; wire rootballed	785.00	5.00	92.50	10.94	932.70	nr	**1036.14**
7 m high, 7 x transplanted; wire rootballed	2140.00	7.50	138.75	16.41	2322.45	nr	**2477.61**
Shrub planting - General							
Preamble: For preparation of planting areas see "Cultivation" at the beginning of the section on planting.							
Shrub planting							
Setting out; selecting planting from holding area; loading to wheelbarrows; planting as plan or as directed; distance from holding area maximum 50 m; plants 2 - 3 litre containers							
plants in groups of 100 nr minimum	-	0.01	0.21	-	-	nr	**0.21**
plants in groups of 10 - 100 nr	-	0.02	0.31	-	-	nr	**0.31**
plants in groups of 3 - 5 nr	-	0.03	0.46	-	-	nr	**0.46**
single plants not grouped	-	0.04	0.74	-	-	nr	**0.74**
Forming planting holes; in cultivated ground (cultivating not included); by mechanical auger; trimming holes by hand; depositing excavated material alongside holes							
250 diameter	-	0.03	0.62	0.04	-	nr	**0.66**
250 x 250 mm	-	0.04	0.74	0.08	-	nr	**0.82**
300 x 300 mm	-	0.08	1.39	0.10	-	nr	**1.48**
Hand excavation; forming planting holes; in cultivated ground (cultivating not included); depositing excavated material alongside holes							
100 mm x 100 mm x 100 mm deep; with mattock or hoe	-	0.01	0.12	-	-	nr	**0.12**
250 mm x 250 mm x 300 mm deep	-	0.04	0.74	-	-	nr	**0.74**
300 mm x 300 mm x 300 mm deep	-	0.06	1.03	-	-	nr	**1.03**
400 mm x 400 mm x 400 mm deep	-	0.13	2.31	-	-	nr	**2.31**
500 mm x 500 mm x 500 mm deep	-	0.25	4.63	-	-	nr	**4.63**
600 mm x 600 mm x 600 mm deep	-	0.43	8.01	-	-	nr	**8.01**
900 mm x 900 mm x 600 mm deep	-	1.00	18.50	-	-	nr	**18.50**
1.00 m x 1.00 m x 600 mm deep	-	1.23	22.75	-	-	nr	**22.75**
1.25 m x 1.25 m x 600 mm deep	-	1.93	35.70	-	-	nr	**35.70**
Hand excavation; forming planting holes; in uncultivated ground; depositing excavated material alongside holes							
100 mm x 100 mm x 100 mm deep with mattock or hoe	-	0.03	0.46	-	-	nr	**0.46**
250 mm x 250 mm x 300 mm deep	-	0.06	1.03	-	-	nr	**1.03**
300 mm x 300 mm x 300 mm deep	-	0.06	1.16	-	-	nr	**1.16**
400 mm x 400 mm x 400 mm deep	-	0.25	4.63	-	-	nr	**4.63**
500 mm x 500 mm x 500 mm deep	-	0.33	6.02	-	-	nr	**6.02**
600 mm x 600 mm x 600 mm deep	-	0.55	10.18	-	-	nr	**10.18**
900 mm x 900 mm x 600 mm deep	-	1.25	23.13	-	-	nr	**23.13**
1.00 m x 1.00 m x 600 mm deep	-	1.54	28.44	-	-	nr	**28.44**
1.25 m x 1.25 m x 600 mm deep	-	2.41	44.63	-	-	nr	**44.63**
Bare root planting; to planting holes (forming holes not included); including backfilling with excavated material (bare root plants not included)							
bare root 1+1; 30 - 90 mm high	-	0.02	0.31	-	-	nr	**0.31**
bare root 1+2; 90 - 120 mm high	-	0.02	0.31	-	-	nr	**0.31**

Q PAVING/PLANTING/FENCING/SITE FURNITURE

Item Excluding site overheads and profit	PC £	Labour hours	Labour £	Plant £	Material £	Unit	Total rate £
Q31 PLANTING - cont'd							
Shrub planting - cont'd							
Containerised planting; to planting holes (forming holes not included); including backfilling with excavated material (shrub or ground cover not included)							
9 cm pot	-	0.01	0.19	-	-	nr	**0.19**
2 litre container	-	0.02	0.37	-	-	nr	**0.37**
3 litre container	-	0.02	0.41	-	-	nr	**0.41**
5 litre container	-	0.03	0.62	-	-	nr	**0.62**
10 litre container	-	0.05	0.93	-	-	nr	**0.93**
15 litre container	-	0.07	1.23	-	-	nr	**1.23**
20 litre container	-	0.08	1.54	-	-	nr	**1.54**
Shrub planting; 2 litre containerised plants; in cultivated ground (cultivating not included); PC £2.60/nr							
average 2 plants per m^2	-	0.06	1.04	-	5.38	m^2	**6.42**
average 3 plants per m^2	-	0.08	1.55	-	8.07	m^2	**9.62**
average 4 plants per m^2	-	0.11	2.07	-	10.76	m^2	**12.83**
average 6 plants per m^2	-	0.17	3.11	-	16.14	m^2	**19.25**
Extra over shrubs for stakes	0.50	0.02	0.31	-	0.50	nr	**0.81**
Composted bark soil conditioners; 20 m^3 loads; on beds by mechanical loader; spreading and rotavating into topsoil; by machine							
50 mm thick	136.50	-	-	6.11	136.50	100 m^2	**142.61**
100 mm thick	273.00	-	-	10.36	286.65	100 m^2	**297.01**
150 mm thick	409.50	-	-	14.78	429.98	100 m^2	**444.75**
200 mm thick	546.00	-	-	19.11	573.30	100 m^2	**592.41**
Mushroom compost; 25 m^3 loads; delivered not further than 25 m from location; cultivating into topsoil by pedestrian operated machine							
50 mm thick	96.00	2.86	52.86	1.58	96.00	100 m^2	**150.44**
100 mm thick	192.00	6.05	111.91	1.58	192.00	100 m^2	**305.49**
150 mm thick	288.00	8.90	164.74	1.58	288.00	100 m^2	**454.32**
200 mm thick	384.00	12.90	238.74	1.58	384.00	100 m^2	**624.32**
Mushroom compost; 25 m^3 loads; on beds by mechanical loader; spreading and rotavating into topsoil by tractor drawn rotavator							
50 mm thick	96.00	-	-	6.11	96.00	100 m^2	**102.11**
100 mm thick	192.00	-	-	10.36	192.00	100 m^2	**202.36**
150 mm thick	288.00	-	-	14.78	288.00	100 m^2	**302.78**
200 mm thick	384.00	-	-	19.11	384.00	100 m^2	**403.11**
Manure; 20 m^3 loads; delivered not further than 25 m from location; cultivating into topsoil by pedestrian operated machine							
50 mm thick	196.25	2.86	52.86	1.58	196.25	100 m^2	**250.69**
100 mm thick	212.50	6.05	111.91	1.58	223.13	100 m^2	**336.61**
150 mm thick	318.75	8.90	164.74	1.58	334.69	100 m^2	**501.01**
200 mm thick	425.00	12.90	238.74	1.58	446.25	100 m^2	**686.57**
Fertilizers (7:7:7); PC £0.62/kg; to beds; by hand							
35 g/m^2	2.16	0.17	3.08	-	2.16	100 m^2	**5.25**
50 g/m^2	3.09	0.17	3.08	-	3.09	100 m^2	**6.17**
70 g/m^2	4.32	0.17	3.08	-	4.32	100 m^2	**7.41**
Fertilizers; "Enmag" PC £1.80/kg slow release fertilizer; to beds; by hand							
35 g/m^2	6.30	0.17	3.08	-	6.30	100 m^2	**9.38**
50 g/m^2	9.00	0.17	3.08	-	9.00	100 m^2	**12.08**
70 g/m^2	13.49	0.17	3.08	-	13.49	100 m^2	**16.58**
Note: For machine incorporation of fertilizers and soil conditioners see "Cultivation".							

Q PAVING/PLANTING/FENCING/SITE FURNITURE

Item Excluding site overheads and profit	PC £	Labour hours	Labour £	Plant £	Material £	Unit	**Total rate £**
Herbaceous and groundcover planting							
Herbaceous plants; PC £1.20/nr; including forming planting holes in cultivated ground (cultivating not included); backfilling with excavated material; 1 litre containers							
average 4 plants per m^2 - 500 mm centres	-	0.09	1.73	-	4.80	m^2	**6.53**
average 6 plants per m^2 - 408 mm centres	-	0.14	2.59	-	7.20	m^2	**9.79**
average 8 plants per m^2 - 354 mm centres	-	0.19	3.46	-	9.60	m^2	**13.06**
Note: For machine incorporation of fertilizers and soil conditioners see "Cultivation".							
Plant support netting; Bridport Gundry; on 50 mm diameter stakes; 750 mm long; driving into ground at 1.50 m centres							
green extruded plastic mesh; 125 mm square	0.38	0.04	0.74	-	0.38	m^2	**1.12**
Bulb planting							
Bulbs; including forming planting holes in cultivated area (cultivating not included); backfilling with excavated material							
small	13.00	0.83	15.42	-	13.00	100 nr	**28.42**
medium	22.00	0.83	15.42	-	22.00	100 nr	**37.42**
large	25.00	0.91	16.82	-	25.00	100 nr	**41.82**
Bulbs; in grassed area; using bulb planter; including backfilling with screened topsoil or peat and cut turf plug							
small	13.00	1.67	30.83	-	13.00	100 nr	**43.83**
medium	22.00	1.67	30.83	-	22.00	100 nr	**52.83**
large	25.00	2.00	37.00	-	25.00	100 nr	**62.00**
Aquatic planting							
Aquatic plants; in prepared growing medium in pool; plant size 2 - 3 litre containerised (plants not included)	-	0.04	0.74	-	-	nr	**0.74**
Operations after planting							
Initial cutting back to shrubs and hedge plants; including disposal of all cuttings	-	1.00	18.50	-	-	100 m^2	**18.50**
Mulch; Melcourt Industries Ltd; "Bark Nuggets®"; to plant beds; delivered in 80 m^3 loads; maximum distance 25 m							
50 mm thick	1.50	0.04	0.82	-	1.57	m^2	**2.39**
75 mm thick	2.24	0.07	1.23	-	2.36	m^2	**3.59**
100 mm thick	2.99	0.09	1.64	-	3.14	m^2	**4.78**
Mulch; Melcourt Industries Ltd; "Bark Nuggets®"; to plant beds; delivered in 25 m^3 loads; maximum distance 25 m							
50 mm thick	1.90	0.03	0.54	-	2.00	m^2	**2.54**
75 mm thick	3.22	0.07	1.23	-	3.38	m^2	**4.61**
100 mm thick	4.29	0.09	1.64	-	4.50	m^2	**6.15**
Mulch; Melcourt Industries Ltd; "Amenity Bark Mulch"; FSC; to plant beds; delivered in 80 m^3 loads; maximum distance 25 m							
50 mm thick	0.98	0.04	0.82	-	1.03	m^2	**1.85**
75 mm thick	1.47	0.07	1.23	-	1.55	m^2	**2.78**
100 mm thick	1.97	0.09	1.64	-	2.06	m^2	**3.71**
Mulch; Melcourt Industries Ltd; "Amenity Bark Mulch"; FSC; to plant beds; delivered in 25 m^3 loads; maximum distance 25 m							
50 mm thick	1.63	0.04	0.82	-	1.71	m^2	**2.54**
75 mm thick	2.45	0.07	1.23	-	2.57	m^2	**3.81**
100 mm thick	3.27	0.09	1.64	-	3.43	m^2	**5.07**

Item Excluding site overheads and profit	PC £	Labour hours	Labour £	Plant £	Material £	Unit	**Total rate £**
Q31 PLANTING - cont'd							
Mulch mats; English Woodlands; lay mulch mat to planted area or plant station; mat secured with metal J pins 240 mm x 3 mm							
Mats to individual plants and plant stations							
Bonflora Biodegradable; 45 cm dia; circular mat	1.60	0.05	0.93	-	1.92	ea	**2.85**
Bonflora Biodegradable; 60 cm dia; circular mat	2.10	0.05	0.93	-	2.42	ea	**3.35**
Hemcore Biodegradable; 50 cm x 50 cm; square mat	0.75	0.05	0.93	-	1.07	ea	**2.00**
Woven Polypropylene; 50 cm x 50 cm; square mat	2.90	0.05	0.93	-	3.22	ea	**4.14**
Woven Polypropylene; 1 m x 1m; square mat	7.30	0.07	1.23	-	7.62	ea	**8.85**
Hedging mats							
Woven Polypropylene; 1 m x 100 m roll; hedge planting	0.43	0.01	0.23	-	0.59	m^2	**0.82**
General planting areas							
Plantex membrane; 1 m x 14 m roll; weed control	1.11	0.01	0.19	-	1.27	m^2	**1.45**
Selective herbicides; in accordance with manufacturer's instructions; PC £29.23/litre; by machine; application rate							
30 ml/m^2	-	-	-	0.30	0.88	100 m^2	**1.18**
35 ml/m^2	-	-	-	0.30	1.02	100 m^2	**1.32**
40 ml/m^2	-	-	-	0.30	1.17	100 m^2	**1.47**
50 ml/m^2	-	-	-	0.30	1.46	100 m^2	**1.76**
3.00 L/ha	-	-	-	30.01	87.67	ha	**117.69**
3.50 L/ha	-	-	-	30.01	102.29	ha	**132.30**
4.00 L/ha	-	-	-	30.01	116.90	ha	**146.91**
Selective herbicides; in accordance with manufacturer's instructions; PC £29.23/litre; by hand; application rate							
30 ml/m^2	-	0.33	6.17	-	0.88	100 m^2	**7.04**
35 ml/m^2	-	0.33	6.17	-	1.02	100 m^2	**7.19**
40 ml/m^2	-	0.33	6.17	-	1.17	100 m^2	**7.34**
50 ml/m^2	-	0.33	6.17	-	1.46	100 m^2	**7.63**
3.00 L/ha	-	33.33	616.67	-	87.67	ha	**704.34**
3.50 L/ha	-	33.33	616.67	-	102.29	ha	**718.95**
4.00 L/ha	-	33.33	616.67	-	116.90	ha	**733.57**
General herbicides; in accordance with manufacturer's instructions; "Knapsack" spray application							
"Casoron G (residual)"; at 1 kg/125 m^2	3.62	0.33	6.17	-	3.98	100 m^2	**10.14**
"Roundup Pro - Bi Active"; systemic herbicide at 50 ml/100 m^2	0.53	0.33	6.17	-	0.53	100 m^2	**6.69**
Dextrone X"; contact herbicide at 50 ml/100 m^2	-	0.33	6.17	-	0.46	100 m^2	**6.62**
Fertilizers; in top 150 mm of topsoil at 35 g/m^2							
fertilizer (18:0:0+Mg+Fe)	7.40	0.12	2.27	-	7.40	100 m^2	**9.67**
"Enmag"	6.30	0.12	2.27	-	6.61	100 m^2	**8.88**
fertilizer (7:7:7)	2.16	0.12	2.27	-	2.16	100 m^2	**4.43**
fertilizer (20:10:10)	2.83	0.12	2.27	-	2.83	100 m^2	**5.09**
"Super Phosphate Powder"	2.13	0.12	2.27	-	2.23	100 m^2	**4.50**
"Bone Meal"	3.23	0.12	2.27	-	3.39	100 m^2	**5.66**
Fertilizers; in top 150 mm of topsoil at 70 g/m^2							
fertilizer (18:0:0+Mg+Fe)	14.80	0.12	2.27	-	14.80	100 m^2	**17.07**
"Enmag"	12.59	0.12	2.27	-	13.22	100 m^2	**15.49**
fertilizer (7:7:7)	4.32	0.12	2.27	-	4.32	100 m^2	**6.59**
fertilizer (20:10:10)	5.65	0.12	2.27	-	5.65	100 m^2	**7.92**
"Super Phosphate Powder"	4.25	0.12	2.27	-	4.47	100 m^2	**6.73**
"Bone meal"	6.46	0.12	2.27	-	6.78	100 m^2	**9.05**

Q PAVING/PLANTING/FENCING/SITE FURNITURE

Item Excluding site overheads and profit	PC £	Labour hours	Labour £	Plant £	Material £	Unit	Total rate £
Maintenance operations (Note: the following rates apply to aftercare maintenance executed as part of a landscaping contract only)							
Weeding and hand forking planted areas; including disposing weeds and debris on site; areas maintained weekly	-	-	0.07	-	-	m^2	**0.07**
Weeding and hand forking planted areas; including disposing weeds and debris on site; areas maintained monthly	-	0.01	0.19	-	-	m^2	**0.19**
Extra over weeding and hand forking planted areas for disposing excavated material off site; to tip not exceeding 13 km; mechanically loaded							
slightly contaminated	-	-	-	1.65	15.00	m^3	**16.65**
rubbish	-	-	-	1.65	15.00	m^3	**16.65**
inert material	-	-	-	1.10	10.00	m^3	**11.10**
Mulch; Melcourt Industries Ltd; "Bark Nuggets®"; to plant beds; delivered in 80 m^3 loads; maximum distance 25 m							
50 mm thick	1.50	0.04	0.82	-	1.57	m^2	**2.39**
75 mm thick	2.24	0.07	1.23	-	2.36	m^2	**3.59**
100 mm thick	2.99	0.09	1.64	-	3.14	m^2	**4.78**
Mulch; Melcourt Industries Ltd; "Bark Nuggets®"; to plant beds; delivered in 25 m^3 loads; maximum distance 25 m							
50 mm thick	2.15	0.04	0.82	-	2.25	m^2	**3.07**
75 mm thick	3.22	0.07	1.23	-	3.38	m^2	**4.61**
100 mm thick	4.29	0.09	1.64	-	4.50	m^2	**6.15**
Mulch; Melcourt Industries Ltd; "Amenity Bark Mulch"; FSC; to plant beds; delivered in 80 m^3 loads; maximum distance 25 m							
50 mm thick	0.98	0.04	0.82	-	1.03	m^2	**1.85**
75 mm thick	1.47	0.07	1.23	-	1.55	m^2	**2.78**
100 mm thick	1.97	0.09	1.65	-	2.06	m^2	**3.71**
Mulch; Melcourt Industries Ltd; "Amenity Bark Mulch"; FSC; to plant beds; delivered in 25 m^3 loads; maximum distance 25 m							
50 mm thick	1.63	0.04	0.82	-	1.71	m^2	**2.54**
75 mm thick	2.45	0.07	1.23	-	2.57	m^2	**3.81**
100 mm thick	3.27	0.09	1.64	-	3.43	m^2	**5.07**
Selective herbicides in accordance with manufacturer's instructions; PC £29.23/litre; by machine; application rate							
30 ml/m^2	-	-	-	0.30	0.88	100 m^2	**1.18**
35 ml/m^2	-	-	-	0.30	1.02	100 m^2	**1.32**
40 ml/m^2	-	-	-	0.30	1.17	100 m^2	**1.47**
50 ml/m^2	-	-	-	0.30	1.46	100 m^2	**1.76**
3.00 L/ha	-	-	-	30.01	87.67	ha	**117.69**
3.50 L/ha	-	-	-	30.01	102.29	ha	**132.30**
4.00 L/ha	-	-	-	30.01	116.90	ha	**146.91**
Selective herbicides; in accordance with manufacturer's instructions; PC £29.23/litre; by hand; application rate							
30 ml/m^2	-	0.17	3.08	-	0.88	100 m^2	**3.96**
35 ml/m^2	-	0.17	3.08	-	1.02	100 m^2	**4.11**
40 ml/m^2	-	0.17	3.08	-	1.17	100 m^2	**4.25**
50 ml/m^2	-	0.17	3.08	-	1.46	100 m^2	**4.55**
3.00 L/ha	-	16.67	308.33	-	87.67	ha	**396.01**
3.50 L/ha	-	16.67	308.33	-	102.29	ha	**410.62**
4.00 L/ha	-	16.67	308.34	-	116.90	ha	**425.24**

Q PAVING/PLANTING/FENCING/SITE FURNITURE

Item Excluding site overheads and profit	PC £	Labour hours	Labour £	Plant £	Material £	Unit	Total rate £
Q31 PLANTING - cont'd							
Maintenance operations - cont'd							
General herbicides; in accordance with manufacturer's instructions; "Knapsack" spray application							
"Casoron G (residual)"; at 1 kg/125 m^2	3.62	0.33	6.17	-	3.98	100 m^2	**10.14**
Dextrone X at 50 ml/100 m^2	-	0.33	6.17	-	0.46	100 m^2	**6.62**
"Roundup Pro - Biactive"; at 50 ml/100 m^2	0.53	0.33	6.17	-	0.53	100 m^2	**6.69**
Fertilizers; at 35 g/m^2							
fertilizer (18:0:0+Mg+Fe)	7.40	0.12	2.27	-	7.40	100 m^2	**9.67**
"Enmag"	6.30	0.12	2.27	-	6.61	100 m^2	**8.88**
fertilizer (7:7:7)	2.16	0.12	2.27	-	2.16	100 m^2	**4.43**
fertilizer (20:10:10)	2.83	0.12	2.27	-	2.83	100 m^2	**5.09**
"Super Phosphate Powder"	2.13	0.12	2.27	-	2.23	100 m^2	**4.50**
"Bone meal"	3.23	0.12	2.27	-	3.39	100 m^2	**5.66**
Fertilizers; at 70 g/m^2							
fertilizer (18:0:0+Mg+Fe)	14.80	0.12	2.27	-	14.80	100 m^2	**17.07**
"Enmag"	12.59	0.12	2.27	-	13.22	100 m^2	**15.49**
fertilizer (7:7:7)	4.32	0.12	2.27	-	4.32	100 m^2	**6.59**
fertilizer (20:10:10)	5.65	0.12	2.27	-	5.65	100 m^2	**7.92**
"Super Phosphate Powder"	4.25	0.12	2.27	-	4.47	100 m^2	**6.73**
"Bone meal"	6.46	0.12	2.27	-	6.78	100 m^2	**9.05**
Watering planting; evenly; at a rate of 5 litre/m^2							
using hand-held watering equipment	-	0.25	4.63	-	-	100 m^2	**4.63**
using sprinkler equipment and with sufficient water pressure to run 1 nr 15 m radius sprinkler	-	0.14	2.57	-	-	100 m^2	**2.57**
using movable spray lines powering 3 nr sprinkler heads with a radius of 15 m and allowing for 60% overlap (irrigation machinery costs not included)	-	0.02	0.29	-	-	100 m^2	**0.29**
Forestry planting							
Deep ploughing rough ground to form planting ridges at							
2.00 m centres	-	0.63	11.56	7.51	-	100 m^2	**19.07**
3.00 m centres	-	0.59	10.88	7.07	-	100 m^2	**17.95**
4.00 m centres	-	0.40	7.40	4.81	-	100 m^2	**12.21**
Notching plant forestry seedlings; "T" or "L" notch	29.00	0.75	13.88	-	29.00	100 nr	**42.88**
Turf planting forestry seedlings	29.00	2.00	37.00	-	29.00	100 nr	**66.00**
Selective herbicides; in accordance with manufacturer's instructions; PC £29.23/litre; by hand; application rate							
30 ml/100 m^2	-	0.17	3.08	-	0.88	100 m^2	**3.96**
35 ml/100 m^2	-	0.17	3.08	-	1.02	100 m^2	**4.11**
40 ml/100 m^2	-	0.17	3.08	-	1.17	100 m^2	**4.25**
50 ml/100 m^2	-	0.17	3.08	-	1.46	100 m^2	**4.55**
3.00 L/ha	-	16.67	308.33	-	87.67	ha	**396.01**
3.50 L/ha	-	16.67	308.33	-	102.29	ha	**410.62**
4.00 L/ha	-	16.67	308.33	-	116.90	ha	**425.23**
Selective herbicides; in accordance with manufacturer's instructions; PC £29.23/litre; by machine; application rate							
30 ml/100 m^2	-	-	-	0.30	0.88	100 m^2	**1.18**
35 ml/100 m^2	-	-	-	0.30	1.02	100 m^2	**1.32**
40 ml/100 m^2	-	-	-	0.30	1.17	100 m^2	**1.47**
50 ml/100 m^2	-	-	-	0.30	1.46	100 m^2	**1.76**
3.00 L/ha	-	-	-	30.01	87.67	ha	**117.69**
3.50 L/ha	-	-	-	30.01	102.29	ha	**132.30**
4.00 L/ha	-	-	-	30.01	116.90	ha	**146.91**
Fertilizers; at 35 g/m^2							
fertilizer (18:0:0+Mg+Fe)	7.40	0.12	2.27	-	7.40	100 m^2	**9.67**
"Enmag"	6.30	0.12	2.27	-	6.61	100 m^2	**8.88**
fertilizer (7:7:7)	2.16	0.12	2.27	-	2.16	100 m^2	**4.43**
fertilizer (20:10:10)	2.83	0.12	2.27	-	2.83	100 m^2	**5.09**

Q PAVING/PLANTING/FENCING/SITE FURNITURE

Item Excluding site overheads and profit	PC £	Labour hours	Labour £	Plant £	Material £	Unit	**Total rate £**
"Super Phosphate Powder"	2.13	0.12	2.27	-	2.23	100 m²	**4.50**
"Bone meal"	3.23	0.12	2.27	-	3.39	100 m²	**5.66**
Fertilizers; at 70 g/m²							
fertilizer (18:0:0+Mg + Fe)	14.80	0.12	2.27	-	14.80	100 m²	**17.07**
"Enmag"	12.59	0.12	2.27	-	13.22	100 m²	**15.49**
fertilizer (7:7:7)	4.32	0.12	2.27	-	4.32	100 m²	**6.59**
fertilizer (20:10:10)	5.65	0.12	2.27	-	5.65	100 m²	**7.92**
"Super Phosphate Powder"	4.25	0.12	2.27	-	4.47	100 m²	**6.73**
"Bone meal"	6.46	0.12	2.27	-	6.78	100 m²	**9.05**
Tree tubes; to young trees	125.00	0.30	5.55	-	125.00	100 nr	**130.55**
Cleaning and weeding around seedlings; once	-	0.50	9.25	-	-	100 nr	**9.25**
Treading in and firming ground around seedlings planted; at 2500 per ha after frost or other ground disturbance; once	-	0.33	6.17	-	-	100 nr	**6.17**
Beating up initial planting; once (including supply of replacement seedlings at 10% of original planting)	2.90	0.25	4.63	-	2.90	100 nr	**7.53**
Work to existing planting							
Cutting and trimming ornamental hedges; to specified profiles; including cleaning out hedge bottoms; hedge cut 2 occasions per annum; by hand							
up to 2.00 m high	-	0.03	0.62	-	0.25	m	**0.87**
2.00 - 4.00 m high	-	0.05	0.93	1.58	0.50	m	**3.01**
Cutting and laying field hedges; including stakes and ethering; removing or burning all debris (Note: Rate at which work executed varies greatly with width and height of hedge; a typical hedge could be cut and laid at rate of 7 m run per man day)	-	1.00	18.50	-	1.62	m	**20.12**
labour charge per man day (specialist daywork)	-	1.00	18.50	-	-	m	**18.50**
Trimming field hedges; to specified heights and shapes							
using cutting bar	-	5.00	92.50	-	-	100 m	**92.50**
using flail	-	0.20	3.70	2.79	-	100 m	**6.49**
Q35 LANDSCAPE MAINTENANCE							
Preamble: Long-term landscape maintenance Maintenance on long-term contracts differs in cost from that of maintenance as part of a landscape contract. In this section the contract period is generally 3 - 5 years. Staff are generally allocated to a single project only and therefore productivity is higher whilst overhead costs are lower. Labour costs in this section are lower than the costs used in other parts of the book. Machinery is assumed to be leased over a 5 year period and written off over the same period. The costs of maintenance and consumables for the various machinery types have been included in the information that follows. Finance costs for the machinery have not been allowed for. The rates shown below are for machines working in unconfined contiguous areas. Users should adjust the times and rates if working in smaller spaces or spaces with obstructions.							

Q PAVING/PLANTING/FENCING/SITE FURNITURE

Item Excluding site overheads and profit	PC £	Labour hours	Labour £	Plant £	Material £	Unit	**Total rate £**
Q35 LANDSCAPE MAINTENANCE - cont'd							
Market prices of landscape chemicals at suggested application rates							
Residual herbicides							
Embargo G; Rigby Taylor Ltd; residual pre and post emergent herbicide							
80 - 125 kg/ha; annual meadow-grass, black grass, charlock, common chickweed, common mouse-ear, common orache, common poppy, corn marigold, corn spurrey, fat-hen, groundsel, hedge mustard, scentless mayweed, small nettle, sow-thistle, stinking chamomile, wild-oat	-	-	-	-	0.48	100 m^2	-
Premiere; Scotts Horticulture; granular weedkiller approved for use in new plantings of trees and shrubs							
Application rate 1 kg/100 m^2 (100 kg/ha)	-	-	-	-	30.38	100 m^2	-
Casoron G; Scotts Horticulture; control of broadleaved weeds; existing weeds and germinating weeds; a wide range of annual and perennial weeds							
Selective; 560 g/100 m^2: control of germinating annual and perennial weeds and light to moderate infestation of established annuals	-	-	-	-	2.53	100 m^2	-
Selective; 1 kg/100 m^2: control more persistent weeds in the above category	-	-	-	-	2.53	100 m^2	-
Selective; 1.25 kg/100 m^2: control established weed	-	-	-	-	5.65	100 m^2	-
Selective; 1 kg/280 m^2	-	-	-	-	1.46	100 m^2	-
Contact herbicides; Scotts Horticulture							
Dextrone X (Diquat Paraquat)							
8.5 L/ha	-	-	-	-	0.77	100 m^2	-
3.0 L/ha	-	-	-	-	0.28	100 m^2	-
Speedway 2 (Diquat Paraquat)							
1220 m^2/4 kg	-	-	-	-	2.25	100 m^2	-
3333 m^2/4 kg	-	-	-	-	0.82	100 m^2	-
Selective herbicides							
Rigby Taylor; junction for broadleaf weeds in turf grasses							
Application rate; 1.2 L/ha	-	-	-	-	0.45	100 m^2	-
Rigby Taylor; Greenor systemic selective treatment of weeds in turfgrass							
Application rate; 4 L/ha	-	-	-	-	0.88	100 m^2	-
Rigby Taylor Bastion-T; control of weeds in both established and newly seeded turf							
Application rate	-	-	-	-	0.88	100 m^2	-
Scotts; Intrepid 2; for broadleaf weed control in grass							
Application rate; 7700 m^2/5 L	-	-	-	-	0.52	100 m^2	-
Scotts; Re-Act; selective herbicide; established managed amenity turf, and newly seeded grass; controls many annual and perennial weeds; will not vaporise in hot conditions							
Application rate; 9090 - 14,285 m^2/5 L	-	-	-	-	0.45	100 m^2	-
Bayer Environmental Science; Asulox; post-emergence translocated herbicide for the control of bracken and docks							
Application rate; docks; 17857 m^2/5 L	-	-	-	-	0.42	100 m^2	-
Application rate; bracken; 4545 m^2/5 L	-	-	-	-	1.65	100 m^2	-

Item Excluding site overheads and profit	PC £	Labour hours	Labour £	Plant £	Material £	Unit	**Total rate £**
Bayer Environmental Science; Clovotox; selective herbicide for the control of clover and other difficult broadleaved weeds in turf							
Application rate; 4545 m^2/5 L	-	-	-	-	0.77	100 m^2	-
Dicotox Extra; Bayer Environmental Science; showerproof herbicide for the control of certain broadleaved weeds on turf and for the control of heather and woody weeds in forestry situations; also controls ragwort							
Application rate; 2.8 L/ha; plantains, hawksweed, buttercups, etc.	-	-	-	-	0.22	100 m^2	-
Application rate; 5.6 L/ha; birdsfoot, clovers, mouse-ear, etc. (1 application)	-	-	-	-	0.43	100 m^2	-
Application rate; 5.6 L/ha; pearlwort, yarrow (2 applications)	-	-	-	-	0.87	100 m^2	-
Application rate; 13 L/ha; woody weeds and heather in established trees and shrub areas	-	-	-	-	1.01	100 m^2	-
Dormone; Bayer Environmental Science; selective herbicide for the control of broadleaved weeds in amenity grassland situations; may also be used in situations in or near water							
Application rate; 2.8 L/ha	-	-	-	-	0.18	100 m^2	-
Application rate; 5.6 L/ha	-	-	-	-	0.36	100 m^2	-
Near water; application rate; 4.5 L/ha	-	-	-	-	0.29	100 m^2	-
Near water; application rate; 9 L/ha	-	-	-	-	0.58	100 m^2	-
Spearhead; Bayer Environmental Science; selective herbicide for the control of broadleaved weeds in established turf							
Application rate; 4.5 L/ha	-	-	-	-	0.84	100 m^2	-
Total herbicides; Note: these application rates will vary dependent on season							
Rigby Taylor Ltd; Gallup Biograde amenity; Glyphosate 360 g/L formulation							
General use; woody weeds; ash, beech, bracken, bramble; 3 L/ha	-	-	-	-	0.27	100 m^2	-
Annual and perennial grasses; heather (peat soils); 4 L/ha	-	-	-	-	0.36	100 m^2	-
Pre-planting; general clearance; 5 L/ha	-	-	-	-	0.45	100 m^2	-
Heather; mineral soils; 6 L/ha	-	-	-	-	0.54	100 m^2	-
Rhododendron; 10 L/ha	-	-	-	-	0.90	100 m^2	-
Gallup Hi-aktiv Amenity; 490 g/L formulation							
General use; woody weeds; ash, beech, bracken, bramble; 2.2 L/ha	-	-	-	-	0.23	100 m^2	-
Annual and perennial grasses; heather (peat soils); 2.9 L/ha	-	-	-	-	0.31	100 m^2	-
Pre-planting; general clearance; 3.7 L/ha	-	-	-	-	0.39	100 m^2	-
Heather; mineral soils; 4.4 L/ha	-	-	-	-	0.47	100 m^2	-
Rhododendron; 7.3 L/ha	-	-	-	-	0.77	100 m^2	-
Roundup Pro Biactive; Scotts							
Application rate; 5 L/ha	-	-	-	-	0.53	100 m^2	-
Casoron G; Scotts							
Total; 2.25 kg/100 m^2	-	-	-	-	10.17	100 m^2	-
Kerb							
"Kerb Flowable (Propyzamide)"	-	-	-	-	2.16	100 m^2	-
"Kerb Granules" (2 x 2000 tree pack)	-	-	-	-	0.06	tree	-

Q PAVING/PLANTING/FENCING/SITE FURNITURE

Item Excluding site overheads and profit	PC £	Labour hours	Labour £	Plant £	Material £	Unit	Total rate £
Q35 LANDSCAPE MAINTENANCE - cont'd							
Woody weed herbicides; Rigby Taylor Ltd							
Timbrel; summer applied; water based Bayer Environmental Science; selective scrub and brushwood herbicide							
2.0 L/ha; bramble, briar, broom, gorse, nettle	-	-	-	-	0.98	100 m^2	-
4.0 L/ha; alder, birch, blackthorn, dogwood, elder, poplar, rosebay willowherb, sycamore	-	-	-	-	1.97	100 m^2	-
6.0 L/ha; beech, box, buckthorn, elm, hazel, hornbeam, horse chestnut, lime, maple, privet, rowan, Spanish chestnut, willow, wild pear	-	-	-	-	2.95	100 m^2	-
8.0 L/ha; ash, oak, rhododendron	-	-	-	-	3.93	100 m^2	-
Timbrel; winter applied; paraffin or diesel based; Bayer Environmental Science; selective scrub and brushwood herbicide							
3.0 L/ha; bramble, briar, broom, gorse, nettle	-	-	-	-	1.47	100 m^2	-
6.0 L/ha; alder, ash, beech, birch, blackthorn, box, buckthorn, dogwood, elder, elm, hazel, hornbeam, horse chestnut, lime, maple, oak, poplar, privet, rowan, Spanish chestnut, sycamore, willow, wild pear	-	-	-	-	2.95	100 m^2	-
10.0 L/ha; hawthorn, laurel, rhododendron	-	-	-	-	4.92	100 m^2	-
Aquatic herbicides							
Casoron G; control of broadleaved weeds; existing weeds and germinating weeds; a wide range of annual and perennial weeds							
Aquatic: 1,665 m^2/25 kg	-	-	-	-	6.78	100 m^2	-
Aquatic: 5,555 m^2/25 kg	-	-	-	-	2.03	100 m^2	-
Fungicides							
Rigby Taylor; Masalon; systemic							
Application rate; 8 L/ha	-	-	-	-	7.18	100 m^2	-
Scotts UK Daconil turf							
Application rate; 333 m^2/L	-	-	-	-	5.13	100 m^2	-
Moss control							
Enforcer; Scotts; surface biocide and mosskiller for the control of mosses in turf or on external hard surfaces							
Turf areas; application rate 1 L/200 m^2	-	-	-	-	6.13	100 m^2	-
Hard surfaces; moss and fungi control; application rate 1 L/588 m^2	-	-	-	-	2.11	100 m^2	-
Insect control							
Merit Turf; insecticide for the control of chafer grubs and leatherjackets; Bayer Environmental Science							
Application rate; 10 kg/3333 m^2	-	-	-	-	5.20	100 m^2	-
Mildothane Turf Liquid; casting worm control and fungicide; Bayer Environmental Science							
Application rate; 5 L/6667 m^2	-	-	-	-	0.07	100 m^2	-
Spray mark indicator							
Rigby Taylor; Trail Blazer; mixed with chemicals; price per 1000 L applied	-	-	-	-	19.75	1000 L	-

Q PAVING/PLANTING/FENCING/SITE FURNITURE

Item Excluding site overheads and profit	PC £	Labour hours	Labour £	Plant £	Material £	Unit	**Total rate £**
Grass cutting - ride-on or tractor drawn equipment; Norris & Gardiner Ltd; works carried out on one site only such as large fields or amenity areas where labour and machinery is present for a full day							
Using multiple-gang mower with cylindrical cutters; contiguous areas such as playing fields and the like larger than 3000 m^2							
3 gang; 2.13 m cutting width	-	0.01	0.16	0.11	-	100 m^2	**0.27**
5 gang; 3.40 m cutting width	-	0.01	0.15	0.14	-	100 m^2	**0.29**
7 gang; 4.65 m cutting width	-	0.01	0.09	0.10	-	100 m^2	**0.19**
Using multiple-gang mower with cylindrical cutters; non-contiguous areas such as verges and general turf areas							
3 gang	-	0.02	0.32	0.11	-	100 m^2	**0.43**
5 gang	-	0.01	0.21	0.14	-	100 m^2	**0.35**
Using multiple-rotary mower with vertical drive shaft and horizontally rotating bar or disc cutters; contiguous areas larger than 3000 m^2							
Cutting grass, overgrowth or the like using flail mower or reaper	-	0.02	0.31	0.24	-	100 m^2	**0.55**
Grass cutting - ride-on or tractor drawn equipment; Norris & Gardiner Ltd; works carried out on multiple sites; machinery moved by trailer between sites							
Using multiple-gang mower with cylindrical cutters; contiguous areas such as playing fields and the like larger than 3000 m^2							
3 gang; 2.13 m cutting width	-	0.02	0.26	0.11	-	100 m^2	**0.37**
5 gang; 3.40 m cutting width	-	0.02	0.24	0.14	-	100 m^2	**0.38**
7 gang; 4.65 m cutting width	-	0.01	0.15	0.10	-	100 m^2	**0.25**
Using multiple-gang mower with cylindrical cutters; non-contiguous areas such as verges and general turf areas							
3 gang	-	0.03	0.52	0.11	-	100 m^2	**0.63**
5 gang	-	0.02	0.33	0.14	-	100 m^2	**0.47**
Using multiple-rotary mower with vertical drive shaft and horizontally rotating bar or disc cutters; contiguous areas larger than 3000 m^2							
Cutting grass, overgrowth or the like using flail mower or reaper	-	0.03	0.50	0.24	-	100 m^2	**0.74**
Cutting grass, overgrowth or the like; using tractor-mounted side-arm flail mower; in areas inaccessible to alternative machine; on surface							
not exceeding 30 deg from horizontal	-	0.02	0.36	0.30	-	100 m^2	**0.66**
30 deg to 50 deg from horizontal	-	0.05	0.72	0.30	-	100 m^2	**1.02**
Grass cutting - 'Ride-On' self-propelled equipment; Norris & Gardiner Ltd; works carried out on one site only such as large fields or amenity areas where labour and machinery is present for a full day							
Using ride-on multiple-cylinder mower							
3 gang; 2.13 m cutting width	-	0.01	0.16	0.11	-	100 m^2	**0.27**
5 gang; 3.40 m cutting width	-	0.01	0.10	0.14	-	100 m^2	**0.24**
Using ride-on multiple-rotary mower with horizontally rotating bar, disc or chain cutters							
cutting width 1.52 m	-	0.01	0.20	0.16	-	100 m^2	**0.36**
cutting width 1.82 m	-	0.01	0.17	0.12	-	100 m^2	**0.29**
cutting width 2.97 m	-	0.01	0.10	0.13	-	100 m^2	**0.23**

Item Excluding site overheads and profit	PC £	Labour hours	Labour £	Plant £	Material £	Unit	Total rate £
Q35 LANDSCAPE MAINTENANCE - cont'd							
Grass cutting - 'Ride-On' self-propelled equipment; Norris & Gardiner Ltd; works carried out on multiple sites; machinery moved by trailer between sites							
Using ride-on multiple-cylinder mower							
3 gang; 2.13 m cutting width	-	0.02	0.26	0.11	-	100 m^2	**0.37**
5 gang; 3.40 m cutting width	-	0.01	0.16	0.14	-	100 m^2	**0.30**
Using ride-on multiple-rotary mower with horizontally rotating bar, disc or chain cutters							
cutting width 1.52 m	-	0.02	0.32	0.16	-	100 m^2	**0.48**
cutting width 1.82 m	-	0.02	0.27	0.12	-	100 m^2	**0.39**
cutting width 2.97 m	-	0.01	0.17	0.13	-	100 m^2	**0.30**
Add for using grass box/collector for removal and depositing of arisings	-	0.04	0.59	0.05	-	100 m^2	**0.64**
Grass cutting - pedestrian operated equipment; Norris & Gardiner Ltd							
Using cylinder lawn mower fitted with not less than five cutting blades, front and rear rollers; on surface not exceeding 30 deg from horizontal; arisings let fly; width of cut							
51 cm	-	0.06	0.94	0.19	-	100 m	**1.13**
61 cm	-	0.05	0.79	0.17	-	100 m	**0.96**
71 cm	-	0.04	0.68	0.17	-	100 m	**0.85**
91 cm	-	0.03	0.53	0.17	-	100 m	**0.70**
Using rotary self-propelled mower; width of cut							
45 cm	-	0.04	0.61	0.06	-	100 m^2	**0.67**
81 cm	-	0.02	0.34	0.07	-	100 m^2	**0.41**
91 cm	-	0.02	0.30	0.06	-	100 m^2	**0.36**
120 cm	-	0.01	0.23	0.24	-	100 m^2	**0.47**
Add for using grass box for collecting and depositing arisings							
removing and depositing arisings	-	0.05	0.78	-	-	100 m^2	**0.78**
Add for 30 to 50 deg from horizontal	-	-	-	-	-	33%	**-**
Add for slopes exceeding 50 deg	-	-	-	-	-	100%	**-**
Cutting grass or light woody undergrowth; using trimmer with nylon cord or metal disc cutter; on surface							
not exceeding 30 deg from horizontal	-	0.20	3.10	0.28	-	100 m^2	**3.38**
30-50 deg from horizontal	-	0.40	6.20	0.56	-	100 m^2	**6.76**
exceeding 50 deg from horizontal	-	0.50	7.75	0.70	-	100 m^2	**8.45**
Grass cutting - collecting arisings; Norris & Gardiner Ltd							
Extra over for tractor drawn and self-propelled machinery using attached grass boxes; depositing arisings							
22 cuts per year	-	0.05	0.78	-	-	100 m^2	**0.78**
18 cuts per year	-	0.08	1.16	-	-	100 m^2	**1.16**
12 cuts per year	-	0.10	1.55	-	-	100 m^2	**1.55**
4 cuts per year	-	0.25	3.88	-	-	100 m^2	**3.88**
bedding plant	-	-	-	-	0.27	each	**-**
Arisings collected by trailed sweepers							
22 cuts per year	-	0.02	0.26	0.23	-	100 m^2	**0.49**
18 cuts per year	-	0.02	0.34	0.23	-	100 m^2	**0.57**
12 cuts per year	-	0.03	0.52	0.23	-	100 m^2	**0.75**
4 cuts per year	-	0.07	1.03	0.23	-	100 m^2	**1.26**
Disposing arisings on site; 100 m distance max.							
22 cuts per year	-	0.01	0.12	0.02	-	100 m^2	**0.14**
18 cuts per year	-	0.01	0.15	0.03	-	100 m^2	**0.18**
12 cuts per year	-	0.01	0.22	0.04	-	100 m^2	**0.27**
4 cuts per year	-	0.04	0.68	0.13	-	100 m^2	**0.80**

Q PAVING/PLANTING/FENCING/SITE FURNITURE

Item Excluding site overheads and profit	PC £	Labour hours	Labour £	Plant £	Material £	Unit	Total rate £
Disposal of arisings off site							
22 cuts per year	-	0.01	0.13	0.04	0.50	100 m²	**0.67**
18 cuts per year	-	0.01	0.16	0.05	0.67	100 m²	**0.87**
12 cuts per year	-	0.01	0.10	0.14	1.20	100 m²	**1.44**
4 cuts per year	-	0.02	0.31	0.82	2.00	100 m²	**3.13**
Harrowing							
Harrowing grassed area with							
drag harrow	-	0.01	0.19	0.13	-	100 m²	**0.32**
chain or light flexible spiked harrow	-	0.02	0.26	0.13	-	100 m²	**0.39**
Scarifying mechanical							
A Plant Hire Co Ltd							
Sisis ARP4; including grass collection box; towed by tractor; area scarified annually	-	0.02	0.36	0.10	-	100 m²	**0.46**
Sisis ARP4; including grass collection box; towed by tractor; area scarified two years previously	-	0.03	0.43	0.12	-	100 m²	**0.55**
Pedestrian operated self-powered equipment	-	0.07	1.08	0.21	-	100 m²	**1.29**
Add for disposal of arisings	-	0.03	0.39	1.02	10.00	100 m²	**11.41**
Scarifying by hand							
hand implement	-	0.50	7.75	-	-	100 m²	**7.75**
add for disposal of arisings	-	0.03	0.39	1.02	10.00	100 m²	**11.41**
Rolling							
Rolling grassed area; equipment towed by tractor; once over; using							
smooth roller	-	0.01	0.22	0.15	-	100 m²	**0.38**
Turf aeration							
By machine; A Plant Hire Co Ltd							
Vertidrain turf aeration equipment towed by tractor to effect a minimum penetration of 100 to 250 mm at 100 mm centres	-	0.04	0.60	1.23	-	100 m²	**1.83**
Ryan GA 30; self-propelled turf aerating equipment; to effect a minimum penetration of 100 mm at varying centres	-	0.04	0.60	1.43	-	100 m²	**2.04**
Groundsman; pedestrian operated; self-powered solid or slitting tine turf aerating equipment to effect a minimum penetration of 100 mm	-	0.14	2.17	1.36	-	100 m²	**3.53**
Cushman core harvester; self-propelled; for collection of arisings	-	0.02	0.30	0.63	-	100 m²	**0.93**
By hand							
hand fork; to effect a minimum penetration of 100 mm and spaced 150 mm apart	-	1.33	20.67	-	-	100 m²	**20.67**
hollow tine hand implement; to effect a minimum penetration of 100 mm and spaced 150 mm apart	-	2.00	31.00	-	-	100 m²	**31.00**
collection of arisings by hand	-	3.00	46.50	-	-	100 m²	**46.50**
Turf areas; surface treatments and top dressing; Boughton Loam Ltd							
Apply screened topdressing to grass surfaces; spread using Tru-Lute							
Sand soil mixes 90/10 to 50/50	0.11	-	0.03	0.03	0.11	m²	**0.17**
Apply screened soil 3 mm, Kettering loam to goal mouths and worn areas							
20 mm thick	0.91	0.01	0.16	-	1.09	m²	**1.24**
10 mm thick	0.45	0.01	0.16	-	0.54	m²	**0.70**

Q PAVING/PLANTING/FENCING/SITE FURNITURE

Item Excluding site overheads and profit	PC £	Labour hours	Labour £	Plant £	Material £	Unit	Total rate £
Q35 LANDSCAPE MAINTENANCE - cont'd							
Leaf clearance; clearing grassed area of leaves and other extraneous debris							
Using equipment towed by tractor							
Large grassed areas with perimeters of mature trees such as sports fields and amenity areas	-	0.01	0.19	0.07	-	100 m^2	**0.26**
Large grassed areas containing ornamental trees and shrub beds	-	-	0.07	0.93	-	100 m^2	**1.00**
Using pedestrian operated mechanical equipment and blowers							
Grassed areas with perimeters of mature trees such as sports fields and amenity areas	-	0.02	0.26	1.49	-	100 m^2	**1.75**
Grassed areas containing ornamental trees and shrub beds	-	0.10	1.55	0.64	-	100 m^2	**2.19**
Verges	-	0.07	1.03	0.08	-	100 m^2	**1.12**
By hand							
Grassed areas with perimeters of mature trees such as sports fields and amenity areas	-	0.10	1.55	0.20	-	100 m^2	**1.75**
grassed areas containing ornamental trees and shrub beds	-	0.17	2.58	0.33	-	100 m^2	**2.92**
verges	-	0.33	5.17	0.66	-	100 m^2	**5.83**
Removal of arisings							
Areas with perimeters of mature trees	-	0.01	0.10	0.09	0.80	100 m^2	**0.99**
Areas containing ornamental trees and shrub beds	-	0.02	0.31	0.34	2.00	100 m^2	**2.65**
Litter clearance							
Collection and disposal of litter from grassed area	-	0.01	0.16	-	0.05	100 m^2	**0.20**
Collection and disposal of litter from isolated grassed area not exceeding 1000 m^2	-	0.04	0.62	-	0.05	100 m^2	**0.67**
Edge maintenance							
Maintain edges where lawn abuts pathway or hard surface using							
Strimmer	-	0.01	0.08	0.01	-	m	**0.08**
Shears	-	0.02	0.26	-	-	m	**0.26**
Maintain edges where lawn abuts plant bed using							
Mechanical edging tool	-	0.01	0.10	0.03	-	m	**0.13**
Shears	-	0.01	0.17	-	-	m	**0.17**
Half moon edging tool	-	0.02	0.31	-	-	m	**0.31**
Tree guards, stakes and ties etc.							
Adjusting existing tree tie	-	0.03	0.52	-	-	nr	**0.52**
Taking up single or double tree stake and ties; removing and disposing	-	0.05	0.93	-	-	nr	**0.93**
Pruning shrubs							
Trimming ground cover planting							
Soft groundcover; vinca ivy and the like	-	1.00	15.50	-	-	100 m^2	**15.50**
Woody groundcover; cotoneaster and the like	-	1.50	23.25	-	-	100 m^2	**23.25**
Pruning massed shrub border (measure ground area)							
Shrub beds pruned annually	-	0.01	0.16	-	-	m^2	**0.16**
Shrub beds pruned hard every 3 years	-	0.03	0.43	-	-	m^2	**0.43**
Cutting off dead heads							
Bush or standard rose	-	0.05	0.78	-	-	nr	**0.78**
Climbing rose	-	0.08	1.29	-	-	nr	**1.29**
Pruning roses							
Bush or standard rose	-	0.05	0.78	-	-	nr	**0.78**
Climbing rose or rambling rose; tying in as required	-	0.07	1.03	-	-	nr	**1.03**

Q PAVING/PLANTING/FENCING/SITE FURNITURE

Item Excluding site overheads and profit	PC £	Labour hours	Labour £	Plant £	Material £	Unit	**Total rate £**
Pruning ornamental shrub; height before pruning (increase these rates by 50% if pruning work has not been executed during the previous two years)							
not exceeding 1 m	-	0.04	0.62	-	-	nr	**0.62**
1 to 2 m	-	0.06	0.86	-	-	nr	**0.86**
exceeding 2 m	-	0.13	1.94	-	-	nr	**1.94**
Removing excess growth etc from face of building etc.; height before pruning							
not exceeding 2 m	-	0.03	0.44	-	-	nr	**0.44**
2 to 4 m	-	0.05	0.78	-	-	nr	**0.78**
4 to 6 m	-	0.08	1.29	-	-	nr	**1.29**
6 to 8 m	-	0.13	1.94	-	-	nr	**1.94**
8 to 10 m	-	0.14	2.21	-	-	nr	**2.21**
Removing epicormic growth from base of shrub or trunk and base of tree: any height; any diameter; number of growths							
not exceeding 10	-	0.05	0.78	-	-	nr	**0.78**
10 to 20	-	0.07	1.03	-	-	nr	**1.03**
Beds, borders and planters							
Lifting							
bulbs	-	0.50	7.75	-	-	100 nr	**7.75**
tubers or corms	-	0.40	6.20	-	-	100 nr	**6.20**
established herbaceous plants; hoeing and depositing for replanting	-	2.00	31.00	-	-	100 nr	**31.00**
Temporary staking and tying in herbaceous plant	-	0.03	0.52	-	0.10	nr	**0.62**
Cutting down spent growth of herbaceous plant; clearing arisings							
unstaked	-	0.02	0.31	-	-	nr	**0.31**
staked; not exceeding 4 stakes per plant; removing stakes and putting into store	-	0.03	0.39	-	-	nr	**0.39**
Hand weeding							
newly planted areas	-	2.00	31.00	-	-	100 m^2	**31.00**
established areas	-	0.50	7.75	-	-	100 m^2	**7.75**
Removing grasses from groundcover areas	-	3.00	46.55	-	-	100 m^2	**46.55**
Hand digging with fork; not exceeding 150 mm deep; breaking down lumps; leaving surface with a medium tilth	-	1.33	20.67	-	-	100 m^2	**20.67**
Hand digging with fork or spade to an average depth of 230 mm; breaking down lumps; leaving surface with a medium tilth	-	2.00	31.00	-	-	100 m^2	**31.00**
Hand hoeing; not exceeding 50 mm deep; leaving surface with a medium tilth	-	0.40	6.20	-	-	100 m^2	**6.20**
Hand raking to remove stones etc.; breaking down lumps: leaving surface with a fine tilth prior to planting	-	0.67	10.33	-	-	100 m^2	**10.33**
Hand weeding; planter, window box; not exceeding 1.00 m^2							
ground level box	-	0.05	0.78	-	-	nr	**0.78**
box accessed by stepladder	-	0.08	1.29	-	-	nr	**1.29**
Spreading only compost, mulch or processed bark to a depth of 75 mm							
on shrub bed with existing mature planting	-	0.09	1.41	-	-	m^2	**1.41**
recently planted areas	-	0.07	1.03	-	-	m^2	**1.03**
groundcover and herbaceous areas	-	0.08	1.16	-	-	m^2	**1.16**
Clearing cultivated area of leaves, litter and other extraneous debris; using hand implement							
weekly maintenance	-	0.13	1.94	-	-	100 m^2	**1.94**
daily maintenance	-	0.02	0.26	-	-	100 m^2	**0.26**

Q PAVING/PLANTING/FENCING/SITE FURNITURE

Item Excluding site overheads and profit	PC £	Labour hours	Labour £	Plant £	Material £	Unit	Total rate £
Q35 LANDSCAPE MAINTENANCE - cont'd							
Bedding							
Lifting							
bedding plants; hoeing and depositing for disposal	-	3.00	46.50	-	-	100 m^2	**46.50**
Hand digging with fork; not exceeding 150 mm deep: breaking down lumps; leaving surface with a medium tilth	-	0.75	11.63	-	-	100 m^2	**11.63**
Hand weeding							
newly planted areas	-	2.00	31.00	-	-	100 m^2	**31.00**
established areas	-	0.50	7.75	-	-	100 m^2	**7.75**
Hand digging with fork or spade to an average depth of 230 mm; breaking down lumps; leaving surface with a medium tilth	-	0.50	7.75	-	-	100 m^2	**7.75**
Hand hoeing: not exceeding 50 mm deep; leaving surface with a medium tilth	-	0.40	6.20	-	-	100 m^2	**6.20**
Hand raking to remove stones etc.; breaking down lumps; leaving surface with a fine tilth prior to planting	-	0.67	10.33	-	-	100 m^2	**10.33**
Hand weeding; planter, window box; not exceeding 1.00 m^2							
ground level box	-	0.05	0.78	-	-	nr	**0.78**
box accessed by stepladder	-	0.08	1.29	-	-	nr	**1.29**
Spreading only; compost, mulch or processed bark to a depth of 75 mm							
on shrub bed with existing mature planting	-	0.09	1.41	-	-	m^2	**1.41**
recently planted areas	-	0.07	1.03	-	-	m^2	**1.03**
groundcover and herbaceous areas	-	0.08	1.16	-	-	m^2	**1.16**
Collecting bedding from nursery	-	3.00	46.50	13.64	-	100 m^2	**60.14**
Setting out							
mass planting single variety	-	0.13	1.94	-	-	m^2	**1.94**
pattern	-	0.33	5.17	-	-	m^2	**5.17**
Planting only							
massed bedding plants	-	0.20	3.10	-	-	m^2	**3.10**
Clearing cultivated area of leaves, litter and other extraneous debris; using hand implement							
weekly maintenance	-	0.13	1.94	-	-	100 m^2	**1.94**
daily maintenance	-	0.02	0.26	-	-	100 m^2	**0.26**
Irrigation and watering							
Hand held hosepipe; flow rate 25 litres per minute; irrigation requirement							
10 litres/m^2	-	0.74	11.42	-	-	100 m^2	**11.42**
15 litres/m^2	-	1.10	17.05	-	-	100 m^2	**17.05**
20 litres/m^2	-	1.46	22.68	-	-	100 m^2	**22.68**
25 litres/m^2	-	1.84	28.47	-	-	100 m^2	**28.47**
Hand held hosepipe; flow rate 40 litres per minute; irrigation requirement							
10 litres/m^2	-	0.46	7.16	-	-	100 m^2	**7.16**
15 litres/m^2	-	0.69	10.74	-	-	100 m^2	**10.74**
20 litres/m^2	-	0.91	14.15	-	-	100 m^2	**14.15**
25 litres/m^2	-	1.15	17.77	-	-	100 m^2	**17.77**
Hedge cutting; field hedges cut once or twice annually							
Trimming sides and top using hand tool or hand held mechanical tools							
not exceeding 2 m high	-	0.10	1.55	0.14	-	10 m^2	**1.69**
2 to 4 m high	-	0.33	5.17	0.47	-	10 m^2	**5.63**

Q PAVING/PLANTING/FENCING/SITE FURNITURE

Item Excluding site overheads and profit	PC £	Labour hours	Labour £	Plant £	Material £	Unit	**Total rate £**
Hedge cutting; ornamental							
Trimming sides and top using hand tool or hand held mechanical tools							
not exceeding 2 m high	-	0.13	1.94	0.18	-	10 m^2	**2.11**
2 to 4 m high	-	0.50	7.75	0.70	-	10 m^2	**8.45**
Hedge cutting; reducing width; hand tool or hand held mechanical tools							
Not exceeding 2 m high							
Average depth of cut not exceeding 300 mm	-	0.10	1.55	0.14	-	10 m^2	**1.69**
Average depth of cut 300 to 600 mm	-	0.83	12.92	1.17	-	10 m^2	**14.08**
Average depth of cut 600 to 900 mm	-	1.25	19.38	1.75	-	10 m^2	**21.13**
2 to 4 m high							
Average depth of cut not exceeding 300 mm	-	0.03	0.39	0.04	-	10 m^2	**0.42**
Average depth of cut 300 to 600 mm	-	0.13	1.94	0.18	-	10 m^2	**2.11**
Average depth of cut 600 to 900 mm	-	2.50	38.75	3.50	-	10 m^2	**42.25**
4 to 6 m high							
Average depth of cut not exceeding 300 mm	-	0.10	1.55	0.14	-	m^2	**1.69**
Average depth of cut 300 to 600 mm	-	0.17	2.58	0.23	-	m^2	**2.82**
Average depth of cut 600 to 900 mm	-	0.50	7.75	0.70	-	m^2	**8.45**
Hedge cutting; reducing width; tractor mounted hedge cutting equipment							
Not exceeding 2 m high							
Average depth of cut not exceeding 300 mm	-	0.04	0.62	0.67	-	10 m^2	**1.29**
Average depth of cut 300 to 600 mm	-	0.05	0.78	0.83	-	10 m^2	**1.61**
Average depth of cut 600 to 900 mm	-	0.20	3.10	3.33	-	10 m^2	**6.43**
2 to 4 m high							
Average depth of cut not exceeding 300 mm	-	0.01	0.19	0.21	-	10 m^2	**0.40**
Average depth of cut 300 to 600 mm	-	0.03	0.39	0.42	-	10 m^2	**0.80**
Average depth of cut 600 to 900 mm	-	0.02	0.31	0.33	-	10 m^2	**0.64**
Hedge cutting; reducing height; hand tool or hand held mechanical tools							
Not exceeding 2 m high							
Average depth of cut not exceeding 300 mm	-	0.07	1.03	0.09	-	10 m^2	**1.13**
Average depth of cut 300 to 600 mm	-	0.13	2.07	0.19	-	10 m^2	**2.25**
Average depth of cut 600 to 900 mm	-	0.40	6.20	0.56	-	10 m^2	**6.76**
2 to 4 m high							
Average depth of cut not exceeding 300 mm	-	0.03	0.52	0.05	-	m^2	**0.56**
Average depth of cut 300 to 600 mm	-	0.07	1.03	0.09	-	m^2	**1.13**
Average depth of cut 600 to 900 mm	-	0.20	3.10	0.28	-	m^2	**3.38**
4 to 6 m high							
Average depth of cut not exceeding 300 mm	-	0.07	1.03	0.09	-	m^2	**1.13**
Average depth of cut 300 to 600 mm	-	0.13	1.94	0.18	-	m^2	**2.11**
Average depth of cut 600 to 900 mm	-	0.25	3.88	0.35	-	m^2	**4.23**
Hedge cutting; removal and disposal of arisings							
Sweeping up and depositing arisings							
300 mm cut	-	0.05	0.78	-	-	10 m^2	**0.78**
600 mm cut	-	0.20	3.10	-	-	10 m^2	**3.10**
900 mm cut	-	0.40	6.20	-	-	10 m^2	**6.20**
Chipping arisings							
300 mm cut	-	0.02	0.31	0.11	-	10 m^2	**0.42**
600 mm cut	-	0.08	1.29	0.48	-	10 m^2	**1.77**
900 mm cut	-	0.20	3.10	1.14	-	10 m^2	**4.24**
Disposal of unchipped arisings							
300 mm cut	-	0.02	0.26	0.51	1.33	10 m^2	**2.10**
600 mm cut	-	0.03	0.52	1.02	2.00	10 m^2	**3.54**
900 mm cut	-	0.08	1.29	2.56	5.00	10 m^2	**8.85**
Disposal of chipped arisings							
300 mm cut	-	-	0.05	0.17	2.00	10 m^2	**2.22**
600 mm cut	-	0.02	0.26	0.17	4.00	10 m^2	**4.43**
900 mm cut	-	0.03	0.52	0.34	10.00	10 m^2	**10.86**

Q PAVING/PLANTING/FENCING/SITE FURNITURE

Item Excluding site overheads and profit	PC £	Labour hours	Labour £	Plant £	Material £	Unit	Total rate £
Q35 LANDSCAPE MAINTENANCE - cont'd							
Herbicide applications; CDA (controlled droplet application); chemical application via low pressure specialised wands to landscape planting; application to maintain 1.00 m diameter clear circles (0.79 m²) around new planting							
Scotts Ltd; Roundup Bi-Active glyphosate; enhanced movement glyphosate; application rate 15 litre/ha							
plants at 1.50 m centres; 4444 nr/ha	66.80	18.52	342.62	-	73.48	ha	**416.10**
plants at 1.75 m centres; 3265 nr/ha	49.14	13.60	251.60	-	54.05	ha	**305.65**
plants at 2.00 m centres; 2500 nr/ha	37.59	10.42	192.77	-	41.35	ha	**234.12**
mass spraying	190.50	10.00	185.00	-	209.55	ha	**394.55**
Bayer Environmental Science; Vanquish Biactive; enhanced movement glyphosate; application rate 15 litre/ha							
plants at 1.50 m centres; 4444 nr/ha	51.28	18.52	342.62	-	56.41	ha	**399.03**
plants at 1.75 m centres; 3265 nr/ha	37.63	13.60	251.60	-	41.40	ha	**293.00**
plants at 2.00 m centres; 2500 nr/ha	28.86	10.42	192.77	-	31.75	ha	**224.52**
mass spraying	146.25	10.00	185.00	-	160.88	ha	**345.88**
Xanadu; glyphosate and diuron for control of emerged weeds at a single application per season; application rate 20 litre/ha							
plants at 1.50 m centres; 4444 nr/ha	103.45	18.52	342.62	-	113.79	ha	**456.41**
plants at 1.75 m centres; 3265 nr/ha	75.89	13.60	251.60	-	83.48	ha	**335.08**
plants at 2.00 m centres; 2500 nr/ha	58.21	10.42	192.77	-	64.03	ha	**256.80**
mass spraying	294.72	10.00	185.00	-	324.19	ha	**509.19**
Herbicide applications; standard backpack spray applicators; application to maintain 1.00 m diameter clear circles (0.79 m²) around new planting							
Scotts Ltd; Roundup Pro Bi- Active; glyphosate; enhanced movement glyphosate; application rate 5 litre/ha							
plants at 1.50 m centres; 4444 nr/ha	18.45	24.69	456.76	-	20.29	ha	**477.05**
plants at 1.75 m centres; 3265 nr/ha	13.49	18.14	335.59	-	14.84	ha	**350.43**
plants at 2.00 m centres; 2500 nr/ha	10.41	13.89	256.96	-	11.45	ha	**268.41**
mass spraying	52.70	13.00	240.50	-	57.97	ha	**298.47**
Q40 FENCING							
Protective fencing							
Cleft chestnut rolled fencing; fixing to 100 mm diameter chestnut posts; driving into firm ground at 3 m centres							
900 mm high	3.32	0.11	1.97	-	6.14	m	**8.11**
1200 mm high	3.90	0.16	2.96	-	6.75	m	**9.71**
1500 mm high; 3 strand	6.18	0.21	3.95	-	8.84	m	**12.79**
Enclosures; Earth Anchors							
"Rootfast" anchored galvanized steel enclosures post ref ADP 20-1000; 1000 mm high x 20 mm diameter with ref AA25-750 socket and padlocking ring	34.00	0.10	1.85	-	36.17	nr	**38.02**
steel cable, orange plastic coated	1.25	-	0.04	-	1.25	m	**1.29**

Q PAVING/PLANTING/FENCING/SITE FURNITURE

Item Excluding site overheads and profit	PC £	Labour hours	Labour £	Plant £	Material £	Unit	Total rate £
Boundary fencing; strained wire and wire mesh; AVS Fencing Supplies Ltd							
Strained wire fencing; concrete posts only at 2750 mm centres, 610 mm below ground; excavating holes; filling with concrete; replacing topsoil; disposing surplus soil off site							
900 mm high	2.91	0.56	10.28	0.60	4.25	m	**15.12**
1200 mm high	4.38	0.56	10.28	0.60	5.71	m	**16.59**
1400 mm high	4.63	0.56	10.28	0.60	5.97	m	**16.84**
1800 mm high	4.99	0.78	14.39	1.49	6.33	m	**22.21**
2400 mm high	7.17	1.67	30.84	0.60	8.50	m	**39.93**
Extra over strained wire fencing for concrete straining posts with one strut; posts and struts 610 mm below ground; struts cleats, stretchers, winders, bolts, and eye bolts; excavating holes; filling to within 150 mm of ground level with concrete (1:12) - 40 mm aggregate; replacing topsoil; disposing surplus soil off site							
900 mm high	23.57	0.67	12.39	1.34	52.57	nr	**66.31**
1200 mm high	26.21	0.67	12.34	1.35	55.86	nr	**69.55**
1400 mm high	30.10	0.67	12.34	1.35	64.05	nr	**77.74**
1800 mm high	37.29	0.67	12.34	1.35	68.87	nr	**82.56**
2400 mm high	51.25	0.83	15.41	1.59	91.14	nr	**108.15**
Extra over strained wire fencing for concrete straining posts with two struts; posts and struts 610 mm below ground; excavating holes; filling to within 150 mm of ground level with concrete (1:12) - 40 mm aggregate; replacing topsoil; disposing surplus soil off site							
900 mm high	33.99	0.91	16.84	1.73	77.00	nr	**95.56**
1200 mm high	37.89	0.85	15.72	1.35	81.99	nr	**99.06**
1400 mm high	43.70	0.91	16.80	1.35	100.42	nr	**118.56**
1800 mm high	54.14	0.85	15.72	1.35	106.12	nr	**123.19**
2400 mm high	51.25	0.67	12.33	1.59	109.46	nr	**123.38**
Strained wire fencing; galvanised steel angle posts only at 2750 mm centres; 610 mm below ground; driving in							
900 mm high; 40 mm x 40 mm x 5 mm	4.97	0.03	0.56	-	4.97	m	**5.53**
1200 mm high; 40 mm x 40 mm x 5 mm	5.83	0.03	0.62	-	5.83	m	**6.45**
1400 mm high; 40 mm x 40 mm x 5 mm	6.96	0.06	1.12	-	6.96	m	**8.08**
1800 mm high; 40 mm x 40 mm x 5 mm	8.46	0.07	1.34	-	8.46	m	**9.81**
2400 mm high; 45 mm x 45 mm x 5 mm	10.26	0.12	2.24	-	10.26	m	**12.50**
Galvanised steel angle straining posts with one strut for strained wire fencing; setting in concrete							
900 mm high; 50 mm x 50 mm x 6 mm	50.89	0.67	12.33	-	50.89	nr	**63.22**
1200 mm high; 50 mm x 50 mm x 6 mm	71.18	0.67	12.33	-	71.18	nr	**83.51**
1400 mm high; 50 mm x 50 mm x 6 mm	82.74	0.67	12.33	-	82.74	nr	**95.07**
1800 mm high; 50 mm x 50 mm x 6 mm	94.08	0.67	12.33	-	94.08	nr	**106.41**
2400 mm high; 60 mm x 60 mm x 6 mm	112.32	0.67	12.33	-	112.32	nr	**124.65**
Galvanised steel straining posts with two struts for strained wire fencing; setting in concrete							
900 mm high; 50 mm x 50 mm x 6 mm	76.15	1.00	18.50	-	76.15	nr	**94.65**
1200 mm high; 50 mm x 50 mm x 6 mm	104.78	1.00	18.50	-	104.78	nr	**123.28**
1400 mm high; 50 mm x 50 mm x 6 mm	124.06	1.00	18.50	-	124.06	nr	**142.56**
1800 mm high; 50 mm x 50 mm x 6 mm	-	1.00	18.50	-	150.83	nr	**169.33**
2400 mm high; 60 mm x 60 mm x 6 mm	187.20	1.00	18.50	-	187.20	nr	**205.70**
Strained wire; to posts (posts not included); 3 mm galvanized wire; fixing with galvanized stirrups							
900 mm high; 2 wire	0.26	0.03	0.61	-	0.37	m	**0.98**
1200 mm high; 3 wire	0.39	0.05	0.86	-	0.50	m	**1.36**
1400 mm high; 3 wire	0.39	0.05	0.86	-	0.50	m	**1.36**
1800 mm high; 3 wire	0.39	0.05	0.86	-	0.50	m	**1.36**

Q PAVING/PLANTING/FENCING/SITE FURNITURE

Item Excluding site overheads and profit	PC £	Labour hours	Labour £	Plant £	Material £	Unit	Total rate £
Q40 FENCING - cont'd							
Boundary fencing - cont'd							
Barbed wire; to posts (posts not included); 3 mm galvanised wire; fixing with galvanised stirrups							
900 mm high; 2 wire	0.24	0.07	1.23	-	0.35	m	**1.58**
1200 mm high; 3 wire	0.36	0.09	1.72	-	0.47	m	**2.19**
1400 mm high; 3 wire	0.36	0.09	1.72	-	0.47	m	**2.19**
1800 mm high; 3 wire	0.36	0.09	1.72	-	0.47	m	**2.19**
Chain link fencing; AVS Fencing Supplies Ltd; to strained wire and posts priced separately; 3 mm galvanised wire; 51 mm mesh; galvanized steel components; fixing to line wires threaded through posts and strained with eye-bolts; posts (not included)							
900 mm high	4.40	0.07	1.23	-	4.51	m	**5.74**
1200 mm high	5.49	0.07	1.23	-	5.60	m	**6.84**
1800 mm high	6.20	0.10	1.85	-	6.31	m	**8.16**
2400 mm high	9.60	0.10	1.85	-	9.71	m	**11.56**
Chain link fencing; to strained wire and posts priced separately; 3.15 mm plastic coated galvanized wire (wire only 2.50 mm); 51 mm mesh; galvanised steel components; fencing with line wires threaded through posts and strained with eye-bolts; posts (not included) (Note: plastic coated fencing can be cheaper than galvanised finish as wire of a smaller cross-sectional area can be used)							
900 mm high	2.60	0.07	1.23	-	2.82	m	**4.05**
1200 mm high	3.60	0.07	1.23	-	3.82	m	**5.05**
1400 mm high	4.40	0.07	1.23	-	4.62	m	**5.85**
1800 mm high	4.96	0.13	2.31	-	5.84	m	**8.15**
Extra over strained wire fencing for cranked arms and galvanized barbed wire							
1 row	2.79	0.02	0.31	-	2.79	m	**3.10**
2 row	2.91	0.05	0.93	-	2.91	m	**3.84**
3 row	3.03	0.05	0.93	-	3.03	m	**3.96**
Field fencing; Jacksons Fencing; welded wire mesh; fixed to posts and straining wires measured separately							
Cattle fence; 1100 m high 114 x 300 mm at bottom to 230 x 300 mm at top	0.97	0.10	1.85	-	1.17	m	**3.02**
Sheep fence; 900 mm high; 140 x 300 mm at bottom to 230 x 300 mm at top	0.86	0.10	1.85	-	1.06	m	**2.91**
Deer fence; 1900 mm high; 89 x 150 mm at bottom to 267 x 300 mm at top	3.22	0.13	2.31	-	3.42	m	**5.73**
Extra for concreting in posts	-	0.50	9.25	-	3.86	nr	**13.11**
Extra for straining post	8.82	0.75	13.88	-	8.82	nr	**22.70**
Rabbit netting; Jacksons Fencing; timber stakes; peeled kiln dried pressure treated; pointed; 1.8 m posts driven 900 mm into ground at 3 m centres (line wires and netting priced separately)							
75 - 100 mm stakes	1.62	0.25	4.63	-	1.62	m	**6.25**
Corner posts or straining posts 150 mm diameter 2.3 m high set in concrete; centres to suit local conditions or changes of direction							
1 strut	10.57	1.00	18.50	3.84	16.89	each	**39.23**
2 strut	12.68	1.00	18.50	3.84	19.00	each	**41.34**
Strained wire; to posts (posts not included); 3 mm galvanized wire; fixing with galvanized stirrups							
900 mm high; 2 wire	0.26	0.03	0.61	-	0.37	m	**0.98**
1200 mm high; 3 wire	0.39	0.05	0.86	-	0.50	m	**1.36**

Q PAVING/PLANTING/FENCING/SITE FURNITURE

Item Excluding site overheads and profit	PC £	Labour hours	Labour £	Plant £	Material £	Unit	**Total rate £**
Rabbit netting; 31 mm 19 gauge 1050 high netting fixed to posts line wires and straining posts or corner posts all priced separately							
900 high turned in	0.70	0.04	0.74	-	0.71	m	**1.45**
900 high buried 150 mm in trench	0.70	0.08	1.54	-	0.71	m	**2.25**
Boundary fencing; strained wire and wire mesh; Jacksons Fencing							
Tubular chain link fencing; galvanized; plastic-coated; 60.3 mm diameter posts at 3.0 m centres; setting 700 mm into ground; choice of ten mesh colours; including excavating holes; backfilling and removing surplus soil; with top rail only							
900 mm high	-	-	-	-	-	m	**27.21**
1200 mm high	-	-	-	-	-	m	**28.68**
1800 mm high	-	-	-	-	-	m	**31.82**
2000 mm high	-	-	-	-	-	m	**33.74**
Tubular chain link fencing; galvanised; plastic coated; 60.3 mm diameter posts at 3.0 m centres; cranked arms and 3 lines barbed wire; setting 700 mm into ground; including excavating holes; backfilling and removing surplus soil; with top rail only							
2000 mm high	-	-	-	-	-	m	**35.33**
1800 mm high	-	-	-	-	-	m	**34.64**
Boundary fencing; Steelway-Fensecure Ltd							
"Classic" 2 rail tubular fencing; top and bottom with strecher bars and straining wires in between; comprising 60.3 mm tubular posts at 3.00 m centres; setting in concrete; 35 mm top rail tied with aluminium and steel fittings; 50 mm x 50 mm x 355/2.5 mm PVC coated chain link; all components galvanized and coated in green nylon							
964 mm high	-	-	-	-	-	m	**29.00**
1269 mm high	-	-	-	-	-	m	**30.79**
1574 mm high	-	-	-	-	-	m	**35.75**
1878 mm high	-	-	-	-	-	m	**36.46**
2188 mm high	-	-	-	-	-	m	**38.44**
2458 mm high	-	-	-	-	-	m	**43.06**
2948 mm high	-	-	-	-	-	m	**55.22**
3562 mm high	-	-	-	-	-	m	**57.42**
End posts; classic range 60.3 mm diameter; setting in concrete							
964 mm high	-	-	-	-	-	nr	**41.69**
1269 mm high	-	-	-	-	-	nr	**44.75**
1574 mm high	-	-	-	-	-	nr	**48.91**
1878 mm high	-	-	-	-	-	nr	**54.20**
2188 mm high	-	-	-	-	-	nr	**58.84**
2458 mm high	-	-	-	-	-	nr	**67.98**
2948 mm high	-	-	-	-	-	nr	**77.31**
3562 mm high	-	-	-	-	-	nr	**90.91**
Corner posts; 60.3 mm diameter; setting in concrete							
964 mm high	-	-	-	-	-	nr	**33.00**
1269 mm high	-	-	-	-	-	nr	**16.79**
1574 mm high	-	-	-	-	-	nr	**37.75**
1878 mm high	-	-	-	-	-	nr	**38.46**
2188 mm high	-	-	-	-	-	nr	**44.94**
2458 mm high	-	-	-	-	-	nr	**47.56**
2948 mm high		-	-	-	-	nr	**57.22**
3562 mm high	-	-	-	-	-	nr	**63.92**

Q PAVING/PLANTING/FENCING/SITE FURNITURE

Item Excluding site overheads and profit	PC £	Labour hours	Labour £	Plant £	Material £	Unit	Total rate £
Q40 FENCING - cont'd							
Boundary fencing; McArthur Group							
"Paladin" welded mesh colour coated green fencing; fixing to metal posts at 2.975 m centres with manufacturer's fixings; setting 600 mm deep in firm ground; including excavating holes; backfilling and removing surplus excavated material; includes 10 year product guarantee							
1800 mm high	48.90	0.66	12.21	-	48.90	m	**61.11**
2000 mm high	49.91	0.66	12.21	-	57.30	m	**69.51**
2400 mm high	57.51	0.66	12.21	-	66.11	m	**78.32**
Extra over welded galvanised plastic coated mesh fencing for concreting in posts	7.72	0.11	2.06	-	7.72	m	**9.77**
Boundary fencing							
"Orsogril" rectangular steel bar mesh fence panels; pleione pattern; bolting to 60 mm x 8 mm uprights at 2 m centres; mesh 62 mm x 66 mm; setting in concrete							
930 mm high panels	-	-	-	-	-	m	**118.00**
1326 mm high panels	-	-	-	-	-	m	**151.00**
1722 mm high panels	-	-	-	-	-	m	**198.00**
"Orsogril" rectangular steel bar mesh fence panels; pleione pattern; bolting to 8 mm x 80 mm uprights at 2 m centres; mesh 62 mm x 66 mm; setting in concrete							
930 mm high panel	-	-	-	-	-	m	**122.00**
1326 mm high panel	-	-	-	-	-	m	**157.00**
1722 mm high panel	-	-	-	-	-	m	**202.00**
"Orsogril" rectangular steel bar mesh fence panels; sterope pattern; bolting to 60 mm x 80 mm uprights at 2.00 m centres; mesh 62 mm x 132 mm; setting in concrete paving (paving not included)							
930 mm high panels	-	-	-	-	-	m	**124.19**
1326 mm high panels	-	-	-	-	-	m	**162.78**
1722 mm high panels	-	-	-	-	-	m	**204.73**
Security fencing; Jacksons Fencing							
"Barbican" galvanized steel paling fencing; on 60 mm x 60 mm posts at 3 m centres; setting in concrete							
1250 mm high	-	-	-	-	-	m	**61.40**
1500 mm high	-	-	-	-	-	m	**65.73**
2000 mm high	-	-	-	-	-	m	**75.77**
2500 mm high	-	-	-	-	-	m	**89.05**
Gates; to match "Barbican" galvanised steel paling fencing							
width 1 m	-	-	-	-	-	nr	**806.82**
width 2 m	-	-	-	-	-	nr	**816.69**
width 3 m	-	-	-	-	-	nr	**838.89**
width 4 m	-	-	-	-	-	nr	**867.26**
width 8 m	-	-	-	-	-	pair	**1767.85**
width 9 m	-	-	-	-	-	pair	**2131.77**
width 10 m	-	-	-	-	-	pair	**2422.93**
Security fencing; Jacksons Fencing; intruder guards							
"Viper Spike Intruder Guards"; to existing structures, including fixing bolts							
ref Viper 1; 40 mm x 5 mm x 1.1 m long with base plate	25.44	1.00	18.50	-	25.44	nr	**43.94**
ref Viper 3; 160 mm x 190 mm wide; U shape; to prevent intruders climbing pipes	31.99	0.50	9.25	-	31.99	nr	**41.24**

Q PAVING/PLANTING/FENCING/SITE FURNITURE

Item Excluding site overheads and profit	PC £	Labour hours	Labour £	Plant £	Material £	Unit	Total rate £
Security fencing; Jacksons Fencing							
"Razor Barb Concertina"; spiral wire security barriers; fixing to 600 mm steel ground stakes							
ref 3275; 450 mm diameter roll - medium barb; galvanized	1.95	0.02	0.37	-	2.08	m	**2.45**
ref 3276; 730 mm diameter roll - medium barb; galvanized	2.49	0.02	0.37	-	2.62	m	**2.99**
ref 3277; 950 mm diameter roll - medium barb; galvanized	2.50	0.02	0.37	-	2.63	m	**3.00**
3 lines "Barbed Tape" medium barb barbed tape; on 50 mm x 50 mm mild steel angle posts; setting in concrete							
ref 3283; barbed tape medium barb; galvanized	0.72	0.21	3.90	-	13.47	m	**17.37**
5 lines "Barbed Tape" medium barb barbed tape; on 50 mm x 50 mm mild steel angle posts; setting in concrete							
ref 3283; barbed tape medium barb; galvanized	1.20	0.22	4.11	-	13.95	m	**18.06**
Trip rails; metal; Broxap Street Furniture							
Double row 48.3 mm internal diameter mild steel tubular rails with sleeved joints; to cast iron posts 525 mm above ground mm long; setting in concrete at 1.20 m centres; all standard painted							
BX 201; cast iron with single rail 42 mm diameter rail	54.83	1.00	18.50	-	60.39	m	**78.89**
22 mm diameter mild steel rails with ferrule joints; fixing through holes in 44 mm x 13 mm mild steel standards 700 mm long; setting in concrete at 1.20 m centres; priming	22.66	1.33	24.67	-	23.02	m	**47.69**
Heavy duty; anti ram 540 high with 100 x 100 mm rail; Birdsmouth style	40.65	1.00	18.50	-	41.01	m	**59.51**
Trip rails; Birdsmouth; Jacksons Fencing							
Diamond rail fencing (Birdsmouth); posts 100 x 100 mm softwood planed at 1.35 m centres set in 1:3:6 concrete; rail 75 x 75 nominal secured with galvanized straps nailed to posts							
Posts 900 mm (600 above ground) at 1.35 m centres	6.73	0.50	9.25	0.26	8.34	m	**17.84**
Posts 1.20 m (900 above ground) at 1.35 m centres	7.72	0.50	9.25	0.26	9.32	m	**18.83**
Posts 900 mm (600 above ground) at 1.80 m centres	5.64	0.38	6.95	0.19	6.86	m	**14.00**
Posts 1.20 m (900 above ground) at 1.80 m centres	6.38	0.38	6.95	0.19	7.58	m	**14.73**
Trip rails; Townscape Products Ltd							
Hollow steel section knee rails; galvanised; 500 mm high; setting in concrete							
1000 mm bays	87.16	2.00	37.00	-	94.06	m	**131.06**
1200 mm bays	18.77	2.00	37.00	-	81.74	m	**118.74**
Timber fencing; AVS Fencing Supplies Ltd							
Palisade fencing; 22 mm x 75 mm softwood vertical palings with flat tops; nailing to 3 nr 50 mm x 100 mm horizontal softwood rails; housing into 100 mm x 100 mm softwood posts with weathered tops at 3.00 m centres; setting in concrete; all treated timber							
1800 mm high	21.61	0.47	8.64	-	25.68	m	**34.32**

Q PAVING/PLANTING/FENCING/SITE FURNITURE

Item Excluding site overheads and profit	PC £	Labour hours	Labour £	Plant £	Material £	Unit	**Total rate £**
Q40 FENCING - cont'd							
Timber fencing - cont'd							
Palisade fencing; 22 mm x 75 mm oak vertical palings with flat tops; nailing to 3 nr 50 mm x 125 mm horizontal oak rails; housing into 100 mm x 100 mm oak posts with weathered tops at 3.00 m centres; setting in concrete							
1800 mm high	40.60	0.47	8.63	-	44.67	m	**53.30**
Post-and-rail fencing; 3 nr 90 mm x 38 mm softwood horizontal rails; fixing with galvanized nails to 150 mm x 75 mm softwood posts; including excavating and backfilling into firm ground at 1.80 m centres; all treated timber							
1200 mm high	14.77	0.35	6.48	-	14.84	m	**21.32**
Post-and-rail fencing; 3 nr 90 mm x 38 mm oak horizontal rails; fixing with galvanised nails to 150 mm x 75 mm oak posts; including excavating and backfilling into firm ground at 1.80 m centres							
1200 mm high	29.66	0.35	6.48	-	29.73	m	**36.21**
Morticed post-and-rail fencing; 3 nr horizontal 90 mm x 38 mm softwood rails; fixing with galvanized nails; 90 mm x 38 mm softwood centre prick posts; to 150 mm x 75 mm softwood posts; including excavating and backfilling into firm ground at 2.85 m centres; all treated timber							
1200 mm high	12.57	0.35	6.48	-	12.64	m	**19.12**
1350 mm high five rails	15.06	0.40	7.40	-	15.13	m	**22.53**
Cleft rail fencing; oak or chestnut adze tapered rails 2.80 m long; morticed into joints; to 125 mm x 100 mm softwood posts 1.95 m long; including excavating and backfilling into firm ground at 2.80 m centres							
two rails	16.72	0.25	4.63	-	16.72	m	**21.34**
three rails	19.37	0.28	5.18	-	19.37	m	**24.55**
four rails	22.02	0.35	6.48	-	22.02	m	**28.50**
Close boarded fencing; 2 nr softwood rails; 150 x 25 mm gravel boards; 2 ex 100 x 22 softwood pales lapped 13 mm; to concrete posts; including excavating and backfilling into firm ground at 3.00 m centres							
900 mm high	14.47	0.47	8.64	-	15.80	m	**24.44**
1050 mm high	11.15	0.47	8.64	-	18.00	m	**26.64**
1200 mm high	12.11	0.47	8.64	-	18.96	m	**27.60**
Close boarded fencing; 3 nr softwood rails; 150 x 25 mm gravel boards; 89 mm x 19 mm softwood pales lapped 13 mm; to concrete posts; including excavating and backfilling into firm ground at 3.00 m centres							
1350 mm high	20.62	0.56	10.36	-	21.96	m	**32.32**
1650 mm high	22.30	0.56	10.36	-	23.64	m	**34.00**
1800 mm high	22.67	0.56	10.36	-	24.00	m	**34.36**
Close boarded fencing; 3 nr softwood rails; 150 x 25 mm gravel boards; 89 mm x 19 mm softwood pales lapped 13 mm; to softwood posts 100 x 100 mm; including excavating and backfilling into firm ground at 3.00 m centres							
1350 mm high	17.32	0.56	10.36	-	18.65	m	**29.01**
1650 mm high	18.86	0.56	10.36	-	20.19	m	**30.56**
1800 mm high	19.51	0.56	10.36	-	20.84	m	**31.21**

Q PAVING/PLANTING/FENCING/SITE FURNITURE

Item Excluding site overheads and profit	PC £	Labour hours	Labour £	Plant £	Material £	Unit	**Total rate £**
Close boarded fencing; 3 nr softwood rails; 150 x 25 mm gravel boards; 89 mm x 19 mm softwood pales lapped 13 mm; to oak posts; including excavating and backfilling into firm ground at 3.00 m centres							
1350 mm high	14.69	0.56	10.36	-	21.53	m	**31.90**
1650 mm high	22.39	0.56	10.36	-	23.72	m	**34.08**
1800 mm high	23.36	0.56	10.36	-	24.69	m	**35.05**
Close boarded fencing; 3 nr softwood rails; 150 x 25 mm gravel boards; 89 mm x 19 mm softwood pales lapped 13 mm; to oak posts; including excavating and backfilling into firm ground at 2.47 m centres							
1800 mm high	24.93	0.68	12.54	-	26.27	m	**38.80**
Close boarded fencing - all oak; 2 nr 76 mm x 38 mm rectangular rails; 152 mm x 25 mm gravel boards; 89 mm x 19 mm oak featheredge boards lapped 13 mm; to oak posts; including excavating and backfilling into firm ground at 2.47 m centres							
1070 mm high	9.42	0.56	10.36	-	39.34	m	**49.70**
1800 mm high	46.24	0.56	10.36	-	47.57	m	**57.94**
Extra over close boarded fencing for 65 mm x 38 mm oak cappings	5.99	0.05	0.93	-	5.99	nr	**6.92**
"Hit and miss" horizontal rail fencing; 87 mm x 38 mm top and bottom rails; 100 mm x 22 mm vertical boards arranged alternately on opposite side of rails; to 100 mm x 100 mm posts; including excavating and backfilling into firm ground; setting in concrete at 1.8 m centres							
treated softwood; 1800 mm high	31.04	1.33	24.67	-	33.57	m	**58.23**
treated softwood; 2000 mm high	32.30	1.33	24.67	-	39.03	m	**63.69**
primed softwood; 1800 mm high	31.04	1.67	30.83	-	33.57	m	**64.40**
primed softwood; 2000 mm high	32.30	1.67	30.83	-	39.03	m	**69.86**
Screen fencing in horizontal wavy edge overlap panels; 1.83 m wide; to 100 mm x 100 mm preservative treated softwood posts; including excavating and backfilling into firm ground							
900 mm high	8.34	0.47	8.64	-	15.46	m	**24.10**
1200 mm high	8.42	0.47	8.64	-	16.32	m	**24.96**
1500 mm high	8.47	0.56	10.39	-	18.50	m	**28.89**
1800 mm high	8.50	0.56	10.39	-	19.24	m	**29.64**
Screen fencing for oak in lieu of softwood posts							
900 mm high	14.61	0.47	8.64	-	18.77	m	**27.41**
1200 mm high	16.27	0.47	8.64	-	20.43	m	**29.07**
1500 mm high	17.89	0.56	10.39	-	23.43	m	**33.82**
1800 mm high	19.43	0.56	10.39	-	24.96	m	**35.36**
Trellis tops to fencing (see more detailed items in the Minor Works section of this book)							
Extra over screen fencing for 300 mm high trellis tops; slats at 100 mm centres; including additional length of posts	3.96	0.10	1.85	-	3.96	m	**5.81**
Concrete fencing							
Panel fencing; to precast concrete posts; in 2 m bays; setting posts 600 mm into ground; sandfaced finish							
900 mm high	9.94	0.25	4.63	-	11.46	m	**16.08**
1200 mm high	13.72	0.25	4.63	-	15.24	m	**19.87**

Q PAVING/PLANTING/FENCING/SITE FURNITURE

Item Excluding site overheads and profit	PC £	Labour hours	Labour £	Plant £	Material £	Unit	Total rate £
Q40 FENCING - cont'd							
Concrete fencing - cont'd							
Panel fencing; to precast concrete posts; in 2 m bays; setting posts 750 mm into ground; sandfaced finish							
1500 mm high	18.01	0.33	6.17	-	19.53	m	**25.69**
1800 mm high	22.02	0.36	6.73	-	23.54	m	**30.27**
2100 mm high	25.63	0.40	7.40	-	27.15	m	**34.55**
2400 mm high	29.60	0.40	7.40	-	31.12	m	**38.52**
extra over capping panel to all of the above	5.06	-	-	-	5.06	m	**-**
Windbreak fencing							
Fencing; English Woodlands; "Shade and Shelter Netting" windbreak fencing; green; to 100 mm diameter treated softwood posts; setting 450 mm into ground; fixing with 50 mm x 25 mm treated softwood battens nailed to posts, including excavating and backfilling into firm ground; setting in concrete at 3 m centres							
1200 mm high	1.24	0.16	2.88	-	5.70	m	**8.58**
1800 mm high	1.76	0.16	2.88	-	6.22	m	**9.10**
Ball stop fencing; Steelway-Fensecure Ltd							
Ball stop net; 30 x 30 mm netting fixed to 60.3 diameter 12 mm solid bar lattice galvanised dual posts, top, middle and bottom rails							
4.5 m high	-	-	-	-	-	m	**75.95**
5 .0 m high	-	-	-	-	-	m	**86.89**
6.0 m high	-	-	-	-	-	m	**95.88**
7.0 m high	-	-	-	-	-	m	**103.82**
8.0 m high	-	-	-	-	-	m	**112.15**
9.0 m high	-	-	-	-	-	m	**120.83**
10.0 m high	-	-	-	-	-	m	**129.90**
Ball stop net; corner posts							
4.5 m high	-	-	-	-	-	m	**53.98**
5.0 m high	-	-	-	-	-	m	**59.08**
6.0 m high	-	-	-	-	-	m	**91.55**
7.0 m high	-	-	-	-	-	m	**81.76**
8.0 m high	-	-	-	-	-	m	**93.10**
9.0 m high	-	-	-	-	-	m	**104.44**
10 m high	-	-	-	-	-	m	**115.78**
Ball stop net; end posts							
4.5 m high	-	-	-	-	-	m	**44.91**
5.0 m high	-	-	-	-	-	m	**50.00**
6.0 m high	-	-	-	-	-	m	**61.35**
7.0 m high	-	-	-	-	-	m	**72.69**
8.0 m high	-	-	-	-	-	m	**84.03**
9.0 m high	-	-	-	-	-	m	**95.37**
10 m high	-	-	-	-	-	m	**107.85**
Railings							
Mild steel bar railings of balusters at 115 mm centres welded to flat rail top and bottom; bays 2.00 m long; bolting to 51 mm x 51 mm hollow square section posts; setting in concrete							
galvanized; 900 mm high	-	-	-	-	-	m	**62.33**
galvanized; 1200 mm high	-	-	-	-	-	m	**75.50**
galvanized; 1500 mm high	-	-	-	-	-	m	**83.91**
galvanized; 1800 mm high	-	-	-	-	-	m	**100.66**
primed; 900 mm high	-	-	-	-	-	m	**58.75**
primed; 1200 mm high	-	-	-	-	-	m	**71.33**
primed; 1500 mm high	-	-	-	-	-	m	**78.86**
primed; 1800 mm high	-	-	-	-	-	m	**93.98**

Item Excluding site overheads and profit	PC £	Labour hours	Labour £	Plant £	Material £	Unit	**Total rate £**
Mild steel blunt top railings of 19 mm balusters at 130 mm centres welded to bottom rail; passing through holes in top rail and welded; top and bottom rails 40 mm x 10 mm; bolting to 51 mm x 51 mm hollow square section posts; setting in concrete							
galvanized; 800 mm high	-	-	-	-	-	m	**67.12**
galvanized; 1000 mm high	-	-	-	-	-	m	**70.48**
galvanized; 1300 mm high	-	-	-	-	-	m	**83.91**
galvanized; 1500 mm high	-	-	-	-	-	m	**90.62**
primed; 800 mm high	-	-	-	-	-	m	**64.61**
primed; 1000 mm high	-	-	-	-	-	m	**67.12**
primed; 1300 mm high	-	-	-	-	-	m	**79.71**
primed; 1500 mm high	-	-	-	-	-	m	**85.60**
Railings; traditional pattern; 16 mm diameter verticals at 127 mm intervals with horizontal bars near top and bottom; balusters with spiked tops; 51 x 20 mm standards; including setting 520 mm into concrete at 2.75 m centres							
primed; 1200 mm high	-	-	-	-	-	m	**75.50**
primed; 1500 mm high	-	-	-	-	-	m	**85.60**
primed; 1800 mm high	-	-	-	-	-	m	**92.29**
Interlaced bow-top mild steel railings; traditional park type; 16 mm diameter verticals at 80 mm intervals, welded at bottom to 50 x 10 mm flat and slotted through 38 mm x 8 mm top rail to form hooped top profile; 50 mm x 10 mm standards; setting 560 mm into concrete at 2.75 m centres							
galvanized; 900 mm high	-	-	-	-	-	m	**114.12**
galvanized; 1200 mm high	-	-	-	-	-	m	**139.29**
galvanized; 1500 mm high	-	-	-	-	-	m	**166.13**
galvanized; 1800 mm high	-	-	-	-	-	m	**192.98**
primed; 900 mm high	-	-	-	-	-	m	**110.75**
primed; 1200 mm high	-	-	-	-	-	m	**134.25**
primed; 1500 mm high	-	-	-	-	-	m	**159.43**
primed; 1800 mm high	-	-	-	-	-	m	**184.58**
Metal estate fencing; Jacksons Fencing; mild steel flat bar angular fencing; galvanized; main posts at 5.00 m centres; fixed with 1:3:6 concrete 430 deep; intermediate posts fixed with fixing claw at 1.00 m centres							
5 nr plain flat or round bar							
1200 high; under 500 m	45.90	0.28	5.14	-	48.07	m	**53.21**
1200 high; over 500 m	39.60	0.28	5.14	-	41.77	m	**46.91**
Five bar horizontal steel fencing; 1.20 m high; 38 mm x 10 mm joiner standards at 4.50 m centres; 38 mm x 8 mm intermediate standards at 900 mm centres; including excavating and backfilling into firm ground; setting in concrete							
galvanized	-	-	-	-	-	m	**89.12**
primed	-	-	-	-	-	m	**73.26**
Extra over five bar horizontal steel fencing for 76 mm diameter end and corner posts							
galvanized	-	-	-	-	-	each	**97.43**
primed	-	-	-	-	-	each	**57.01**

Q PAVING/PLANTING/FENCING/SITE FURNITURE

Item Excluding site overheads and profit	PC £	Labour hours	Labour £	Plant £	Material £	Unit	Total rate £
Q40 FENCING - cont'd							
Five bar horizontal steel fencing with mild steel copings; 38 mm square hollow section mild steel standards at 1.80 m centres; including excavating and backfilling into firm ground; setting in concrete							
galvanized; 900 mm high	-	-	-	-	-	m	**116.63**
galvanized; 1200 mm high	-	-	-	-	-	m	**122.69**
galvanized; 1500 mm high	-	-	-	-	-	m	**131.35**
galvanized; 1800 mm high	-	-	-	-	-	m	**140.01**
primed; 900 mm high	-	-	-	-	-	m	**103.49**
primed; 1200 mm high	-	-	-	-	-	m	**106.23**
primed; 1500 mm high	-	-	-	-	-	m	**112.00**
primed; 1800 mm high	-	-	-	-	-	m	**120.53**
Gates - General							
Preamble: Gates in fences; see specification for fencing, as gates in traditional or proprietary fencing systems are usually constructed of the same materials and finished as the fencing itself.							
Gates, hardwood; Longlyf Timber Products Ltd							
Hardwood entrance gate, five bar diamond braced, curved hanging stile, planed iroko; fixed to 150 mm x 150 mm softwood posts; inclusive of hinges and furniture							
ref 1100 040; 0.9 m wide	199.57	5.00	92.50	-	253.77	nr	**346.27**
ref 1100 041; 1.2 m wide	213.06	5.00	92.50	-	267.26	nr	**359.76**
ref 1100 042; 1.5 m wide	276.00	5.00	92.50	-	330.20	nr	**422.70**
ref 1100 043; 1.8 m wide	292.39	5.00	92.50	-	346.59	nr	**439.09**
ref 1100 044; 2.1 m wide	320.70	5.00	92.50	-	374.90	nr	**467.40**
ref 1100 045; 2.4 m wide	339.54	5.00	92.50	-	393.74	nr	**486.24**
ref 1100 047; 3.0 m wide	369.86	5.00	92.50	-	424.06	nr	**516.56**
ref 1100 048; 3.3 m wide	386.71	5.00	92.50	-	440.91	nr	**533.41**
ref 1100 049; 3.6 m wide	400.16	5.00	92.50	-	454.36	nr	**546.86**
Hardwood field gate, five bar diamond braced, planed iroko; fixed to 150 mm x 150 mm softwood posts; inclusive of hinges and furniture							
ref 1100 100; 0.9 m wide	144.00	4.00	74.00	-	198.20	nr	**272.20**
ref 1100 101; 1.2 m wide	156.70	4.00	74.00	-	210.90	nr	**284.90**
ref 1100 102; 1.5 m wide	186.84	4.00	74.00	-	241.04	nr	**315.04**
ref 1100 103; 1.8 m wide	201.09	4.00	74.00	-	255.29	nr	**329.29**
ref 1100 104; 2.1 m wide	247.44	4.00	74.00	-	301.64	nr	**375.64**
ref 1100 105; 2.4 m wide	262.43	4.00	74.00	-	316.63	nr	**390.63**
ref 1100 106; 2.7 m wide	277.43	4.00	74.00	-	331.63	nr	**405.63**
ref 1100 108; 3.3 m wide	307.41	4.00	74.00	-	361.61	nr	**435.61**
ref 1100 109; 3.6 m wide	323.20	4.00	74.00	-	377.40	nr	**451.40**
Gates, softwood; Jacksons Fencing							
Timber field gates, including wrought iron ironmongery; five bar type; diamond braced; 1.80 m high; to 200 mm x 200 mm posts; setting 750 mm into firm ground							
treated softwood; width 2400 mm	72.58	10.00	185.00	-	157.87	nr	**342.87**
treated softwood; width 2700 mm	79.58	10.00	185.00	-	164.86	nr	**349.86**
treated softwood; width 3000 mm	82.89	10.00	185.00	-	168.17	nr	**353.17**
treated softwood; width 3300 mm	88.95	10.00	185.00	-	174.23	nr	**359.23**
Featherboard garden gates, including ironmongery; to 100 mm x 120 mm posts; 1 nr diagonal brace							
treated softwood; 1.0 m x 1.2 m high	15.03	3.00	55.50	-	87.83	nr	**143.33**
treated softwood; 1.0 m x 1.5 m high	19.85	3.00	55.50	-	94.43	nr	**149.93**
treated softwood; 1.0 m x 1.8 m high	70.54	3.00	55.50	-	100.75	nr	**156.25**

Q PAVING/PLANTING/FENCING/SITE FURNITURE

Item Excluding site overheads and profit	PC £	Labour hours	Labour £	Plant £	Material £	Unit	Total rate £
Picket garden gates, including ironmongery; to match picket fence; width 1000 mm; to 100 mm x 120 mm posts; 1 nr diagonal brace							
treated softwood; 950 mm high	60.43	3.00	55.50	-	83.45	nr	**138.95**
treated softwood; 1200 mm high	63.20	3.00	55.50	-	86.22	nr	**141.72**
treated softwood; 1800 mm high	72.98	3.00	55.50	-	103.19	nr	**158.69**
Gates, tubular steel; Jacksons Fencing							
Tubular mild steel field gates, including ironmongery; diamond braced; 1.80 m high; to tubular steel posts; setting in concrete							
galvanized; width 3000 mm	109.48	5.00	92.50	-	200.99	nr	**293.49**
galvanized; width 3300 mm	115.84	5.00	92.50	-	207.35	nr	**299.85**
galvanized; width 3600 mm	122.45	5.00	92.50	-	213.97	nr	**306.47**
galvanized; width 4200 mm	135.66	5.00	92.50	-	227.17	nr	**319.67**
Gates, sliding; Jacksons Fencing							
"Sliding Gate"; including all galvanized rails and vertical rail infill panels; special guide and shutting frame posts (Note: foundations installed by suppliers)							
access width 4.00 m; 1.5 m high gates	-	-	-	-	-	nr	**3867.54**
access width 4.00 m; 2.0 m high gates	-	-	-	-	-	nr	**3921.83**
access width 4.00 m; 2.5 m high gates	-	-	-	-	-	nr	**3974.87**
access width 6.00 m; 1.5 m high gates	-	-	-	-	-	nr	**4418.99**
access width 6.00 m; 2.0 m high gates	-	-	-	-	-	nr	**4484.38**
access width 6.00 m; 2.5 m high gates	-	-	-	-	-	nr	**4547.30**
access width 8.00 m; 1.5 m high gates	-	-	-	-	-	nr	**5035.83**
access width 8.00 m; 2.0 m high gates	-	-	-	-	-	nr	**5111.07**
access width 8.00 m; 2.5 m high gates	-	-	-	-	-	nr	**5182.64**
access width 10.00 m; 2.0 m high gates	-	-	-	-	-	nr	**6210.28**
access width 10.00 m; 2.5 m high gates	-	-	-	-	-	nr	**6302.80**
Stiles and kissing gates							
Stiles; Jacksons Fencing; "Jacksons Nr 2007" stiles; 2 nr posts; setting into firm ground; 3 nr rails; 2 nr treads	65.16	3.00	55.50	-	75.45	nr	**130.95**
Kissing gates; Jacksons Fencing; in galvanised metal bar; fixing to fencing posts (posts not included); 1.65 m x 1.30 m x 1.00 m high	185.63	5.00	92.50	-	185.63	nr	**278.13**
Pedestrian guard rails and barriers							
Mild steel pedestrian guard rails; Broxap Street Furniture; to BS 3049:1976; 1.00 m high with 150 mm toe space; to posts at 2.00 m centres, galvanized finish							
vertical "in-line" bar infill panel	37.00	2.00	37.00	-	43.32	m	**80.32**
vertical staggered bar infill with 230 mm visibility gap at top	42.00	2.00	37.00	-	48.32	m	**85.32**

Q PAVING/PLANTING/FENCING/SITE FURNITURE

Item Excluding site overheads and profit	PC £	Labour hours	Labour £	Plant £	Material £	Unit	**Total rate £**
Q50 SITE/STREET FURNITURE/EQUIPMENT							
Standing stones; CED Ltd; erect standing stones; vertical height above ground; in concrete base; including excavation setting in concrete to 1/3 depth and crane offload into position							
Purple schist							
1.00 m high	75.00	1.50	27.75	14.00	89.91	nr	**131.66**
1.25 m high	95.00	1.50	27.75	14.00	104.75	nr	**146.51**
1.50 m high	152.00	2.00	37.00	16.80	164.05	nr	**217.85**
2.00 m high	250.00	3.00	55.50	28.00	284.84	nr	**368.34**
2.50 m high	380.00	1.50	27.75	56.00	447.47	nr	**531.22**
Cattle grids - General							
Preamble: Cattle grids are not usually prefabricated, owing to their size and weight. Specification must take into account the maximum size and weight of vehicles likely to cross the grid. Drainage and regular clearance of the grid is essential. Warning signs should be erected on both approaches to the grid. For the cost of a complete grid and pit installation see "Approximate Estimates" section (book only).							
Cattle grids; Jacksons Fencing							
Cattle grids; supply only							
to carry 12 tonnes evenly distributed; 2.90 m x 2.36 m grid; galvanized	627.92	-	-	-	627.92	nr	-
to carry 12 tonnes evenly distributed; 3.66 m x 2.50 m grid; galvanized	1094.71	-	-	-	1094.71	nr	-
Barriers - General							
Preamble: The provision of car and lorry control barriers may form part of the landscape contract. Barriers range from simple manual counterweighted poles to fully automated remote-control security gates, and the exact degree of control required must be specified. Complex barriers may need special maintenance and repair.							
Barriers; Autopa Ltd							
Manually operated pole barriers; counterbalance; to tubular steel supports; bolting to concrete foundation (foundation not included); aluminium boom; various finishes							
clear opening up to 3.00 m	817.00	6.00	111.00	-	838.32	nr	**949.32**
clear opening up to 4.00 m	879.00	6.00	111.00	-	900.32	nr	**1011.32**
clear opening 5.00	941.00	6.00	111.00	-	962.32	nr	**1073.32**
clear opening 6.00	1003.00	6.00	111.00	-	1024.32	nr	**1135.32**
clear opening 7.00	1063.00	6.00	111.00	-	1084.32	nr	**1195.32**
catch pole; arm rest for all manual barriers	116.00	-	-	-	118.67	nr	-
Electrically operated pole barriers; enclosed fan-cooled motor; double worm reduction gear; overload clutch coupling with remote controls; aluminium boom; various finishes (exclusive of electrical connections by electrician)							
Autopa AU 3.0, 3 m boom	2310.00	6.00	111.00	-	2331.32	nr	**2442.32**
Autopa AU 4.5, 4.5 m boom with catchpost	2410.00	6.00	111.00	-	2431.32	nr	**2542.32**
Autopa AU 6.0, 6.0 m boom with catchpost	2510.00	6.00	111.00	-	2531.32	nr	**2642.32**

Q PAVING/PLANTING/FENCING/SITE FURNITURE

Item Excluding site overheads and profit	PC £	Labour hours	Labour £	Plant £	Material £	Unit	**Total rate £**
Vehicle crash barriers - General							
Preamble: See Department of Environment Technical Memorandum BE5.							
Vehicle crash barriers							
Steel corrugated beams; untensioned; Broxap Street Furniture; effective length 3.20 m 310 mm deep x 85 mm corrugations							
steel posts; Z-section; roadside posts	73.50	0.33	6.17	0.70	81.17	m	**88.04**
steel posts; Z-section; off highway	65.63	0.33	6.17	0.70	73.30	m	**80.16**
steel posts RSJ 760 high; for anchor fixing	85.75	0.33	6.17	-	99.81	m	**105.98**
steel posts RSJ 560 high; for anchor fixing to car parks	82.69	0.33	6.17	-	96.75	m	**102.92**
extra over for curved rail 6.00 m radius	23.63	-	-	-	23.63	m	-
Bases for street furniture							
Excavating; filling with concrete 1:3:6; bases for street furniture							
300 mm x 450 mm x 500 mm deep	-	0.75	13.88	-	6.24	nr	**20.11**
300 mm x 600 mm x 500 mm deep	-	1.11	20.54	-	8.31	nr	**28.85**
300 mm x 900 mm x 500 mm deep	-	1.67	30.89	-	12.47	nr	**43.37**
1750 mm x 900 mm x 300 mm deep	-	2.33	43.10	-	43.65	nr	**86.75**
2000 mm x 900 mm x 300 mm deep	-	2.67	49.40	-	49.89	nr	**99.28**
2400 mm x 900 mm x 300 mm deep	-	3.20	59.20	-	59.87	nr	**119.06**
2400 mm x 1000 mm x 300 mm deep	-	3.56	65.86	-	66.51	nr	**132.37**
Precast concrete flags; to BS 7263; to concrete bases (not included); bedding and jointing in cement:mortar (1:4)							
450 mm x 600 mm x 50 mm	8.52	1.17	21.58	-	13.11	m^2	**34.69**
Precast concrete paving blocks; to concrete bases (not included); bedding in sharp sand; butt joints							
200 mm x 100 mm x 65 mm	6.74	0.50	9.25	-	8.65	m^2	**17.90**
200 mm x 100 mm x 80 mm	7.50	0.50	9.25	-	9.30	m^2	**18.55**
Engineering paving bricks; to concrete bases (not included); bedding and jointing in sulphate-resisting cement:lime:sand mortar (1:1:6)							
over 300 mm wide	10.50	0.56	10.39	-	19.32	m^2	**29.72**
Edge restraints to pavings; haunching in concrete (1:3:6)							
200 mm x 300 mm	4.82	0.10	1.85	-	4.82	m	**6.67**
Bases; Earth Anchors Ltd							
"Rootfast" ancillary anchors; ref A1; 500 mm long 25 mm diameter and top strap ref F2; including bolting to site furniture (site furniture not included)	9.00	0.50	9.25	-	9.00	set	**18.25**
installation tool for above	22.00	-	-	-	22.00	nr	-
"Rootfast" ancillary anchors; ref A4; heavy duty 40 mm square fixed head anchors; including bolting to site furniture (site furniture not included)	28.00	0.33	6.17	-	28.00	set	**34.17**
"Rootfast" ancillary anchors; F4; vertical socket; including bolting to site furniture (site furniture not included)	12.45	0.05	0.93	-	12.45	set	**13.38**
"Rootfast" ancillary anchors; ref F3; horizontal socket; including bolting to site furniture (site furniture not included)	13.70	0.05	0.93	-	13.70	set	**14.63**
installation tools for the above	62.00	-	-	-	62.00	nr	-
"Rootfast" ancillary anchors; ref A3 Anchored bases; including bolting to site furniture (site furniture not included)	44.00	0.33	6.17	-	44.00	set	**50.17**
installation tools for the above	62.00	-	-	-	62.00	nr	-

Q PAVING/PLANTING/FENCING/SITE FURNITURE

Item Excluding site overheads and profit	PC £	Labour hours	Labour £	Plant £	Material £	Unit	Total rate £
Q50 SITE/STREET FURNITURE/EQUIPMENT - cont'd							
Furniture/equipment - General							
Preamble: The following items include fixing to manufacturer's instructions; holding down bolts or other fittings and making good (excavating, backfilling and tarmac, concrete or paving bases not included).							
Dog waste bins; all-steel							
Earth Anchors Ltd							
HG45A; 45l; earth anchored; post mounted	167.00	0.33	6.17	-	167.00	nr	**173.17**
HG45A; 45l; as above with pedal operation	196.00	0.33	6.17	-	196.00	nr	**202.17**
Dog waste bins; cast iron							
Bins; Furnitubes International Ltd							
ref PED 701; "Pedigree"; post mounted cast iron dog waste bins; 1250 mm total height above ground; 400 mm square bin	786.00	0.75	13.88	-	792.32	nr	**806.19**
Litter bins; precast concrete in textured white or exposed aggregate finish; with wire baskets and drainage holes							
Bins; Marshalls Plc							
"Boulevard 700" concrete circular litter bin	328.70	0.50	9.25	-	328.70	nr	**337.95**
Bins; Neptune Outdoor Furniture Ltd							
ref SF 16 - 42 L	205.00	0.50	9.25	-	205.00	nr	**214.25**
ref SF 14 - 100 L	286.00	0.50	9.25	-	286.00	nr	**295.25**
Bins; Townscape Products Ltd							
"Sutton"; 750 mm high x 500 mm diameter, 70 litre capacity including GRP canopy	166.56	0.33	6.17	-	285.58	nr	**291.75**
"Braunton"; 750 mm high x 500 mm diameter, 70 litre capacity including GRP canopy	176.57	1.50	27.75	-	295.59	nr	**323.34**
Litter bins; metal; stove-enamelled perforated metal for holder and container							
Bins; Townscape Products Ltd							
"Metro"; 440 x 420 x 800 high; 62 litre capacity	573.00	0.33	6.17	-	573.00	nr	**579.17**
"Voltan" large round; 460 diameter x 780 high; 56 litre capacity	493.08	0.33	6.17	-	493.08	nr	**499.25**
"Voltan" small round with pedestal; 410 diameter x 760 high; 31 litre capacity	477.42	0.33	6.17	-	477.42	nr	**483.59**
Litter bins; all-steel							
Bins; Marshall Plc							
"MSF Central"; steel litter bin	237.50	1.00	18.50	-	237.50	nr	**256.00**
"MSF Central"; stainless steel litter bin	491.77	0.67	12.33	-	491.77	nr	**504.10**
Bins; Earth Anchors Ltd							
"Ranger", 100 L litter bin, 107l, pedestal mounted	825.00	0.50	9.25	-	825.00	nr	**834.25**
"Big Ben", 82 L, steel frame and liner, colour coated finish, earth anchored	257.00	1.00	18.50	-	257.00	nr	**275.50**
"Beau", 42 L, steel frame and liner, colour coated finish, earth anchored	205.00	1.00	18.50	-	205.00	nr	**223.50**
Bins; Townscape Products Ltd							
"Baltimore Major" with GRP canopy; 560 diameter x 960 high, 140 litre capacity	827.95	1.00	18.50	-	827.95	nr	**846.45**
Litter bins; cast iron							
Bins; Marshalls Plc							
"MSF5501" Heritage, cast iron litter bin	441.75	1.00	18.50	4.69	441.75	nr	**464.94**

Q PAVING/PLANTING/FENCING/SITE FURNITURE

Item Excluding site overheads and profit	PC £	Labour hours	Labour £	Plant £	Material £	Unit	Total rate £
Bins; Furnitubes International Ltd							
"Wave Bin"; ref WVB 440; free standing 55 litre liners; 440 mm diameter x 850 mm high	343.00	0.50	9.25	-	346.98	nr	**356.23**
"Wave Bin" ref WVB 520; free standing 85 litre liners; 520 mm diameter x 850 mm high; cast iron plinth	390.00	0.50	9.25	-	393.98	nr	**403.23**
"Covent Garden"; ref COV 702; side opening, 500 mm square x 1050 mm high; 105 litre capacity	290.00	0.50	9.25	-	296.65	nr	**305.90**
"Covent Garden"; ref COV 803; side opening, 500 mm diameter x 1025 mm high; 85 litre capacity	247.00	0.50	9.25	-	253.65	nr	**262.90**
"Covent Garden"; ref COV 912; open top; 500 mm A/F octagonal x 820 mm high; 85 litre capacity	307.00	0.50	9.25	-	313.65	nr	**322.90**
"Albert"; ref ALB 800; open top; 400 mm diameter x 845 mm high; 55 litre capacity	507.00	0.50	9.25	-	513.65	nr	**522.90**
Bins; Broxap Street Furniture							
"Chester"; pedestal mounted	348.00	1.00	18.50	-	354.65	nr	**373.15**
Bins; Townscape Products Ltd							
"York Major"; 650 diameter x 1060 high, 140 litre capacity	991.29	1.00	18.50	-	993.15	nr	**1011.65**
Litter bins; timber faced; hardwood slatted casings with removable metal litter containers; ground or wall fixing							
Bins; Lister Lutyens Co Ltd							
"Monmouth"; 675 mm high x 450 mm wide; freestanding	114.00	0.33	6.17	-	114.00	nr	**120.17**
"Monmouth"; 675 mm high x 450 mm wide; bolting to ground (without legs)	105.00	1.00	18.50	-	113.64	nr	**132.14**
Bins; Woodscape Ltd							
Square, 580 mm x 580 mm x 950 mm high, with lockable lid	595.00	0.50	9.25	-	595.00	nr	**604.25**
Round, 580 mm diameter by 950 mm high, with lockable lid	595.00	0.50	9.25	-	595.00	nr	**604.25**
Plastic litter and grit bins; glassfibre reinforced polyester grit bins; yellow body; hinged lids							
Bins; Wybone Ltd; Victoriana glass fibre, cast iron effect litter bins, including lockable liner							
ref LBV/2; 521 mm x 521 mm x 673 mm high; open top; square shape; 0.078 m^3 capacity	179.91	1.00	18.50	-	186.56	nr	**205.06**
ref LVC/3; 457 mm diameter x 648 mm high; open top; drum shape; 0.084 m^3 capacity; with lockable liner	196.26	1.00	18.50	-	202.91	nr	**221.41**
Bins; Amberol Ltd; floor standing double walled to accept stabilizing ballast							
"Enviro Bin"; 150 litre	185.00	0.33	6.17	-	185.00	nr	**191.17**
"Enviro Bin"; 90 litre	139.00	0.33	6.17	-	139.00	nr	**145.17**
"Westminster" hooded; 90 litre	120.00	0.33	6.17	-	120.00	nr	**126.17**
"Westminster"; 90 litre	95.00	0.33	6.17	-	95.00	nr	**101.17**
Bins; Amberol Ltd; pole mounted							
Envirobin 50 litre top emptying with liner	82.50	0.50	9.25	-	82.50	nr	**91.75**
Grit bins; Furnitubes International Ltd							
Grit and salt bin; yellow glass fibre; hinged lid; 310 litre	273.00	-	-	-	273.00	ea	-
Grit and salt bin; yellow glass fibre; hinged lid; 170 litre	274.00	-	-	-	274.00	ea	-

Q PAVING/PLANTING/FENCING/SITE FURNITURE

Item Excluding site overheads and profit	PC £	Labour hours	Labour £	Plant £	Material £	Unit	**Total rate £**
Q50 SITE/STREET FURNITURE/EQUIPMENT - cont'd							
Lifebuoy stations							
Lifebuoy stations; Earth Anchors Ltd							
"Rootfast" lifebuoy station complete with post ref AP44; SOLAS approved lifebouys 590 mm and lifeline	162.00	2.00	37.00	-	171.87	nr	**208.87**
installation tool for above	-	-	-	-	79.00	nr	**-**
Outdoor Seats							
CED Ltd; stone bench; "Sinuous bench"; stone type bench to organic "S" pattern; laid to concrete base; 500 high x 500 wide (not included)							
2.00 m long	1000.00	4.00	74.00	-	1000.00	nr	**1074.00**
5.00 m long	2500.00	8.00	148.00	-	2500.00	nr	**2648.00**
10.00 m long	5000.00	11.00	203.50	-	5000.00	nr	**5203.50**
Outdoor seats; concrete framed - General Preamble: Prices for the following concrete framed seats and benches with hardwood slats include for fixing (where necessary) by bolting into existing paving or concrete bases (bases not included) or building into walls or concrete foundations (walls and foundations not included).							
Outdoor seats; concrete framed							
Outdoor seats; Townscape Products Ltd							
"Oxford" benches; 1800 mm x 430 mm x 440 mm	242.95	2.00	37.00	-	255.58	nr	**292.58**
"Maidstone" seats; 1800 mm x 610 mm x 785 mm	361.23	2.00	37.00	-	373.86	nr	**410.86**
Outdoor seats; concrete							
Outdoor seats; Marshalls Ltd							
"Boulevard 2000" concrete seat	632.86	2.00	37.00	-	653.45	nr	**690.45**
Outdoor seats; metal framed - General Preamble: Metal framed seats with hardwood backs to various designs can be bolted to ground anchors.							
Outdoor seats; metal framed							
Outdoor seats; Furnitubes International Ltd							
ref NS 6; "Newstead"; steel standards with iroko slats; 1.80 m long	223.00	2.00	37.00	-	243.59	nr	**280.59**
ref NEB 6; "New Forest Single Bench"; cast iron standards with iroko slats; 1.83 m long	210.00	2.00	37.00	-	230.59	nr	**267.59**
ref EA 6; "Eastgate"; cast iron standards with iroko slats; 1.86 m long	253.00	2.00	37.00	-	273.59	nr	**310.59**
ref NE 6; "New Forest Seat"; cast iron standards with iroko slats; 1.83 m long	285.00	2.00	37.00	-	305.59	nr	**342.59**
Outdoor seats; Columbia Cascade Ltd; range of 10 nr colours							
ref 2604-6; double Boulevard metal benches; 1800 mm x 1500 mm x 860 mm high	995.00	3.00	55.50	-	1008.29	nr	**1063.79**
Outdoor seats; Orchard Street Furniture Ltd; "Bramley"; broad iroko slats to steel frame							
1.20 m long	161.44	2.00	37.00	-	174.73	nr	**211.73**
1.80 m long	198.94	2.00	37.00	-	212.23	nr	**249.23**
2.40 m long	229.79	2.00	37.00	-	243.08	nr	**280.08**
Outdoor seats; Orchard Street Furniture Ltd; "Laxton"; narrow iroko slats to steel frame							
1.20 m long	182.45	2.00	37.00	-	195.74	nr	**232.74**
1.80 m long	224.20	2.00	37.00	-	237.49	nr	**274.49**
2.40 m long	254.79	2.00	37.00	-	268.08	nr	**305.08**

Q PAVING/PLANTING/FENCING/SITE FURNITURE

Item Excluding site overheads and profit	PC £	Labour hours	Labour £	Plant £	Material £	Unit	**Total rate £**
Outdoor seats; Orchard Street Furniture Ltd; "Lambourne"; iroko slats to cast iron frame							
1.80 m long	473.94	2.00	37.00	-	487.23	nr	**524.23**
2.40 m long	557.71	2.00	37.00	-	571.00	nr	**608.00**
Outdoor seats; SMP Playgrounds Ltd; "Datchet", steel and timber							
backless benches; ref OFDB6; 1.80 m long	378.00	2.00	37.00	-	391.29	nr	**428.29**
bench seats with backrests; ref OFDS6; 1.80 m long	549.00	2.00	37.00	-	562.29	nr	**599.29**
Outdoor seats; Broxap Street Furniture; metal and timber							
"Eastgate"	379.00	2.00	37.00	-	392.29	nr	**429.29**
"Serpent"	389.00	2.00	37.00	-	402.29	nr	**439.29**
"Rotherham"	589.00	2.00	37.00	-	602.29	nr	**639.29**
Outdoor seats; Earth Anchors Ltd; "Forest-Saver", steel frame and recycled slats							
bench; 1.8 m	189.00	1.00	18.50	-	189.00	nr	**207.50**
seat; 1.8 m	292.00	1.00	18.50	-	292.00	nr	**310.50**
Outdoor seats; Earth Anchors Ltd; "Evergreen", ci frame and recycled slats							
bench; 1.8 m	396.00	1.00	18.50	-	396.00	nr	**414.50**
seat; 1.8 m	487.00	1.00	18.50	-	487.00	nr	**505.50**
Outdoor seats; all-steel							
Outdoor seats; Marshalls Plc							
"MSF Central" stainless steel seat	950.00	0.50	9.25	-	950.00	nr	**959.25**
"MSFCentral" steel seat	475.00	2.00	37.00	-	475.00	nr	**512.00**
"MSF 502" cast iron Heritage seat	443.65	2.00	37.00	-	443.65	nr	**480.65**
Outdoor seats; Earth Anchors Ltd, "Ranger"							
bench; 1.8 m	266.00	1.00	18.50	-	266.00	nr	**284.50**
seat; 1.8 m	418.00	1.00	18.50	-	418.00	nr	**436.50**
Outdoor seats; all timber							
Outdoor seats; Broxap Street Furniture; teak seats with back and armrests							
ref BX49 4081-2; Cambridge Seat; two seater; 1.3 m	298.00	2.00	37.00	-	318.59	nr	**355.59**
ref BX49 4081-3; Cambridge Seat; three seater; 1.8 m	329.00	2.00	37.00	-	349.59	nr	**386.59**
ref BX49 4080; Milano Seat; three seater; 1.8 m	358.00	2.00	37.00	-	378.59	nr	**415.59**
ref BX49 4082; Lutyens Seat; three seater; 1.96 m	399.00	2.00	37.00	-	419.59	nr	**456.59**
Outdoor seats; Lister Lutyens Co Ltd; "Mendip"; teak							
1.524 m long	465.00	1.00	18.50	-	489.54	nr	**508.04**
1.829 m long	561.00	1.00	18.50	-	585.54	nr	**604.04**
2.438 m long (inc. centre leg)	703.00	1.00	18.50	-	727.54	nr	**746.04**
Outdoor seats; Lister Lutyens Co Ltd; "Sussex"; hardwood							
1.5 m	185.00	2.00	37.00	-	209.54	nr	**246.54**
Outdoor seats; Woodscape Ltd; solid hardwood							
seat type "3" with back; 2.00 m long; freestanding	675.00	2.00	37.00	-	699.54	nr	**736.54**
seat type "3" with back; 2.00 m long; building in	715.00	4.00	74.00	-	739.54	nr	**813.54**
seat type "4" with back; 2.00 m long; fixing to wall	435.00	3.00	55.50	-	446.94	nr	**502.44**
seat type "5" with back; 2.50 m long; freestanding	845.00	2.00	37.00	-	869.54	nr	**906.54**
seat type "4"; 2.00 m long; fixing to wall	325.00	2.00	37.00	-	349.54	nr	**386.54**
seat type "5" with back; 2.00 m long; freestanding	690.00	2.00	37.00	-	714.54	nr	**751.54**
seat type "4" with back; 2.50 m long; fixing to wall	505.00	3.00	55.50	-	516.94	nr	**572.44**

Item Excluding site overheads and profit	PC £	Labour hours	Labour £	Plant £	Material £	Unit	Total rate £
Q50 SITE/STREET FURNITURE/EQUIPMENT - cont'd							
Outdoor seats; Woodscape Ltd; solid hardwood - cont'd							
seat type "5" with back; building in	730.00	2.00	37.00	-	754.54	nr	**791.54**
seat type "5" with back; 2.50 m long; building in	885.00	3.00	55.50	-	909.54	nr	**965.04**
bench type "1"; 2.00 m long; freestanding	535.00	2.00	37.00	-	559.54	nr	**596.54**
bench type "1"; 2.00 m long; building in	575.00	4.00	74.00	-	599.54	nr	**673.54**
bench type "2"; 2.00 m long; freestanding	510.00	2.00	37.00	-	534.54	nr	**571.54**
bench type "2"; 2.00 m long; building in	550.00	4.00	74.00	-	574.54	nr	**648.54**
bench type "2"; 2.50 m long; freestanding	580.00	2.00	37.00	-	604.54	nr	**641.54**
bench type "2"; 2.50 m long; building in	620.00	4.00	74.00	-	644.54	nr	**718.54**
bench type "2"; 2.00 m long overall; curved to 5 m radius, building in	710.00	4.00	74.00	-	734.54	nr	**808.54**
Outdoor seats; Orchard Street Furniture Ltd; "Allington"; all iroko							
1.20 m long	254.26	2.00	37.00	-	278.80	nr	**315.80**
1.80 m long	292.29	2.00	37.00	-	316.83	nr	**353.83**
2.40 m long	355.59	2.00	37.00	-	380.13	nr	**417.13**
Outdoor seats; SMP Playgrounds Ltd; "Lowland range"; all timber							
bench seats with backrest; ref OFDS6; 1.50 m long	549.00	2.00	37.00	-	562.29	nr	**599.29**
Outdoor seats; tree benches/seats							
Tree bench; Neptune Outdoor Furniture Ltd, "Beaufort" hexagonal, timber							
SF34-15A; 1500 mm diameter	944.00	0.50	9.25	-	944.00	nr	**953.25**
Tree seat; Neptune Outdoor Furniture Ltd, "Beaufort" hexagonal, timber, with back							
SF32-10A; 720 mm diameter	1132.00	0.50	9.25	-	1132.00	nr	**1141.25**
SF32-20A; 1720 mm diameter	1324.00	0.50	9.25	-	1324.00	nr	**1333.25**
Street furniture ranges							
Townscape Products Ltd; "Belgrave"; natural grey concrete							
bollards; 250 mm diameter x 500 mm high	76.90	1.00	18.50	-	80.22	nr	**98.72**
seats; 1800 mm x 600 mm x 736 mm high	282.96	2.00	37.00	-	282.96	nr	**319.96**
Picnic benches - General							
Preamble: The following items include for fixing to ground in to manufacturer's instructions or concreting in.							
Picnic tables and benches							
Picnic tables and benches; Broxap Street Furniture "Eastgate" picnic unit	569.00	2.00	37.00	-	582.29	nr	**619.29**
Picnic tables and benches; Woodscape Ltd Table and benches built in, 2 m long	1725.00	2.00	37.00	-	1738.29	nr	**1775.29**
Market prices of containers							
Plant containers; terracotta							
Capital Garden Products Ltd							
Large Pot LP63; weathered terracotta - 1170 x 1600 dia.	-	-	-	-	633.00	nr	-
Large Pot LP38; weathered terracotta - 610 x 970 dia.	-	-	-	-	276.00	nr	-
Large Pot LP23; weathered terracotta - 480 x 580 dia.	-	-	-	-	147.00	nr	-
Manhole Cover Planter 3022 760 x 560 x 210 high	-	-	-	-	102.00	nr	-
Apple Basket 2215; 380 x 560 dia.	-	-	-	-	105.00	nr	-
Indian style Shimmer Pot 2322; 585 x 560 dia.	-	-	-	-	141.00	nr	-
Indian style Shimmer Pot 1717; 430 x 430 dia.	-	-	-	-	84.00	nr	-
Indian style Shimmer Pot 1314; 330 x 355 dia.	-	-	-	-	74.00	nr	-

Q PAVING/PLANTING/FENCING/SITE FURNITURE

Item Excluding site overheads and profit	PC £	Labour hours	Labour £	Plant £	Material £	Unit	**Total rate £**
Plant containers; faux lead							
Capital Garden Products Ltd							
Trough 2508 'Tudor Rose'; 620 x 220 x 230 high	-	-	-	-	58.00	nr	-
Tub 2004 'Elizabethan'; 510 mm square	-	-	-	-	74.00	nr	-
Tub 1513 'Elizabethan'; 380 mm square	-	-	-	-	46.00	nr	-
Tub 1601 'Tudor Rose'; 420 x 400 dia.	-	-	-	-	48.00	nr	-
Plant containers; window boxes							
Capital Garden Products Ltd							
Adam 5401; faux lead; 1370 x 270 x 210 h	-	-	-	-	96.00	nr	-
Wheatsheaf WH54; faux lead; 1370 x 270 x 210 h	-	-	-	-	96.00	nr	-
Oakleaf OAK24; terracotta; 610 x 230 x 240 h	-	-	-	-	62.00	nr	-
Swag 2402; faux lead; 610 x 200 x 210 h	-	-	-	-	34.00	nr	-
Plant containers; timber							
Plant containers; hardwood; Neptune Outdoor Furniture Ltd							
'Beaufort' T38-4D, 1500 x 1500 x 900 mm high	-	-	-	-	1057.00	nr	-
'Beaufort' T38-3C, 1000 x 1500 x 700 mm high	-	-	-	-	786.00	nr	-
'Beaufort' T38-2A, 1000 x 500 x 500 mm high	-	-	-	-	410.00	nr	-
'Kara' T42-4D, 1500 x 1500 x 900 mm high	-	-	-	-	1140.00	nr	-
'Kara' T42-3C, 1000 x 1500 x 700 mm high	-	-	-	-	850.00	nr	-
'Kara' T42-2A, 1000 x 500 x 500 mm high	-	-	-	-	444.00	nr	-
Plant containers; hardwood; Woodscape Ltd							
Square, 900 x 900 x 420 mm high	-	-	-	-	283.20	nr	-
Plant containers; precast concrete							
Plant containers; Marshalls Plc							
"Boulevard 700" circular base and ring	479.51	2.00	37.00	16.50	479.51	nr	**533.01**
"Boulevard 1200" circular base and ring	601.64	1.00	18.50	16.50	601.64	nr	**636.64**
Cycle holders							
Cycle stands; Marshalls Plc							
"Sheffield" steel cycle stand; RCS1	41.80	0.50	9.25	-	41.80	nr	**51.05**
"Sheffield" stainless steel cycle stand; RSCS1	114.95	0.50	9.25	-	114.95	nr	**124.20**
Cycle holders; Autopa; "VELOPA" galvanized steel							
ref R; fixing to wall or post; making good	28.80	1.00	18.50	-	42.09	nr	**60.59**
ref SR(V); fixing in ground; making good	35.10	1.00	18.50	-	48.39	nr	**66.89**
Cycle holders; Townscape Products Ltd							
"Guardian" cycle holders; tubular steel frame; setting in concrete; 1250 mm x 550 mm x 775 mm high; making good	271.84	1.00	18.50	-	285.13	nr	**303.63**
"Penny" cycle stands; 600 mm diameter; exposed aggregate bollards with 8 nr cycle holders; in galvanized steel; setting in concrete; making good	541.93	1.00	18.50	-	555.22	nr	**573.72**
Cycle holders; Broxap Street Furniture							
"Neath" cycle rack; ref BX/MW/AG; for 6 nr cycles; semi-vertical; galvanised and polyester powder coated; 1.320 m wide x 2.542 m long x 1.80 m high	385.00	10.00	185.00	-	411.58	nr	**596.58**
"Premier Senior" combined shelter and rack; ref BX/MW/AW; for 10 nr cycles; horizontal; galvanised only; 2.13 m wide x 3.05 m long x 2.15 m high	860.00	10.00	185.00	-	886.58	nr	**1071.58**
"Toast Rack" double sided free standing cycle rack; ref BX/MW/GH; for 10 nr cycles; galvanised and polyester coated; 3.250 m long	430.00	4.00	74.00	-	440.66	nr	**514.66**

Q PAVING/PLANTING/FENCING/SITE FURNITURE

Item Excluding site overheads and profit	PC £	Labour hours	Labour £	Plant £	Material £	Unit	Total rate £
Q50 SITE/STREET FURNITURE/EQUIPMENT - cont'd							
Directional signage; cast aluminium							
Signage; Furnitubes International Ltd							
ref FFL 1; "Lancer"; cast aluminium finials	58.00	0.07	1.23	-	58.00	nr	**59.23**
ref FAA IS; arrow end type cast aluminium directional arms, single line; 90 mm wide	123.00	2.00	37.00	-	123.00	nr	**160.00**
ref FAA ID; arrow end type cast aluminium directional arms; double line; 145 mm wide	149.00	0.13	2.47	-	149.00	nr	**151.47**
ref FAA IS; arrow end type cast aluminium directional arms; treble line; 200 mm wide	172.00	0.20	3.70	-	172.00	nr	**175.70**
ref FCK1 211 G; "Kingston"; composite standard root columns	325.00	2.00	37.00	-	341.25	nr	**378.25**
Excavating; for bollards and barriers; by hand							
Holes for bollards							
400 mm x 400 mm x 400 mm; disposing off site	-	0.42	7.71	-	0.77	nr	**8.48**
600 mm x 600 mm x 600 mm; disposing off site	-	0.97	17.95	-	2.16	nr	**20.11**
Concrete bollards - General							
Preamble: Precast concrete bollards are available in a very wide range of shapes and sizes. The bollards listed here are the most commonly used sizes and shapes; manufacturer's catalogues should be consulted for the full range. Most manufacturers produce bollards to match their suites of street furniture, which may include planters, benches, litter bins and cycle stands. Most parallel sided bollards can be supplied in removable form, with a reduced shank, precast concrete socket and lifting hole to permit removal with a bar.							
Concrete bollards							
Marshalls Plc; cylinder; straight or tapered; 200 mm - 400 mm diameter; plain grey concrete; setting into firm ground (excavating and backfilling not included)							
"Bridgford" 915 mm high above ground	66.49	2.00	37.00	-	76.54	nr	**113.54**
Marshalls Plc; cylinder; straight or tapered; 200 mm - 400 mm diameter; "Beadalite" reflective finish; setting into firm ground (excavating and backfilling not included)							
"Wexham" concrete bollard; exposed silver grey	133.05	2.00	37.00	-	138.45	nr	**175.45**
"Wexham Major" concrete bollard; exposed silver grey	214.20	2.00	37.00	-	224.25	nr	**261.25**
Precast concrete verge markers; various shapes; 450 mm high							
plain grey concrete	34.00	1.00	18.50	-	37.86	nr	**56.36**
white concrete	35.65	1.00	18.50	-	39.51	nr	**58.01**
exposed aggregate	37.75	1.00	18.50	-	41.61	nr	**60.11**
Other bollards							
Removable parking posts; Dee-Organ Ltd							
"Spacekeeper"; ref 3014203; folding plastic coated galvanized steel parking posts with key; reflective bands; including 300 mm x 300 mm x 300 mm concrete foundations and fixing bolts; 850 mm high	120.59	2.00	37.00	-	132.78	nr	**169.78**
"Spacesaver" ref 1506101; hardwood removable bollards including key and base-plate; 150 mm x 150 mm; 800 mm high	140.75	2.00	37.00	-	144.98	nr	**181.98**

Q PAVING/PLANTING/FENCING/SITE FURNITURE

Item Excluding site overheads and profit	PC £	Labour hours	Labour £	Plant £	Material £	Unit	**Total rate £**
Removable parking posts; Marshalls Plc							
"RT\RD4" domestic telescopic bollard	156.28	2.00	37.00	-	161.68	nr	**198.68**
"RT\R8" heavy duty telescopic bollard	210.90	2.00	37.00	-	216.30	nr	**253.30**
Plastic bollards; Marshalls Plc							
"Lismore" 3 ring recycled plastic bollard	71.10	2.00	37.00	-	76.50	nr	**113.50**
Cast iron bollards - General							
Preamble: The following bollards are particularly suitable for conservation areas. Logos for civic crests can be incorporated to order.							
Cast iron bollards							
Bollards; Furnitubes International Ltd (excavating and backfilling not included)							
"Doric Round"; 920 mm high x 170 mm diameter	99.00	2.00	37.00	-	104.14	nr	**141.14**
"Gunner round"; 750 mm high x 165 mm diameter	60.00	2.00	37.00	-	65.14	nr	**102.14**
"Manchester Round"; 975 mm high; 225 mm square base	102.00	2.00	37.00	-	107.14	nr	**144.14**
"Cannon"; 1140 mm x 210 mm diameter	125.00	2.00	37.00	-	130.14	nr	**167.14**
"Kenton"; heavy duty galvanized steel; 900 mm high; 350 mm diameter	192.00	2.00	37.00	-	197.14	nr	**234.14**
Bollards; Marshalls Plc; (excavating and backfilling not included)							
"MSF103" cast iron bollard - "Small Manchester"	118.75	2.00	37.00	-	124.15	nr	**161.15**
"MSF102" cast iron bollard - "Manchester"	128.25	2.00	37.00	-	133.65	nr	**170.65**
Cast iron bollards with rails - General							
Preamble: The following cast iron bollards are suitable for conservation areas.							
Cast iron bollards with rails							
Cast iron posts with steel tubular rails; Broxap Street Furniture; setting into firm ground (excavating not included)							
"Sheffield Short"; 420 mm high; one rail, type A	74.00	1.50	27.75	-	76.67	nr	**104.42**
"Mersey"; 1085 mm high above ground; two rails, type D	115.00	1.50	27.75	-	117.67	nr	**145.42**
"Promenade"; 1150 mm high above ground; square; three rails, type C	129.00	1.50	27.75	-	131.67	nr	**159.42**
"Type A" mild steel tubular rail including connector	4.13	0.03	0.62	-	5.52	m	**6.14**
"Type C" mild steel tubular rail including connector	4.97	0.03	0.62	-	6.50	m	**7.11**
Steel bollards							
Steel bollards; Marshalls Plc (excavating and backfilling not included)							
"SSB01" stainless steel bollard; 101 x 1250	109.25	2.00	37.00	-	114.65	nr	**151.65**
"RS001" stainless steel bollard; 114 x 1500	123.50	2.00	37.00	-	128.90	nr	**165.90**
"RB119 - Brunel" steel bollard; 168 x 1500	116.85	2.00	37.00	-	126.90	nr	**163.90**
Timber bollards							
Woodscape Ltd; durable hardwood							
RP 250/1500; 250 mm diameter x 1500 mm long	200.10	1.00	18.50	-	200.33	nr	**218.83**
SP 250/1500; 250 mm square x 1500 mm long	200.10	1.00	18.50	-	200.33	nr	**218.83**
SP 150/1200; 150 mm square x 1200 mm long	72.10	1.00	18.50	-	72.33	nr	**90.83**
SP 125/750; 125 mm square x 750 mm long	39.80	1.00	18.50	-	40.03	nr	**58.53**
RP 125/750; 125 mm diameter x 750 mm long	39.80	1.00	18.50	-	40.03	nr	**58.53**

Q PAVING/PLANTING/FENCING/SITE FURNITURE

Item Excluding site overheads and profit	PC £	Labour hours	Labour £	Plant £	Material £	Unit	Total rate £
Q50 SITE/STREET FURNITURE/EQUIPMENT - cont'd							
Deterrent bollards							
Semi-mountable vehicle deterrent and kerb protection bollards; Furnitubes International Ltd (excavating and backfilling not included)							
"Bell decorative"	301.00	2.00	37.00	-	308.32	nr	**345.32**
"Half bell"	312.00	2.00	37.00	-	319.32	nr	**356.32**
"Full bell"	401.00	2.00	37.00	-	408.32	nr	**445.32**
"Three quarter bell"	365.00	2.00	37.00	-	372.32	nr	**409.32**
Security bollards							
Security bollards; Furnitubes International Ltd (excavating and backfilling not included)							
"Gunner"; reinforced with steel insert and tie bars; 750 mm high above ground; 600 mm below ground	132.00	2.00	37.00	-	139.72	nr	**176.72**
"Burr Bloc Type 6" removable steel security bollard; 750 mm high above ground; 410 mm x 285 mm	440.00	2.00	37.00	-	457.36	nr	**494.36**
Tree grilles; cast iron							
Cast iron tree grilles; Furnitubes International Ltd							
ref GS 1070 Greenwich; two part; 1000 mm square; 700 mm diameter tree hole	75.00	2.00	37.00	-	75.00	nr	**112.00**
ref GC 1270 Greenwich; two part, 1200 mm diameter; 700 mm diameter tree hole	92.00	2.00	37.00	-	92.00	nr	**129.00**
Cast iron tree grilles; Marshalls Plc							
"Heritage" cast iron grille plus frame; 1 m x 1 m	323.00	3.00	55.50	-	333.32	nr	**388.82**
Cast iron tree grilles; Townscape Products Ltd							
"Baltimore" 1200 mm square x 460 mm diameter tree hole	391.46	2.00	37.00	-	391.46	nr	**428.46**
"Baltimore" hexagonal, maximum width 1440 mm nominal, 600 mm diameter tree hole	629.87	2.00	37.00	-	629.87	nr	**666.87**
Tree grilles; steel							
Steel tree grilles; Furnitubes International Ltd							
ref GSF 102G Greenwich; steel tree grille frame for GS 1045, one part	197.00	2.00	37.00	-	197.00	nr	**234.00**
ref GSF 122G Greenwich; steel tree grille frame for GS 1270 and GC 1245, one part	266.00	2.00	37.00	-	266.00	nr	**303.00**
Note: Care must be taken to ensure that tree grids and guards are removed when trees grow beyond the specified diameter of guard.							
Playground equipment - General							
Preamble: The range of equipment manufactured or available in the UK is so great that comprehensive coverage would be impossible, especially as designs, specifications and prices change fairly frequently. The following information should be sufficient to give guidance to anyone designing or equipping a playground. In comparing prices note that only outline specification details are given here and that other refinements which are not mentioned may be the reason for some difference in price between two apparently identical elements. The fact that a particular manufacturer does not appear under one item heading does not necessarily imply that he does not make it. Landscape designers are advised to check that equipment complies with ever more stringent safety standards before specifying.							

Item Excluding site overheads and profit	PC £	Labour hours	Labour £	Plant £	Material £	Unit	**Total rate £**
Playground equipment - Installation The rates below include for installation of the specified equipment by the manufacturers. Most manufactures will offer an option to install the equipment supplied by them.							
Play systems; Kompan Ltd; "Galaxy"; multiple play activity systems for non-prescribed play; for children 6-14 years; galvanized steel and high density polyethylene							
Adara GXY 906; 14 different play activities	-	-	-	-	-	nr	**13610.00**
Sirius GXY 8011; 10 different play activities	-	-	-	-	-	nr	**8420.00**
Sports and social areas; multi-use games areas (MUGA); Kompan Ltd; enclosed sports areas; complete with surfacing boundary and goals and targets; galvanized steel framework with high density polyethylene panels, galvanized steel goals and equipment; surfacing priced separately							
Pitch complete; suitable for multiple ball sports; suitable for use with natural artificial or hard landscape surfaces; fully enclosed including 2 nr end sports walls							
FRE 2202 end sports wall multigoal only	-	-	-	-	-	nr	**9750.00**
FRE 2000 12.0 x 20.0 m	-	-	-	-	-	nr	**23820.00**
FRE 2116 19.00 x 36.00	-	-	-	-	-	nr	**42170.00**
FRE 3000 meeting point shelter or social area inclusive of 2 nr modular S benches, 3.920 m x 1.40 m	-	-	-	-	-	nr	**4040.00**
Swings - General Preamble: Prices for the following vary considerably. Those given represent the middle of the range and include multiple swings with tubular steel frames and timber or tyre seats; ground fixing and priming only.							
Swings							
Swings; Wicksteed Ltd							
traditional swings; 1850 mm high; 1 bay; 2 seat	-	-	-	-	-	nr	**1861.00**
traditional swings; 1850 mm high; 2 bay; 4 seat	-	-	-	-	-	nr	**2981.00**
traditional swings; 2450 mm high; 1 bay; 2 seat	-	-	-	-	-	nr	**1799.00**
traditional swings; 2450 mm high; 2 bay; 4 seat	-	-	-	-	-	nr	**2831.00**
traditional swings; 3050 mm high; 1 bay; 2 seat	-	-	-	-	-	nr	**1909.00**
traditional swings; 3050 mm high; 2 bay; 4 seat	-	-	-	-	-	nr	**2967.00**
double arch swing; cradle safety seats; 1850 mm high	-	-	-	-	-	nr	**1882.00**
twin double arch swing; cradle safety seat; 1850 mm high	-	-	-	-	-	nr	**3148.00**
single arch swing; flat rubber safety seat; 2450 mm high	-	-	-	-	-	nr	**1452.00**
double arch swing; flat rubber safety seats; 2450 mm high	-	-	-	-	-	nr	**1779.00**
Swings; Lappset UK Ltd							
ref 020414M; swing frame with two flat seats	-	-	-	-	-	nr	**1360.00**
Swings; Kompan Ltd							
ref M947-52; double swings	-	-	-	-	-	nr	**2810.00**
ref M948-52; double swings	-	-	-	-	-	nr	**2300.00**
ref M951; "Sunflower" swings	-	-	-	-	-	nr	**1320.00**

Q PAVING/PLANTING/FENCING/SITE FURNITURE

Item Excluding site overheads and profit	PC £	Labour hours	Labour £	Plant £	Material £	Unit	Total rate £
Q50 SITE/STREET FURNITURE/EQUIPMENT - cont'd							
Slides							
Slides; Wicksteed Ltd							
"Pedestal" slides; 3.40 m	-	-	-	-	-	nr	**2571.00**
"Pedestal" slides; 4.40 m	-	-	-	-	-	nr	**3343.00**
"Pedestal" slides; 5.80 m	-	-	-	-	-	nr	**3890.00**
"Embankment" slides; 3.40 m	-	-	-	-	-	nr	**1970.00**
"Embankment" slides; 4.40 m	-	-	-	-	-	nr	**2559.00**
"Embankment" slides; 5.80 m	-	-	-	-	-	nr	**3256.00**
"Embankment" slides; 7.30 m	-	-	-	-	-	nr	**4137.00**
"Embankment" slides; 9.10 m	-	-	-	-	-	nr	**5180.00**
"Embankment" slides; 11.00 m	-	-	-	-	-	nr	**6196.00**
Slides; Lappset UK Ltd							
ref 142015M; slide	-	-	-	-	-	nr	**2940.00**
ref 141115M; "Jumbo" slide	-	-	-	-	-	nr	**4030.00**
Slides; Kompan Ltd							
ref M351; slides	-	-	-	-	-	nr	**2250.00**
ref M322; slide and cave	-	-	-	-	-	nr	**2650.00**
Moving equipment - General							
Preamble: The following standard items of playground equipment vary considerably in quality and price; the following prices are middle of the range.							
Moving equipment							
Roundabouts; Wicksteed Ltd							
"Turnstile"	-	-	-	-	-	nr	**391.00**
"Speedway" (without restrictor)	-	-	-	-	-	nr	**2825.00**
"Spiro Whirl" (without restrictor)	-	-	-	-	-	nr	**3216.00**
Roundabouts; Kompan Ltd							
Supernova GXY916 multifunctional spinning and balancing disc; capacity approximately 15 children	-	-	-	-	-	nr	**3380.00**
Seesaws							
Seesaws; Lappset UK Ltd							
ref 010300; seesaws	-	-	-	-	-	nr	**825.00**
ref 010237; seesaws	-	-	-	-	-	nr	**1950.00**
Seesaws; Wicksteed Ltd							
"Seesaw" (non-bump)	-	-	-	-	-	nr	**2130.00**
"Jolly Gerald" (non-bump)	-	-	-	-	-	nr	**2350.00**
"Rocking Rockette" (with motion restrictor)	-	-	-	-	-	nr	**2990.00**
"Rocking Horse" (with motion restrictor)	-	-	-	-	-	nr	**3560.00**
Play sculptures - General							
Preamble: Many variants on the shapes of playground equipment are available, simulating spacecraft, trains, cars, houses etc., and these designs are frequently changed. The basic principles remain constant but manufacturer's catalogues should be checked for the latest styles.							

Q PAVING/PLANTING/FENCING/SITE FURNITURE

Item Excluding site overheads and profit	PC £	Labour hours	Labour £	Plant £	Material £	Unit	Total rate £
Climbing equipment and play structures; General							
Preamble: Climbing equipment generally consists of individually designed modules. Play structures generally consist of interlinked and modular pieces of equipment and sculptures. These may consist of climbing, play and skill based modules, nets and various other activities. Both are set into either safety surfacing or defined sand pit areas. The equipment below outlines a range from various manufacturers. Individual catalogues should be consulted in each instance. Safety areas should be allowed round all equipment.							
Climbing equipment and play structures							
Climbing equipment and play structures; Kompan Ltd							
ref M480 "Castle"	-	-	-	-	-	nr	**26370.00**
Climbing equipment; Wicksteed Ltd							
Funrun Fitness Trail "Under Starter's Orders" set of 12 units	-	-	-	-	-	nr	**13281.00**
Climbing equipment; Lappset UK Ltd							
ref 138401M; "Storks Nest"	-	-	-	-	-	nr	**3685.00**
ref 122457M; "Playhouse"	-	-	-	-	-	nr	**3460.00**
ref 120100M; "Activity Tower"	-	-	-	-	-	nr	**12195.00**
ref 120124M; "Tower and Climbing Frame"	-	-	-	-	-	nr	**11865.00**
SMP Playgrounds							
Action Pack - Cape Horn; 8 module	-	-	-	-	-	nr	**9595.00**
Spring equipment							
Spring based equipment for 1 - 8 year olds; Kompan Ltd							
ref M101; "Crazy Hen"	-	-	-	-	-	nr	**590.00**
ref M128; "Crazy Daisy"	-	-	-	-	-	nr	**800.00**
ref M141; "Crazy Seesaw"	-	-	-	-	-	nr	**1460.00**
ref M155; "Quartet Seesaw"	-	-	-	-	-	nr	**1110.00**
Spring based equipment for under 12's; Lappset UK Ltd							
ref 010501; "Horse Springer"	-	-	-	-	-	nr	**1010.00**
Sand pits							
Market prices							
Play pit sand; Boughton Loam Ltd	-	-	-	-	46.98	m^3	-
Kompan Ltd							
"Basic 500" 2900 mm x 1570 mm x 310 mm deep	-	-	-	-	-	nr	**970.00**
Flagpoles							
Flagpoles; Harrison External Display Systems; in glass fibre; smooth white finish; terylene halyards with mounting accessories; setting in concrete; to manufacturers recommendations (excavating not included)							
6.00 m high, plain	159.00	3.00	55.50	-	180.32	nr	**235.82**
6.00 m high, hinged baseplate, external halyard	198.00	4.00	74.00	-	219.32	nr	**293.32**
6.00 m high, hinged baseplate, internal halyard	303.00	4.00	74.00	-	324.32	nr	**398.32**
10.00 m high, plain	315.00	5.00	92.50	-	336.32	nr	**428.82**
10.00 m high, hinged baseplate, external halyard	771.00	5.00	92.50	-	792.32	nr	**884.82**
10.00 m high, hinged baseplate, internal halyard	448.00	5.00	92.50	-	469.32	nr	**561.82**
15.00 m high, plain	721.00	6.00	111.00	-	742.32	nr	**853.32**
15.00 m high, hinged baseplate, external halyard	771.00	6.00	111.00	-	792.32	nr	**903.32**
15.00 m high, hinged baseplate, internal halyard	970.00	6.00	111.00	-	991.32	nr	**1102.32**

Q PAVING/PLANTING/FENCING/SITE FURNITURE

Item Excluding site overheads and profit	PC £	Labour hours	Labour £	Plant £	Material £	Unit	**Total rate £**
Q50 SITE/STREET FURNITURE/EQUIPMENT - cont'd							
Flagpoles - cont'd							
Flagpoles; Harrison External Display Systems; tapered hollow steel section; base plate flange; galvanized; white painted finish; lowering gear; including bolting to concrete (excavating not included)							
7.00 m high	131.00	3.00	55.50	-	151.30	nr	**206.80**
10.00 m high	315.00	3.50	64.75	-	338.29	nr	**403.04**
13.00 m high	342.00	6.00	111.00	-	369.70	nr	**480.70**
Flagpoles; Harrison External Display Systems; in glass fibre; smooth white finish; terylene halyards for wall mounting							
3.0 m pole	180.00	2.00	37.00	-	180.00	nr	**217.00**
Flagpoles; Harrison External Display Systems; parallel galvanised steel banner flagpole - guide price							
6 m high	1000.00	4.00	74.00	-	1021.32	nr	**1095.32**
Sports equipment; Edwards Sports Products Ltd							
Tennis courts; steel posts and sockets for hardcourt							
round	246.23	1.00	18.50	-	246.23	set	**264.73**
square	235.01	1.00	18.50	-	235.01	set	**253.51**
Tennis nets; not including posts or fixings							
ref 5001; "Championship"	124.57	3.00	55.50	-	131.68	set	**187.18**
ref 5021; "Matchplay"	89.27	3.00	55.50	-	96.38	set	**151.88**
ref 5075; "Club"	69.72	3.00	55.50	-	76.83	set	**132.33**
Football goals; senior; socketed; including backirons and 2.5 mm nets							
ref 2711/2197/20261; aluminium	1042.75	4.00	74.00	-	1049.86	set	**1123.86**
ref 2760/2190/20261; steel	696.99	0.14	2.66	-	894.41	set	**897.07**
Freestanding football goals; senior; including nets							
ref 2719/2000A; aluminium	1644.00	4.00	74.00	-	1651.11	set	**1725.11**
ref 2890/20261; steel	935.89	4.00	74.00	-	943.00	set	**1017.00**
Freestanding mini-soccer goals; including nets							
ref 2717/CN; aluminium	492.94	4.00	74.00	-	500.05	set	**574.05**
ref 2961/CN; steel	465.28	4.00	74.00	-	472.39	set	**546.39**
Rugby goal posts; including sockets							
ref 2800; 7.32 m high	671.52	6.00	111.00	-	690.47	set	**801.47**
ref 2802; 10.70 m high	927.38	6.00	111.00	-	946.33	set	**1057.33**
Hockey goals; steel; including backboards and nets							
ref 2906/2310; freestanding	1185.14	1.00	18.50	-	1185.14	set	**1203.64**
ref 2851/2853/2841/2310; socketed	902.39	1.00	18.50	-	902.39	set	**920.89**
Cricket cages; steel; including netting							
ref 5454; freestanding	858.33	12.00	222.00	-	858.33	set	**1080.33**
ref 5454/W; wheelaway	1104.66	12.00	222.00	-	1104.66	set	**1326.66**

R DISPOSAL SYSTEMS

Item Excluding site overheads and profit	PC £	Labour hours	Labour £	Plant £	Material £	Unit	**Total rate £**
R12 DRAINAGE BELOW GROUND							
Silt pits and inspection chambers							
Excavating pits; starting from ground level; by machine							
maximum depth not exceeding 1.00 m	-	0.50	9.25	16.50	-	m^3	**25.75**
maximum depth not exceeding 2.00 m	-	0.50	9.25	24.75	-	m^3	**34.00**
maximum depth not exceeding 4.00 m	-	0.50	9.25	49.50	-	m^3	**58.75**
Disposal of excavated material; depositing on site in permanent spoil heaps; average 50 m	-	0.04	0.77	1.61	-	m^3	**2.38**
Filling to excavations; obtained from on site spoil heaps; average thickness not exceeding 0.25 m	-	0.13	2.47	4.40	-	m^3	**6.86**
Surface treatments; compacting; bottoms of excavations	-	0.05	0.93	-	-	m^2	**0.93**
Earthwork support; distance between opposing faces not exceeding 2.00 m							
maximum depth not exceeding 1.00 m	-	0.20	3.70	-	4.28	m^3	**7.98**
maximum depth not exceeding 2.00 m	-	0.30	5.55	-	4.28	m^3	**9.83**
maximum depth not exceeding 4.00 m	-	0.67	12.33	-	1.69	m^3	**14.02**
Silt pits and inspection chambers; in situ concrete							
Beds; plain in situ concrete; 11.50 N/mm^2 - 40 mm aggregate							
thickness not exceeding 150 mm	-	1.00	18.50	-	98.71	m^3	**117.21**
thickness 150 mm - 450 mm	-	0.67	12.33	-	98.71	m^3	**111.04**
Benchings in bottoms; plain in situ concrete; 25.50 N/mm^2 - 20 mm aggregate							
thickness 150 mm - 450 mm	-	2.00	37.00	-	98.71	m^3	**135.71**
Isolated cover slabs; reinforced in situ concrete; 21.00 N/mm^2 - 20 mm aggregate							
thickness not exceeding 150 mm	98.71	4.00	74.00	-	98.71	m^3	**172.71**
Fabric reinforcement; BS 4483; A193 (3.02 kg/m^2) in cover slabs	3.16	0.06	1.16	-	3.48	m^2	**4.63**
Formwork to reinforced in situ concrete; isolated cover slabs							
soffits; horizontal	-	3.28	60.68	-	4.40	m^2	**65.08**
height not exceeding 250 mm	-	0.97	17.95	-	1.87	m	**19.81**
Silt pits and inspection chambers; precast concrete units							
Precast concrete inspection chamber units; BS 5911; FP McCann Ltd; bedding, jointing and pointing in cement mortar (1:3); 600 mm x 450 mm internally							
600 mm deep	35.55	6.00	111.00	-	39.86	nr	**150.86**
900 mm deep	43.95	7.00	129.50	-	48.47	nr	**177.97**
Drainage chambers; FP McCann Ltd; 1200 mm x 750 mm reducing to 600 mm x 600 mm; no base unit; depth of invert							
1050 mm deep	229.90	9.00	166.50	-	241.35	nr	**407.85**
1650 mm deep	331.10	11.00	203.50	-	348.51	nr	**552.01**
2250 mm deep	432.30	12.50	231.25	-	454.52	nr	**685.77**
Cover slabs for chambers or shaft sections; FP McCann Ltd; heavy duty							
900 mm diameter internally	39.10	0.67	12.32	-	39.10	nr	**51.42**
1050 mm diameter internally	47.15	2.00	37.00	11.00	47.15	nr	**95.15**
1200 mm diameter internally	53.59	1.00	18.50	11.00	53.59	nr	**83.09**
1500 mm diameter internally	108.00	1.00	18.50	11.00	108.00	nr	**137.50**
1800 mm diameter internally	129.60	2.00	37.00	24.75	129.60	nr	**191.35**

R DISPOSAL SYSTEMS

Item Excluding site overheads and profit	PC £	Labour hours	Labour £	Plant £	Material £	Unit	Total rate £
R12 DRAINAGE BELOW GROUND - cont'd							
Brickwork							
Walls to manholes; bricks; PC £300.00/1000; in cement mortar (1:3)							
one brick thick	36.00	3.00	55.50	-	44.65	m^2	**100.15**
one and a half brick thick	54.00	4.00	74.00	-	66.97	m^2	**140.97**
two brick thick projection of footing or the like	72.00	4.80	88.80	-	89.30	m^2	**178.10**
Walls to manholes; engineering bricks; PC £260.00/1000; in cement mortar (1:3)							
one brick thick	31.20	3.00	55.50	-	39.61	m^2	**95.11**
one and a half brick thick	46.80	4.00	74.00	-	59.41	m^2	**133.41**
two brick thick projection of footing or the like	62.40	4.80	88.80	-	79.22	m^2	**168.02**
Extra over common or engineering bricks in any mortar for fair face; flush pointing as work proceeds; English bond walls or the like	-	0.13	2.47	-	-	m^2	**2.47**
In situ finishings; cement:sand mortar (1:3); steel trowelled; 13 mm one coat work to manhole walls; to brickwork or blockwork base; over 300 mm wide	-	0.80	14.80	-	2.28	m^2	**17.08**
Building into brickwork; ends of pipes; making good facings or renderings							
small	-	0.20	3.70	-	-	nr	**3.70**
large	-	0.30	5.55	-	-	nr	**5.55**
extra large	-	0.40	7.40	-	-	nr	**7.40**
extra large; including forming ring arch cover	-	0.50	9.25	-	-	nr	**9.25**
Inspection Chambers; Hepworth Plc							
Polypropylene up to 600 mm deep							
300 mm diameter; 960 mm deep; heavy duty round covers and frames; double seal recessed with 6 nr 110 mm outlets/inlets	178.46	3.00	55.50	-	180.87	nr	**236.37**
Polypropylene up to 1200 mm deep							
475 mm diameter; 1030 mm deep; heavy duty round covers and frames; double seal recessed with 5 nr 110 mm outlets/inlets	59.14	4.00	74.00	-	237.95	nr	**311.95**
Cast iron inspection chambers; to BS 437; St Gobain Pipelines Plc; drainage systems bolted flat covers; bedding in cement mortar (1:3); mechanical coupling joints							
100 mm x 100 mm							
one branch each side	177.36	2.20	40.70	-	180.18	nr	**220.88**
two branches each side	323.07	2.40	44.40	-	324.21	nr	**368.61**
100 mm x 150 mm							
one branch each side	225.03	1.80	33.30	-	225.26	nr	**258.56**
150 mm x 150 mm							
one branch	231.96	1.50	27.75	-	232.19	nr	**259.94**
one branch each side	267.44	2.75	50.88	-	267.67	nr	**318.55**
Step irons; to BS 1247; Ashworth Ltd drainage systems; malleable cast iron; galvanized; building into joints							
General purpose pattern; for one brick walls	4.34	0.17	3.15	-	4.34	nr	**7.48**
Best quality vitrified clay half section channels; Hepworth Plc; bedding and jointing in cement:mortar (1:2)							
Channels; straight							
100 mm	5.15	0.80	14.80	-	8.31	m	**23.11**
150 mm	8.57	1.00	18.50	-	11.73	m	**30.23**
225 mm	19.26	1.35	24.98	-	22.42	m	**47.40**
300 mm	39.55	1.80	33.30	-	42.71	m	**76.01**

R DISPOSAL SYSTEMS

Item Excluding site overheads and profit	PC £	Labour hours	Labour £	Plant £	Material £	Unit	**Total rate £**
Bends; 15, 30, 45 or 90 degrees							
100 mm bends	4.43	0.75	13.88	-	6.01	nr	**19.89**
150 mm bends	7.66	0.90	16.65	-	10.04	nr	**26.69**
225 mm bends	29.72	1.20	22.20	-	32.88	nr	**55.08**
300 mm bends	60.60	1.10	20.35	-	64.56	nr	**84.91**
Best quality vitrified clay three quarter section channels; Hepworth Plc; bedding and jointing in cement:mortar (1:2)							
Branch bends; 115, 140 or 165 degrees; left or right hand							
100 mm	4.43	0.75	13.88	-	6.01	nr	**19.89**
150 mm	7.66	0.90	16.65	-	10.04	nr	**26.69**
Intercepting traps							
Vitrified clay; inspection arms; brass stoppers; iron levers; chains and staples; galvanized; staples cut and pinned to brickwork; cement:mortar (1:2) joints to vitrified clay pipes and channels; bedding and surrounding in concrete; 11.50 N/mm^2 - 40 mm aggregate; cutting and fitting brickwork; making good facings							
100 mm inlet; 100 mm outlet	65.79	3.00	55.50	-	95.24	nr	**150.74**
150 mm inlet; 150 mm outlet	94.86	2.00	37.00	-	130.16	nr	**167.16**
Excavating trenches; using 3 tonne tracked excavator; to receive pipes; grading bottoms; earthwork support; filling with excavated material to within 150 mm of finished surfaces and compacting; completing fill with topsoil; disposal of surplus soil							
Services not exceeding 200 mm nominal size							
average depth of run not exceeding 0.50 m	0.83	0.12	2.22	1.00	1.00	m	**4.22**
average depth of run not exceeding 0.75 m	0.83	0.16	3.01	1.38	1.00	m	**5.40**
average depth of run not exceeding 1.00 m	0.83	0.28	5.24	2.41	1.00	m	**8.65**
average depth of run not exceeding 1.25 m	0.83	0.38	7.09	3.24	0.83	m	**11.17**
Granular beds to trenches; lay granular material, to trenches excavated separately, to receive pipes (not included)							
300 wide x 100 thick							
reject sand	-	0.05	0.93	0.23	1.68	m	**2.83**
reject gravel	-	0.05	0.93	0.23	1.57	m	**2.73**
shingle 40 mm aggregate	-	0.05	0.93	0.23	3.50	m	**4.65**
sharp sand	3.05	0.05	0.93	0.23	3.36	m	**4.51**
300 wide x 150 thick							
reject sand	-	0.08	1.39	0.34	2.52	m	**4.24**
reject gravel	-	0.08	1.39	0.34	2.36	m	**4.09**
shingle 40 mm aggregate	-	0.08	1.39	0.34	5.24	m	**6.97**
sharp sand	4.58	0.08	1.39	0.34	5.04	m	**6.76**
Excavating trenches; using 3 tonne tracked excavator; to receive pipes; grading bottoms; earthwork support; filling with imported granular material type 2 and compacting; disposal of surplus soil							
Services not exceeding 200 mm nominal size							
average depth of run not exceeding 0.50 m	4.86	0.09	1.60	0.70	6.66	m	**8.96**
average depth of run not exceeding 0.75 m	7.29	0.11	2.00	0.87	9.99	m	**12.87**
average depth of run not exceeding 1.00 m	9.72	0.14	2.58	1.14	13.32	m	**17.04**
average depth of run not exceeding 1.25 m	12.15	0.23	4.23	1.94	16.65	m	**22.82**

R DISPOSAL SYSTEMS

Item Excluding site overheads and profit	PC £	Labour hours	Labour £	Plant £	Material £	Unit	**Total rate £**
R12 DRAINAGE BELOW GROUND - cont'd							
Excavating trenches, using 3 tonne tracked excavator, to receive pipes; grading bottoms; earthwork support; filling with concrete, ready mixed ST2; disposal of surplus soil							
Services not exceeding 200 mm nominal size							
average depth of run not exceeding 0.50 m	11.34	0.11	1.97	0.36	13.43	m	**15.76**
average depth of run not exceeding 0.75 m	17.01	0.13	2.41	0.45	20.13	m	**22.99**
average depth of run not exceeding 1.00 m	22.68	0.17	3.08	0.60	26.84	m	**30.53**
average depth of run not exceeding 1.25 m	28.35	0.23	4.16	0.90	33.56	m	**38.62**
Earthwork support; providing support to opposing faces of excavation; moving along as work proceeds; A Plant Acrow							
Maximum depth not exceeding 2.00 m							
distance between opposing faces not exceeding 2.00 m	-	0.80	14.80	17.70	-	m	**32.50**
Clay pipes and fittings: to BS EN295:1:1991; Hepworth Plc; Supersleeve							
100 mm clay pipes; polypropylene slip coupling; in trenches (trenches not included)							
laid straight	2.36	0.25	4.63	-	4.10	m	**8.72**
short runs under 3.00 m	2.36	0.31	5.78	-	4.10	m	**9.88**
Extra over 100 mm clay pipes for							
bends; 15 - 90 degree; single socket	7.85	0.25	4.63	-	7.85	nr	**12.47**
junction; 45 or 90 degree; double socket	16.54	0.25	4.63	-	16.54	nr	**21.17**
slip couplings polypropylene	2.78	0.08	1.54	-	2.78	nr	**4.32**
gully with "P" trap; 100 mm; 154 mm x 154 mm plastic grating	26.86	1.00	18.50	-	39.46	nr	**57.96**
150 mm vitrified clay pipes; polypropylene slip coupling; in trenches (trenches not included)							
laid straight	9.41	0.30	5.55	-	9.41	m	**14.96**
short runs under 3.00 m	7.24	0.33	6.17	-	7.24	m	**13.41**
Extra over 150 mm vitrified clay pipes for							
bends; 15 - 90 degree	10.48	0.28	5.18	-	15.54	nr	**20.71**
junction; 45 or 90 degree; 100 x 150	14.03	0.40	7.40	-	24.13	nr	**31.53**
junction; 45 or 90 degree; 150 x 150	15.39	0.40	7.40	-	25.49	nr	**32.89**
slip couplings polypropylene	5.05	0.05	0.93	-	5.05	nr	**5.98**
taper pipe 100 - 150 mm	15.77	0.50	9.25	-	15.77	nr	**25.02**
taper pipe 150 - 225 mm	40.49	0.50	9.25	-	40.49	nr	**49.74**
socket adaptor; connection to traditional pipes and fittings	10.63	0.33	6.11	-	15.68	nr	**21.79**
Accessories in clay							
150 mm access pipe	39.53	-	-	-	49.63	nr	-
150 mm rodding eye	36.04	0.50	9.25	-	40.05	nr	**49.30**
gully with "P" traps; 150 mm; 154 mm x 154 mm plastic grating	44.62	1.00	18.50	-	51.27	nr	**69.77**
PVC-u pipes and fittings; to BS EN1401; Wavin Plastics Ltd; OsmaDrain							
110 mm PVC-u pipes; in trenches (trenches not included)							
laid straight	6.47	0.08	1.48	-	6.47	m	**7.95**
short runs under 3.00 m	6.47	0.12	2.22	-	7.28	m	**9.50**
Extra over 110 mm PVC-u pipes for							
bends; short radius	12.61	0.25	4.63	-	12.61	nr	**17.23**
bends; long radius	22.43	0.25	4.63	-	22.43	nr	**27.05**
junctions; equal; double socket	15.05	0.25	4.63	-	15.05	nr	**19.68**
slip couplings	7.30	0.25	4.63	-	7.30	nr	**11.93**
adaptors to clay	13.49	0.50	9.25	-	13.49	nr	**22.74**

R DISPOSAL SYSTEMS

Item Excluding site overheads and profit	PC £	Labour hours	Labour £	Plant £	Material £	Unit	**Total rate £**
160 mm PVC-u pipes; in trenches (trenches not included)							
laid straight	14.90	0.08	1.48	-	14.90	m	**16.38**
short runs under 3.00 m	26.12	0.12	2.22	-	27.98	m	**30.20**
Extra over 160 mm PVC-u pipes for							
socket bend double 90 or 45 degrees	45.28	0.20	3.70	-	45.28	nr	**48.98**
socket bend double 15 or 30 degrees	42.34	0.20	3.70	-	42.34	nr	**46.04**
socket bend single 87.5 or 45 degrees	25.81	0.20	3.70	-	25.81	nr	**29.51**
socket bend single 15 or 30 degrees	22.87	0.20	3.70	-	22.87	nr	**26.57**
bends; short radius	33.64	0.25	4.63	-	46.25	nr	**50.88**
bends; long radius	87.67	0.25	4.63	-	87.67	nr	**92.30**
junctions; single	82.90	0.33	6.17	-	82.90	nr	**89.07**
pipe coupler	16.82	0.05	0.93	-	16.82	nr	**17.75**
slip couplings PVC-u	8.85	0.05	0.93	-	8.85	nr	**9.78**
adaptors to clay	36.27	0.50	9.25	-	36.27	nr	**45.52**
level invert reducer	14.55	0.50	9.25	-	14.55	nr	**23.80**
spiggot	31.37	0.20	3.70	-	31.37	nr	**35.07**
Accessories in PVC-u							
110 mm screwed access cover	-	-	-	-	-	nr	**12.31**
110 mm rodding eye	25.79	0.50	9.25	-	29.81	nr	**39.06**
gully with "P" traps; 110 mm; 154 mm x 154 mm grating	47.35	1.00	18.50	-	48.96	nr	**67.46**
Kerbs; to gullies; in one course Class B engineering bricks; to 4 nr sides; rendering in cement:mortar (1:3); dished to gully gratings	2.08	1.00	18.50	-	2.87	nr	**21.37**
Gullies; concrete; FP McCann							
Concrete road gullies; to BS 5911; trapped with rodding eye and stoppers; 450 mm diameter x 1.07 m deep	35.00	6.00	111.00	-	57.26	nr	**168.25**
Gullies; vitrified clay; Hepworth Plc; bedding in concrete; 11.50 N/mm^2 - 40 mm aggregate							
Vitrified clay yard gullies (mud); trapped; domestic duty (up to 1 tonne)							
RGP5; 100 mm outlet; 100 mm dia; 225 mm internal width 585 mm internal depth	87.49	3.50	64.75	-	88.08	nr	**152.83**
RGP7; 150 mm outlet; 100 mm dia; 225 mm internal width 585 mm internal depth	87.49	3.50	64.75	-	88.08	nr	**152.83**
Vitrified clay yard gullies (mud); trapped; medium duty (up to 5 tonnes)							
100 mm outlet; 100 mm dia; 225 mm internal width 585 mm internal depth	123.69	3.50	64.75	-	124.28	nr	**189.03**
150 mm outlet; 100 mm dia; 225 mm internal width 585 mm internal depth	123.62	3.50	64.75	-	124.21	nr	**188.96**
Combined filter and silt bucket for yard gullies							
225 mm wide	28.76	-	-	-	28.76	nr	-
Vitrified clay road gullies; trapped with rodding eye							
100 mm outlet; 300 mm internal dia, 600 mm internal depth	76.80	3.50	64.75	-	77.40	nr	**142.15**
150 mm outlet; 300 mm internal dia, 600 mm internal depth	78.65	3.50	64.75	-	79.24	nr	**143.99**
150 mm outlet; 400 mm internal dia, 750 mm internal depth	91.22	3.50	64.75	-	91.81	nr	**156.56**
150 mm outlet; 450 mm internal dia, 900 mm internal depth	-	3.50	64.75	-	124.01	nr	**188.76**
Hinged gratings and frames for gullies; alloy							
135 mm for 100 mm dia gully	-	-	-	-	10.54	nr	-
193 mm for 150 mm dia gully	-	-	-	-	18.22	nr	-
120 mm x 120 mm	-	-	-	-	6.47	nr	-
150 mm x 150 mm	-	-	-	-	11.72	nr	-
230 mm x 230 mm	-	-	-	-	21.43	nr	-
316 mm x 316 mm	-	-	-	-	56.83	nr	-

R DISPOSAL SYSTEMS

Item Excluding site overheads and profit	PC £	Labour hours	Labour £	Plant £	Material £	Unit	Total rate £
R12 DRAINAGE BELOW GROUND - cont'd							
Gullies; vitrified clay - cont'd							
Hinged gratings and frames for gullies; cast iron							
265 mm for 225 mm dia gully	-	-	-	-	36.41	nr	-
150 mm x 150 mm	-	-	-	-	11.72	nr	-
230 mm x 230 mm	-	-	-	-	21.43	nr	-
316 mm x 316 mm	-	-	-	-	56.83	nr	-
Universal gully trap PVC-u; Wavin Plastics Ltd; OsmaDrain system; bedding in concrete; 11.50 N/mm^2 - 40 mm aggregate							
Universal gully fitting; comprising gully trap only							
110 mm outlet; 110 mm dia; 205 mm internal depth	10.70	3.50	64.75	-	11.29	nr	**76.04**
Vertical inlet hopper c\w plastic grate							
272 x 183 mm	14.38	0.25	4.63	-	14.38	nr	**19.01**
Sealed access hopper							
110 x 110 mm	31.24	0.25	4.63	-	31.24	nr	**35.86**
Universal gully PVC-u; Wavin Plastics Ltd; OsmaDrain system; accessories to universal gully trap							
Hoppers; backfilling with clean granular material; tamping; surrounding in lean mix concrete							
plain hopper with 110 spigot 150 mm long	11.75	0.40	7.40	-	12.09	nr	**19.49**
vertical inlet hopper with 110 spigot 150 mm long	14.38	0.40	7.40	-	14.38	nr	**21.78**
sealed access hopper with 110 spigot 150 mm long	31.24	0.40	7.40	-	31.24	nr	**38.64**
plain hopper; solvent weld to trap	8.03	0.40	7.40	-	8.03	nr	**15.43**
vertical inlet hopper; solvent wed to trap	13.08	0.40	7.40	-	13.08	nr	**20.48**
sealed access cover; PVC-U	15.91	0.10	1.85	-	15.91	nr	**17.76**
Gullies PVC-u; Wavin Plastics Ltd; OsmaDrain system; bedding in concrete; 11.50 N/mm^2 - 40 mm aggregate							
Bottle gully; providing access to the drainage system for cleaning							
bottle gully; 228 x 228 x 317 mm deep	26.38	0.50	9.25	-	26.86	nr	**36.11**
sealed access cover; PVC-u 217 x 217 mm	20.51	0.10	1.85	-	20.51	nr	**22.36**
grating; ductile iron 215 x 215 mm	15.81	0.10	1.85	-	15.81	nr	**17.66**
bottle gully riser; 325 mm	3.49	0.50	9.25	-	4.53	nr	**13.78**
Yard gully; trapped 300 mm diameter 600 mm deep; including catchment bucket and ductile iron cover and frame, medium duty loading							
305 mm diameter 600 mm deep	163.27	2.50	46.25	-	166.40	nr	**212.65**
Kerbs to gullies							
One course Class B engineering bricks to 4 nr sides; rendering in cement:mortar (1:3); dished to gully gratings							
150 mm x 150 mm	1.04	0.33	6.17	-	1.59	nr	**7.75**
Linear Drainage							
Marshalls Plc "Mini Beany" combined kerb and channel drainage system; to trenches (not included)							
Precast concrete drainage channel base; 185 mm - 385 mm deep; bedding, jointing and pointing in cement mortar (1:3); on 150 mm deep concrete (ready mixed) foundation; including haunching with in situ concrete; 11.50 N/mm^2 - 40 aggregate one side; channels 250 mm wide x 1.00 long							
straight; 1.00 m long	18.32	1.00	18.50	-	21.53	m	**40.03**
straight; 500 mm long	18.32	1.05	19.43	-	21.53	m	**40.96**
radial; 30 -10 m or 9-6 m internal or external	16.93	1.33	24.67	-	20.14	m	**44.81**
angles 45 or 90 degree	50.50	1.00	18.50	-	53.71	nr	**72.21**

R DISPOSAL SYSTEMS

Item Excluding site overheads and profit	PC £	Labour hours	Labour £	Plant £	Material £	Unit	**Total rate £**
Mini Beany Top Block; perforated kerb unit to drainage channel above; natural grey							
straight	9.71	0.33	6.17	-	10.92	m	**17.08**
radial; 30-10 m or 9-6 m internal or external	12.11	0.50	9.25	-	13.32	m	**22.57**
angles 45 or 90 degree	28.20	0.50	9.25	-	29.41	nr	**38.66**
Mini Beany; outfalls; two section concrete trapped outfall with Mini Beany cast iron access cover and frame; to concrete foundation							
High capacity outfalls; silt box 150/225 outlet; two section trapped outfall silt box and cast iron access cover	34.22	1.00	18.50	-	211.93	nr	**230.43**
Inline side or end outlet outfall; 150 mm; 2 section concrete trapped outfall; cast iron Mini Beany access cover and frame	183.20	1.00	18.50	-	183.77	nr	**202.27**
Ancillaries to Mini Beany							
End cap	10.26	0.25	4.63	-	10.26	nr	**14.88**
End cap outlets	26.65	0.25	4.63	-	26.65	nr	**31.27**
Precast concrete channels; Charcon Hard Landscaping; on 150 mm deep concrete foundations; including haunching with in situ concrete; 21.00 N/mm² - 20 aggregate; both sides							
"Charcon Safeticurb"; slotted safety channels; for pedestrians and light vehicles							
ref DBJ; 305 x 305 mm	56.74	0.67	12.33	-	73.52	m	**85.85**
ref DBA; 250 x 250 mm	29.21	0.67	12.33	-	45.99	m	**58.32**
"Charcon Safeticurb"; slotted safety channels; for heavy vehicles							
ref DBM; 248 x 248 mm	52.01	0.80	14.80	-	68.79	m	**83.59**
ref Clearway; 324 x 257 mm	59.65	0.80	14.80	-	76.43	m	**91.23**
"Charcon Safeticurb"; inspection units; ductile iron lids; including jointing to drainage channels							
248 x 248 x 914 mm	70.78	1.50	27.75	-	71.77	nr	**99.52**
Silt box tops; concrete frame; cast iron grid lids; type 1; set over gully							
457 x 610 mm	317.25	2.00	37.00	-	318.24	nr	**355.24**
Manhole covers; type K; cast iron; providing inspection to blocks and back gullies	403.07	2.00	37.00	-	405.24	nr	**442.24**
Linear drainage							
Loadings for slot drains							
A15 - 1.5 tonne; pedestrian							
B125 - 12.5 tonne; domestic use							
C250 - 25 tonne; car parks, supermarkets, industrial units							
D400 - 40 tonne; highways							
E600 - 60 tonne; forklifts							
Linear drainage; channels; Aco Building Products; laid to concrete bed C25 on compacted granular base on 200 mm deep concrete bed; haunched with 200 mm concrete surround; all in 750 x 430 wide trench with compacted 200 granular base surround (excavation and sub-base not included)							
Slot drains							
ACO Brickslot; Load class C250; discreet slot drainage system ideal for visually sensitive paved areas	69.06	0.33	6.17	-	70.48	m	**76.65**
Extra over for laying of paving (PC £300.00/1000) adjacent to channel on epoxy mortar bed; stretcher bond	0.36	0.33	6.17	-	3.44	m	**0.00**

R DISPOSAL SYSTEMS

Item Excluding site overheads and profit	PC £	Labour hours	Labour £	Plant £	Material £	Unit	**Total rate £**
R12 DRAINAGE BELOW GROUND - cont'd							
Linear drainage; channels - cont'd							
Channel footpath drain							
Aco Channel drain to pedestrian footpaths; connection to existing kerbdrain; laying on 150 mm concrete base and surround; excavation and sub-base not included	23.56	2.00	37.00	-	32.31	m	**69.31**
NK100; to C250; 100 mm bore; polymer concrete drain units; tapered to falls; on 150 mm concrete with 150 concrete surround							
Constant or pre sloped invert	33.67	0.80	14.80	-	44.41	m	**59.21**
Accessories for NK100							
Sump unit - complete with sediment bucket	60.59	1.00	18.50	-	60.59	nr	**79.09**
End cap - closing piece	-	-	-	-	7.94	nr	-
End cap - inlet/outlet	-	-	-	-	16.92	nr	-
Step connectors	-	-	-	-	12.97	nr	-
Aco Building Products; grating covers for NK100							
Load class A15 (pedestrian areas)							
Slotted galvanised steel grating. Ref: 423QL/424QL	-	-	-	-	34.15	m	-
Slotted stainless steel grating. Ref: 470QL	-	-	-	-	99.70	m	-
1433 Load Class C250 (Cars and light vans)							
Slotted galvanised steel grating. Ref: 405/406.	-	-	-	-	34.15	m	-
Galvanised steel mesh grating. Ref: 413/414.	-	-	-	-	25.60	m	-
Heelguard™' composite 522QL	-	-	-	-	33.40	m	-
'Heelguard™' ductile iron grating. Ref: 508QL.	-	-	-	-	35.24	m	-
Wade Ltd; stainless steel channel drains; specialised applications; bespoke manufacture							
Drain in stainless steel; c/w tie in lugs and inbuilt falls to 100 mm spigot outlet; stainless steel gratings							
"NE" Channel; Ref 12430 secured gratings	252.00	1.20	22.20	-	262.74	m	**284.94**
Linear drainage; channels; Ensor Building Products; on 100 mm deep concrete bed; 100 mm concrete fill both sides							
Polymer concrete drain units; tapered to falls							
channel units; Grade 'A', ref Stora-drain 100; 100 mm wide including galvanised grate and locks	21.88	0.80	14.80	-	28.30	m	**43.10**
channel units; Grade 'B', ref Stora-drain 100; 100 mm wide including galvanised grate and locks	26.75	0.80	14.80	-	33.17	m	**47.97**
channel units; Grade 'C', ref Stora-drain 100; 100 mm wide including galvanised grate and locks	28.50	0.80	14.80	-	34.92	m	**49.72**
channel units; Grade 'C', ref Stora-drain 100; 100 mm wide including anti-heel grate and locks	29.69	0.80	14.80	-	36.11	m	**50.91**
Stora Drain sump unit, steel bucket, Class A15 galv. slotted grating	44.59	2.00	37.00	-	51.01	nr	**88.01**
Stora Drain sump unit, steel bucket, Class B125 galv. mesh grating	46.33	2.00	37.00	-	52.75	nr	**89.75**
Stora Drain sump unit, steel bucket, Class C250 galv. slotted grating	47.62	2.00	37.00	-	54.04	nr	**91.04**
Stora Drain sump unit, steel bucket, C250 'Heelsafe' slotted grating	44.14	2.00	37.00	-	50.56	nr	**87.56**

R DISPOSAL SYSTEMS

Item Excluding site overheads and profit	PC £	Labour hours	Labour £	Plant £	Material £	Unit	**Total rate £**
Access covers and frames							
Loading information; Note - Groups 5 - 6 are not covered within the scope of this book, please refer to the Spon's Civil Engineering Price Book							
Group 1 - Class A15; pedestrian access only - 1.5 tonne maximum weight							
Group 2 - Class B125; car parks and pedestrian areas with occasional vehicle use - 12.5 tonne maximum weight limit							
Group 3 - Class B250; gully tops in areas extending more than 500 mm from the kerb into the carriageway							
Group 4 - Class D400; carriageways of roads							
Access covers and frames; BSEN124; St Gobain Pipelines Plc; bedding frame in cement mortar (1:3); cover in grease and sand; light duty; clear opening sizes; base size shown in brackets							
Group 1; solid top single seal; ductile iron; plastic frame							
Pedestrian; 450 mm diameter (525) x 35 deep	15.31	3.00	55.50	-	17.57	nr	**73.07**
Driveway; 450 mm diameter (550) x 35 deep	23.87	3.00	55.50	-	26.13	nr	**81.63**
Group 1; solid top single seal; ductile iron; steel frame							
450 mm x 450 mm (565 x 565) x 35 deep	62.10	1.50	27.75	-	63.87	nr	**91.62**
450 mm x 600 mm (710 x 510) x 35 deep	56.64	1.80	33.30	-	59.08	nr	**92.38**
600 mm x 600 mm (710 x 710) x 35 deep	56.70	2.00	37.00	-	59.82	nr	**96.82**
Group 1; coated; double seal solid top; fabricated steel							
450 mm x 450 mm	44.10	1.50	27.75	-	45.87	nr	**73.62**
600 mm x 450 mm	51.30	1.80	33.30	-	53.75	nr	**87.05**
600 mm x 600 mm	56.70	2.00	37.00	-	59.82	nr	**96.82**
Group 2; single seal solid top							
450 diameter x 41 deep	42.18	1.50	27.75	-	43.95	nr	**71.70**
600 diameter x 75 deep	80.33	1.50	27.75	-	82.10	nr	**109.85**
600 x 450 (710 x 560) x 41 deep	56.64	1.50	27.75	-	58.41	nr	**86.16**
600 x 600 (711 x 711) x 44 deep	60.69	1.50	27.75	-	62.46	nr	**90.21**
600 x 600 (755 x 755) x 75 deep	75.12	1.50	27.75	-	76.89	nr	**104.64**
Recessed covers and frames; Bripave; galvanized; single seal; clear opening sizes shown in brackets							
450 x 450 x 93 (596 x 596) 1.5 tonne load	69.71	1.50	27.75	-	71.49	nr	**99.24**
450 x 450 x 93 (596 x 596) 5 tonne load	87.30	1.50	27.75	-	89.07	nr	**116.82**
450 x 450 x 93 (596 x 596) 11 tonne load	105.30	1.50	27.75	-	107.07	nr	**134.82**
600 x 450 x 93 (746 x 596) 1.5 tonne load	211.69	2.00	37.00	-	213.46	nr	**250.46**
600 x 450 x 93 (746 x 596) 5 tonne load	227.54	2.00	37.00	-	229.31	nr	**266.31**
600 x 450 x 93 (746 x 596) 11 tonne load	234.17	2.00	37.00	-	235.94	nr	**272.94**
600 x 600 x 93 (746 x 596) 1.5 tonne load	221.09	2.33	43.10	-	224.21	nr	**267.31**
600 x 600 x 93 (746 x 596) 5 tonne load	231.83	2.33	43.10	-	234.95	nr	**278.06**
600 x 600 x 93 (746 x 596) 11 tonne load	243.44	2.33	43.10	-	246.56	nr	**289.67**
750 x 600 x 93 (896 x 746) 1.5 tonne load	251.95	3.00	55.50	-	253.72	nr	**309.22**
750 x 600 x 93 (896 x 746) 5 tonne load	280.17	3.00	55.50	-	281.94	nr	**337.44**
750 x 600 x 93 (896 x 746) 11 tonne load	306.10	3.00	55.50	-	307.87	nr	**363.37**
Extra over to the above for double seal to covers							
450 x 450 x 93 (596 x 596)	-	-	-	-	50.40	nr	-
600 x 450 x 93 (746 x 596)	-	-	-	-	53.10	nr	-
600 x 600 x 93 (746 x 596)	-	-	-	-	56.70	nr	-
750 x 600 x 93 (896 x 746)	-	-	-	-	62.10	nr	-

R DISPOSAL SYSTEMS

Item Excluding site overheads and profit	PC £	Labour hours	Labour £	Plant £	Material £	Unit	Total rate £
R12 DRAINAGE BELOW GROUND - cont'd							
Access covers and frames; Jones of Oswestry; bedding frame in cement:mortar (1:3); cover in grease and sand; clear opening sizes							
Access covers and frames; "Suprabloc"; to paved areas; filling with blocks cut and fitted to match surrounding paving; BS standard size manholes available							
pedestrian weight; 300 mm x 300 mm	94.43	2.50	46.25	-	99.91	nr	**146.16**
pedestrian weight; 450 mm x 450 mm	115.52	2.50	46.25	-	115.52	nr	**161.77**
pedestrian weight; 450 mm x 600 mm	126.86	2.50	46.25	-	132.57	nr	**178.82**
light vehicular weight; 300 mm x 300 mm	95.46	2.40	44.40	-	100.94	nr	**145.34**
light vehicular weight; 450 mm x 450 mm	118.62	3.00	55.50	-	124.90	nr	**180.40**
light vehicular weight; 450 mm x 600 mm	130.58	4.00	74.00	-	136.06	nr	**210.06**
heavy vehicular weight; 300 mm x 300 mm	98.44	3.00	55.50	-	104.15	nr	**159.65**
heavy vehicular weight; 450 mm x 600 mm	121.46	3.00	55.50	-	127.74	nr	**183.24**
heavy vehicular weight; 600 mm x 600 mm	133.99	4.00	74.00	-	143.12	nr	**217.12**
Extra over "Suprabloc" manhole frames and covers for							
filling recessed manhole covers with brick paviors; PC £305.00/1000	12.20	1.00	18.50	-	13.95	m^2	**32.45**
filling recessed manhole covers with vehicular paving blocks; PC £6.74/m^2	6.74	0.75	13.88	-	7.27	m^2	**21.14**
filling recessed manhole covers with concrete paving flags; PC £10.20/m^2	10.21	0.35	6.47	-	10.52	m^2	**17.00**
R13 LAND DRAINAGE							
Ditching; clear silt and bottom ditch not exceeding 1.50 m deep; strim back vegetation; disposing to spoil heaps; by machine							
up to 1.50 m wide at top	-	6.00	111.00	27.00	-	100 m	**138.00**
1.50 - 2.50 m wide at top	-	8.00	148.00	36.00	-	100 m	**184.00**
2.50 - 4.00 m wide at top	-	9.00	166.50	40.50	-	100 m	**207.00**
Ditching; clear only vegetation from ditch not exceeding 1.50 m deep; disposing to spoil heaps; by strimmer							
up to 1.50 m wide at top	-	5.00	92.50	7.01	-	100 m	**99.51**
1.50 - 2.50 m wide at top	-	6.00	111.00	12.61	-	100 m	**123.61**
2.50 - 4.00 m wide at top	-	7.00	129.50	19.62	-	100 m	**149.12**
Ditching; clear silt from ditch not exceeding 1.50 m deep; trimming back vegetation; disposing to spoil heaps; by hand							
up to 1.50 m wide at top	-	15.00	277.50	7.01	-	100 m	**284.51**
1.50 - 2.50 m wide at top	-	27.00	499.50	12.61	-	100 m	**512.11**
2.50 - 4.00 m wide at top	-	42.00	777.00	19.62	-	100 m	**796.62**
Ditching; excavating and forming ditch and bank to given profile (normally 45 degrees); in loam or sandy loam; by machine							
Width 300 mm							
depth 600 mm	-	3.70	68.45	33.30	-	100 m	**101.75**
depth 900 mm	-	5.20	96.20	46.80	-	100 m	**143.00**
depth 1200 mm	-	7.20	133.20	64.80	-	100 m	**198.00**
depth 1500 mm	-	9.40	173.90	74.03	-	100 m	**247.93**
Width 600 mm							
depth 600 mm	-	-	-	89.51	-	100 m	**89.51**
depth 900 mm	-	-	-	132.82	-	100 m	**132.82**
depth 1200 mm	-	-	-	180.47	-	100 m	**180.47**
depth 1500 mm	-	-	-	259.88	-	100 m	**259.88**

R DISPOSAL SYSTEMS

Item Excluding site overheads and profit	PC £	Labour hours	Labour £	Plant £	Material £	Unit	**Total rate £**
Width 900 mm							
depth 600 mm	-	-	-	315.00	-	100 m	**315.00**
depth 900 mm	-	-	-	555.00	-	100 m	**555.00**
depth 1200 mm	-	-	-	745.80	-	100 m	**745.80**
depth 1500 mm	-	-	-	924.00	-	100 m	**924.00**
Width 1200 mm							
depth 600 mm	-	-	-	495.00	-	100 m	**495.00**
depth 900 mm	-	-	-	735.00	-	100 m	**735.00**
depth 1200 mm	-	-	-	973.50	-	100 m	**973.50**
depth 1500 mm	-	-	-	1247.40	-	100 m	**1247.40**
Width 1500 mm							
depth 600 mm	-	-	-	610.50	-	100 m	**610.50**
depth 900 mm	-	-	-	930.60	-	100 m	**930.60**
depth 1200 mm	-	-	-	1237.50	-	100 m	**1237.50**
depth 1500 mm	-	-	-	1551.00	-	100 m	**1551.00**
Extra for ditching in clay	-	-	-	-	-	20%	-
Ditching; excavating and forming ditch and bank to given profile (normal 45 degrees); in loam or sandy loam; by hand							
Width 300 mm							
depth 600 mm	-	36.00	666.00	-	-	100 m	**666.00**
depth 900 mm	-	42.00	777.00	-	-	100 m	**777.00**
depth 1200 mm	-	56.00	1036.00	-	-	100 m	**1036.00**
depth 1500 mm	-	70.00	1295.00	-	-	100 m	**1295.00**
Width 600 mm							
depth 600 mm	-	56.00	1036.00	-	-	100 m	**1036.00**
depth 900 mm	-	84.00	1554.00	-	-	100 m	**1554.00**
depth 1200 mm	-	112.00	2072.00	-	-	100 m	**2072.00**
depth 1500 mm	-	140.00	2590.00	-	-	100 m	**2590.00**
Width 900 mm							
depth 600 mm	-	84.00	1554.00	-	-	100 m	**1554.00**
depth 900 mm	-	126.00	2331.00	-	-	100 m	**2331.00**
depth 1200 mm	-	168.00	3108.00	-	-	100 m	**3108.00**
depth 1500 mm	-	210.00	3885.00	-	-	100 m	**3885.00**
Width 1200 mm							
depth 600 mm	-	112.00	2072.00	-	-	100 m	**2072.00**
depth 900 mm	-	168.00	3108.00	-	-	100 m	**3108.00**
depth 1200 mm	-	224.00	4144.00	-	-	100 m	**4144.00**
depth 1500 mm	-	280.00	5180.00	-	-	100 m	**5180.00**
Extra for ditching in clay	-	-	-	-	-	50%	-
Extra for ditching in compacted soil	-	-	-	-	-	90%	-
Piped ditching							
Jointed concrete pipes; to BS 5911 pt.3; FP McCann Ltd; including bedding, haunching and topping with 150 mm concrete; 11.50 N/mm^2 - 40 mm aggregate; to existing ditch							
300 mm diameter	12.99	0.67	12.33	6.60	29.83	m	**48.76**
450 mm diameter	19.25	0.67	12.33	6.60	44.51	m	**63.45**
600 mm diameter	31.28	0.67	12.33	6.60	67.15	m	**86.08**
900 mm diameter	76.33	1.00	18.50	9.90	97.65	m	**126.05**
Jointed concrete pipes; to BS 5911 pt.1 class S; FP McCann Ltd; including bedding, haunching and topping with 150 mm concrete; 11.50 N/mm^2 - 40 mm aggregate; to existing ditch							
1200 mm diameter	131.14	1.00	18.50	9.90	159.03	m	**187.43**
Extra over jointed concrete pipes for bends to 45 deg	194.85	0.67	12.33	8.25	203.35	nr	**223.94**
Extra over jointed concrete pipes for single junctions 300 mm dia	94.52	0.67	12.33	8.25	103.02	nr	**123.61**
Extra over jointed concrete pipes for single junctions 450 mm dia	115.98	0.67	12.33	7.50	124.48	nr	**144.32**

R DISPOSAL SYSTEMS

Item Excluding site overheads and profit	PC £	Labour hours	Labour £	Plant £	Material £	Unit	Total rate £
R13 LAND DRAINAGE - cont'd							
Piped ditching - cont'd							
Extra over jointed concrete pipes for single junctions 600 mm dia	135.23	0.67	12.33	7.50	143.73	nr	**163.57**
Extra over jointed concrete pipes for single junctions 900 mm dia	256.06	0.67	12.33	7.50	264.56	nr	**284.40**
Extra over jointed concrete pipes for single junctions 1200 mm dia	166.45	0.67	12.33	7.50	174.95	nr	**194.79**
Concrete road gullies; to BS 5911; trapped; cement:mortar (1:2) joints to concrete pipes; bedding and surrounding in concrete; 11.50 N/mm^2 - 40 mm aggregate; 450 mm diameter x 1.07 m deep; rodding eye; stoppers	35.00	6.00	111.00	-	54.05	nr	**165.05**
Mole drainage; White Horse Contractors Ltd							
Drain by mole plough; 50 mm diameter mole set at depth of 450 mm in parallel runs							
1.20 m centres	-	-	-	-	-	ha	**576.87**
1.50 m centres	-	-	-	-	-	ha	**521.64**
2.00 m centres	-	-	-	-	-	ha	**479.13**
2.50 m centres	-	-	-	-	-	ha	**426.69**
3.00 m centres	-	-	-	-	-	ha	**383.79**
Drain by mole plough; 75 mm diameter mole set at depth of 450 mm in parallel runs							
1.20 m centres	-	-	-	-	-	ha	**691.30**
1.50 m centres	-	-	-	-	-	ha	**622.11**
2.00 m centres	-	-	-	-	-	ha	**589.02**
2.50 m centres	-	-	-	-	-	ha	**510.12**
3.00 m centres	-	-	-	-	-	ha	**443.38**
Sand slitting; White Horse Contractors Ltd							
Drainage slits; at 1.00 m centres; using spinning disc trenching machine; backfilling to 100 mm of surface with pea gravel; blind with sharp sand; arisings to be loaded, hauled and tipped onsite							
250 mm depth	-	-	-	-	-	100 m^2	**54.97**
300 mm depth	-	-	-	-	-	100 m^2	**61.76**
400 mm depth	-	-	-	-	-	100 m^2	**74.25**
450 mm depth	-	-	-	-	-	100 m^2	**79.75**
250 mm depth	-	-	-	-	-	ha	**15127.75**
300 mm depth	-	-	-	-	-	ha	**15671.13**
400 mm depth	-	-	-	-	-	ha	**16214.50**
450 mm depth	-	-	-	-	-	ha	**17389.50**
Trenchless drainage system; White Horse Contractors Ltd; insert perforated pipes by means of laser graded deep plough machine; backfill with gravel (not included)							
Laterals 80 mm							
depth 700 mm	-	-	-	-	-	m	**2.63**
Main 100 mm							
depth 900 mm	-	-	-	-	-	m	**3.07**
Main 160 mm							
depth 1000 mm	-	-	-	-	-	m	**4.11**
Gravel backfill							
laterals 80 mm	-	-	-	-	-	m	**4.21**
main 100 mm	-	-	-	-	-	m	**6.12**
main 160 mm	-	-	-	-	-	m	**7.44**

R DISPOSAL SYSTEMS

Item Excluding site overheads and profit	PC £	Labour hours	Labour £	Plant £	Material £	Unit	**Total rate £**
Agricultural drainage; calculation table							
For calculation of drainage per hectare, the following table can be used; rates show the lengths of drains per unit and not the value							
Lateral drains							
10.00 m centres	-	-	-	-	-	m/ha	**1000.00**
15.00 m centres	-	-	-	-	-	m/ha	**650.00**
25.00 m centres	-	-	-	-	-	m/ha	**400.00**
30.00 m centres	-	-	-	-	-	m/ha	**330.00**
Main drains (per hectare)							
1 nr (at 100 m centres)	-	-	-	-	-	m/ha	**100.00**
2 nr (at 50 m centres)	-	-	-	-	-	m/ha	**200.00**
3 nr (at 33.333 m centres)	-	-	-	-	-	m/ha	**300.00**
4 nr (at 25 m centres)	-	-	-	-	-	m/ha	**400.00**
Agricultural drainage; excavating							
Removing 150 mm depth of topsoil; 300 mm wide; depositing beside trench; by machine	-	1.50	27.75	13.50	-	100 m	**41.25**
Removing 150 mm depth of topsoil; 300 mm wide; depositing beside trench; by hand	-	8.00	148.00	-	-	100 m	**148.00**
Disposing on site; to spoil heaps; by machine							
not exceeding 100 m distance	-	0.07	1.22	2.34	-	m^3	**3.56**
average 100 - 150 m distance	-	0.08	1.47	2.81	-	m^3	**4.27**
average 150 - 200 m distance	-	0.09	1.71	3.28	-	m^3	**5.00**
Removing excavated material from site to tip not exceeding 13 km; mechanically loaded							
excavated material and clean hardcore rubble	-	-	-	1.10	10.00	m^3	**11.10**
Agricultural drainage; White Horse Contractors Ltd; excavating trenches (with minimum project size of 500 m) by trenching machine; spreading arisings on site; excluding backfill							
Width 150 mm							
depth 450 mm	-	-	-	-	-	100 m	**195.00**
depth 600 mm	-	-	-	-	-	100 m	**227.70**
depth 750 mm	-	-	-	-	-	100 m	**258.50**
Width 225 mm							
depth 450 mm	-	-	-	-	-	100 m	**183.70**
depth 600 mm	-	-	-	-	-	100 m	**266.20**
depth 750 mm	-	-	-	-	-	100 m	**302.50**
Width 300 mm							
depth 600 mm	-	-	-	-	-	100 m	**303.60**
depth 750 mm	-	-	-	-	-	100 m	**343.20**
depth 900 mm	-	-	-	-	-	100 m	**382.80**
depth 1000 mm	-	-	-	-	-	100 m	**414.70**
Width 375 mm							
depth 600 mm	-	-	-	-	-	100 m	**367.36**
depth 750 mm	-	-	-	-	-	100 m	**413.28**
depth 900 mm	-	-	-	-	-	100 m	**439.20**
depth 1000 mm	-	-	-	-	-	100 m	**515.20**
Sports or amenity drainage; White Horse Contractors Ltd; excavating trenches (with minimum project size of 500 m) by trenching machine; arisings to spoil heap on site maximum 100 m							
Width 150 mm							
depth 450 mm	-	-	-	-	-	100 m	**355.64**
depth 600 mm	-	-	-	-	-	100 m	**413.59**
depth 750 mm	-	-	-	-	-	100 m	**469.53**
Width 225 mm							
depth 450 mm	-	-	-	-	-	100 m	**388.50**
depth 600 mm	-	-	-	-	-	100 m	**[illegible]**
depth 750 mm	-	-	-	-	-	100 m	**549.45**

R DISPOSAL SYSTEMS

Item Excluding site overheads and profit	PC £	Labour hours	Labour £	Plant £	Material £	Unit	Total rate £
R13 LAND DRAINAGE - cont'd							
Sports or amenity drainage - cont'd							
Width 300 mm							
depth 600 mm	-	-	-	-	-	100 m	**551.45**
depth 750 mm	-	-	-	-	-	100 m	**623.38**
depth 900 mm	-	-	-	-	-	100 m	**695.30**
depth 1000 mm	-	-	-	-	-	100 m	**746.46**
Width 375 mm							
depth 600 mm	-	-	-	-	-	100 m	**655.34**
depth 750 mm	-	-	-	-	-	100 m	**737.26**
depth 900 mm	-	-	-	-	-	100 m	**885.60**
depth 1000 mm	-	-	-	-	-	100 m	**993.60**
Agricultural drainage; excavating for drains; by backacter excavator JCB C3X Sitemaster; including disposing spoil to spoil heaps not exceeding 100 m; boning to levels by laser							
Width 150 mm - 225 mm							
depth 450 mm	-	9.67	178.83	260.00	-	100 m	**438.84**
depth 600 mm	-	13.00	240.50	390.00	-	100 m	**630.50**
depth 700 mm	-	15.50	286.75	487.50	-	100 m	**774.25**
depth 900 mm	-	23.00	425.50	780.00	-	100 m	**1205.50**
Width 300 mm							
depth 450 mm	-	9.67	178.83	260.00	-	100 m	**438.84**
depth 700 mm	-	14.00	259.00	429.00	-	100 m	**688.00**
depth 900 mm	-	23.00	425.50	780.00	-	100 m	**1205.50**
depth 1000 mm	-	25.00	462.50	858.00	-	100 m	**1320.50**
depth 1200 mm	-	25.00	462.50	975.00	-	100 m	**1437.50**
depth 1500 mm	-	31.00	573.50	1092.00	-	100 m	**1665.50**
Agricultural drainage; excavating for drains; by 7 tonne tracked excavator; including disposing spoil to spoil heaps not exceeding 100 m							
Width 225 mm							
depth 450 mm	-	7.00	129.50	81.75	-	100 m	**211.25**
depth 600 mm	-	7.17	132.58	85.16	-	100 m	**217.74**
depth 700 mm	-	7.35	135.93	88.86	-	100 m	**224.79**
depth 900 mm	-	7.76	143.60	97.32	-	100 m	**240.92**
depth 1000 mm	-	8.00	148.00	102.19	-	100 m	**250.19**
depth 1200 mm	-	8.26	152.87	107.57	-	100 m	**260.44**
Width 300 mm							
depth 450 mm	-	7.35	135.93	88.86	-	100 m	**224.79**
depth 600 mm	-	7.76	143.60	97.32	-	100 m	**240.92**
depth 700 mm	-	8.26	152.87	107.57	-	100 m	**260.44**
depth 900 mm	-	9.25	171.13	127.73	-	100 m	**298.86**
depth 1000 mm	-	10.14	187.64	145.98	-	100 m	**333.63**
depth 1200 mm	-	11.00	203.50	163.50	-	100 m	**367.00**
Width 600 mm							
depth 450 mm	-	13.00	240.50	204.38	-	100 m	**444.88**
depth 600 mm	-	14.11	261.06	227.08	-	100 m	**488.14**
depth 700 mm	-	16.33	302.17	272.50	-	100 m	**574.67**
depth 900 mm	-	17.29	319.79	291.96	-	100 m	**611.75**
depth 1000 mm	-	18.38	340.12	314.42	-	100 m	**654.54**
depth 1200 mm	-	21.18	391.86	371.59	-	100 m	**763.45**
Agricultural drainage; excavating for drains; by hand, including disposing spoil to spoil heaps not exceeding 100 m							
Width 150 mm							
depth 450 mm	-	22.03	407.56	-	-	100 m	**407.56**
depth 600 mm	-	29.38	543.53	-	-	100 m	**543.53**
depth 700 mm	-	34.27	634.00	-	-	100 m	**634.00**
depth 900 mm	-	44.06	815.11	-	-	100 m	**815.11**

R DISPOSAL SYSTEMS

Item Excluding site overheads and profit	PC £	Labour hours	Labour £	Plant £	Material £	Unit	**Total rate £**
Width 225 mm							
depth 450 mm	-	33.05	611.42	-	-	100 m	**611.42**
depth 600 mm	-	44.06	815.11	-	-	100 m	**815.11**
depth 700 mm	-	51.41	951.09	-	-	100 m	**951.09**
depth 900 mm	-	66.10	1222.85	-	-	100 m	**1222.85**
depth 1000 mm	-	73.44	1358.64	-	-	100 m	**1358.64**
Width 300 mm							
depth 450 mm	-	44.06	815.11	-	-	100 m	**815.11**
depth 600 mm	-	58.75	1086.88	-	-	100 m	**1086.88**
depth 700 mm	-	68.54	1267.99	-	-	100 m	**1267.99**
depth 900 mm	-	88.13	1630.40	-	-	100 m	**1630.40**
depth 1000 mm	-	97.92	1811.52	-	-	100 m	**1811.52**
Width 375 mm							
depth 450 mm	-	55.08	1018.98	-	-	100 m	**1018.98**
depth 600 mm	-	73.44	1358.64	-	-	100 m	**1358.64**
depth 700 mm	-	85.68	1585.08	-	-	100 m	**1585.08**
depth 900 mm	-	110.16	2037.96	-	-	100 m	**2037.96**
depth 1000 mm	-	122.40	2264.40	-	-	100 m	**2264.40**
Width 450 mm							
depth 450 mm	-	66.10	1222.85	-	-	100 m	**1222.85**
depth 600 mm	-	88.13	1630.40	-	-	100 m	**1630.40**
depth 700 mm	-	102.82	1902.17	-	-	100 m	**1902.17**
depth 900 mm	-	132.19	2445.51	-	-	100 m	**2445.51**
depth 1000 mm	-	146.88	2717.28	-	-	100 m	**2717.28**
Width 600 mm							
depth 450 mm	-	88.13	1630.40	-	-	100 m	**1630.40**
depth 600 mm	-	117.50	2173.75	-	-	100 m	**2173.75**
depth 700 mm	-	137.09	2536.16	-	-	100 m	**2536.16**
depth 900 mm	-	176.26	3260.81	-	-	100 m	**3260.81**
depth 1000 mm	-	195.84	3623.04	-	-	100 m	**3623.04**
Width 900 mm							
depth 450 mm	-	132.19	2445.51	-	-	100 m	**2445.51**
depth 600 mm	-	176.26	3260.81	-	-	100 m	**3260.81**
depth 700 mm	-	205.63	3804.16	-	-	100 m	**3804.16**
depth 900 mm	-	264.38	4891.03	-	-	100 m	**4891.03**
depth 1000 mm	-	293.76	5434.56	-	-	100 m	**5434.56**
Earthwork support; moving along as work proceeds							
Maximum depth not exceeding 2.00 m							
distance between opposing faces not exceeding 2.00 m	-	0.80	14.80	17.70	-	m	**32.50**
Agricultural drainage; pipe laying							
Hepworth; agricultural clay drain pipes; to BS 1196; 300 mm length; butt joints; in straight runs							
75 mm diameter	351.65	8.00	148.00	-	360.44	100 m	**508.44**
100 mm diameter	607.39	9.00	166.50	-	622.58	100 m	**789.08**
150 mm diameter	1241.42	10.00	185.00	-	1272.46	100 m	**1457.46**
Extra over clay drain pipes for filter-wrapping pipes with "Terram" or similar filter fabric							
"Terram 700"	0.21	0.04	0.74	-	0.21	m^2	**0.95**
"Terram 1000"	0.21	0.04	0.74	-	0.21	m^2	**0.95**
Junctions between drains in clay pipes							
75 mm x 75 mm	12.81	0.25	4.63	-	13.20	nr	**17.82**
100 mm x 100 mm	16.00	0.25	4.63	-	16.63	nr	**21.26**
100 mm x 150 mm	19.70	0.25	4.63	-	20.64	nr	**25.27**

R DISPOSAL SYSTEMS

Item Excluding site overheads and profit	PC £	Labour hours	Labour £	Plant £	Material £	Unit	**Total rate £**
R13 LAND DRAINAGE - cont'd							
Wavin Plastics Ltd; flexible plastic perforated pipes in trenches (not included); to a minimum depth of 450 mm (couplings not included)							
"OsmaDrain"; flexible plastic perforated pipes in trenches (not included); to a minimum depth of 450 mm (couplings not included)							
80 mm diameter; available in 100 m coil	56.50	2.00	37.00	-	57.91	100 m	**94.91**
100 mm diameter; available in 100 m coil	97.00	2.00	37.00	-	99.42	100 m	**136.43**
160 mm diameter; available in 35 m coil	235.00	2.00	37.00	-	240.88	100 m	**277.88**
"WavinCoil"; plastic pipe junctions							
80 mm x 80 mm	2.50	0.05	0.93	-	2.50	nr	**3.42**
100 mm x 100 mm	2.79	0.05	0.93	-	2.79	nr	**3.71**
100 mm x 60 mm	2.75	0.05	0.93	-	2.75	nr	**3.67**
100 mm x 80 mm	2.66	0.05	0.93	-	2.66	nr	**3.58**
160 mm x 160 mm	6.83	0.05	0.93	-	6.83	nr	**7.75**
"WavinCoil"; couplings for flexible pipes							
80 mm diameter	0.95	0.03	0.62	-	0.95	nr	**1.57**
100 mm diameter	1.05	0.03	0.62	-	1.05	nr	**1.67**
160 mm diameter	1.42	0.03	0.62	-	1.42	nr	**2.04**
Market prices of backfilling materials							
Sand	15.25	-	-	-	18.30	m^3	-
Gravel rejects	15.00	-	-	-	16.50	m^3	-
Topsoil; allowing for 20% settlement	18.50	-	-	-	22.20	m^3	-
Agricultural drainage; backfilling trench after laying pipes with gravel rejects or similar; blind filling with ash or sand; topping with 150 mm topsoil from dumps not exceeding 100 m; by machine							
Width 150 mm							
depth 450 mm	-	3.30	61.05	53.40	114.82	100 m	**229.27**
depth 600 mm	-	4.30	79.55	69.90	151.31	100 m	**300.76**
depth 750 mm	-	4.96	91.76	80.79	171.07	100 m	**343.62**
depth 900 mm	-	6.30	116.55	102.90	216.07	100 m	**435.52**
Width 225 mm							
depth 450 mm	-	4.95	91.58	80.10	176.50	100 m	**348.17**
depth 600 mm	-	6.45	119.33	104.85	227.05	100 m	**451.22**
depth 750 mm	-	7.95	147.08	129.60	277.75	100 m	**554.42**
depth 900 mm	-	9.45	174.83	154.35	328.30	100 m	**657.47**
Width 375 mm							
depth 450 mm	-	8.25	152.63	133.50	294.06	100 m	**580.18**
depth 600 mm	-	10.75	198.88	174.75	378.36	100 m	**751.98**
depth 750 mm	-	13.25	245.13	216.00	462.81	100 m	**923.93**
depth 900 mm	-	15.75	291.38	257.25	547.11	100 m	**1095.73**
Agricultural drainage; backfilling trench after laying pipes with gravel rejects or similar, blind filling with ash or sand, topping with 150 mm topsoil from dumps not exceeding 100 m; by hand							
Width 150 mm							
depth 450 mm	-	18.63	344.65	-	106.49	100 m	**451.15**
depth 600 mm	-	24.84	459.54	-	151.31	100 m	**610.85**
depth 750 mm	-	31.05	574.42	-	129.44	100 m	**703.87**
depth 900 mm	-	37.26	689.31	-	216.07	100 m	**905.38**
Width 225 mm							
depth 450 mm	-	27.94	516.89	-	176.50	100 m	**693.38**
depth 600 mm	-	37.26	689.31	-	227.05	100 m	**916.35**
depth 750 mm	-	46.57	861.54	-	277.75	100 m	**1139.29**
depth 900 mm	-	55.89	1033.96	-	328.30	100 m	**1362.26**

R DISPOSAL SYSTEMS

Item Excluding site overheads and profit	PC £	Labour hours	Labour £	Plant £	Material £	Unit	**Total rate £**
Width 375 mm							
depth 450 mm	-	46.57	861.54	-	294.06	100 m	**1155.60**
depth 600 mm	-	61.10	1130.35	-	378.36	100 m	**1508.71**
depth 750 mm	-	77.63	1436.15	-	462.81	100 m	**1898.96**
depth 900 mm	-	93.15	1723.28	-	547.11	100 m	**2270.38**
Catchwater or french drains; laying pipes to drain excavated separately; 100 mm diameter non-coilable perforated plastic pipes; to BS 4962; including straight jointing; pipes laid with perforations uppermost; lining trench; wrapping pipes with filter fabric; backfilling with shingle							
Width 300 mm							
depth 450 mm	1019.94	9.10	168.35	30.00	1465.58	100 m	**1663.93**
depth 600 mm	1023.56	9.84	182.04	46.86	1619.59	100 m	**1848.49**
depth 750 mm	1027.34	10.60	196.10	59.40	1773.79	100 m	**2029.29**
depth 900 mm	1031.04	11.34	209.79	71.61	1927.90	100 m	**2209.30**
depth 1000 mm	1033.51	11.84	219.04	79.86	2030.64	100 m	**2329.54**
depth 1200 mm	1038.44	12.84	237.54	96.36	2236.11	100 m	**2570.01**
Width 450 mm							
depth 450 mm	1025.49	10.44	193.14	86.46	1696.74	100 m	**1976.34**
depth 600 mm	1031.04	11.34	209.79	71.61	1927.90	100 m	**2209.30**
depth 750 mm	1036.59	12.46	230.51	90.09	2159.05	100 m	**2479.65**
depth 900 mm	1042.14	13.60	251.60	108.90	2390.21	100 m	**2750.71**
depth 1000 mm	1045.84	14.34	265.29	121.11	2544.32	100 m	**2930.72**
depth 1200 mm	1053.24	15.84	293.04	145.86	2852.53	100 m	**3291.43**
depth 1500 mm	1064.34	18.10	334.85	183.15	3314.84	100 m	**3832.84**
depth 2000 mm	1082.84	21.84	404.04	244.86	4085.36	100 m	**4734.26**
Width 600 mm							
depth 450 mm	1031.04	11.24	207.94	71.61	1927.90	100 m	**2207.45**
depth 600 mm	1038.44	12.84	237.54	96.36	2236.11	100 m	**2570.01**
depth 750 mm	1045.84	14.34	265.29	121.11	2544.32	100 m	**2930.72**
depth 900 mm	1053.24	15.84	293.04	145.86	2852.53	100 m	**3291.43**
depth 1000 mm	1060.29	16.84	311.54	162.36	3049.51	100 m	**3523.41**
depth 1200 mm	1068.04	18.84	348.54	195.36	3468.94	100 m	**4012.84**
depth 1500 mm	1082.84	21.84	404.04	514.86	4085.36	100 m	**5004.26**
depth 2000 mm	1107.51	19.84	367.04	442.86	5112.73	100 m	**5922.63**
Width 900 mm							
depth 450 mm	1042.14	13.60	251.60	108.90	2390.21	100 m	**2750.71**
depth 600 mm	1053.24	7.92	146.52	145.86	2852.53	100 m	**3144.91**
depth 750 mm	1064.34	9.05	167.43	183.15	3314.84	100 m	**3665.41**
depth 900 mm	1075.44	10.17	188.15	220.11	3777.15	100 m	**4185.41**
depth 1000 mm	1082.84	10.92	202.02	244.86	4085.36	100 m	**4532.24**
depth 1200 mm	1097.64	12.42	229.77	294.36	4701.78	100 m	**5225.91**
depth 1500 mm	1119.84	14.67	271.39	368.61	5626.41	100 m	**6266.42**
depth 2000 mm	1156.84	18.92	350.02	582.36	7167.46	100 m	**8099.84**
depth 2500 mm	1193.84	22.17	410.14	706.11	8708.51	100 m	**9824.77**
depth 3000 mm	1230.84	25.92	479.52	739.86	10249.56	100 m	**11468.94**
Catchwater or french drains; laying pipes to d rains excavated separately; 160 mm diameter non-coilable perforated plastic pipes; to BS 4962; including straight jointing; pipes laid with perforations uppermost; lining trench; wrapping pipes with filter fabric; backfilling with shingle							
Width 600 mm							
depth 450 mm	1861.07	11.20	207.20	69.30	2752.93	100 m	**3029.43**
depth 600 mm	1868.47	12.70	234.95	94.05	3061.14	100 m	**3390.14**
depth 750 mm	1875.87	14.20	262.70	118.80	3369.35	100 m	**3750.85**
depth 900 mm	1883.27	15.70	290.45	143.55	3677.56	100 m	**4111.56**
depth 1000 mm	1888.21	16.70	308.95	160.05	3883.03	100 m	**4352.03**
depth 1200 mm	1898.07	18.70	345.95	193.15	4293.98	100 m	**4833.08**
depth 1500 mm	1912.87	21.70	401.45	242.55	4910.40	100 m	**5554.40**
depth 2000 mm	1937.54	26.70	493.95	325.05	5937.77	100 m	**6756.77**
depth 2500 mm	1962.21	31.70	586.45	407.55	6965.13	100 m	**7959.13**
depth 3000 mm	1986.87	36.70	678.95	490.05	7992.50	100 m	**9161.50**

Item Excluding site overheads and profit	PC £	Labour hours	Labour £	Plant £	Material £	Unit	**Total rate £**
R13 LAND DRAINAGE - cont'd							
Catchwater or french drains - cont'd							
Width 900 mm							
depth 450 mm	1042.14	13.60	251.60	108.90	2390.21	100 m	**2750.71**
depth 600 mm	1053.24	15.84	293.04	145.86	2852.53	100 m	**3291.43**
depth 750 mm	1064.34	18.10	334.85	183.15	3314.84	100 m	**3832.84**
depth 900 mm	1075.44	20.34	376.29	220.11	3777.15	100 m	**4373.55**
depth 1000 mm	1082.84	21.84	404.04	244.86	4085.36	100 m	**4734.26**
depth 1200 mm	1097.64	24.84	459.54	294.36	4701.78	100 m	**5455.68**
depth 1500 mm	1119.84	29.34	542.79	368.61	5626.41	100 m	**6537.81**
depth 2000 mm	1156.84	36.84	681.54	492.36	7167.46	100 m	**8341.36**
depth 2500 mm	1193.84	44.34	820.29	616.11	8708.51	100 m	**10144.91**
depth 3000 mm	1230.84	51.84	959.04	739.86	10249.56	100 m	**11948.46**
Catchwater or french drains; Exxon Chemical Geopolymers Ltd; "Filtram" filter drain; in trenches (trenches not included); comprising filter fabric, liquid conducting core and 110 mm PVC-u slitpipes; all in accordance with manufacturer's instructions; backfilling with shingle							
Width 600 mm							
depth 1000 mm	1058.18	16.84	311.54	162.36	3058.00	100 m	**3531.90**
depth 1200 mm	1068.04	18.84	348.54	195.36	3468.94	100 m	**4012.84**
depth 1500 mm	1082.84	21.84	404.04	514.86	4085.36	100 m	**5004.26**
depth 2000 mm	1107.51	19.84	367.04	442.86	5112.73	100 m	**5922.63**
Width 900 mm							
depth 450 mm	1042.14	13.60	251.60	108.90	2390.21	100 m	**2750.71**
depth 600 mm	1053.24	15.84	293.04	145.86	2852.53	100 m	**3291.43**
depth 750 mm	1064.34	18.10	334.85	183.15	3314.84	100 m	**3832.84**
depth 900 mm	1075.44	20.34	376.29	220.11	3777.15	100 m	**4373.55**
depth 1000 mm	1082.84	21.84	404.04	244.86	4085.36	100 m	**4734.26**
depth 1200 mm	1097.64	24.84	459.54	294.36	4701.78	100 m	**5455.68**
depth 1500 mm	1119.84	29.34	542.79	368.61	5626.41	100 m	**6537.81**
depth 2000 mm	1156.84	37.84	700.04	582.36	7167.46	100 m	**8449.86**
depth 2500 mm	1193.84	44.34	820.29	706.11	8708.51	100 m	**10234.91**
depth 3000 mm	1230.84	51.84	959.04	829.86	10249.56	100 m	**12038.46**
Outfalls							
Reinforced concrete outfalls to water course; flank walls; for 150 mm drain outlets; overall dimensions							
900 mm x 1050 mm x 900 mm high	-	-	-	-	-	m	**487.60**
GRC outfall headwalls	-	-	-	-	-	nr	**105.00**
JKH unit, standard small	-	-	-	-	-	nr	**85.00**
Preamble: Soakaway design based on BS EN 752-4. Flat rate hourly rainfall = 50 mm/hr and assumes 100 impermeability of the run-off area. A storage capacity of the soakaway should be 1/3 of the hourly rainfall. Formulae for calculating soakaway depths are provided in the publications mentioned below and in the Memoranda section of this publication. The design of soakaways is dependent on, amongst other factors, soil conditions, permeability, groundwater level and runoff. The definitive documents for design of soakaways are CIRIA 156 and BRE Digest 365 dated September 1991. The suppliers of the systems below will assist through their technical divisions. Excavation earthwork support of pits and disposal not included.							

R DISPOSAL SYSTEMS

Item Excluding site overheads and profit	PC £	Labour hours	Labour £	Plant £	Material £	Unit	**Total rate £**
Soakaways							
Excavating mechanical; to reduce levels							
maximum depth not exceeding 1.00 m; JCB sitemaster	-	0.05	0.93	1.65	-	m^3	**2.58**
maximum depth not exceeding 1.00 m; 360 Tracked excavator	-	0.04	0.74	1.65	-	m^3	**2.39**
maximum depth not exceeding 2.00 m; 360 Tracked excavator	-	0.06	1.11	2.48	-	m^3	**3.58**
Disposal							
Excavated material; off site; to tip; mechanically loaded (JCB)							
inert	-	0.04	0.77	1.25	19.80	m^3	**21.82**
In situ concrete ring beam foundations to base of soakaway; 300 mm wide x 250 deep; poured on or against earth or unblinded hardcore							
Internal diameters of rings							
900 mm	17.05	4.00	74.00	-	18.75	nr	**92.75**
1200 mm	22.72	4.50	83.25	-	25.00	nr	**108.25**
1500 mm	28.41	5.00	92.50	-	31.24	nr	**123.74**
2400 mm	45.45	5.50	101.75	-	50.00	nr	**151.75**
Concrete soakaway rings; Milton Pipes Ltd; perforations and step irons to concrete rings at manufacturers recommended centres; placing of concrete ring soakaways to in situ concrete ring beams (1:3:6) (not included); filling and surrounding base with gravel 225 deep (not included)							
Ring diameter 900 mm							
1.00 deep; volume 636 litres	42.50	1.25	23.13	18.05	134.00	nr	**175.17**
1.50 deep; volume 954 litres	63.75	1.65	30.52	47.64	198.35	nr	**276.52**
2.00 deep; volume 1272 litres	85.00	1.65	30.52	47.64	262.70	nr	**340.87**
Ring diameter 1200 mm							
1.00 deep; volume 1131 litres	56.00	3.00	55.50	43.31	161.91	nr	**260.72**
1.50 deep; volume 1696 litres	84.00	4.50	83.25	64.97	238.16	nr	**386.38**
2.00 deep; volume 2261 litres	112.00	4.50	83.25	64.97	314.41	nr	**462.63**
2.50 deep; volume 2827 litres	140.00	6.00	111.00	86.63	354.61	nr	**552.23**
Ring diameter 1500 mm							
1.00 deep; volume 1767 litres	96.00	4.50	83.25	43.31	227.81	nr	**354.37**
1.50 deep; volume 2651 litres	144.00	6.00	111.00	60.64	334.36	nr	**506.00**
2.00 deep; volume 3534 litres	192.00	6.00	111.00	60.64	440.91	nr	**612.55**
2.50 deep; volume 4418 litres	140.00	7.50	138.75	108.28	447.46	nr	**694.49**
Ring diameter 2400 mm							
1.00 deep; volume 4524 litres	324.00	6.00	111.00	43.31	502.16	nr	**656.47**
1.50 deep; volume 6786 litres	486.00	7.50	138.75	60.64	743.31	nr	**942.70**
2.00 deep; volume 9048 litres	648.00	7.50	138.75	60.64	989.61	nr	**1189.00**
2.50 deep; volume 11310 litres	810.00	9.00	166.50	108.28	1230.76	nr	**1505.54**
Extra over for							
250 mm depth chamber ring	-	-	-	-	-	100 %	-
500 mm depth chamber ring	-	-	-	-	-	50 %	-
Cover slabs to soakaways							
Heavy duty precast concrete							
900 diameter	62.70	1.00	18.50	21.66	62.70	nr	**102.86**
1200 diameter	81.70	1.00	18.50	21.66	81.70	nr	**121.86**
1500 diameter	125.70	1.00	18.50	21.66	125.70	nr	**165.86**
2400 diameter	496.70	1.00	18.50	21.66	496.70	nr	**536.86**
Step irons to concrete chamber rings	24.40	-	-	-	24.40	m	-

R DISPOSAL SYSTEMS

Item Excluding site overheads and profit	PC £	Labour hours	Labour £	Plant £	Material £	Unit	Total rate £
R13 LAND DRAINAGE - cont'd							
Cover slabs to soakaways - cont'd Extra over soakaways for filter wrapping with a proprietary filter membrane							
900 mm diameter x 1.00 m deep	1.16	1.00	18.50	-	1.45	nr	**19.95**
900 mm diameter x 2.00 m deep	2.33	1.50	27.75	-	2.91	nr	**30.66**
1050 mm diameter x 1.00 m deep	1.36	1.50	27.75	-	1.70	nr	**29.45**
1050 mm diameter x 2.00 m deep	2.71	2.00	37.00	-	3.39	nr	**40.39**
1200 mm diameter x 1.00 m deep	1.55	2.00	37.00	-	1.94	nr	**38.94**
1200 mm diameter x 2.00 m deep	3.10	2.50	46.25	-	3.87	nr	**50.12**
1500 mm diameter x 1.00 m deep	1.94	2.50	46.25	-	2.42	nr	**48.67**
1500 mm diameter x 2.00 m deep	3.86	2.50	46.25	-	4.83	nr	**51.08**
1800 mm diameter x 1.00 m deep	2.32	3.00	55.50	-	2.90	nr	**58.40**
1800 mm diameter x 2.00 m deep	4.65	3.25	60.13	-	5.81	nr	**65.93**
Gravel surrounding to concrete ring soakaway 40 mm aggregate backfilled to vertical face of soakaway wrapped with geofabric (not included)							
250 thick	33.30	0.20	3.70	8.66	33.30	m^3	**45.66**
Backfilling to face of soakaway; carefully compacting as work proceeds Arising from the excavations							
average thickness exceeding 0.25 m; depositing in layers 150 mm maximum thickness	-	0.03	0.62	4.33	-	m^3	**4.95**
"Aquacell" soakaway; Wavin Plastics Ltd; preformed polypropylene soakaway infiltration crate units; to trenches; surrounded by geotextile and 40 mm aggregate laid 100 thick in trenches (excavation, disposal and backfilling not included) 1.00 x 500 x 400; internal volume 190 litres							
4 crates; 2.00 x 1.00 x 400; 760 litres	153.12	1.60	29.60	13.86	197.75	nr	**241.21**
8 crates; 2.00 x 1.00 x 800; 1520 litres	306.24	3.20	59.20	17.82	361.64	nr	**438.66**
12 crates; 6.00 x 500 x 800; 2280 litres	459.36	4.80	88.80	36.96	566.84	nr	**692.60**
16 crates; 4.00 x 1.00 x 800; 3040 litres	612.48	6.40	118.40	33.00	709.19	nr	**860.59**
20 crates; 5.00 x 1.00 x 800; 3800 litres	765.60	8.00	148.00	32.67	868.33	nr	**1049.00**
30 crates; 15.00 x 1.00 x 400; 5700 litres	1148.40	12.00	222.00	89.10	1404.66	nr	**1715.76**
60 crates; 15.00 x 1.00 x 800; 11400 litres	2296.80	20.00	370.00	116.49	2641.43	nr	**3127.92**
Geofabric surround to Aquacell units; Terram Ltd "Terram" synthetic fibre filter fabric; to face of concrete rings (not included); anchoring whilst backfilling (not included)							
"Terram 1000", 0.70 mm thick; mean water flow 50 litre/m^2/s	0.41	0.05	0.93	-	0.49	m^2	**1.42**
Fin drains; Cooper Clarke Ltd; Geofin shallow drain; drainage of sports fields and grassed landscaped areas Geofin shallow drain; to trenches; excavation by trenching machine; backfilled with single size aggregate 20 mm and covered with sharp sand rootzone 200 thick							
Geofin 25 mm thick x 150 deep	2.85	0.05	0.93	0.70	11.03	m	**12.65**
Geofin 25 mm thick x 450 deep	3.09	0.07	1.23	0.93	15.00	m	**17.16**
Geofin 25 mm thick x 900 deep	3.80	0.10	1.85	0.93	22.40	m	**25.18**
Fin drains; Cooper Clarke Ltd; Geofin composite fin drain laid to slabs Geofin fin drain laid horizontally to slab or blinded ground; covered with 20 mm shingle 200 mm thick							
Geofin 25 mm thick x 900 wide laid flat to falls	4.22	0.02	0.37	0.14	10.88	m^2	**11.39**

S PIPED SUPPLY SYSTEMS

Item Excluding site overheads and profit	PC £	Labour hours	Labour £	Plant £	Material £	Unit	**Total rate £**
S10 COLD WATER							
Borehole drilling; White Horse Contractors Ltd							
Drilling operations to average 100 m depth; lining with 150 mm diameter sieve							
drilling to 100 m	-	-	-	-	-	nr	**12000.00**
rate per m over 100 m	-	-	-	-	-	m	**120.00**
pump, rising main, cable and kiosk	-	-	-	-	-	nr	**5000.00**
Blue MDPE polythene pipes; type 50; for cold water services; with compression fittings; bedding on 100 mm DOT type 1 granular fill material							
Pipes							
20 mm diameter	0.31	0.08	1.48	-	3.98	m	**5.46**
25 mm diameter	0.38	0.08	1.48	-	4.05	m	**5.53**
32 mm diameter	0.93	0.08	1.48	-	4.62	m	**6.10**
50 mm diameter	1.62	0.10	1.85	-	5.32	m	**7.17**
60 mm diameter	3.50	0.10	1.85	-	7.25	m	**9.10**
Hose union bib taps; to BS 5412; including fixing to wall; making good surfaces							
15 mm	11.09	0.75	13.88	-	12.95	nr	**26.82**
22 mm	15.65	0.75	13.88	-	18.44	nr	**32.31**
Stopcocks; to BS 1010; including fixing to wall; making good surfaces							
15 mm	6.58	0.75	13.88	-	8.44	nr	**22.31**
22 mm	10.25	0.75	13.88	-	12.11	nr	**25.98**
Standpipes; to existing 25 mm water mains							
1.00 m high	-	-	-	-	-	nr	**225.00**
Hose junction bib taps; to standpipes							
19 mm	-	-	-	-	-	nr	**80.00**
S14 IRRIGATION							
Revaho Ltd; water recycling							
Fully automatic self cleaning water recycle system for use with domestic grey water including sub surface tank self flushing filter pump and control system; excludes plumbing of the drainage system from the house into the tank	-	-	-	-	-	nr	**2500.00**
Revaho UK Ltd; main or ring main supply							
Excavate and lay mains supply pipe to supply irrigated area							
PE 80 20 mm	-	-	-	-	-	m	**4.38**
PE 80 25 mm	-	-	-	-	-	m	**4.55**
Sports Pro 63 mm PE	-	-	-	-	-	m	**7.81**
Sports Pro 50 mm PE	-	-	-	-	-	m	**6.38**
Sports Pro 40 mm PE	-	-	-	-	-	m	**5.50**
Sports Pro 32 mm PE	-	-	-	-	-	m	**4.94**
Revaho UK Ltd; head control to sprinkler stations							
Valves installed at supply point on irrigation supply manifold; includes for pressure control and filtration							
2 valve unit	-	-	-	-	-	nr	**118.83**
4 valve unit	-	-	-	-	-	nr	**185.50**
6 valve unit	-	-	-	-	-	nr	**262.34**
12 valve unit	-	-	-	-	-	nr	**503.03**
Solenoid valve; 25 mm with chamber; extra over for each active station	-	-	-	-	-	nr	**95.54**

S PIPED SUPPLY SYSTEMS

Item Excluding site overheads and profit	PC £	Labour hours	Labour £	Plant £	Material £	Unit	Total rate £
S14 IRRIGATION - cont'd							
Revaho UK Ltd; head control to sprinkler stations - cont'd							
Cable to solenoid valves (alternative to valves on manifold as above)							
4 core	-	-	-	-	-	m	**1.41**
6 core	-	-	-	-	-	m	**6.87**
12 core	-	-	-	-	-	m	**2.87**
Revaho UK Ltd; supply irrigation infrastructure for irrigation system							
Header tank and submersible pump and pressure stat							
1000 litre (500 gallon 25 mm/10 days to 1500 m^2) 2 m^3/hr pump	-	-	-	-	-	nr	**440.70**
4540 litre (1000 gallon 25 mm/10 days to 3500 m^2)	-	-	-	-	-	nr	**892.70**
9080 litre (2000 gallon 25 mm/10 days to 7000 m^2)	-	-	-	-	-	nr	**1478.04**
Electric multistation controllers 240 V							
"Nelson" EZ Pro Junior; 4 station standard	-	-	-	-	-	nr	**134.29**
"Nelson" EZ Pro Junior; 6 station standard	-	-	-	-	-	nr	**150.86**
"Nelson" EZ Pro Junior; 12 station standard	-	-	-	-	-	nr	**255.76**
"Nelson" EZ Pro Junior; 12 station; radio controlled	-	-	-	-	-	nr	**525.42**
"Nelson" EZ Pro Junior; 12 station; with moisture sensor	-	-	-	-	-	nr	**883.29**
"Nelson" EZ Pro Junior; 12 station; radio controlled and moisture sensor	-	-	-	-	-	nr	**1060.75**
Revaho UK Ltd; station consisting of multiple sprinklers; inclusive of all trenching wiring and connections to ring main or main supply							
Sprayheads "Nelson" including nozzle 21 m^2 coverage							
100 mm (4")	-	-	-	-	-	nr	**13.97**
150 mm (6")	-	-	-	-	-	nr	**19.73**
300 mm (12")	-	-	-	-	-	nr	**22.75**
Matched Precipitation Sprinklers "Nelson"							
MP1000; 16 m^2 coverage	-	-	-	-	-	nr	**32.01**
MP2000; 30 m^2 coverage	-	-	-	-	-	nr	**32.01**
MP3000; 81 m^2 coverage	-	-	-	-	-	nr	**32.01**
Gear drive sprinklers; "Nelson" placed to provide head to head (100%) overlap; average							
Nelson 6000; 121 m^2 coverage	-	-	-	-	-	nr	**42.05**
Nelson 7000; 255 m^2 coverage	-	-	-	-	-	nr	**74.12**
Nelson 7500; 361 m^2 coverage	-	-	-	-	-	nr	**82.31**
Revaho UK Ltd; driplines; inclusive of connections and draindown systems							
"Netafim" Techline 30 cm emitter							
fixed on soil surface	-	-	-	-	-	m	**0.96**
sub-surface	-	-	-	-	-	m	**1.41**
"Netafim" Techline 50 cm emitter							
fixed on soil surface	-	-	-	-	-	m	**0.81**
sub-surface	-	-	-	-	-	m	**1.27**
Commissioning and testing of irrigation system							
per station	-	-	-	-	-	nr	**35.60**
Annual maintenance costs of irrigation system							
Call out charge per visit	-	-	-	-	-	nr	**296.63**
Extra over per station	-	-	-	-	-	nr	**11.87**

S PIPED SUPPLY SYSTEMS

Item Excluding site overheads and profit	PC £	Labour hours	Labour £	Plant £	Material £	Unit	**Total rate £**
Revaho UK Ltd; EZ Pro Max moisture sensing control system							
Large garden comprising 7000 square meters and 24 stations with 4 moisture sensors							
Turf only	-	-	-	-	-	nr	**12000.00**
Turf/shrub beds 70/30	-	-	-	-	-	nr	**7000.00**
Medium garden comprising 3500 square meters and 12 stations with 2 moisture sensors							
Turf only	-	-	-	-	-	nr	**8000.00**
Turf/shrub beds 70/30	-	-	-	-	-	nr	**10400.00**
Smaller gardens garden comprising 1000 square meters and 6 stations with 1 moisture sensors							
Turf only	-	-	-	-	-	nr	**4900.00**
Turf/shrub beds 70/30	-	-	-	-	-	nr	**5400.00**
Turf/shrub beds 50/50	-	-	-	-	-	nr	**5800.00**
Leaky Pipe Systems Ltd; " Leaky Pipe"; moisture leaking pipe irrigation system							
Main supply pipe inclusive of machine excavation; exclusive of connectors							
20 mm LDPE polytubing	0.84	0.05	0.93	0.58	0.86	m	**2.37**
16 mm LDPE polytubing	0.56	0.05	0.93	0.58	0.57	m	**2.08**
Water filters and cartridges							
No 10; 20 mm	-	-	-	-	41.90	nr	-
Big Blue and RR30 cartridge; 25 mm	-	-	-	-	107.20	nr	-
Water filters and pressure regulator sets; complete assemblies							
No 10; flow rate 3.1 - 82 litres per minute	-	-	-	-	71.50	nr	-
Leaky pipe hose; placed 150 mm sub surface for turf irrigation							
Distance between laterals 350 mm; excavation and backfilling priced separately							
LP12L low leak	3.65	0.04	0.74	-	3.65	m^2	**4.39**
LP12H high leak	3.16	0.04	0.74	-	3.16	m^2	**3.90**
LP12UH ultra high leak	3.85	0.04	0.74	-	3.85	m^2	**4.59**
Leaky pipe hose; laid to surface for landscape irrigation; distance between laterals 600 mm							
LP12L low leak	2.12	0.03	0.46	-	2.18	m^2	**2.64**
LP12H high leak	1.84	0.03	0.46	-	1.89	m^2	**2.35**
LP12UH ultra high leak	2.24	0.03	0.46	-	2.30	m^2	**2.76**
Leaky pipe hose; laid to surface for landscape irrigation; distance between laterals 900 mm							
LP12L low leak	1.42	0.02	0.31	-	1.46	m^2	**1.77**
LP12H high leak	1.23	0.02	0.31	-	1.26	m^2	**1.57**
LP12UH ultra high leak	1.50	0.02	0.31	-	1.54	m^2	**1.84**
Leaky pipe hose; laid to surface for tree irrigation laid around circumference of tree pit							
LP12L low leak	1.96	0.13	2.31	-	2.15	nr	**4.47**
LP12H high leak	1.70	0.13	2.31	-	1.87	nr	**4.18**
LP12UH ultra high leak	2.07	0.13	2.31	-	2.27	nr	**4.58**
Accessories							
Automatic multi-station controller stations inclusive of connections	289.00	2.00	80.00	-	289.00	nr	**369.00**
Solenoid valves inclusive of wiring and connections to a multi-station controller; nominal distance from controller 25 m	48.00	0.50	9.25	-	398.00	nr	**407.25**

S PIPED SUPPLY SYSTEMS

Item Excluding site overheads and profit	PC £	Labour hours	Labour £	Plant £	Material £	Unit	Total rate £
S15 FOUNTAINS/WATER FEATURES							
Lakes and ponds - General							
Preamble: The pressure of water against a retaining wall or dam is considerable, and where water retaining structures form part of the design of water features, the landscape architect is advised to consult a civil engineer. Artificially contained areas of water in raised reservoirs over 25,000 m^3 have to be registered with the local authority, and their dams will have to be covered by a civil engineer's certificate of safety.							
Typical linings - General							
Preamble: In addition to the traditional methods of forming the linings of lakes and ponds in puddled clay or concrete, there are a number of lining materials available. They are mainly used for reservoirs but can also help to form comparatively economic water features especially in soil which is not naturally water retentive. Information on the construction of traditional clay puddle ponds can be obtained from the British Trust for Conservation Volunteers, 36 St. Mary's Street, Wallingford, Oxfordshire, OX10 0EU. Tel: (01491) 39766. The cost of puddled clay ponds depends on the availability of suitable clay, the type of hand or machine labour that can be used, and the use to which the pond is to be put.							
Lake liners; Fairwater Ltd; to evenly graded surface of excavations (excavating not included); all stones over 75 mm; removing debris; including all welding and jointing of liner sheets							
Geotextile underlay; inclusive of spot welding to prevent dragging							
to water features	-	-	-	-	-	m^2	**2.06**
to lakes or large features	-	-	-	-	-	1000 m^2	**2055.00**
Butyl rubber liners; "Varnamo" inclusive of site vulcanising laid to geotextile above							
0.75 mm thick	-	-	-	-	-	m^2	**7.21**
0.75 mm thick	-	-	-	-	-	1000 m^2	**7138.60**
1.00 mm thick	-	-	-	-	-	m^2	**8.21**
1.00 mm thick	-	-	-	-	-	1000 m^2	**8209.30**
Lake liners; Landline Ltd; "Landflex" or "Alkorplan" geomembranes; to prepared surfaces (surfaces not included); all joints fully welded; installation by Landline employees							
"Landflex HC" polyethylene geomembranes							
0.50 mm thick	-	-	-	-	-	1000 m^2	**3230.00**
0.75 mm thick	-	-	-	-	-	1000 m^2	**3700.00**
1.00 mm thick	-	-	-	-	-	1000 m^2	**4160.00**
1.50 mm thick	-	-	-	-	-	1000 m^2	**4730.00**
2.00 mm thick	-	-	-	-	-	1000 m^2	**5260.00**
"Alkorplan PVC" geomembranes							
0.80 mm thick	-	-	-	-	-	1000 m	**5190.00**
1.20 mm thick	-	-	-	-	-	1000 m	**7800.00**
Lake liners; Monarflex							
Polyethylene lake and reservoir lining system; welding on site by Monarflex technicians (surface preparation and backfilling not included)							
"Blackline"; 500 micron	-	-	-	-	-	100 m^2	**371.00**
"Blackline"; 750 micron	-	-	-	-	-	100 m^2	**431.00**
"Blackline"; 1000 micron	-	-	-	-	-	100 m^2	**498.00**

S PIPED SUPPLY SYSTEMS

Item Excluding site overheads and profit	PC £	Labour hours	Labour £	Plant £	Material £	Unit	**Total rate £**
Operations over surfaces of lake liners							
Dug ballast; evenly spread over excavation already brought to grade							
150 mm thick	499.50	2.00	37.00	28.88	549.45	100 m^2	**615.33**
200 mm thick	666.00	3.00	55.50	43.31	732.60	100 m^2	**831.41**
300 mm thick	999.00	3.50	64.75	50.53	1098.90	100 m^2	**1214.18**
Imported topsoil; evenly spread over excavation							
100 mm thick	185.00	1.50	27.75	21.66	222.00	100 m^2	**271.41**
150 mm thick	277.50	2.00	37.00	28.88	333.00	100 m^2	**398.88**
200 mm thick	370.00	3.00	55.50	43.31	444.00	100 m^2	**542.81**
Blinding existing subsoil with 50 mm sand	152.60	1.00	18.50	28.88	167.86	100 m^2	**215.24**
Topsoil from excavation; evenly spread over excavation							
100 mm thick	-	-	-	21.66	-	100 m^2	**21.66**
200 mm thick	-	-	-	28.88	-	100 m^2	**28.88**
300 mm thick	-	-	-	43.31	-	100 m^2	**43.31**
Extra over for screening topsoil using a "Powergrid" screener; removing debris	-	-	-	4.75	0.50	m^3	**5.25**
Lake construction; Fairwater Ltd; lake construction at ground level; excavation; forming of lake; commissioning; excavated material spread on site							
Lined with existing site clay; natural edging with vegetation meeting the water							
500 m^2	-	-	-	-	-	nr	**8400.00**
1000 m^2	-	-	-	-	-	nr	**14600.00**
1500 m^2	-	-	-	-	-	nr	**20000.00**
2000 m^2	-	-	-	-	-	nr	**25000.00**
Lined with 0.75 mm butyl on geotextile underlay; natural edging with vegetation meeting the water							
500 m^2	-	-	-	-	-	nr	**13500.00**
1000 m^2	-	-	-	-	-	nr	**23400.00**
1500 m^2	-	-	-	-	-	nr	**32000.00**
2000 m^2	-	-	-	-	-	nr	**40000.00**
Extra over to the above for hard block edging to secure and protect liner; measured at lake perimeter	-	-	-	-	-	m	**42.00**
Lined with imported puddling clay; natural edging with vegetation meeting the water							
500 m^2	-	-	-	-	-	nr	**22000.00**
1000 m^2	-	-	-	-	-	nr	**38000.00**
1500 m^2	-	-	-	-	-	nr	**52000.00**
2000 m^2	-	-	-	-	-	nr	**65000.00**
Waterfall construction; Fairwater Ltd							
Stone placed on top of butyl liner; securing with concrete and dressing to form natural rock pools and edgings							
Portland stone m^3 rate	-	-	-	-	-	m^3	**630.89**
Portland stone tonne rate	-	-	-	-	-	tonne	**349.35**
Balancing tank; blockwork construction; inclusive of recirculation pump and pond level control; 110 mm balancing pipe to pond; waterproofed with butyl rubber membrane; pipework mains water top-up and overflow							
450 x 600 x 1000 mm	-	-	-	-	-	nr	**1312.14**
extra for pump							
2000 gallons per hour; submersible	-	-	-	-	-	nr	**200.36**

S PIPED SUPPLY SYSTEMS

Item Excluding site overheads and profit	PC £	Labour hours	Labour £	Plant £	Material £	Unit	**Total rate £**
S15 FOUNTAINS/WATER FEATURES - cont'd							
Ornamental pools - General Preamble: Small pools may be lined with one of the materials mentioned under lakes and ponds, or may be in rendered brickwork, puddled clay or, for the smaller sizes, fibreglass. Most of these tend to be cheaper than waterproof concrete. Basic prices for various sizes of concrete pools are given in the Approximate Estimates section (book only). Prices for excavation, grading, mass concrete, and precast concrete retaining walls are given in the relevant sections. The manufacturers should be consulted before specifying the type and thickness of pool liner, as this depends on the size, shape and proposed use of the pool. The manufacturer's recommendation on foundations and construction should be followed.							
Ornamental pools Pool liners; to 50 mm sand blinding to excavation (excavating not included); all stones over 50 mm; removing debris from surfaces of excavation; including all welding and jointing of liner sheets							
black polythene; 1000 gauge	-	-	-	-	-	m^2	**2.45**
blue polythene; 1000 gauge	-	-	-	-	-	m^2	**2.73**
coloured PVC; 1500 gauge	-	-	-	-	-	m^2	**3.31**
black PVC; 1500 gauge	-	-	-	-	-	m^2	**2.82**
black butyl; 0.75 mm thick	-	-	-	-	-	m^2	**6.85**
black butyl; 1.00 mm thick	-	-	-	-	-	m^2	**7.80**
black butyl; 1.50 mm thick	-	-	-	-	-	m^2	**26.70**
Fine gravel; 100 mm; evenly spread over area of pool; by hand	3.33	0.13	2.31	-	3.33	m^2	**5.64**
Selected topsoil from excavation; 100 mm; evenly spread over area of pool; by hand	-	0.13	2.31	-	-	m^2	**2.31**
Extra over selected topsoil for spreading imported topsoil over area of pool; by hand	18.50	-	-	-	22.20	m^3	**-**
Pool surrounds and ornament; Haddonstone Ltd; Portland Bath or terracotta cast stone Pool surrounds; installed to pools or water feature construction priced separately; surrounds and copings to 112.5 mm internal brickwork							
C4HSKVP half small pool surround; internal diameter 1780 mm; kerb features continuous moulding enriched with ovolvo and palmette designs; inclusive of plinth and integral conch shell vases flanked by dolphins	822.13	16.00	296.00	-	834.35	nr	**1130.35**
C4SKVP small pool surround as above but with full circular construction; internal diameter 1780 mm	1644.26	48.00	888.00	-	1658.82	nr	**2546.82**
C4MKVP medium pool surround; internal diameter 2705 mm; inclusive of plinth and integral vases	2466.38	48.00	888.00	-	2496.72	nr	**3384.72**
C4XLKVP extra large pool surround; internal diameter 5450 mm	4932.77	140.00	2590.00	-	4968.24	nr	**7558.24**
Pool centre pieces and fountains; inclusive of plumbing and pumps							
HC350 Lotus bowl; 1830 wide with C1700 triple dolphin fountain; HD2900 Doric pedestal	2542.13	8.00	148.00	-	2544.42	nr	**2692.42**
C251 Gothic Fountain and Gothic Upper Base A350; freestanding fountain	690.21	4.00	74.00	-	692.50	nr	**766.50**
HC521 Romanesque fountain; freestanding bowl with self-circulating fountain; filled with cobbles; 815 mm diameter; 348 mm high	317.45	2.00	37.00	-	348.18	nr	**385.18**
C300 Lion fountain 610 mm high on fountain base C305; 280 mm high	188.09	2.00	37.00	-	190.38	nr	**227.38**

S PIPED SUPPLY SYSTEMS

Item Excluding site overheads and profit	PC £	Labour hours	Labour £	Plant £	Material £	Unit	Total rate £
Wall fountains, watertanks and fountains; inclusive of installation drainage, automatic top up, pump, and balancing tank							
Capital Garden Products Ltd							
Lion Wall Fountain F010; 970 x 940 x 520	-	4.00	74.00	-	1897.07	nr	**1971.07**
Dolphin Wall Fountain F001; 740 x 510	-	4.00	74.00	-	1759.07	nr	**1833.07**
Dutch Master Wall Fountain F012; 740 x 410	-	4.00	74.00	-	1745.07	nr	**1819.07**
James II Watertank 2801; 730 x 730 x 760 h - 405 litres	-	4.00	74.00	-	1895.07	nr	**1969.07**
James II Fountain 4901 bp; 710 x 1300 x 1440 h - 655 litres	-	4.00	74.00	-	1759.07	nr	**1833.07**
Marble wall fountains; Architectural Heritage Ltd; inclusive of installation drainage, automatic top up, pump, and balancing tank							
Breccia Pernice Marble Wall Fountain supported by two Dolphins, 1770 high, 1000 wide, 640 deep	16000.00	4.00	74.00	-	17671.07	nr	**17745.07**
'The River God', Verona Marble wall mask fountain, 890 high, 810 wide, 230 deep	8000.00	4.00	74.00	-	9671.07	nr	**9745.07**
Fountain kits; typical prices of submersible units comprising fountain pumps, fountain nozzles, underwater spotlights, nozzle extension armatures, underwater terminal boxes and electrical control panels							
Single aerated white foamy water columns; ascending jet 70 mm diameter; descending water up to four times larger; jet height adjustable between 1.00 m and 1.70 m	3900.00	-	-	-	4470.00	nr	**4470.00**
Single aerated white foamy water columns, ascending jet 110 mm diameter; descending water up to four times larger; jet height adjustable between 1.50 m and 3.00 m	6500.00	-	-	-	7380.02	nr	**7380.02**
Natural swimming pools; Fairwater Ltd							
Natural swimming pool of 50 m^2 area; shingle regeneration zone 50 m^2 planted with marginal and aquatic plants	-	-	-	-	-	nr	**55000.00**
Swimming pools; Thompson Landscapes							
Conventional swimming pool construction inclusive of excavation reinforced gunnite construction, heating, lighting, pumping, filtration and mosaic lined							
rectangular construction 14 x 6 m	-	-	-	-	-	nr	**51000.00**
Duraglide cover inclusive of all electrical works	-	-	-	-	-	nr	**22000.00**
automatic cleaning system with main drains	-	-	-	-	-	nr	**13000.00**

V ELECTRICAL SUPPLY/POWER/LIGHTING SYSTEMS

Item Excluding site overheads and profit	PC £	Labour hours	Labour £	Plant £	Material £	Unit	Total rate £
V41 STREET/AREA/FLOOD LIGHTING							
Street area floodlighting - Generally Preamble: There are an enormous number of luminaires available which are designed for small scale urban and garden projects. The designs are continually changing and the landscape designer is advised to consult the manufacturer's latest catalogue. Most manufacturers supply light fittings suitable for column, bracket, bulkhead, wall or soffit mounting. Highway lamps and columns for trafficked roads are not included in this section as the design of highway lighting is a very specialised subject outside the scope of most landscape contracts. The IP reference number refers to the waterproof properties of the fitting; the higher the number the more waterproof the fitting. Most items can be fitted with time clocks or PIR controls.							
Market prices of lamps							
Lamps							
70 w HQIT-S	-	-	-	-	9.22	nr	-
70 w HQIT	-	-	-	-	18.87	nr	-
70 w SON	-	-	-	-	4.22	nr	-
70 w SONT	-	-	-	-	4.22	nr	-
100 w SONT	-	-	-	-	6.34	nr	-
150 w SONT	-	-	-	-	7.71	nr	-
28 w 2D	-	-	-	-	22.00	nr	-
100 w GLS\E27	-	-	-	-	2.07	nr	-
Bulkhead and canopy fittings; including fixing to wall and light fitting (lamp, final painting, electric wiring, connections or related fixtures such as switch gear and time clock mechanisms not included unless otherwise indicated)							
Bulkhead and canopy fittings; Louis Poulsen "Nyhavn Wall" small domed top conical shade with rings; copper wall lantern; finished untreated copper to achieve verdigris finish; also available in white aluminium; 310 mm diameter shade; with wall mounting arm; to IP 44	422.00	0.50	9.25	-	424.79	nr	**434.04**
Bulkhead and canopy fittings; Sugg Lighting "Princess" IP54 backlamp; 375 mm x 229 mm; copper frame; stove painted in black; with chimney and lampholder	264.51	0.50	9.25	-	267.30	nr	**276.55**
"Victoria" IP54 backlamp; 502 mm x 323 mm; copper frame; polished copper finish; with chimney, door and lampholder	353.89	0.50	9.25	-	356.68	nr	**365.93**
"Palace" IP54 backlamp; 457 mm x 321 mm; copper frame; stove painted black finish; with chimney, door and lampholder	353.13	0.50	9.25	-	355.92	nr	**365.17**
"Windsor" IP54 backlamp; 650 mm x 306 mm; copper frame; polished and lacquered finish; with door and lampholder	345.67	0.50	9.25	-	348.46	nr	**357.71**
"Windsor" IP54 gas backlamp, 650 mm x 306 mm, copper frame, polished and lacquered finish, hinged door, double inverted cluster mantle with permanent pilot and mains solenoid	619.20	0.50	9.25	-	621.99	nr	**631.24**

V ELECTRICAL SUPPLY/POWER/LIGHTING SYSTEMS

Item Excluding site overheads and profit	PC £	Labour hours	Labour £	Plant £	Material £	Unit	**Total rate £**
Floodlighting; ground, wall or pole mounted; including fixing (lamp, final painting, electric wiring, connections or related fixtures such as switch gear and time clock mechanisms not included unless otherwise indicated)							
Floodlighting; Louis Poulsen - Landscape Division							
SPR-12; multi purpose ground/spike mounted wide angle floodlight; 70 w HIT or SON; to IP65	219.00	1.00	18.50	-	224.76	nr	**243.26**
Floodlight accessories; Louis Poulsen - Landscape Division							
earth spike for SPR-12	25.00	-	-	-	25.00	nr	-
louvre for SPR-12	41.00	-	-	-	41.00	nr	-
cowl for SPR-12	41.00	-	-	-	41.00	nr	-
barn doors for SPR-12	89.00	-	-	-	89.00	nr	-
Large area/pitch floodlighting; CU Phosco							
ref FL444 1000 w SON-T; floodlights with lamp and loose gear; narrow asymmetric beam	408.20	1.00	40.00	-	408.20	nr	**448.20**
ref FL444 2.0 kw MBIOS; floodlight with lamp and loose gear; projector beam	613.00	1.00	40.00	-	613.00	nr	**653.00**
Large area floodlighting; CU Phosco							
ref FL345/G/250S; floodlight with lamp and integral gear	314.80	1.00	40.00	-	314.80	nr	**354.80**
ref FL345/G/400 MBI; floodlight with lamp and integral gear	340.00	1.00	40.00	-	340.00	nr	**380.00**
Small area floodlighting; Sugg Lighting; floodlight for feature lighting; clear or toughened glass							
"Scenario"; lamp 150 w HPS-T; lamp	582.82	1.00	40.00	-	582.82	nr	**622.82**
"Scenario"; lamp 150 w HQI-T; lamp	582.82	1.00	40.00	-	582.82	nr	**622.82**
Spotlights for uplighting and for illuminating signs and notice boards, statuary and other features; ground, wall or pole mounted; including fixing, light fitting and priming (lamp, final painting, electric wiring, connections or related fixtures such as switch gear and time clock mechanisms not included); all mains voltage (240 v) unless otherwise stated							
Spotlights; Louis Poulsen - Landscape Division							
"Maxispotter" miniature solid copper spotlight designed to patternate and age naturally; with mounting bracket; and internal anti glare louvre; low voltage halogen reflector; 20/35/50 w; to IP56	79.95	1.00	18.50	-	85.71	nr	**104.21**
"WeeBee Spot"; ref SP-05 for low voltage halogen reflector; 20/35/50 w; c/w integral transformer and wall mounting box; to IP65	152.00	1.00	18.50	-	157.76	nr	**176.26**
Spotlight accessories; Louis Poulsen - Landscape Division							
"ES"; earth spike for WeeBee Spot	16.00	-	-	-	16.00	nr	-
Spotlighters, uplighters and cowl lighting; Marlin Lighting; TECHNO - SHORT ARM; head adjustable 130 deg rotation 350 deg; projection 215 mm on 210 mm base plate; 355 high; integral gear; PG16 cable gland; black							
ref SIM 3518-09; 150 w CDMT/HQIT	397.45	1.00	40.00	-	397.45	nr	**437.45**
ref SIM 3517-09; 70 w CDMT/HQIT	357.76	1.00	40.00	-	357.76	nr	**397.76**

V ELECTRICAL SUPPLY/POWER/LIGHTING SYSTEMS

Item Excluding site overheads and profit	PC £	Labour hours	Labour £	Plant £	Material £	Unit	Total rate £
V41 STREET/AREA/FLOOD LIGHTING - cont'd							
Recessed uplighting; including walk/drive over fully recessed uplighting; excavating, ground fixing, concreting in and making good surfaces (electric wiring, connections or related fixtures such as switch gear and time-clock mechanisms not included unless otherwise stated) (Note: transformers will power multiple lights dependent on the distance between the light units); all mains voltage (240 v) unless otherwise stated							
Recessed uplighting; Louis Poulsen - Landscape Division; "WeeBee Up"; ultra-compact recessed halogen uplight in diecast aluminium and S/Steel top plate and toughened safety glass; 20/35 w; complete with installation sleeve; low voltage requires transformer; to IP67							
ref BU-01	105.00	2.00	37.00	-	110.76	nr	**147.76**
Recessed uplighting; Louis Poulsen - Landscape Division; 266 mm diameter; diecast aluminium with stainless steel top plate 10 mm toughened safety glass; 2000 kg drive over; integral control gear; to IP67							
IPR-14; HIT Metal Halide; white light; spot or flood or wall wash distribution	384.00	1.50	27.75	-	389.76	nr	**417.51**
IRR-14 HIT Compact fluorescent; white low power consumption; flood distribution	384.00	1.50	27.75	-	389.76	nr	**417.51**
Accessories for IPR-14 uplighters; Louis Poulsen - Landscape Division							
Rockguard	110.00	-	-	-	110.00	nr	-
Stainless steel installation sleeve	47.00	-	-	-	47.00	nr	-
Anti glare louvre (internal tilt)	69.00	-	-	-	69.00	nr	-
Recessed uplighting; Louis Poulsen - Landscape Division; Nimbus 125-150 mm diameter uplighters; manufactured from diecast aluminium with stainless steel top plate 8 mm toughened safety glass; 2000 kg drive over; integral control gear; to IP67							
BU-02; 95 mm deep; 20/35/50 w low voltage halogen; requires remote transformer	112.11	2.00	37.00	-	117.87	nr	**154.87**
BU-03; 172 mm deep; 20/35/50 w low voltage halogen; with integral transformer	154.60	2.00	37.00	-	160.36	nr	**197.36**
Accessories for Nimbus uplighters; Louis Poulsen - Landscape Division							
"IS" installation sleeve	41.00	-	-	-	41.00	nr	-
Wall recessed; Marlin Lighting; "EOS" range; integral gear; asymmetric reflector; toughened reeded glass; to IP55							
ref SIM 4629-09; 10 w TCD "MINI EOS" 145 x 90 mm	92.18	1.00	40.00	-	92.18	nr	**132.18**
ref SIM 4639-09; "MEGA EOS" wall recessing box	15.57	0.25	10.00	-	15.57	nr	**25.57**
ref SIM 4619-09; 26 w TCD "RECTANGULAR EOS" 270 x 145 mm	150.96	1.00	40.00	-	150.96	nr	**190.96**
ref SIM 4532-12; "RECTANGULAR EOS" wall recessing box	15.57	0.25	10.00	-	15.57	nr	**25.57**
ref SIM 4628-09; 12 v 20 w QT9 "MINI EOS" 145 x 90 mm	138.16	1.00	40.00	-	138.16	nr	**178.16**
ref SIM 4623-12; "MINI EOS" wall recessing box	9.38	0.25	10.00	-	9.38	nr	**19.38**

V ELECTRICAL SUPPLY/POWER/LIGHTING SYSTEMS

Item Excluding site overheads and profit	PC £	Labour hours	Labour £	Plant £	Material £	Unit	**Total rate £**
Underwater lighting							
Underwater lighting; Louis Poulsen - Landscape Division							
UW-05 "Minipower" cast bronze underwater floodlight c\w mounting bracket; 50 w low voltage reflector lamp; requires remote transformer	206.24	1.00	40.00	-	206.24	nr	**246.24**
Low-level lighting; positioned to provide glare free light wash to pathways, steps and terraces; including forming post holes, concreting in, making good to surfaces (final painting, electric wiring, connections or related fixtures such as switch gear and time-clock mechanisms not included)							
Low-level lighting; Louis Poulsen - Landscape Division							
"Sentry"; ref AM-08; single-sided bollard type sculptural pathway light; aluminium; 670 mm high; for 35 w HIT; to IP65	365.00	2.00	37.00	-	370.76	nr	**407.76**
"Footliter"; ref GL-05; low voltage pathlighter; 311 mm high; 6.00 m light distribution; in aluminium or solid copper; c\w earth spike; requires remote transformer; 20 w halogen; to IP44; copper finish	79.07	2.00	37.00	-	84.83	nr	**121.83**
Lighted bollards; including excavating, ground fixing, concreting in and making good surfaces (lamp, final painting, electric wiring, connections or related fixtures such as switch gear and time-clock mechanisms not included) (Note: all illuminated bollards must be earthed); heights given are from ground level to top of bollards							
Lighted bollards; Louis Poulsen							
"Orbiter" vandal resistant; head of cast aluminium; domed top; anti-glare rings; pole extruded aluminium; diffuser clear UV stabilised polycarbonate; powder coated; 1040 mm high; 255 mm diameter; with root or base plate; IP44	499.00	2.50	46.25	-	505.32	nr	**551.57**
"Waterfront" solidly proportioned; head of cast silumin; domed top; symmetrical distribution; pole extruded aluminium sandblasted or painted white; internal diffuser clear UV stabilised polycarbonate; 865 mm high; diameter 260 mm; IP55	583.00	2.50	46.25	-	589.32	nr	**635.57**
"Bysted" concentric louvred bollard; head cast iron; post COR-TEN steel; externally untreated to provide natural aging effect of uniform oxidised red surface finish; internal painted white; lamp diffuser rings of clear polycarbonate; 1130 mm high 280 mm diameter; to IP44	826.00	2.50	46.25	-	832.32	nr	**878.57**
Lighted bollards; Woodscape Ltd							
Illuminated bollard; in "Very durable hardwood"; integral die cast aluminium lighting unit; 165 x 165 x 1.00 high	212.80	2.50	46.25	-	225.14	nr	**271.39**

V ELECTRICAL SUPPLY/POWER/LIGHTING SYSTEMS

Item Excluding site overheads and profit	PC £	Labour hours	Labour £	Plant £	Material £	Unit	Total rate £
V41 STREET/AREA/FLOOD LIGHTING - cont'd							
Reproduction street lanterns and columns; including excavating, concreting in, backfilling and disposing of spoil, or fixing to ground or wall, making good surfaces, light fitting and priming (final painting, electric wiring, connections or related fixtures such as switch gear and time-clock mechanisms not included) (Note: lanterns up to 14 in and columns up to 7 ft are suitable for residential lighting)							
Reproduction street lanterns; Sugg Lighting; reproduction lanterns handmade to original designs; all with ES lamp holder							
"Westminster" IP54 hexagonal hinged top lantern with door, two piece folded polycarbonate glazing; lamp 100 w HQI T; integral photo electric cell							
"Small" 1016 mm high x 356 mm wide	587.62	0.50	9.25	-	587.62	nr	**596.87**
"Large" 1124 mm high x 760 mm wide	923.45	0.50	9.25	-	923.45	nr	**932.70**
"Guildhall" IP54; handcrafted copper frame; clear polycarbonate glazing circular tapered lantern with hemispherical top; integral photo electric cell; with door							
"small" 711 mm high x 330 mm wide	516.58	0.50	9.25	-	519.37	nr	**528.62**
"medium" 1150 mm high x 432 mm wide	709.72	0.50	9.25	-	712.51	nr	**721.76**
"large" 1370 mm high x 550 mm wide	950.88	0.50	9.25	-	953.67	nr	**962.92**
"Grosvenor" circular lantern with door; copper frame; polished copper finish; polycarbonate glazing							
"small" 790 mm x 330 mm; IP54	574.70	0.50	9.25	-	577.49	nr	**586.74**
"medium" 1080 mm x 435 mm; IP65	544.33	0.50	9.25	-	547.12	nr	**556.37**
"Classic Globe" IP54 lantern with hinged outer frame; cast aluminium frame; black polyester powder coating							
"medium" 965 mm x 483 mm	521.48	0.50	9.25	-	524.27	nr	**533.52**
"Windsor" lantern; copper frame; polished copper finish							
"small" 905 x 356 mm; IP54; with door	410.33	0.50	9.25	-	413.12	nr	**422.37**
"small" 905 x 356 mm; IP65; without door	408.65	0.50	9.25	-	411.44	nr	**420.69**
"medium" 1124 mm x 420 mm; IP54; with door	445.65	0.50	9.25	-	448.44	nr	**457.69**
"large" 1124 mm x 470 mm; IP54; with door	506.20	0.50	9.25	-	508.99	nr	**518.24**
"medium" gas lantern, 1124 mm x 420 mm; IP54; with door; integral solenoid and pilot	791.62	0.50	9.25	-	794.41	nr	**803.66**
Reproduction brackets and suspensions; Sugg Lighting							
Iron brackets							
"Bow" bracket - 6'0"	734.71	2.00	37.00	-	737.50	nr	**774.50**
"Ornate" iron bracket - large	329.64	2.00	37.00	-	332.43	nr	**369.43**
"Ornate" iron bracket - medium	316.94	2.00	37.00	-	319.73	nr	**356.73**
"Swan neck" iron bracket - large	285.60	2.00	37.00	-	288.39	nr	**325.39**
"Swan neck" iron bracket - medium	161.86	2.00	37.00	-	164.65	nr	**201.65**
Cast brackets							
"Universal" cast bracket	216.74	2.00	37.00	-	219.53	nr	**256.53**
"Abbey" bracket	112.64	2.00	37.00	-	115.43	nr	**152.43**
"Short Abbey" bracket	98.61	2.00	37.00	-	101.40	nr	**138.40**
"Plaza" cast bracket	112.65	2.00	37.00	-	115.44	nr	**152.44**
Base mountings							
"Universal" pedestal	214.98	2.00	37.00	-	217.77	nr	**254.77**
"Universal" plinth	121.45	2.00	37.00	-	124.24	nr	**161.24**

V ELECTRICAL SUPPLY/POWER/LIGHTING SYSTEMS

Item Excluding site overheads and profit	PC £	Labour hours	Labour £	Plant £	Material £	Unit	**Total rate £**
Reproduction lighting columns							
Reproduction lighting columns; Sugg Lighting							
"Harborne" C11; fabricated iron/steel heavy duty post with integral cast root; 3 - 5 m	822.90	8.00	148.00	-	830.80	nr	**978.80**
"Aylesbury" C12; base fabricated heavy gauge 89 mm aluminium post; 3 - 5 m	1327.45	8.00	148.00	-	1335.35	nr	**1483.35**
"Cannonbury" C13; fabricated heavy gauge 89 mm aluminium post; 3 - 5 m	827.07	8.00	148.00	-	834.97	nr	**982.97**
Standard column C14; rooted British steel 168/89 mm embellished post; 5 - 8 m	444.91	8.00	148.00	-	452.81	nr	**600.81**
"Large Constitution Hill" C22X; cast aluminium, steel cored post with extended spigot; 4.3 m	3948.36	6.00	111.00	-	3956.26	nr	**4067.26**
"Cardiff" C29; cast aluminium post; 3.5 m	1721.69	6.00	111.00	-	1729.59	nr	**1840.59**
"Seven Dials" C36; cast aluminium post; 3.8 m	1291.80	8.00	148.00	-	1299.70	nr	**1447.70**
"Royal Exchange" C42; traditional cast aluminium post; welded multi arm construction; 2.1 m	2049.55	6.00	111.00	-	2057.45	nr	**2168.45**
Precinct lighting lanterns; ground, wall or pole mounted; including fixing and light fitting (lamps, poles, brackets, final painting, electric wiring, connections or related fixtures such as switch gear and time-clock mechanisms not included)							
Precinct lighting lanterns; Louis Poulsen							
"Nyhavn Park" side entry 90 degree curved arm mounted; steel rings over domed top; housing cast silumin sandblasted with integral gear; protected by UV stabilized clear polycarbonate diffuser; IP55	615.00	1.00	18.50	-	615.00	nr	**633.50**
"Kipp"; bottom entry pole-top; shot blasted or lacquered Hanover design award; hinged diffuser for simple maintenance; indirect lighting technique ensures low glare; IP55	402.00	1.00	18.50	-	402.00	nr	**420.50**
Precinct lighting lanterns; Sugg Lighting							
Juno Dome; IP66; molded GRP body; graphite finish	393.45	1.00	18.50	-	393.45	nr	**411.95**
Juno Cone; IP66; molded GRP body; graphite finish	396.45	1.00	18.50	-	396.45	nr	**414.95**
Sharkon; IP65; cast aluminium body; silver and blue finish	374.72	1.00	18.50	-	374.72	nr	**393.22**
Transformers for low voltage lighting; Louis Poulsen - Landscape Division; distance from electrical supply 25 m; trenching and backfilling measured separately							
Boxed transformer							
50 Va for maximum 50 w of lamp	64.42	2.00	80.00	-	91.92	nr	**171.92**
100Va for maximum 100 w of lamp	68.15	2.00	80.00	-	95.65	nr	**175.65**
150 Va for maximum 150 w of lamp	68.16	2.00	80.00	-	95.66	nr	**175.66**
200 Va for maximum 200 w of lamp	84.95	2.00	80.00	-	112.45	nr	**192.45**
250 Va for maximum 50 w of lamp	99.36	2.00	80.00	-	126.86	nr	**206.86**
Buried transformers							
50 Va for maximum 50 w of lamp	64.37	2.00	80.00	-	91.87	nr	**171.87**
Installation; electric cable in trench 500 mm deep (trench not included); all in accordance with IEE regulations							
Light duty 600 volt grade armoured							
3 core 2.5 mm	1.10	0.01	0.09	-	1.10	m	**1.19**
4 core 2.5 mm	1.50	0.01	0.09	-	1.50	m	**1.59**
Twin and earth PVC cable in plastic conduit (conduit not included)							
2.50 mm^2	0.35	0.01	0.19	-	0.35	m	**0.53**
4.00 mm^2	0.72	0.01	0.19	-	0.72	m	**0.91**
16 mm heavy gauge high impact PVC conduit	3.60	0.03	0.62	-	3.60	m	**4.22**
Main switch and fuse unit; 30 A	-	-	-	-	70.23	nr	
Weatherproof switched socket; 13 A	-	2.00	80.00	-	14.00	nr	**94.00**

ESSENTIAL READING FROM TAYLOR AND FRANCIS

The Dynamic Landscape
Naturalistic Planting in an Urban Context

Edited by
Nigel Dunnett and James Hitchmough

The last quarter of the twentieth-century witnessed a burgeoning of interest in ecological or naturally inspired use of vegetation in the designed landscape. More recently a strong aesthetic element has been added to what was formerly a movement aimed at creating nature-like landscapes.

This book advances a fusion of scientific and ecological planting design philosophy that can address the need for more sustainable designed landscapes. It is a major statement on the design, implementation and management of ecologically-inspired landscape vegetation. With contributions from people at the forefront of developments in this field, in both Europe and North America, it provides a valuable synthesis of current thinking.

Contents: 1. Introduction 2. The Historical Development of Naturalistic Planting 3. A Contemporary Overview of Naturalistic Planting 4. The Dynamic Nature of Plant Communities 5. A Naturalistic Design Process 6. Herbaceous Plantings 7. Exploring Woodland Design 8. Wetlands and Water Bodies 9. Communicating Naturalistic Plantings: Plans and Specifications 10. The Creative Management of Naturalistic Plantings 11. The Social and Cultural Context of Naturalistic Plantings. Index.

June 2004: 238x225mm: 336 pages
20 tables, 40 line drawings, 67 b+w photos and 85 colour photos
HB: 0-415-25620-8: £63.00

To Order: Tel: +44 (0) 1264 343071 Fax: +44 (0) 1264 343005, or
Post: Taylor and Francis Customer Services, Thomson Publishing Services, Cheriton House, Andover, Hants, SP10 5BE, UK Email: book.orders@tandf.co.uk

For a complete listing of all our titles visit:
www.sponpress.com

Taylor & Francis
Taylor & Francis Group

PART 4

Approximate Estimates- Minor Works

APPROXIMATE ESTIMATES

Prices in this section are based upon the Prices for Measured Works, but allow for incidentals which would normally be measured separately in a Bill of Quantities. They do not include for Preliminaries which are priced elsewhere in this book.

Items shown as sub-contract or specialist rates would normally include the specialist's overhead and profit. All other items which could fall within the scope of works of general landscape and external works contractors would not include profit.

Based on current commercial rates, profits of 15% to 35% may be added to these rates to indicate the likely "with profit" values of the tasks below. The variation quoted above is dependent on the sector in which the works are taking place - domestic, public or commercial.

ESSENTIAL READING FROM TAYLOR AND FRANCIS

Visualization in Landscape and Environmental Planning
Technology and Applications

Ian Bishop and Eckart Lange

This major reference presents the challenges, issues and directions of computer-based visualization of the natural and built environment and the role of such visualization in landscape and environmental planning. It offers a uniquely systematic approach to the potential of visualization and the writers are acknowledged experts in their field of specialization. Case studies are presented to illustrate many aspects of landscape management including forestry, agriculture, ecology, mining and urban development.

Foreword: Stephen Ervin. Part One. 1. Communication, perception and visualization. 2. Visualization Classified. 3. Data sources for three-dimensional models. 4. Visualization Technology. 5. Validity, reliability, and ethics in visualization. Part Two - Applications. 6 - Application in the Forest Landscape. 7. Applications in the agricultural landscape. 8. Applications in Energy, Industry and Infrastructure. 9 Applications in the urban landscape Part Three - Prospects. 10. Visualization Prospects. Index.

May 2005: 246x189 mm: 340 pages
HB: 0-415-30510-1: £58.00

To Order: Tel: +44 (0) 1264 343071 Fax: +44 (0) 1264 343005, or
Post: Taylor and Francis Customer Services, Thomson Publishing Services, Cheriton House, Andover, Hants, SP10 5BE, UK Email: book.orders@tandf.co.uk

For a complete listing of all our titles visit:
www.sponpress.com

Taylor & Francis
Taylor & Francis Group

DEMOLITION AND SITE CLEARANCE

Item Excluding site overheads and profit	Unit	**Total rate £**
DEMOLITION AND SITE CLEARANCE		
Demolish existing surfaces by hand held electric breaker; removal by grab		
Break up plain concrete slab; remove to licensed tip		
150 thick	m^2	**20.00**
200 thick	m^2	**20.00**
Break up reinforced concrete slab and remove to licensed tip		
150 thick	m^2	**20.00**
200 thick	m^2	**25.00**
300 thick	m^2	**45.00**
Break out existing surface and associated 150 mm thick granular base load to remove off site by grab		
macadam 70 mm thick	m^2	**13.76**
block paving 50 thick	m^2	**18.53**
block paving 80 thick	m^2	**20.95**
Demolish existing free standing walls; grub out foundations; remove arisings to tip		
Brick wall; 112 mm thick		
300 mm high	m	**19.62**
500 mm high	m	**21.23**
Brick wall; 225 mm thick		
300 mm high	m	**23.65**
500 mm high	m	**35.06**
1.00 m high	m	**44.74**
1.20 m high	m	**47.96**
1.50 m high	m	**61.00**
1.80 m high	m	**66.64**
Site clearance - generally		
Clear away light fencing and gates (chain link, chestnut paling, light boarded fence or similar) and remove to licensed tip	100 m	**402.00**

GROUNDWORK

Item Excluding site overheads and profit	Unit	Total rate £
GROUNDWORK		
EXCAVATION AND FILLING		
Cut and strip by machine turves 50 thick		
Load to barrows and stack on site not exceeding 25 m travel to stack	100 m²	**510.00**
Load to barrows and disposal off site by skip; distance 25 m	100 m²	**1140.00**
Excavate to reduce levels; mechanical		
Removal to spoil heaps		
excavated directly to loading position	m^3	**7.90**
transporting to loading position 25 m distance	m^3	**14.00**
Excavate to reduce levels; mechanical		
Removal off site by grab		
excavated directly to loading position	m^3	**46.00**
transporting to loading position 25 m distance	m^3	**52.00**
Excavate to reduce levels; mechanical		
Removal off site by skip		
excavated directly to loading position	m^3	**53.00**
transporting to loading position 25 m distance	m^3	**59.00**
Excavate to reduce levels; hand		
Removal off site by skip		
excavated directly to loading position	m^3	**180.00**
excavated directly to loading position	m^3	**240.00**
filled to bags and transporting to loading position 25 m distant	m^3	**360.00**
Excavation and filling; mechanical		
Excavate existing soil on proposed turf or planting area to reduce levels; grade to levels; fill excavated area with topsoil from spoil heaps; removal of and excavated material by skip		
100 mm deep	m^2	**6.80**
200 mm deep	m^2	**14.00**
300 mm deep	m^2	**21.00**
Spread excavated material to levels in layers not exceeding 150 mm; grade to finished levels to receive surface treatments		
By machine		
average thickness 100 mm	m^2	**2.00**
average thickness 100 mm but with imported topsoil	m^2	**4.00**
average thickness 200 mm	m^2	**2.00**
average thickness 200 mm but with imported topsoil	m^2	**7.00**
average thickness 250 mm	m^2	**2.00**
average thickness 250 mm but with imported topsoil	m^2	**9.00**
extra for work to banks exceeding 30 slope	30%	**-**
Filling by hand		
Excavate existing soil on proposed turf or planting area to reduce levels; grade to levels; fill excavated area with topsoil from spoil heaps; removal of excavated material by skip		
100 mm deep	m^2	**9.00**
200 mm deep	m^2	**18.00**
300 mm deep	m^2	**27.00**
500 mm deep	m^2	**45.00**

GROUNDWORK

Item Excluding site overheads and profit	Unit	**Total rate £**
TRENCHES		
Excavate trenches; remove excavated material off site by grab 25 m distance; fill trench to ground level with site mixed concrete 1:3:6; allow for movement of concrete and excavated material		
By machine		
300 mm wide x 250 mm deep	m	**14.34**
500 mm wide x 250 mm deep	m	**20.78**
750 mm wide x 350 mm deep	m	**41.73**
1200 mm wide x 600 mm deep	m	**115.44**
By hand		
300 mm wide x 250 mm deep	m	**23.60**
500 mm wide x 250 mm deep	m	**37.84**
750 mm wide x 350 mm deep	m	**83.44**
1200 mm wide x 600 mm deep	m	**229.67**

IN SITU CONCRETE

Item Excluding site overheads and profit	Unit	Total rate £
IN SITU CONCRETE		
Mix concrete on site; aggregates delivered in 10 tonne loads; deliver mixed concrete to location by mechanical dumper distance 25 m		
1:3:6	m^3	**102.00**
1:2:4	m^3	**113.00**
As above but ready mixed concrete		
10 N/mm²	m^3	**116.00**
15 N/mm²	m^3	**123.00**
Mix concrete on site; aggregates delivered in 10 tonne loads; deliver mixed concrete to location by barrow distance 25 m		
1:3:6	m^3	**147.00**
1:2:4	m^3	**158.00**
As above but aggregates delivered in 1 tonne bags		
1:3:6	m^3	**177.00**
1:2:4	m^3	**188.00**
As above but concrete discharged directly from ready mix lorry to required location		
10 N/mm²	m^3	**120.00**
15 N/mm²	m^3	**120.00**
Excavate foundation trench mechanically; remove spoil offsite by grab; lay 1:3:6 site mixed concrete foundations; distance from mixer 25 m; depth of trench to be 225 mm deeper than foundation to allow for 3 underground brick courses priced separately		
Foundation size		
200 mm deep x 400 mm wide	m	**20.00**
300 mm deep x 500 mm wide	m	**35.00**
400 mm deep x 400 mm wide	m	**35.00**
400 mm deep x 600 mm wide	m	**52.00**
600 mm deep x 600 mm wide	m	**74.00**
As above but hand excavation and disposal to spoil heap 25 m by barrow; disposal off site by grab		
200 mm deep x 400 mm wide	m	**43.00**
300 mm deep x 500 mm wide	m	**71.00**
400 mm deep x 400 mm wide	m	**70.00**
400 mm deep x 600 mm wide	m	**110.00**
600 mm deep x 600 mm wide	m	**150.00**

BRICK/BLOCK WALLING

Item Excluding site overheads and profit	Unit	Total rate £
BRICK/BLOCK WALLING		
BRICK WALLING		
Excavate foundation trench 500 deep; remove spoil to dump off site (all by machine); lay site mixed concrete foundations 1:3:6 350 x 150 thick; construct half brick wall with one brick piers at 2.0 m centres; laid in cement:lime:sand (1:1:6) mortar with flush joints; fair-face one side; DPC two courses underground; engineering brick in cement:sand (1:3) mortar; coping of headers on end		
Wall 900 high above DPC		
in engineering brick (class B) - £300.00/1000	m	**237.00**
in reclaimed bricks - £800.00/1000	m	**348.00**
Excavate foundation trench 400 deep; remove spoil to dump off site; lay GEN 1 concrete foundations 450 wide x 250 thick; construct one brick wall with one and a half brick piers at 3.0 m centres; all in English garden wall bond; laid in cement:lime:sand (1:1:6) mortar with flush joints, fair-face one side; DPC two courses engineering brick in cement:sand (1:3) mortar; engineering brick coping		
Wall 900 high above DPC		
in engineering brick (class B) - £300.00/1000	m	**363.00**
in reclaimed bricks - £800.00/1000	m	**422.00**
Wall 1200 high above DPC		
in engineering brick (class B) - £300.00/1000	m	**424.00**
in reclaimed bricks - £800.00/1000	m	**494.00**
Wall 1800 high above DPC		
in engineering brick (class B) - £300.00/1000	m	**624.00**
in reclaimed bricks - £800.00/1000	m	**731.00**
BLOCK WALLING		
Excavate foundation trench 450 deep; remove spoil to dump off site; lay GEN 1 concrete foundations 600 x 300 thick; construct wall of concrete block; 2 courses belowground		
Solid blocks 7 N/mm^2; wall 1.00 m high		
100 mm thick	m	**110.00**
Hollow blocks filled with concrete; wall 1.00 m high		
215 mm thick	m	**170.00**
Hollow blocks but with steel bar cast into the foundation wall 1.00 m high		
215 mm thick	m	**180.00**
Solid blocks 7 N/mm^2; wall 1.80 m high		
100 mm thick	m	**150.00**
Hollow blocks filled with concrete; wall 1.80 m high		
215 mm thick	m	**250.00**
Hollow blocks but with steel bar cast into the foundation wall 1.80 m high		
215 mm thick	m	**280.00**

ROADS AND PAVINGS

Item Excluding site overheads and profit	Unit	**Total rate £**
ROADS AND PAVINGS		
BASES FOR PAVING		
Excavate ground and reduce levels to receive 38 mm thick slab and 25 mm mortar bed; dispose of excavated material off site; treat substrate with total herbicide		
Lay granular fill Type 1 150 thick laid to falls and compacted		
all by machine	m²	**18.00**
all by hand except disposal by grab	m²	**37.00**
Excavate ground and reduce levels to receive 65 mm thick surface and bed (not included); dispose of excavated material off site; treat substrate with total herbicide		
Lay 100 compacted hardcore; lay 1:2:4 concrete base 150 thick laid to falls		
all by machine	m²	**40.00**
all by hand except disposal by grab	m²	**82.00**
As above but inclusive of reinforcement fabric A142		
all by machine	m²	**44.00**
all by hand except disposal by grab	m²	**86.00**
KERBS AND EDGINGS		
Note: excavation is by machine unless otherwise mentioned		
Excavate trench and construct concrete foundation 150 mm wide x 150 mm deep; lay precast concrete kerb units bedded in semi-dry concrete; slump 35 mm maximum; haunching one side; disposal of arisings off site		
Edgings laid straight		
50 x 150 mm	m	**32.00**
125 mm high x 150 mm wide; bullnosed	m	**32.00**
50 x 200 mm	m	**33.00**
Second hand granite setts		
100 x 100 mm	m	**66.00**
Single course; brick or block edgings laid to stretcher		
concrete blocks 200 x 100	m	**26.00**
engineering bricks	m	**32.00**
paving bricks (PC £520.00 per 1000)	m	**32.00**
Double course; brick or block edgings laid to stretcher		
concrete blocks 200 x 100	m	**29.00**
engineering bricks	m	**37.00**
paving bricks (PC £520.00 per 1000)	m	**39.00**
Single course; brick or block edgings laid to header course (soldier course)		
blocks 200 x 100 x 60; PC £7.09/m^2; butt jointed	m	**29.00**
bricks 200 x 100 x 50; PC £450.00/1000; butt jointed	m	**32.00**
bricks 200 x 100 x 50; PC £450.00/1000; mortar joints	m	**37.00**
Sawn yorkstone edgings; excavate for ground beam; lay concrete 1:2:4 150 mm deep x 33.3% wider than the edging; on 35 mm thick mortar bed; inclusive of haunching one side		
Yorkstone 50 mm thick		
100 mm wide x random lengths	m	**38.00**
100 mm x 100 mm	m	**41.00**
200 mm wide x 100 mm long	m	**50.00**
250 mm wide x random lengths	m	**47.00**
500 mm wide x random lengths	m	**83.00**

ROADS AND PAVINGS

Item Excluding site overheads and profit	Unit	Total rate £
INTERLOCKING BLOCK PAVING		
Excavate ground; supply and lay granular fill Type 1 150 thick laid to falls and compacted; supply and lay block pavers; laid on 50 compacted sharp sand; vibrated; joints filled with loose sand excluding edgings or kerbs measured separately		
Concrete blocks		
200 x 100 x 60	m²	**60.00**
200 x 100 x 80	m²	**63.00**
Reduce levels; lay 150 granular material Type 1; lay precast concrete edging 50 x 150; on concrete foundation 1:2:4; lay 200 x 100 x 60 vehicular block paving to 90 degree herringbone pattern; on 50 compacted sand bed; vibrated; jointed in sand and vibrated		
1.0 m wide clear width between edgings	m	**100.00**
1.5 m wide clear width between edgings	m	**130.00**
2.0 m wide clear width between edgings	m	**160.00**
3.0 m wide clear width between edgings	m	**220.00**
Works by hand; reduce levels; lay 150 granular material Type 1; lay edge restraint of block paving 200 wide on 150 thick concrete foundation 1:2:4 haunched; lay 200 x 100 x 60 vehicular block paving to 90 degree herringbone pattern; on 50 compacted sand bed; vibrated; jointed in sand and vibrated		
1.0 m wide clear width between edgings	m	**94.00**
1.5 m wide clear width between edgings	m	**120.00**
2.0 m wide clear width between edgings	m	**150.00**
3.0 m wide clear width between edgings	m	**210.00**
Works by hand; reduce levels; lay 150 granular material Type 1; lay edge restraint of block paving 200 wide on 150 thick concrete foundation 1:2:4 haunched; lay 200 x 100 x 60 vehicular block paving to 90 degree herringbone pattern; on 50 compacted sand bed; vibrated; jointed in sand and vibrated		
1.0 m wide clear width between edgings	m	**120.00**
1.5 m wide clear width between edgings	m	**140.00**
2.0 m wide clear width between edgings	m	**180.00**
3.0 m wide clear width between edgings	m	**250.00**
BRICK PAVING		
WORKS BY MACHINE		
Excavate and lay base Type 1 150 thick remove arisings; all by machine; lay clay brick paving		
200 x 100 x 50 thick; butt jointed on 50 mm sharp sand bed		
PC £300.00/1000	m²	**72.00**
PC £600.00/1000	m²	**87.00**
200 x 100 x 50 thick; 10 mm mortar joints on 35 mm mortar bed		
PC £300.00/1000	m²	**86.00**
PC £600.00/1000	m²	**100.00**
Excavate and lay 150 mm site mixed concrete base 1:3:6 reinforced with A142 mesh; all by machine; remove arisings; lay clay brick paving		
200 x 100 x 50 thick; 10 mm mortar joints on 35 mm mortar bed; running or stretcher bond		
PC £300.00/1000	m²	**120.00**
PC £600.00/1000	m²	**130.00**
200 x 100 x 50 thick; 10 mm mortar joints on 35 mm mortar bed; butt jointed; herringbone bond		
PC £300.00/1000	m²	**95.00**
PC £600.00/1000	m²	**120.00**
Excavate and lay base readymix concrete base 150 mm thick reinforced with A393 mesh; all by machine; remove arisings; lay clay brick paving		
215 x 102.5 x 50 thick; 10 mm mortar joints on 35 mm mortar bed		
PC £300.00/1000	m²	**85.00**
PC £600.00/1000	m²	**110.00**

ROADS AND PAVINGS

Item Excluding site overheads and profit	Unit	Total rate £
BRICK PAVING - cont'd		
WORKS BY HAND		
Excavate and lay base Type 1 150 thick by hand; arisings barrowed to spoil heap maximum distance 25 m and removal off site by grab; lay clay brick paving		
200 x 100 x 50 thick; butt jointed on 50 mm sharp sand bed		
PC £300.00/1000	m^2	**83.90**
PC £600.00/1000	m^2	**99.20**
Excavate and lay 150 mm concrete base; 1:3:6: site mixed concrete reinforced with A142 mesh; remove arisings to stockpile and then off site by grab; lay clay brick paving		
215 x 102.5 x 50 thick; 10 mm mortar joints on 35 mm mortar bed		
PC £300.00/1000	m^2	**104.00**
PC £600.00/1000	m^2	**120.00**
NATURAL STONE/SLAB PAVING		
WORKS BY MACHINE		
Excavate ground by machine and reduce levels, to receive 65 mm thick slab and 35 mm mortar bed; dispose of excavated material off site; treat substrate with total herbicide; lay granular fill Type 1 150 thick laid to falls and compacted; lay to random rectangular pattern on 35 mm mortar bed		
New riven slabs		
laid random rectangular	m^2	**133.00**
New riven slabs; but to 150 mm plain concrete base		
laid random rectangular	m^2	**148.00**
Reclaimed Cathedral grade riven slabs		
laid random rectangular	m^2	**171.00**
Reclaimed Cathedral grade riven slabs; but to 150 mm plain concrete base		
laid random rectangular	m^2	**186.00**
New slabs sawn 6 sides		
laid random rectangular	m^2	**126.00**
3 sizes, laid to coursed pattern	m^2	**130.00**
New slabs sawn 6 sides; but to 150 mm plain concrete base		
laid random rectangular	m^2	**139.00**
3 sizes, laid to coursed pattern	m^2	**144.00**
WORKS BY HAND		
Excavate ground by hand and reduce levels, to receive 65 mm thick slab and 35 mm mortar bed; barrow all materials and arisings 25 m; dispose of excavated material off site by grab; treat substrate with total herbicide; lay granular fill Type 1 150 thick laid to falls and compacted; lay to random rectangular pattern on 35 mm mortar bed		
New riven slabs		
laid random rectangular	m^2	**149.00**
New riven slabs laid random rectangular; but to 150 mm plain concrete base		
laid random rectangular	m^2	**163.00**
Reclaimed Cathedral grade riven slabs		
laid random rectangular	m^2	**188.00**
Reclaimed Cathedral grade riven slabs; but to 150 mm plain concrete base		
laid random rectangular	m^2	**199.00**
New slabs sawn 6 sides		
laid random rectangular	m^2	**132.00**
3 sizes, sawn 6 sides laid to coursed pattern	m^2	**141.00**
New slabs sawn 6 sides; but to 150 mm plain concrete base		
laid random rectangular	m^2	**144.00**
3 sizes, sawn 6 sides laid to coursed pattern	m^2	**153.00**

PREPARATION FOR SEEDING/TURFING

Item Excluding site overheads and profit	Unit	**Total rate £**
PREPARATION FOR SEEDING/TURFING		
SURFACE PREPARATIONS		
Cultivate existing ground by pedestrian operated rotavator; spread and lightly consolidate topsoil brought from spoil heap 25 m distance in layers not exceeding 150; grade to specified levels; remove stones over 25 mm; rake and grade to a fine tilth		
By machine		
100 mm thick	100 m^2	**138.00**
200 mm thick	100 m^2	**234.00**
300 mm thick	100 m^2	**330.00**
500 mm thick	100 m^2	**892.00**
By hand		
100 mm thick	100 m^2	**462.00**
150 mm thick	100 m^2	**673.00**
300 mm thick	100 m^2	**1300.00**
500 mm thick	100 m^2	**2750.00**
Lift and remove existing turf to skip 25 m distance; cultivate surface to receive new turf; rake to a fine tilth		
Works by hand		
normal turfed area	100 m^2	**654.00**
compacted turfed area	100 m^2	**663.00**
Lift and remove existing turf to skip 25 m distance; cultivate surface to receive new turf; rake to a fine tilth; fill area to receive turf with 50 mm imported topsoil		
Works by hand		
normal turfed area	100 m^2	**826.00**
compacted turfed area	100 m^2	**835.00**
SEEDING AND TURFING		
Domestic lawn areas		
Cultivate recently filled topsoil area; grade to levels and falls and rake to remove stones and debris; add fertilizers; add surface treatment as specified		
Turf areas		
Rolawn medallion	m^2	**5.84**
Seeded areas		
grass seed PC £3.00/kg; application rate 35 g/m^2	m^2	**0.99**
grass seed PC £3.00/kg; application rate 50 g/m^2	m^2	**1.04**
grass seed PC £4.00/kg; application rate 35 g/m^2	m^2	**1.03**
grass seed PC £4.00/kg; application rate 50 g/m^2	m^2	**1.09**
Cultivate recently filled topsoil area; grade to levels and falls and rake to remove stones and debris; add fertilizers; add surface treatment as specified; maintain for 1 year watering and cutting 26 times during the summer; pedestrian mower with grass box; arisings removed off site		
Turf areas		
Rolawn medallion	m^2	**8.08**
Seeded areas		
grass seed PC £3.00/kg; application rate 35 g/m^2	m^2	**3.23**
grass seed PC £3.00/kg; application rate 50 g/m^2	m^2	**3.28**
grass seed PC £4.00/kg; application rate 35 g/m^2	m^2	**3.27**
grass seed PC £4.00/kg; application rate 50 g/m^2	m^2	**3.33**

PLANTING

Item Excluding site overheads and profit	Unit	Total rate £
PLANTING		
TREE PLANTING		
Excavate tree pit by hand; fork over bottom of pit; plant tree with roots well spread out; backfill with excavated material, incorporating tree planting compost at 1 m³ per 3 m³ of soil, one tree stake and two ties; tree pits square in sizes shown		
Light standard bare root tree in pit; PC £8.40		
600 x 600 deep	each	**40.00**
900 x 900 deep	each	**52.00**
Standard tree bare root tree in pit; PC £10.75		
600 x 600 deep	each	**42.73**
900 x 600 deep	each	**55.00**
Standard root balled tree in pit; PC £25.75		
600 x 600 deep	each	**60.00**
900 x 600 deep	each	**76.00**
Selected standard bare root tree, in pit; PC £16.90		
900 x 900 deep	each	**69.00**
1.00 m x 1.00 m x 600 mm deep	each	**97.00**
Selected standard root ball tree in pit; PC £31.90		
900 x 600 deep	each	**84.00**
1.00 m x 1.00 m x 600 mm deep	each	**112.00**
Heavy standard bare root tree, in pit; PC £28.00		
900 x 900 deep	each	**86.00**
1.00 m x 1.00 m x 600 mm deep	each	**94.00**
Heavy standard root ball tree, in pit; PC £41.50		
900 x 600 deep	each	**99.00**
1.00 m x 1.00 m x 600 mm deep	each	**127.00**
Extra heavy standard bare root tree in pit; PC £33.50		
1.00 m x 1.00 deep	each	**122.00**
Extra heavy standard root ball tree in pit; PC £48.50		
1.00 m x 1.00 deep	each	**145.00**
1.50 m x 750 mm deep	each	**174.00**
SHRUB PLANTING		
Treat recently filled ground with systemic weedkiller; Cultivate ground and clear arisings; add mushroom composts and fertilizers;		
cultivation by rotavator; all other works by hand		
compost 50 mm; general purpose fertilizer 35 g/m^2	m^2	**3.91**
compost 100 mm; general purpose fertilizer 35 g/m^2	m^2	**6.73**
compost 100 mm; Enmag 35 g/m^2	m^2	**6.77**
all works by hand		
compacted ground; compost 50 mm; general purpose fertilizer 35 g/m^2	m^2	**4.77**
ground previously planted but cleared of vegetable matter; compost 50 mm; general purpose fertilizer 35 g/m^2	m^2	**4.60**
ground previously planted but cleared of vegetable matter; compost 50 mm; general purpose fertilizer 35 g/m^2	m^2	**4.47**
Excavate planting holes on 300 x 300 mm x 300 deep to area previously prepared; Plant shrubs PC £2.70 each in groups of 3 to 5 inclusive of transport from holding area setting out and final mulching 50 thick		
by hand		
300 mm centres (11.11 plants per m^2)	m^2	**63.00**
400 mm centres (6.26 plants per m^2)	m^2	**40.00**
500 mm centres (4 plants per m^2)	m^2	**25.00**
750 mm centres (1.78 plants per m^2)	m^2	**13.00**

PLANTING

Item Excluding site overheads and profit	Unit	Total rate £
Cultivate and grade shrub bed; bring top 300 mm of topsoil to a fine tilth, incorporating mushroom compost at 50 mm and Enmag slow release fertilizer; rake and bring to given levels; remove all stones and debris over 50 mm; dig planting holes average 300 x 300x 300 mm deep; supply and plant specified shrubs in quantities as shown below; backfill with excavated material as above; water to field capacity and mulch 50 mm bark chips 20 - 40 mm size;		
Shrubs - 3L PC £2.80; ground covers - 9 cm PC £1.50		
Groundcover 30% /shrubs 70% at the distances shown below		
200mm/300 mm	m^2	**61.68**
300 mm/400 mm	m^2	**33.54**
300 mm/500 mm	m^2	**23.50**
Groundcover 50%/shrubs 50% at the distances shown below		
200mm/300 mm	m^2	**65.30**
300 mm/400 mm	m^2	**33.73**
300 mm/500 mm	m^2	**28.56**
400 mm/500 mm	m^2	**23.08**

DRAINAGE

Item Excluding site overheads and profit	Unit	**Total rate £**
DRAINAGE		
Pipe Laying		
Excavate trench by excavator 600 deep; lay bedding and backfill as per material specification below; lay non woven geofabric and fill with topsoil to ground level		
100 mm vitrified clay; laid on earth with excavated backfill	m	**15.00**
110 PVC-U drainpipe; laid on sand bed with gravel backfill	m	**18.00**
Excavate trench by hand 600 deep; lay bedding and backfill as per material specification below; lay non woven geofabric and fill with topsoil to ground level		
100 mm vitrified clay; laid on earth with excavated backfill	m	**20.00**
110 PVC-U drainpipe; laid on sand bed with gravel backfill	m	**22.00**
LINEAR DRAINAGE		
Linear drainage to design sensitive areas		
Excavate trench by machine; Lay Aco Brickslot channel drain on concrete base and surround to falls; all to manufacturers specifications;		
paving surround to both sides of channel	m	**120.00**
Linear drainage to pedestrian area		
Excavate trench by machine; Lay Aco NK100 channel drain on concrete base and surround to falls; all to manufacturers specifications; Paving surround to channel with brick paving PC £300.00/1000		
slotted galvanised grating; paving surround to one side of channel	m	**150.00**
slotted galvanised grating; paving surround to both sides of channel	m	**160.00**
stainless steel grating; paving surround to one side of channel	m	**220.00**
slotted galvanised grating; paving surround to both sides of channel	m	**230.00**
Linear drainage to light vehicular area		
Excavate trench by machine; Lay Aco NK100 channel drain on concrete base and surround to falls; all to manufacturers specifications; Paving surround to channel with brick paving PC £300.00/1000		
slotted galvanised grating; paving surround to one side of channel	m	**140.00**
slotted galvanised grating; paving surround to both sides of channel	m	**149.00**
"Heelguard" composite grating; paving surround to one side of channel	m	**139.00**
"Heelguard" ductile grating; paving surround to both sides of channel	m	**141.00**
Accessories for Channel drain		
Sump unit with sediment bucket	nr	**120.00**
End cap; inlet / outlet	nr	**19.00**
MANHOLES		
Inspection Chambers; brick manhole; excavate pit for inspection chamber including earthwork support and disposal of spoil to dump on site not exceeding 100 m; lay concrete (1:2:4) base 1500 dia. x 200 thick; 110 vitrified clay channels; benching in concrete (1:3:6) allowing one outlet and two inlets for 110 dia pipe; construct inspection chamber 1 brick thick walls of engineering brick Class B; backfill with excavated material; complete with 2 no. cast iron step irons; cover 600 x 450 mm ductile iron; light vehicle loading		
1200 x 1200 x 1200 mm		
excavation by machine	each	**975.00**
excavation by hand	each	**1970.00**
1200 x 1200 x 1200 by machine but with recessed cover		
600 x 450 mm; 5 tonne loading; filled with block pavers	each	**1160.00**
1200 x 1200 x 1500 mm		
excavation by machine	each	**1140.00**
excavation by hand	each	**2270.00**

DRAINAGE

Item Excluding site overheads and profit	Unit	**Total rate £**
Inspection Chambers; Polypropylene; excavate pit for inspection chamber including earthwork support and disposal of spoil to dump on site not exceeding 100 m; lay concrete (1:2:4) base 700 dia. x 200 thick;		
Polypropylene 600 mm deep		
excavation by machine	each	**408.00**
excavation by hand	each	**370.00**
Polypropylene 1200 mm		
excavation by machine	each	**392.00**
excavation by hand	each	**460.00**
GULLIES		
Clay Gully		
Excavate hole; supply and set in concrete vitrified clay trapped mud (dirt) gully complete with galvanized bucket and cast iron hinged locking grate and frame; lay kerb to gully		
1 tonne loading; RGP5; 100 mm outlet; 100 mm dia; 225 mm internal width 585 mm internal depth	each	**311.00**
Gullies PVC-u		
Excavate hole and lay 100 concrete (C20P) base 150 x 150 to suit given invert level of drain; connect to drainage system; backfill with Type 1 granular fill; install gully; complete with cast grate and frame; brick kerb to gully surround		
PVC-u gully 100 mm with P trap; grid and frame included	each	**150.00**
PVC-u gully 100 mm with P trap; grid and frame included	each	**200.00**
Yard Gully trapped; 300 diameter 600 deep; sediment bucket and ductile iron cover	each	**320.00**
SOAKAWAYS		
Soakaway "Aquacell" Wavin Plastics Ltd; preformed polypropylene soakaway infiltration crate units		
Excavate pits or trenches for soakaway units; place Aquacell polypropylene soakaway units in recommended configurations laid on 100 mm shingle; cover with terram and backfill to sides and top with 100 mm shingle; backfill with 150 mm topsoil		
4 crates; 760 litres	nr	**290.00**
8 crates; 1520 litres	nr	**518.00**
12 crates; 2280 litres	nr	**862.00**
16 crates; 3050 litres	nr	**1040.00**

DAVIS LANGDON
Constructing the best and most valued relationships in the industry

www.davislangdon.com

Offices in Europe & the Middle East, Africa, Asia Pacific, Australasia and the USA

Project Management | Cost Management | Management Consulting | Legal Support
Specification Consulting | Engineering Services | Property Tax & Finance

ESSENTIAL READING FROM TAYLOR AND FRANCIS

Spon's First Stage Estimating Handbook
2nd Edition

Bryan Spain

The ultimate paradox of construction project management is that the earlier a decision has to be made, the more difficult the task is of estimating the cost implications. Yet the most important decisions are taken earliest - when the cost estimates are the most difficult to obtain.

Spon's First Stage Estimating Handbook is the only comprehensive and reliable source of first stage estimating costs. Covering the whole spectrum of building, civil engineering, mechanical and electrical work, and landscaping work, vital cost data is presented as costs per square metre, elemental cost analyses, principal rates and composite rates. Compact and clear, Spon's First Stage Estimating Handbook is ideal for meetings with clients when the inevitable question 'just an idea ... a ballpark figure ...' needs an accurate response.

With additional sections on whole life costing and general information, this is an essential reference for all construction professionals and clients making early judgements on the viability of new projects.

April 2006: 216x138 mm: 240 pages
PB: 0-415-38619-5: £42.00

To Order: Tel: +44 (0) 1264 343071 Fax: +44 (0) 1264 343005, or
Post: Taylor and Francis Customer Services, Thomson Publishing Services, Cheriton House, Andover, Hants, SP10 5BE, UK Email: book.orders@tandf.co.uk

For a complete listing of all our titles visit:
www.tandf.co.uk

Taylor & Francis
Taylor & Francis Group

PART 5

Prices for Measured Works - Minor Works

INTRODUCTION

Typical Project Profile

Contract value	£10 000.00 - £70 000.00
Labour rate (see page 4)	£21.00 per hour
Labour rate for maintenance contracts	-
Number of site staff	6 - 9
Project area	1200 m^2
Project location	Outer London
Project components	20% hard landscape 80% soft landscape and planting
Access to works areas	Very good
Contract	Main contract
Delivery of materials	Part loads
Profit and site overheads	Excluded

ESSENTIAL READING FROM TAYLOR AND FRANCIS

4th Edition

Rock Slope Engineering

Civil and Mining

Duncan C. Wyllie and Christopher W. Mah

The stability of rock slopes is an important issue in both civil and mining engineering. On civil projects, rock cuts must be safe from rock falls and large-scale slope instability during both construction and operation. In open pit mining, where slope heights can be many hundreds of meters, the economics of the operation are closely related to the steepest stable slope angle that can be mined.

This extensively updated version of the classic text, *Rock Slope Engineering* by Hoek and Bray, deals comprehensively with the investigation, design and operation of rock slopes. New material contained in this book includes the latest developments in earthquake engineering related to slope stability, probabilistic analysis, numerical analysis, blasting, slope movement monitoring and stabilization methods.

***Rock Slopes Engineering: Civil and Mining* contains both worked examples illustrating data interpretation and design methods, and chapters on civil and mining case studies. It also includes an introduction by Dr. Evert Hoek.**

August 2004: 246x174 mm: 480 pages
242 line drawings, 32 b+w photos and 42 tables
HB: 0-415-28000-1: £112.00
PB: 0-415-28001-X: £40.00

To Order: Tel: +44 (0) 1264 343071 Fax: +44 (0) 1264 343005, or
Post: Taylor and Francis Customer Services, Thomson Publishing Services, Cheriton House, Andover, Hants, SP10 5BE, UK Email: book.orders@tandf.co.uk

For a complete listing of all our titles visit:
www.sponpress.com

Taylor & Francis
Taylor & Francis Group

D GROUNDWORK

Item Excluding site overheads and profit	PC £	Labour hours	Labour £	Plant £	Material £	Unit	Total rate £
D11 SOIL STABILIZATION							
Timber log retaining walls; Longlyf Timber Products Ltd							
Machine rounded softwood logs to trenches priced separately; disposal of excavated material priced separately; inclusive of 75 mm hardcore blinding to trench and backfilling trench with site mixed concrete 1:3:6; geofabric pinned to rear of logs; heights of logs above ground							
500 mm (constructed from 1.80 m lengths)	19.70	1.50	31.50	-	30.59	m	**62.09**
1.20 m (constructed from 1.80 m lengths)	39.40	1.30	27.30	-	58.62	m	**85.92**
1.60 m (constructed from 2.40 m lengths)	51.30	2.50	52.50	-	76.78	m	**129.28**
2.00 m (constructed from 3.00 m lengths)	72.60	3.50	73.50	-	104.37	m	**177.87**
As above but with 150 mm machine rounded timbers							
500 mm	28.52	2.50	52.50	-	39.41	m	**91.91**
1.20 m	94.97	1.75	36.75	-	114.19	m	**150.94**
1.60 m	95.07	3.00	63.00	-	120.58	m	**183.58**
As above but with 200 mm machine rounded timbers							
1.80 m (constructed from 2.40 m lengths)	128.60	4.00	84.00	-	160.42	m	**244.42**
2.40 m (constructed from 3.60 m lengths)	192.90	4.50	94.50	-	224.91	m	**319.41**
Railway sleeper walls; Sleeper Supplies Ltd							
Construct retaining wall from railway sleepers; fixed with steel galvanised pins 12 mm driven into the ground; sleepers laid flat							
Grade 1 softwood; 2590 x 250 x 150 mm							
150 mm; 1 sleeper high	8.73	0.50	10.50	-	13.73	m	**24.23**
300 mm; 2 sleepers high	17.46	1.00	21.00	-	18.46	m	**39.46**
450 mm; 3 sleepers high	25.99	1.50	31.50	-	27.26	m	**58.76**
600 mm; 4 sleepers high	35.01	2.00	42.00	-	36.28	m	**78.28**
Grade 1 softwood as above but with 2 nr galvanised angle iron stakes set into concrete internally and screwed to the inside face of the sleepers							
750 mm; 5 sleepers high	43.31	2.50	52.50	-	84.30	m^2	**136.80**
900 mm; 6 sleepers high	51.84	2.75	57.75	-	95.03	m^2	**152.78**
Grade 1 hardwood; 2590 x 250 x 150 mm							
150 mm; 1 sleeper high	7.36	0.50	10.50	-	7.84	m	**18.34**
300 mm; 2 sleepers high	14.72	1.00	21.00	-	15.68	m	**36.68**
450 mm; 3 sleepers high	21.90	1.50	31.50	-	23.17	m	**54.67**
600 mm; 4 sleepers high	29.50	2.00	42.00	-	30.77	m	**72.77**
Grade 1 hardwood as above but with 2 nr galvanised angle iron stakes set into concrete internally and screwed to the inside face of the sleepers							
750 mm; 5 sleepers high	36.50	2.50	52.50	-	77.48	m^2	**129.98**
900 mm; 6 sleepers high	43.68	2.75	57.75	-	86.88	m^2	**144.63**
New pine softwood; 2500 x 245 x 120 mm							
120 mm; 1 sleeper high	9.08	0.50	10.50	-	9.57	m	**20.07**
240 mm; 2 sleepers high	18.17	1.00	21.00	-	19.14	m	**40.14**
360 mm; 3 sleepers high	27.25	1.50	31.50	-	28.52	m	**60.02**
480 mm; 4 sleepers high	36.34	2.00	42.00	-	37.61	m	**79.61**
New pine softwood as above but with 2 nr galvanised angle iron stakes set into concrete internally and screwed to the inside face of the sleepers							
600 mm; 5 sleepers high	45.42	2.50	52.50	-	86.41	m^2	**138.91**
720 mm; 6 sleepers high	54.50	2.75	57.75	-	97.70	m^2	**155.45**

D GROUNDWORK

Item Excluding site overheads and profit	PC £	Labour hours	Labour £	Plant £	Material £	Unit	Total rate £
D11 SOIL STABILIZATION - cont'd							
Railway sleeper walls - cont'd							
New oak hardwood; 2500 x 200 x 130 mm							
130 mm; 1 sleeper high	10.08	0.50	10.50	-	10.57	m	**21.07**
260 mm; 2 sleepers high	20.17	1.00	21.00	-	21.14	m	**42.14**
390 mm; 3 sleepers high	30.25	1.50	31.50	-	31.52	m	**63.02**
520 mm; 4 sleepers high	40.34	2.00	42.00	-	41.61	m	**83.61**
New oak hardwood as above but with 2 nr galvanised angle iron stakes set into concrete internally and screwed to the inside face of the sleepers							
640 mm; 5 sleepers high	50.42	2.50	52.50	-	91.41	m^2	**143.91**
760 mm; 6 sleepers high	60.50	2.75	57.75	-	103.70	m^2	**161.45**
Excavate foundation trench; set railway sleepers vertically on end in concrete 1:3:6 continuous foundation to 33.3% of their length to form retaining wall							
Grade 1 hardwood; finished height above ground level							
300 mm	13.05	3.00	63.00	2.40	24.79	m	**90.19**
500 mm	18.91	3.00	63.00	2.40	33.58	m	**98.98**
600 mm	26.28	3.00	63.00	2.40	49.41	m	**114.81**
750 mm	28.36	3.50	73.50	2.40	52.53	m	**128.43**
1.00 m	38.77	3.75	78.75	2.40	59.39	m	**140.54**
Excavate and place vertical steel universal beams 165 wide in concrete base at 2.590 centres; fix railway sleepers set horizontally between beams to form horizontal fence or retaining wall							
Grade 1 hardwood; bay length 2.590 m							
0.50 m high; (2 sleepers)	15.03	2.50	52.50	-	56.31	bay	**108.81**
750 m high; (3 sleepers)	22.09	2.60	54.60	-	83.63	bay	**138.23**
1.00 m high; (4 sleepers)	29.60	3.00	63.00	-	107.84	bay	**170.84**
1.25 m high; (5 sleepers)	37.80	3.50	73.50	-	139.09	bay	**212.59**
1.50 m high; (6 sleepers)	44.40	3.00	63.00	-	165.56	bay	**228.56**
1.75 m high; (7 sleepers)	51.69	3.00	63.00	-	172.85	bay	**235.85**
Retaining walls; Maccaferri Ltd							
Wire mesh gabions; galvanized mesh 80 mm x 100 mm; filling with broken stones 125 mm - 200 mm size; wire down securely to manufacturer's instructions; filling front face by hand							
2 x 1 x 0.50 m	13.65	2.00	42.00	6.70	84.87	nr	**133.57**
2 x 1 x 1.00 m	22.57	4.00	84.00	13.41	165.01	nr	**262.42**
PVC coated gabions							
2 x 1 x 0.5 m	22.00	2.00	42.00	6.70	93.22	nr	**141.92**
2 x 1 x 1.00 m	31.00	4.00	84.00	13.41	173.44	nr	**270.85**

D GROUNDWORK

Item Excluding site overheads and profit	PC £	Labour hours	Labour £	Plant £	Material £	Unit	Total rate £
D20 EXCAVATING AND FILLING							
MACHINE SELECTION TABLE							
Road Equipment Ltd; machine volumes for excavating/filling only and placing excavated material alongside or to a dumper; no bulkages are allowed for in the material volumes; these rates should be increased by user-preferred percentages to suit prevailing site conditions; the figures in the next section for "Excavation, mechanical" and filling allow for the use of banksmen within the rates shown below							
1.5 tonne excavators; digging volume							
1 cycle/minute; 0.04 m^3	-	0.42	8.75	2.54	-	m^3	**11.29**
2 cycles/minute; 0.08 m^3	-	0.21	4.37	1.50	-	m^3	**5.87**
3 cycles/minute; 0.12 m^3	-	0.14	2.92	1.16	-	m^3	**4.08**
3 tonne excavators; digging volume							
1 cycle/minute; 0.13 m^3	-	0.13	2.69	1.56	-	m^3	**4.25**
2 cycles/minute; 0.26 m^3	-	0.06	1.35	0.97	-	m^3	**2.32**
3 cycles/minute; 0.39 m^3	-	0.04	0.90	0.87	-	m^3	**1.76**
5 tonne excavators; digging volume							
1 cycle/minute; 0.28 m^3	-	0.06	1.25	0.91	-	m^3	**2.16**
2 cycles/minute; 0.56 m^3	-	0.03	0.63	0.49	-	m^3	**1.12**
3 cycles/minute; 0.84 m^3	-	0.02	0.42	0.42	-	m^3	**0.83**
Dumpers; Road Equipment Ltd							
1 tonne high tip skip loader; volume 0.485 m^3 (775 kg)							
5 loads per hour	-	0.41	8.66	1.64	-	m^3	**10.30**
7 loads per hour	-	0.29	6.19	1.21	-	m^3	**7.39**
10 loads per hour	-	0.21	4.33	0.88	-	m^3	**5.21**
3 tonne dumper; max. volume 2.40 m^3 (3.38 t); available volume 1.9 m^3							
4 loads per hour	-	0.14	2.92	0.63	-	m^3	**3.55**
5 loads per hour	-	0.11	2.33	0.51	-	m^3	**2.85**
7 loads per hour	-	0.08	1.67	0.38	-	m^3	**2.04**
10 loads per hour	-	0.06	1.17	0.28	-	m^3	**1.44**
Lifting turf for preservation							
hand lift and stack	-	0.08	1.75	-	-	m^2	**1.75**
Load to wheelbarrows and dispose offsite by skip							
25 m distance	-	0.02	0.44	2.85	-	m^2	**3.29**
Note: The figures in this section relate to the machine capacities shown in the Major Works section of this book. The figures below allow for dig efficiency based on depth, bulkages of 25% on loamy soils (adjustments should be made for different soil types) and for a banksman.							
Excavating; mechanical; topsoil for preservation; depositing alongside							
3 tonne tracked excavator (bucket volume 0.13 m^3)							
average depth 100 mm	-	0.02	0.50	0.78	-	m^2	**1.29**
average depth 150 mm	-	0.03	0.71	1.10	-	m^2	**1.80**
average depth 200 mm	-	0.04	0.84	1.30	-	m^2	**2.14**
average depth 250 mm	-	0.04	0.92	1.43	-	m^2	**2.36**
average depth 300 mm	-	0.05	1.01	1.56	-	m^2	**2.57**

D GROUNDWORK

Item Excluding site overheads and profit	PC £	Labour hours	Labour £	Plant £	Material £	Unit	Total rate £
D20 EXCAVATING AND FILLING - cont'd							
Excavating; mechanical; topsoil for preservation; depositing to spoil heaps by wheeled dumper; spoil heaps maximum 50 m distance							
3 tonne tracked excavator (bucket volume 0.13 m^3)							
average depth 100 mm	-	0.02	0.50	1.23	-	m^2	**1.73**
average depth 150 mm	-	0.03	0.71	1.76	-	m^2	**2.47**
average depth 200 mm	-	0.04	0.84	2.19	-	m^2	**3.03**
average depth 250 mm	-	0.04	0.92	2.54	-	m^2	**3.47**
average depth 300 mm	-	0.05	1.01	2.89	-	m^2	**3.90**
Excavating; mechanical; to reduce levels							
3 tonne excavator (bucket volume 0.13 m^3)							
maximum depth not exceeding 0.25 m	-	0.13	2.69	5.17	-	m^3	**7.87**
maximum depth not exceeding 1.00 m	-	0.13	2.69	4.91	-	m^3	**7.61**
Pits; 3 tonne tracked excavator							
maximum depth not exceeding 0.25 m	-	0.58	12.18	1.03	-	m^3	**13.21**
maximum depth not exceeding 1.00 m	-	0.50	10.50	5.78	-	m^3	**16.27**
maximum depth not exceeding 2.00 m	-	0.60	12.60	6.93	-	m^3	**19.53**
Trenches; width not exceeding 0.30 m; 3 tonne excavator							
maximum depth not exceeding 0.25 m	-	1.33	28.00	6.40	-	m^3	**34.40**
maximum depth not exceeding 1.00 m	-	0.69	14.38	3.29	-	m^3	**17.67**
maximum depth not exceeding 2.00 m	-	0.60	12.60	2.88	-	m^3	**15.48**
Trenches; width exceeding 0.30 m; 3 tonne excavator							
maximum depth not exceeding 0.25 m	-	0.60	12.60	2.88	-	m^3	**15.48**
maximum depth not exceeding 1.00 m	-	0.50	10.50	2.40	-	m^3	**12.90**
maximum depth not exceeding 2.00 m	-	0.38	8.08	1.85	-	m^3	**9.92**
Extra over any types of excavating irrespective of depth for breaking out existing materials; heavy duty 110 volt breaker tool							
hard rock	-	5.00	105.00	37.08	-	m^3	**142.08**
concrete	-	3.00	63.00	22.25	-	m^3	**85.25**
reinforced concrete	-	4.00	84.00	35.29	-	m^3	**119.29**
brickwork, blockwork or stonework	-	1.50	31.50	11.13	-	m^3	**42.63**
Extra over any types of excavating irrespective of depth for breaking out existing hard pavings; 1600 watt, 110 volt breaker							
concrete; 100 mm thick	-	0.20	4.20	0.83	-	m^2	**5.03**
concrete; 150 mm thick	-	0.25	5.25	1.04	-	m^2	**6.29**
concrete; 200 mm thick	-	0.29	6.00	1.19	-	m^2	**7.19**
concrete; 300 mm thick	-	0.40	8.40	1.67	-	m^2	**10.07**
reinforced concrete; 100 mm thick	-	0.33	7.00	4.20	-	m^2	**11.20**
reinforced concrete; 150 mm thick	-	0.40	8.40	5.04	-	m^2	**13.44**
reinforced concrete; 200 mm thick	-	0.50	10.50	6.30	-	m^2	**16.80**
reinforced concrete; 300 mm thick	-	1.00	21.00	12.60	-	m^2	**33.60**
tarmacadam; 75 mm thick	-	0.14	3.00	0.60	-	m^2	**3.60**
tarmacadam and hardcore; 150 mm thick	-	0.20	4.20	0.83	-	m^2	**5.03**
Extra over any types of excavating irrespective of depth for taking up							
precast concrete paving slabs	-	0.08	1.75	0.35	-	m^2	**2.10**
natural stone paving	-	0.13	2.63	0.52	-	m^2	**3.15**
cobbles	-	0.17	3.50	0.69	-	m^2	**4.20**
brick paviors	-	0.17	3.50	0.69	-	m^2	**4.20**
Excavating; hand							
Topsoil for preservation; loading to barrows							
average depth 100 mm	-	0.30	6.30	-	-	m^2	**6.30**
average depth 200 mm	-	0.60	12.60	-	-	m^2	**12.60**
average depth 300 mm	-	0.90	18.90	-	-	m^2	**18.90**
over 300 mm	-	3.00	63.00	-	-	m^3	**63.00**

D GROUNDWORK

Item Excluding site overheads and profit	PC £	Labour hours	Labour £	Plant £	Material £	Unit	Total rate £
Excavating; hand							
Topsoil to reduce levels							
maximum depth not exceeding 0.25 m	-	2.52	52.92	-	-	m^3	**52.92**
maximum depth not exceeding 1.00 m	-	3.12	65.52	-	-	m^3	**65.52**
Pits							
maximum depth not exceeding 0.25 m	-	2.67	56.00	-	-	m^3	**56.00**
maximum depth not exceeding 1.00 m	-	3.47	72.80	-	-	m^3	**72.80**
maximum depth not exceeding 2.00 m (includes earthwork support)	-	6.93	145.60	-	-	m^3	**200.67**
Trenches; width not exceeding 0.30 m							
maximum depth not exceeding 0.25 m	-	2.86	60.00	-	-	m^3	**60.00**
maximum depth not exceeding 1.00 m	-	3.72	78.14	-	-	m^3	**78.14**
maximum depth not exceeding 1.00 m	-	3.72	78.14	-	-	m^3	**105.67**
Trenches; width exceeding 0.30 m wide							
maximum depth not exceeding 0.25 m	-	2.86	60.00	-	-	m^3	**60.00**
maximum depth not exceeding 1.00 m	-	4.00	84.00	-	-	m^3	**84.00**
maximum depth not exceeding 2.00 m (includes earthwork support)	-	6.00	126.00	-	-	m^3	**181.07**
Filling to make up levels; mechanical (3 tonne tracked excavator); depositing in layers 150 mm maximum thickness							
Arising from the excavations							
maximum thickness less than 0.25 m	-	0.10	2.18	3.39	-	m^3	**5.57**
average thickness exceeding 0.25 m thick	-	0.08	1.68	2.61	-	m^3	**4.29**
Obtained from on site spoil heaps; average 25 m distance; multiple handling							
maximum thickness less than 0.25 m	-	0.16	3.37	6.09	-	m^3	**9.46**
average thickness exceeding 0.25 m thick	-	0.13	2.73	4.90	-	m^3	**7.63**
Obtained off site; planting quality topsoil PC £18.50/m^3							
maximum thickness less than 0.25 m	18.50	0.14	3.01	5.54	23.13	m^3	**31.67**
average thickness exceeding 0.25 m thick	18.50	0.13	2.73	5.11	23.13	m^3	**30.97**
Obtained off site; crushed concrete hardcore; PC £22.00/tonne							
maximum thickness less than 0.25 m	22.00	0.14	2.94	12.99	26.84	m^3	**42.76**
maximum thickness exceeding 0.25 m	22.00	0.13	2.67	12.47	26.84	m^3	**41.98**
Filling to make up levels; hand; depositing in layers 150 mm maximum thickness							
Arising from the excavations							
average thickness exceeding 0.25 m	-	0.75	15.75	-	-	m^3	**15.75**
Obtained from on site spoil heaps; average 25 m distance; multiple handling							
average thickness exceeding 0.25 m thick	-	1.25	26.25	-	-	m^3	**26.25**
Obtained off site; planting quality topsoil PC £31.90/m^3 (10 tonne loads)							
average thickness exceeding 0.25 m thick	52.63	1.25	26.25	-	65.79	m^3	**92.04**
Disposal; mechanical							
Light soils and loams (bulking factor - 1.25)							
Excavated material; off site; to tip; mechanically loaded by grab; capacity of load 7.25 m^3							
inert material	-	-	-	-	-	m^3	**37.95**
soil (sandy and loam) dry	-	-	-	-	-	m^3	**37.95**
soil (sandy and loam) wet	-	-	-	-	-	m^3	**40.25**
broken out compacted materials such as road bases and the like	-	-	-	-	-	m^3	**42.17**
soil (clay) dry	-	-	-	-	-	m^3	**49.26**
soil (clay) wet	-	-	-	-	-	m^3	**51.18**
Disposal by skip; 6 yd^3 (4.6 m^3)							
Excavated material; off site; to tip							
by machine	-	-	-	44.81	-	m^3	**44.81**
by hand	-	1.00	04.00	42.79	-	m^3	**126.79**

D GROUNDWORK

Item Excluding site overheads and profit	PC £	Labour hours	Labour £	Plant £	Material £	Unit	Total rate £
D20 EXCAVATING AND FILLING - cont'd							
Disposal on site							
Excavated material to spoil heaps							
average 25 m distance	-	-	-	5.33	-	m^3	**5.33**
average 50 m distance	-	-	-	5.86	-	m^3	**5.86**
average 100 m distance	-	-	-	6.39	-	m^3	**6.39**
average 200 m distance	-	-	-	7.99	-	m^3	**7.99**
Excavated material; spreading on site							
average 25 m distance	-	-	-	6.71	-	m^3	**6.71**
average 50 m distance	-	-	-	7.38	-	m^3	**7.38**
average 100 m distance	-	-	-	8.48	-	m^3	**8.48**
average 200 m distance	-	-	-	9.36	-	m^3	**9.36**
Disposal; hand							
Excavated material; on site; in spoil heaps							
average 25 m distance	-	2.40	50.40	-	-	m^3	**50.40**
average 50 m distance	-	2.64	55.44	-	-	m^3	**55.44**
average 100 m distance	-	3.00	63.00	-	-	m^3	**63.00**
average 200 m distance	-	3.60	75.60	-	-	m^3	**75.60**
Disposal hand; excavated material in bags							
Filling bags 25 kg and barrowing through buildings							
average 25 m distance	-	8.40	176.40	-	-	m^3	**176.40**
average 50 m distance	-	9.20	193.20	-	-	m^3	**193.20**
average 100 m distance	-	10.10	212.10	-	-	m^3	**212.10**
average 200 m distance	-	11.60	243.60	-	-	m^3	**243.60**
Excavated material; spreading on site							
average 25 m distance	-	2.64	55.44	-	-	m^3	**55.44**
average 50 m distance	-	3.00	63.00	-	-	m^3	**63.00**
average 100 m distance	-	3.60	75.60	-	-	m^3	**75.60**
average 200 m distance	-	4.20	88.20	-	-	m^3	**88.20**
Surface treatments							
Compacting							
bottoms of excavations	-	0.01	0.11	0.02	-	m^2	**0.13**
Grading operations; surface previously excavated to reduce levels to prepare to receive subsequent treatments; grading to accurate levels and falls 20 mm tolerances							
Clay or heavy soils or hardcore							
3 tonne excavator	-	0.02	0.42	0.62	-	m^2	**1.04**
5 tonne excavator	-	0.04	0.84	0.24	-	m^2	**1.08**
by hand	-	0.10	2.10	-	-	m^2	**2.10**
Loamy topsoils							
3 tonne excavator	-	0.01	0.28	0.41	-	m^2	**0.69**
5 tonne excavator	-	0.03	0.56	0.16	-	m^2	**0.72**
by hand	-	0.05	1.05	-	-	m^2	**1.05**
Sand or graded granular materials							
3 tonne excavator	-	0.01	0.21	0.31	-	m^2	**0.52**
5 tonne excavator	-	0.02	0.42	0.12	-	m^2	**0.54**
by hand	-	0.03	0.70	-	-	m^2	**0.70**
Excavating; by hand							
Topsoil for preservation; loading to barrows							
average depth 100 mm	-	0.24	4.44	-	-	m^2	**4.44**
average depth 150 mm	-	0.36	6.66	-	-	m^2	**6.66**
average depth 200 mm	-	0.58	10.66	-	-	m^2	**10.66**
average depth 250 mm	-	0.72	13.32	-	-	m^2	**13.32**
average depth 300 mm	-	0.86	15.98	-	-	m^2	**15.98**

D GROUNDWORK

Item Excluding site overheads and profit	PC £	Labour hours	Labour £	Plant £	Material £	Unit	**Total rate £**
Grading operations; surface recently filled to raise levels to prepare to receive subsequent treatments; grading to accurate levels and falls 20 mm tolerances							
Clay or heavy soils or hardcore							
3 tonne excavator	-	0.02	0.32	0.47	-	m^2	**0.79**
by hand	-	0.08	1.75	-	-	m^2	**1.75**
Loamy topsoils							
3 tonne excavator	-	0.01	0.22	0.32	-	m^2	**0.53**
by hand	-	0.04	0.84	-	-	m^2	**0.84**
Sand or graded granular materials							
3 tonne excavator	-	0.01	0.16	0.23	-	m^2	**0.39**
by hand	-	0.03	0.60	-	-	m^2	**0.60**
Surface preparation							
Trimming surfaces of cultivated ground to final levels, removing roots stones and debris exceeding 50 mm in any direction to tip offsite; slopes less than 15 degrees							
clean ground with minimal stone content	-	0.25	5.25	-	-	100 m^2	**5.25**
slightly stony - 0.5 kg stones per m^2	-	0.33	7.00	0.02	-	100 m^2	**7.02**
very stony - 1.0 - 3.00 kg stones per m^2	-	0.50	10.50	0.03	-	100 m^2	**10.53**
clearing mixed slightly contaminated rubble inclusive of roots and vegetation	-	0.50	10.50	8.56	-	100 m^2	**19.06**
clearing brick-bats stones and clean rubble	-	0.60	12.60	0.15	-	100 m^2	**12.75**
Hand cultivation; preparation for turfing/seeding areas							
Cultivate surface to receive landscape surface treatments using hand fork or implement; 1 spit (300 mm deep), break up ground and bring to a fine tilth by rake							
Compacted ground	-	0.07	1.40	-	-	m^2	**1.40**
Ground previously supporting turf; lifting of turf not included	-	0.06	1.31	-	-	m^2	**1.31**
Ground previously planted with small to medium shrubs; removal of planting not included	-	0.06	1.17	-	-	m^2	**1.17**
Soft ground	-	0.04	0.84	-	-	m^2	**0.84**
Hand cultivation; preparation for planting							
Cultivate surface to receive landscape surface treatments using hand fork or implement; 1 spit (300 mm deep), break up ground and bring to a fine tilth by rake							
Compacted ground	-	0.05	1.05	-	-	m^2	**1.05**
Ground previously supporting turf; lifting of turf not included	-	0.04	0.88	-	-	m^2	**0.88**
Ground previously planted with small to medium shrubs; removal of planting not included	-	0.04	0.75	-	-	m^2	**0.75**
Soft ground	-	0.03	0.60	-	-	m^2	**0.60**

E IN SITU CONCRETE/LARGE PRECAST CONCRETE

Item Excluding site overheads and profit	PC £	Labour hours	Labour £	Plant £	Material £	Unit	Total rate £
E10 MIXING/CASTING/CURING IN SITU CONCRETE							
General							
Please see the notes on concrete mix designations in the Major Works section of this book under E10 In Situ Concrete.							
Concrete mixes; mixed on site; costs for producing concrete; prices for commonly used mixes for various types of work; based on an 85 litre concrete mixer							
Roughest type mass concrete such as footings, road haunchings 300 mm thick; aggregates delivered in 10 tonne loads							
1:3:6	37.80	1.51	31.78	-	64.46	m^3	**96.24**
1:3:6 sulphate resisting	100.58	1.51	31.78	-	100.58	m^3	**132.36**
As above but aggregates delivered in 900 kg bulk bags							
1:3:6	26.66	1.51	31.78	-	94.66	m^3	**126.44**
1:3:6 sulphate resisting	62.78	1.51	31.78	-	130.78	m^3	**162.56**
Most ordinary use of concrete such as mass walls above ground, road slabs etc. and general reinforced concrete work; aggregates delivered in 10 tonne loads							
1:2:4	75.50	1.51	31.78	-	75.50	m^3	**107.27**
1:2:4 sulphate resisting	126.57	1.51	31.78	-	126.57	m^3	**158.35**
As above but aggregates delivered in 900 kg bulk bags							
1:2:4	37.70	1.51	31.78	-	105.70	m^3	**137.47**
1:2:4 sulphate resisting	88.77	1.51	31.78	-	156.77	m^3	**188.55**
Watertight floors, pavements and walls, tanks, pits, steps, paths, surface of two course roads; reinforced concrete where extra strength is required (aggregates delivered in 10 tonne loads)							
1:1.5:3	86.04	1.51	31.78	-	86.04	m^3	**117.81**
As above but aggregates delivered in 900 kg bulk bags							
1:1.5:3	48.24	1.51	31.78	-	116.24	m^3	**148.01**
Plain in situ concrete; site mixed; 10 N/mm² - 40 aggregate (1:3:6) (aggregate delivery indicated)							
Foundations							
ordinary Portland cement; 10 tonne ballast loads	96.24	1.00	21.00	-	96.24	m^3	**117.24**
ordinary Portland cement; 900 kg bulk bags	126.04	1.00	21.00	-	126.04	m^3	**147.04**
sulphate resistant cement; 10 tonne ballast loads	132.36	1.00	21.00	-	132.36	m^3	**153.36**
sulphate resistant cement; 900 kg bulk bags	162.56	1.00	21.00	-	162.56	m^3	**183.56**
Foundations; poured on or against earth or unblinded hardcore							
ordinary Portland cement; 10 tonne ballast loads	96.24	1.00	21.00	-	101.05	m^3	**122.05**
ordinary Portland cement; 900 kg bulk bags	126.04	1.00	21.00	-	132.34	m^3	**153.34**
sulphate resistant cement; 10 tonne ballast loads	132.36	1.00	21.00	-	138.98	m^3	**159.98**
sulphate resistant cement; 900 kg bulk bags	162.56	1.00	21.00	-	170.69	m^3	**191.69**
Isolated foundations							
ordinary Portland cement; 10 tonne ballast loads	96.24	2.00	42.00	-	98.65	m^3	**140.65**
ordinary Portland cement; 900 kg bulk bags	126.04	2.00	42.00	-	129.19	m^3	**171.19**
sulphate resistant cement; 10 tonne ballast loads	132.36	2.00	42.00	-	135.67	m^3	**177.67**
sulphate resistant cement; 900 kg bulk bags	162.56	2.00	42.00	-	166.62	m^3	**208.62**

E IN SITU CONCRETE/LARGE PRECAST CONCRETE

Item Excluding site overheads and profit	PC £	Labour hours	Labour £	Plant £	Material £	Unit	Total rate £
Plain in situ concrete; site mixed; 21 N/mm² - 20 aggregate (1:2:4)							
Foundations							
ordinary Portland cement; 10 tonne ballast loads	107.27	1.00	21.00	-	109.95	m³	**130.95**
ordinary Portland cement; 900 kg bulk bags	-	1.00	21.00	-	137.83	m³	**158.83**
sulphate resistant cement; 10 tonne ballast loads	158.35	1.00	21.00	-	162.31	m³	**183.31**
sulphate resistant cement; 900 kg bulk bags	188.55	1.00	21.00	-	193.26	m³	**214.26**
Foundations; poured on or against earth or unblinded hardcore							
ordinary Portland cement; 10 tonne ballast loads	107.27	1.00	21.00	-	112.63	m³	**133.63**
ordinary Portland cement; 900 kg bulk bags	107.27	1.50	31.50	-	112.63	m³	**144.13**
sulphate resistant cement; 10 tonne ballast loads	158.35	1.10	23.10	-	166.27	m³	**189.37**
sulphate resistant cement; 900 kg bulk bags	188.55	1.50	31.50	-	197.98	m³	**229.48**
Isolated foundations							
ordinary Portland cement	158.35	2.00	42.00	-	158.35	m³	**200.35**
sulphate resistant cement	158.35	2.00	42.00	-	162.31	m³	**204.31**
Reinforced in situ concrete; site mixed; 21 N/mm² - 20 aggregate (1:2:4); aggregates delivered in 10 tonne loads							
Foundations							
ordinary Portland cement	107.27	1.10	23.10	-	109.95	m³	**133.05**
sulphate resistant cement	158.35	1.10	23.10	-	162.31	m³	**185.41**
Foundations; poured on or against earth or unblinded hardcore							
ordinary Portland cement	158.35	1.10	23.10	-	166.27	m³	**189.37**
sulphate resistant cement	158.35	1.10	23.10	-	166.27	m³	**189.37**
Isolated foundations							
ordinary Portland cement	107.27	2.20	46.20	-	109.95	m³	**156.15**
sulphate resistant cement	158.35	2.20	46.20	-	166.27	m³	**212.47**
Plain in situ concrete; ready mixed; Tarmac Southern, 10 N/mm mixes; suitable for mass concrete fill and blinding							
Foundations							
GEN1; Designated mix	77.24	1.50	31.50	-	79.17	m³	**110.67**
ST2; Standard mix	83.14	1.50	31.50	-	85.22	m³	**116.72**
Foundations; poured on or against earth or unblinded hardcore							
GEN1; Designated mix	77.24	1.57	33.08	-	81.10	m³	**114.18**
ST2; Standard mix	83.14	1.57	33.08	-	87.30	m³	**120.37**
Isolated foundations							
GEN1; Designated mix	77.24	2.00	42.00	-	79.17	m³	**121.17**
ST2; Standard mix	83.14	2.00	42.00	-	85.22	m³	**127.22**
Plain in situ concrete; ready mixed; Tarmac Southern, 15 N/mm mixes; suitable for oversite below suspended slabs and strip footings in non aggressive soils							
Foundations							
GEN 2; Designated mix	80.50	1.50	31.50	-	82.51	m³	**114.01**
ST3; Standard mix	75.90	1.50	31.50	-	79.69	m³	**111.19**
Foundations; poured on or against earth or unblinded hardcore							
GEN2; Designated mix	80.50	1.57	33.08	-	82.51	m³	**115.59**
ST3; Standard mix	83.35	1.57	33.08	-	85.43	m³	**118.51**
Isolated foundations							
GEN2; Designated mix	80.50	2.00	42.00	-	82.51	m³	**124.51**
ST3; Standard mix	-	2.00	42.00	-	85.43	m³	**127.43**
Ready mix concrete; extra for loads of less than 6 m³; Euromix							
Concrete mixer on lorry; max. standing time 3 m³/hr	-	1.00	21.00	-	79.50	m³	**100.50**

E IN SITU CONCRETE/LARGE PRECAST CONCRETE

Item Excluding site overheads and profit	PC £	Labour hours	Labour £	Plant £	Material £	Unit	Total rate £
E20 FORMWORK FOR IN SITU CONCRETE							
Plain vertical formwork; basic finish							
Sides of foundations							
height exceeding 1.00 m	-	2.00	42.00		4.35	m^2	**46.35**
height not exceeding 250 mm	-	1.00	21.00		1.22	m	**22.22**
height 250 - 500 mm	-	1.00	21.00		2.17	m	**23.17**
height 500 mm - 1.00 m	-	1.50	31.50		4.35	m	**35.85**
Sides of foundations; left in							
height over 1.00 m	-	2.00	42.00		9.10	m^2	**51.10**
height not exceeding 250 mm	-	1.00	21.00		4.72	m	**25.72**
height 250 - 500 mm	-	1.00	21.00		9.18	m	**30.18**
height 500 mm - 1.00 m	-	1.50	31.50		18.36	m	**49.86**
E30 REINFORCEMENT FOR IN SITU CONCRETE							
For reinforcement bar prices please see the Major Works section of this book.							
Reinforcement fabric; BS 4483; lapped; in beds or suspended slabs							
Fabric 3.6 x 2.0 m^2							
ref A142 (2.22 kg/m^2)	2.01	0.22	4.62	-	2.22	m^2	**6.84**
ref A393 (6.16 kg/m^2)	5.39	0.28	5.88	-	5.93	m^2	**11.81**
Fabric 2.4 x 4.8 m^2							
ref A98 (1.54 kg/m^2)	1.32	0.22	4.62	-	1.45	m^2	**6.07**
ref A142 (2.22 kg/m^2)	1.89	0.22	4.62	-	2.08	m^2	**6.70**
ref A193 (3.02 kg/m^2)	3.48	0.22	4.62	-	3.83	m^2	**8.45**
ref A252 (3.95 kg/m^2)	4.51	0.24	5.04	-	4.96	m^2	**10.00**
ref A393 (6.16 kg/m^2)	6.16	0.28	5.88	-	6.78	m^2	**12.66**

F MASONRY

Item Excluding site overheads and profit	PC £	Labour hours	Labour £	Plant £	Material £	Unit	Total rate £
F10 BRICK BLOCK WALLING							
Batching quantities for these mortar mixes may be found in the Memoranda section of this book.							
Mortar mixes; common mixes for various types of work; mortar mixed on site; prices based on builders merchant rates for cement; mechanically mixed							
Aggregates delivered in 900 kg bulk bags							
1:3	-	0.75	15.75	-	141.76	m³	**157.51**
1:4	-	0.75	15.75	-	112.16	m³	**127.91**
1:1:6	-	0.75	15.75	-	116.74	m³	**132.49**
1:1:6 sulphate resisting	-	0.75	15.75	-	162.10	m³	**177.85**
Aggregates delivered in 10 tonne loads							
1:2	-	0.75	15.75	-	132.53	m³	**148.28**
1:3	-	0.75	15.75	-	110.52	m³	**126.27**
1:4	-	0.75	15.75	-	85.90	m³	**101.65**
1:1:6	-	0.75	15.75	-	90.48	m³	**106.23**
1:1:6 sulphate resisting	-	0.75	15.75	-	135.84	m³	**151.59**
Mortar mixes; common mixes for various types of work; mortar mixed on site; prices based on builders merchant rates for cement; hand mixed							
Aggregates delivered in 900 kg bulk bags							
1:3	-	2.00	42.00	-	143.63	m³	**185.63**
1:4	-	2.00	42.00	-	113.72	m³	**155.72**
1:1:6	-	2.00	42.00	-	118.30	m³	**160.30**
1:1:6 sulphate resisting	-	2.00	42.00	-	163.66	m³	**205.66**
Variation in brick prices							
Add or subtract the following amounts for every £1.00/1000 difference in the PC price of the measured items below							
half brick thick	-	-	-	-	0.06	m²	-
one brick thick	-	-	-	-	0.13	m²	-
one and a half brick thick	-	-	-	-	0.19	m²	-
two brick thick	-	-	-	-	0.25	m²	-
Movement of materials							
Loading to wheelbarrows and transporting to location; per 215 mm thick walls; maximum 25 m distance	-	0.42	8.75	-	-	m²	**8.75**
Class B engineering bricks; PC £300.00/1000; double Flemish bond in cement mortar (1:3)							
Mechanically offloading; maximum 25 m distance; loading to wheelbarrows; transporting to location; per 215 mm thick walls	-	0.42	8.75	-	-	m²	**8.75**
Walls							
half brick thick	-	3.19	67.03	-	19.89	m²	**86.93**
one brick thick	-	6.38	134.06	-	39.79	m²	**173.85**
one and a half brick thick	-	9.58	201.13	-	59.68	m²	**260.81**
two brick thick	-	12.77	268.13	-	79.58	m²	**347.70**
Walls; curved; mean radius 6 m							
half brick thick	-	4.81	100.91	-	20.53	m²	**121.44**
one brick thick	-	9.61	201.73	-	39.79	m²	**241.52**
Walls; curved; mean radius 1.50 m							
half brick thick	-	6.40	134.36	-	20.53	m²	**154.88**
one brick thick	-	12.80	268.89	-	39.79	m²	**308.67**
Walls; tapering; one face battering; average							
one and a half brick thick	-	10.34	217.24	-	59.68	m²	**276.93**
two brick thick	-	15.52	325.86	-	79.58	m²	**405.44**

Item Excluding site overheads and profit	PC £	Labour hours	Labour £	Plant £	Material £	Unit	Total rate £
F10 BRICK BLOCK WALLING - cont'd							
Class B engineering bricks - cont'd							
Walls; battering (retaining)							
one and a half brick thick	-	11.49	241.38	-	59.68	m^2	**301.06**
two brick thick	-	17.24	362.07	-	79.58	m^2	**441.65**
Isolated piers							
one brick thick	-	9.31	195.51	-	43.58	m^2	**239.09**
one and a half brick thick	-	11.97	251.37	-	65.36	m^2	**316.73**
two brick thick	-	13.30	279.30	-	88.42	m^2	**367.72**
three brick thick	-	16.49	346.33	-	130.73	m^2	**477.06**
Projections; vertical							
one brick x half brick	-	0.70	14.70	-	4.43	m	**19.13**
one brick x one brick	-	1.40	29.40	-	8.88	m	**38.28**
one and a half brick x one brick	-	2.10	44.10	-	13.33	m	**57.43**
two brick x one brick	-	2.30	48.30	-	17.77	m	**66.07**
Walls; half brick thick							
in honeycomb bond	-	1.80	37.80	-	14.53	m^2	**52.33**
in quarter bond	-	1.67	35.00	-	19.26	m^2	**54.26**
Facing bricks; PC £300.00/1000; English garden wall bond; in gauged mortar (1:1:6); facework one side							
Mechanically offloading; maximum 25 m distance; loading to wheelbarrows and transporting to location; per 215 mm thick walls	-	0.42	8.75	-	-	m^2	**8.75**
Walls							
half brick thick	-	3.19	67.03	-	20.49	m^2	**87.53**
half brick thick half brick thick (using site cut snap headers to form bond)	-	4.99	104.74	-	20.49	m^2	**125.23**
one brick thick	-	6.38	134.06	-	39.39	m^2	**173.46**
one and a half brick thick	-	9.58	201.10	-	61.48	m^2	**262.58**
two brick thick	-	12.77	268.13	-	81.97	m^2	**350.10**
Walls; curved; mean radius 6 m							
half brick thick	-	4.81	100.91	-	21.92	m^2	**122.84**
one brick thick	-	9.61	201.73	-	42.79	m^2	**244.52**
Walls; curved; mean radius 1.50 m							
half brick thick	-	6.40	134.36	-	21.47	m^2	**155.83**
one brick thick	-	12.80	268.89	-	41.89	m^2	**310.77**
Walls; tapering; one face battering; average							
one and a half brick thick	-	11.49	241.33	-	64.18	m^2	**305.51**
two brick thick	-	17.24	362.07	-	85.57	m^2	**447.64**
Walls; battering (retaining)							
one and a half brick thick	-	10.34	217.20	-	64.18	m^2	**281.38**
two brick thick	-	15.52	325.86	-	85.57	m^2	**411.44**
Isolated piers; English bond; facework all round							
one brick thick	-	9.31	195.51	-	44.17	m^2	**239.68**
one and a half brick thick	-	11.97	251.37	-	66.26	m^2	**317.63**
two brick thick	-	13.30	279.30	-	89.41	m^2	**368.71**
three brick thick	-	16.49	346.33	-	132.52	m^2	**478.85**
Projections; vertical							
one brick x half brick	-	0.93	19.55	-	4.57	m	**24.12**
one brick x one brick	-	1.86	39.10	-	9.14	m	**48.24**
one and a half brick x one brick	-	2.79	58.65	-	13.72	m	**72.37**
two brick x one brick	-	3.06	64.24	-	18.29	m	**82.53**
Brickwork fair faced both sides; facing bricks in gauged mortar (1:1:6)							
Extra for fair face both sides; flush, struck, weathered, or bucket-handle pointing	-	0.89	18.62	-	-	m^2	**18.62**

F MASONRY

Item Excluding site overheads and profit	PC £	Labour hours	Labour £	Plant £	Material £	Unit	Total rate £
Reclaimed bricks; PC £800.00/1000; English garden wall bond; in gauged mortar (1:1:6)							
Walls							
half brick thick (stretcher bond)	-	3.19	67.03	-	51.99	m^2	**119.03**
half brick thick (using site cut snap headers to form bond)	-	4.99	104.74	-	51.99	m^2	**156.73**
one brick thick	-	6.38	134.06	-	102.07	m^2	**236.13**
one and a half brick thick	-	9.58	201.13	-	153.10	m^2	**354.23**
two brick thick	-	12.77	268.13	-	204.13	m^2	**472.26**
Walls; curved; mean radius 6 m							
half brick thick	-	4.81	100.91	-	54.92	m^2	**155.84**
one brick thick	-	9.61	201.73	-	108.79	m^2	**310.51**
Walls; curved; mean radius 1.50 m							
half brick thick	-	6.40	134.36	-	57.32	m^2	**191.68**
one brick thick	-	12.77	268.13	-	113.59	m^2	**381.71**
Walls; stretcher bond; wall ties at 450 centres vertically and horizontally							
one brick thick	-	6.96	146.25	-	102.54	m^2	**248.79**
two brick thick	-	13.93	292.46	-	205.55	m^2	**498.00**
Brickwork fair faced both sides; facing bricks in gauged mortar (1:1:6)							
extra for fair face both sides; flush, struck, weathered, or bucket-handle pointing	-	0.67	14.00	-	-	m^2	**14.00**
Brick copings							
Copings; all brick headers-on-edge; to BS 4729; two angles rounded 53 mm radius; flush pointing top and both sides as work proceeds; one brick wide; horizontal							
machine-made specials	25.65	0.71	14.93	-	27.17	m	**42.10**
hand-made specials	25.65	0.71	14.93	-	27.18	m	**42.11**
Extra over copings for two courses machine-made tile creasings, projecting 25 mm each side; 260 mm wide copings; horizontal	5.54	0.50	10.50	-	6.20	m	**16.70**
Copings; all brick headers-on-edge; flush pointing top and both sides as work proceeds; one brick wide; horizontal							
bricks PC £300.00/1000	4.00	0.71	14.93	-	4.55	m	**19.48**
bricks PC £500.00/1000	6.67	0.71	14.93	-	7.10	m	**22.03**
Dense aggregate concrete blocks; "Tarmac Topblock" or other equal and approved; in gauged mortar (1:2:9)							
Solid blocks 7 N/mm^2							
440 x 215 x 100 mm thick	11.26	1.20	25.20	-	12.27	m^2	**37.47**
440 x 215 x 140 mm thick	16.22	1.30	27.30	-	17.36	m^2	**44.66**
Solid blocks 7 N/mm^2 laid flat							
440 x 100 x 215 mm thick	22.97	3.22	67.62	-	25.69	m^2	**93.31**
Hollow concrete blocks							
440 x 215 x 215 mm thick	26.00	1.30	27.30	-	27.14	m^2	**54.44**
Filling of hollow concrete blocks with concrete as work proceeds; tamping and compacting							
440 x 215 x 215 mm thick	19.61	0.20	4.20	-	19.61	m^2	**23.81**

G STRUCTURAL/CARCASSING METAL/TIMBER

Item Excluding site overheads and profit	PC £	Labour hours	Labour £	Plant £	Material £	Unit	Total rate £
G TIMBER DECKING							
Timber decking							
Supports for timber decking; softwood joists to receive decking boards; joists at 400 mm centres; Southern yellow pine							
38 x 88 mm	15.75	1.00	21.00	-	19.18	m^2	**40.17**
50 x 150 mm	10.85	1.00	21.00	-	13.79	m^2	**34.78**
50 x 125 mm	9.05	1.00	21.00	-	11.80	m^2	**32.80**
Hardwood decking yellow Balau; grooved or smooth; 6 mm joints							
deck boards; 90 mm wide x 19 mm thick	19.05	1.00	21.00	-	21.16	m^2	**42.16**
deck boards; 145 mm wide x 21 mm thick	21.25	1.00	21.00	-	25.49	m^2	**46.49**
deck boards; 145 mm wide x 28 mm thick	28.73	1.00	21.00	-	33.72	m^2	**54.72**
Hardwood decking Ipe; smooth; 6 mm joints							
deck boards; 90 mm wide x 19 mm thick	40.60	1.00	21.00	-	42.71	m^2	**63.71**
deck boards; 145 mm wide x 19 mm thick	29.79	1.00	21.00	-	31.90	m^2	**52.90**
Western Red Cedar; 6 mm joints							
Prime deck grade; 90 mm wide x 40 mm thick	32.27	1.00	21.00	-	37.61	m^2	**58.61**
Prime deck grade; 142 mm wide x 40 mm thick	34.56	1.00	21.00	-	40.13	m^2	**61.13**
Handrails and base rail; fixed to posts at 2.00 m centres							
Posts 100 x 100 x 1370 high	6.70	1.00	21.00	-	7.75	m	**28.75**
Posts turned 1220 high	14.25	1.00	21.00	-	15.30	m	**36.30**
Handrails; balusters							
Square balusters at 100 mm centres	38.00	0.50	10.50	-	38.50	m	**49.00**
Square balusters at 300 mm centres	12.65	0.33	7.00	-	13.05	m	**20.05**
Turned balusters at 100 mm centres	60.00	0.50	10.50	-	60.50	m	**71.00**
Turned balusters at 300 mm centres	19.98	0.33	7.00	-	20.38	m	**27.38**

J WATERPROOFING

Item Excluding site overheads and profit	PC £	Labour hours	Labour £	Plant £	Material £	Unit	**Total rate £**
J30 LIQUID APPLIED TANKING/DAMP PROOFING							
Tanking and damp proofing; Ruberoid Building Products, "Synthaprufe" cold applied bituminous emulsion waterproof coating							
"Synthaprufe"; to smooth finished concrete or screeded slabs; flat; blinding with sand							
two coats	4.76	0.22	4.67	-	5.41	m²	**10.08**
three coats	7.14	0.31	6.51	-	8.03	m²	**14.54**
"Synthaprufe"; to fair faced brickwork with flush joints, rendered brickwork, or smooth finished concrete walls; vertical							
two coats	5.36	0.29	6.00	-	6.07	m²	**12.07**
three coats	7.86	0.40	8.40	-	8.82	m²	**17.22**
Tanking and damp proofing; RIW Ltd							
Liquid asphaltic composition; to smooth finished concrete screeded slabs or screeded slabs; flat							
two coats	6.15	0.33	7.00	-	6.76	m²	**13.76**
Liquid asphaltic composition; fair faced brickwork with flush joints, rendered brickwork, or smooth finished concrete walls; vertical							
two coats	6.15	0.50	10.50	-	6.76	m²	**17.27**
"Heviseal"; to smooth finished concrete or screeded slabs; to surfaces of ponds, tanks, planters; flat							
two coats	7.04	0.33	7.00	-	7.74	m²	**14.74**
"Heviseal"; to fair faced brickwork with flush joints, rendered brickwork, or smooth finished concrete walls; to surfaces of retaining walls, ponds, tanks, planters; vertical							
two coats	7.04	0.50	10.50	-	7.74	m²	**18.24**

M SURFACE FINISHES

Item Excluding site overheads and profit	PC £	Labour hours	Labour £	Plant £	Material £	Unit	Total rate £
M60 PAINTING/CLEAR FINISHING							
Prepare; touch up primer; two undercoats and one finishing coat of gloss oil paint; on metal surfaces							
General surfaces							
girth exceeding 300 mm	0.87	0.33	7.00	-	0.87	m^2	**7.87**
isolated surfaces; girth not exceeding 300 mm	0.24	0.13	2.80	-	0.24	m	**3.04**
isolated areas not exceeding 0.50 m^2 irrespective of girth	0.46	0.13	2.80	-	0.46	nr	**3.26**
Ornamental railings; each side measured overall							
girth exceeding 300 mm	0.87	0.75	15.75	-	0.87	m^2	**16.62**
Prepare; one coat primer; two undercoats and one finishing coat of gloss oil paint; on wood surfaces							
General surfaces							
girth exceeding 300 mm	1.17	0.40	8.40	-	1.17	m^2	**9.57**
isolated areas not exceeding 0.50 m^2 irrespective of girth	0.45	0.36	7.64	-	0.45	nr	**8.09**
isolated surfaces; girth not exceeding 300 mm	0.38	0.20	4.20	-	0.38	m	**4.58**
Prepare; two coats of creosote; on wood surfaces							
General surfaces							
girth exceeding 300 mm	0.20	0.21	4.36	-	0.20	m^2	**4.55**
isolated surfaces; girth not exceeding 300 mm	0.06	0.12	2.42	-	0.06	m	**2.48**
Prepare; proprietary solution primer; two coats of dark stain; on wood surfaces							
General surfaces							
girth exceeding 300 mm	0.14	0.10	2.10	-	0.14	m^2	**2.24**
isolated surfaces; girth not exceeding 300 mm	0.02	0.05	1.05	-	0.02	m	**1.07**
Three coats "Dimex Shield"; to clean, dry surfaces; in accordance with manufacturer's instructions							
Brick or block walls							
girth exceeding 300 mm	1.06	0.28	5.88	-	1.06	m^2	**6.94**
Cement render or concrete walls							
girth exceeding 300 mm	0.83	0.25	5.25	-	0.83	m^2	**6.08**
Two coats resin based paint; "Sandtex Matt"; in accordance with manufacturer's instructions							
Brick or block walls							
girth exceeding 300 mm	7.11	0.20	4.20	-	7.11	m^2	**11.31**
Cement render or concrete walls							
girth exceeding 300 mm	4.84	0.17	3.50	-	4.84	m^2	**8.34**

Q PAVING/PLANTING/FENCING/SITE FURNITURE

Item Excluding site overheads and profit	PC £	Labour hours	Labour £	Plant £	Material £	Unit	Total rate £
Q10 KERBS/EDGINGS/CHANNELS/PAVING ACCESSORIES							
Foundations to kerbs							
Excavating trenches; width 300 mm; 3 tonne excavator							
depth 300 mm	-	0.10	2.10	0.48	2.84	m	**5.42**
depth 400 mm	-	0.11	2.34	0.53	3.79	m	**6.66**
By hand	-	0.60	12.65	-	2.46	m	**15.12**
Excavating trenches; width 450 mm; 3 tonne excavator; disposal off site							
depth 300 mm	-	0.13	2.63	0.56	2.22	m	**5.41**
depth 400 mm	-	0.14	3.00	0.64	2.97	m	**6.61**
Disposal							
Excavated material; off site; to tip							
grab loaded 7.25 m^3	-	-	-	-	-	m^3	**2.84**
Disposal by skip; 6 yd^3 (4.6 m^3)							
by machine	-	-	-	37.96	-	m^3	**37.96**
by hand	-	4.00	84.00	41.08	-	m^3	**125.08**
Foundations; to kerbs, edgings, or channels; in situ concrete; 21 N/mm^2 - 20 aggregate (1:2:4) site mixed; one side against earth face, other against formwork (not included); site mixed concrete							
Site mixed concrete							
150 wide x 100 mm deep	-	0.13	2.80	-	1.61	m	**4.41**
150 wide x 150 mm deep	-	0.17	3.50	-	2.41	m	**5.91**
200 wide x 150 mm deep	-	0.20	4.20	-	3.22	m	**7.42**
300 wide x 150 mm deep	-	0.23	4.90	-	4.83	m	**9.73**
600 wide x 200 mm deep	-	0.29	6.00	-	12.87	m	**18.87**
Ready mixed concrete							
150 wide x 100 mm deep	-	0.13	2.80	-	1.22	m	**4.02**
150 wide x 150 mm deep	-	0.17	3.50	-	1.83	m	**5.33**
200 wide x 150 mm deep	-	0.20	4.20	-	2.44	m	**6.64**
300 wide x 150 mm deep	-	0.23	4.90	-	3.66	m	**8.56**
600 wide x 200 mm deep	-	0.29	6.00	-	9.75	m	**15.75**
Formwork; sides of foundations (this will usually be required to one side of each kerb foundation adjacent to road sub-bases)							
100 mm deep	-	0.04	0.87		0.18	m	**1.05**
150 mm deep	-	0.04	0.87		0.27	m	**1.14**
Precast concrete edging units; including haunching with in situ concrete; 11.50 N/mm^2 - 40 aggregate both sides							
Edgings; rectangular, bullnosed, or chamfered							
50 x 150 mm	1.28	0.33	7.00	-	6.09	m	**13.09**
125 x 150 mm bullnosed	1.39	0.33	7.00	-	6.21	m	**13.21**
50 x 200 mm	1.92	0.33	7.00	-	6.73	m	**13.73**
50 x 250 mm	2.22	0.33	7.00	-	7.04	m	**14.04**
50 x 250 mm flat top	2.74	0.33	7.00	-	7.55	m	**14.56**

Q PAVING/PLANTING/FENCING/SITE FURNITURE

Item Excluding site overheads and profit	PC £	Labour hours	Labour £	Plant £	Material £	Unit	Total rate £
Q10 KERBS/EDGINGS/CHANNELS/PAVING ACCESSORIES - cont'd							
Marshalls Mono; small element precast concrete kerb system; Keykerb Large (KL) upstand of 100 - 125 mm; on 150 mm concrete foundation including haunching with in situ concrete 1:3:6 1 side							
Bullnosed or half battered; 100 x 127 x 200							
laid straight	10.37	0.80	16.80	-	11.38	m	**28.18**
radial blocks laid to curve; 8 blocks/1/4 circle - 500 mm radius	7.98	0.80	16.80	-	8.99	m	**25.79**
radial blocks laid to curve; 8 radial blocks, alternating 8 standard blocks/1/4 circle - 1000 mm radius	13.16	1.25	26.25	-	14.18	m	**40.43**
radial blocks, alternating 16 standard blocks/1/4 circle - 1500 mm radius	12.11	1.50	31.50	-	13.12	m	**44.62**
internal angle 90 degree	4.38	0.20	4.20	-	4.38	nr	**8.58**
external angle	4.38	0.20	4.20	-	4.38	nr	**8.58**
drop crossing kerbs; KL half battered to KL splay; LH and RH	9.16	1.00	21.00	-	10.17	pair	**31.17**
drop crossing kerbs; KL half battered to KS bullnosed	8.91	1.00	21.00	-	9.92	pair	**30.92**
Marshalls Mono; small element precast concrete kerb system; Keykerb Small (KS) upstand of 25 - 50 mm; on 150 mm concrete foundation including haunching with in situ concrete 1:3:6 1 side							
Half battered							
laid straight	7.28	0.80	16.80	-	8.29	m	**25.09**
radial blocks laid to curve; 8 blocks/1/4 circle - 500 mm radius	5.99	1.00	21.00	-	7.00	m	**28.00**
radial blocks laid to curve; 8 radial blocks, alternating 8 standard blocks/1/4 circle - 1000 mm radius	9.63	1.25	26.25	-	10.64	m	**36.89**
radial blocks, alternating 16 standard blocks/1/4 circle - 1500 mm radius	10.30	1.50	31.50	-	11.31	m	**42.81**
internal angle 90 degree	4.38	0.20	4.20	-	4.38	nr	**8.58**
external angle	4.38	0.20	4.20	-	4.38	nr	**8.58**
Second-hand granite setts; 100 x 100 mm; bedding in cement mortar (1:4); on 150 mm deep concrete foundations; including haunching with in situ concrete; 11.50 N/mm^2 - 40 aggregate one side							
Edgings							
300 mm wide	15.16	1.33	28.00	-	19.13	m	**47.13**
Brick or block stretchers; bedding in cement mortar (1:4); on 150 mm deep concrete foundations, including haunching with in situ concrete; 11.50 N/mm^2 - 40 aggregate one side							
Single course							
concrete paving blocks; PC £7.70/m^2; 200 x 100 x 60 mm	1.54	0.31	6.46	-	4.25	m	**10.71**
engineering bricks; PC £300.00/1000; 215 x 102.5 x 65 mm	1.33	0.40	8.40	-	4.11	m	**12.51**
paving bricks; PC £520.00/1000; 215 x 102.5 x 65 mm	2.31	0.40	8.40	-	5.03	m	**13.43**
Two courses							
concrete paving blocks; PC £7.70/m^2; 200 x 100 x 60 mm	3.08	0.40	8.40	-	6.25	m	**14.65**

Q PAVING/PLANTING/FENCING/SITE FURNITURE

Item Excluding site overheads and profit	PC £	Labour hours	Labour £	Plant £	Material £	Unit	Total rate £
engineering bricks; PC £300.00/1000; 215 x 102.5 x 65 mm	2.67	0.57	12.00	-	5.97	m	**17.97**
paving bricks; PC £520.00/1000; 215 x 102.5 x 65 mm	4.62	0.57	12.00	-	7.79	m	**19.79**
Three courses							
concrete paving blocks; PC £7.70/m²; 200 x 100 x 60 mm	4.62	0.44	9.33	-	7.79	m	**17.12**
engineering bricks; PC £300.00/1000; 215 x 102.5 x 65 mm	4.00	0.67	14.00	-	7.37	m	**21.37**
paving bricks; PC £520.00/1000; 215 x 102.5 x 65 mm	6.93	0.67	14.00	-	10.10	m	**24.10**
Bricks on edge; bedding in cement mortar (1:4); on 150 mm deep concrete foundations; including haunching with in situ concrete; 11.50 N/mm² - 40 aggregate one side							
One brick wide							
engineering bricks; 215 x 102.5 x 65 mm	4.00	0.57	12.00	-	7.48	m	**19.48**
paving bricks; 215 x 102.5 x 65 mm	6.93	0.57	12.00	-	10.22	m	**22.21**
Two courses; stretchers laid on edge; 225 mm wide							
engineering bricks; 215 x 102.5 x 65 mm	8.00	1.20	25.20	-	12.70	m	**37.90**
paving bricks; 215 x 102.5 x 65 mm	13.86	1.20	25.20	-	18.16	m	**43.36**
Extra over bricks on edge for standard kerbs to one side; haunching in concrete							
125 x 255 mm; ref HB2; SP	3.27	0.44	9.33	-	6.16	m	**15.49**
Channels; bedding in cement mortar (1:3); joints pointed flush; on concrete foundations (not included)							
Three courses stretchers; 350 mm wide; quarter bond to form dished channels							
engineering bricks; PC £300.00/1000; 215 x 102.5 x 65 mm	4.00	1.00	21.00	-	5.23	m	**26.23**
paving bricks; PC £520.00/1000; 215 x 102.5 x 65 mm	6.93	1.00	21.00	-	7.96	m	**28.96**
Three courses granite setts; 340 mm wide; to form dished channels							
340 mm wide	15.16	2.00	42.00	-	17.52	m	**59.52**
Permaloc "AshphaltEdge"; RTS Ltd; extruded aluminium alloy L shaped edging with 5.33 mm exposed upper lip; edging fixed to roadway base and edge profile with 250 mm steel fixing spike; laid to straight or curvilinear road edge; subsequently filled with macadam (not included)							
Depth of macadam							
38 mm	7.72	0.02	0.35	-	7.91	m	**8.26**
51 mm	8.43	0.02	0.36	-	8.64	m	**9.00**
64 mm	9.09	0.02	0.38	-	9.32	m	**9.70**
76 mm	10.55	0.02	0.42	-	10.81	m	**11.23**
102 mm	11.92	0.02	0.44	-	12.22	m	**12.65**
Permaloc "Cleanline"; RTS Ltd; heavy duty straight profile edging; for edgings to soft landscape beds or turf areas; 3.2 mm x 102 high; 3.2 mm thick with 4.75 mm exposed upper lip; fixed to form straight or curvilinear edge with 305 mm fixing spike							
Milled aluminium							
100 deep	6.64	0.02	0.38	-	6.81	m	**7.19**
Black							
100 deep	7.31	0.02	0.38	-	7.49	m	**7.87**

Q PAVING/PLANTING/FENCING/SITE FURNITURE

Item Excluding site overheads and profit	PC £	Labour hours	Labour £	Plant £	Material £	Unit	Total rate £
Q10 KERBS/EDGINGS/CHANNELS/PAVING ACCESSORIES - cont'd							
Permaloc "Permastrip"; RTS Ltd; heavy duty L shaped profile maintenance strip; 3.2 mm x 89 mm high with 5.2 mm exposed top lip; for straight or gentle curves on paths or bed turf interfaces; fixed to form straight or curvilinear edge with standard 305 mm stake; other stake lengths available							
Milled aluminium							
89 deep	6.64	0.02	0.38	-	6.81	m	**7.19**
Black							
89 deep	7.31	0.02	0.38	-	7.49	m	**7.87**
Permaloc "Proline"; RTS Ltd; medium duty straight profiled maintenance strip; 3.2 mm x 102 mm high with 3.18 mm exposed top lip; for straight or gentle curves on paths or bed turf interfaces; fixed to form straight or curvilinear edge with standard 305 mm stake; other stake lengths available							
Milled aluminium							
89 deep	5.03	0.02	0.38	-	5.16	m	**5.54**
Black							
89 deep	5.80	0.02	0.38	-	5.95	m	**6.33**
Q20 GRANULAR SUB-BASES TO ROADS/PAVINGS							
Hardcore bases; obtained off site; PC £20.00/m³							
By machine							
100 mm thick	2.00	0.05	1.05	1.75	2.00	m²	**4.80**
150 mm thick	3.00	0.07	1.40	2.30	3.00	m²	**6.70**
over 150 thick	20.00	0.50	10.50	17.52	20.00	m³	**48.02**
By hand							
100 mm thick	2.00	0.20	4.20	0.12	2.00	m²	**6.32**
150 mm thick	3.00	0.30	6.31	0.12	3.60	m²	**10.03**
Hardcore; difference for each £1.00 increase/decrease in PC price per m³; price will vary with type and source of hardcore							
average 75 mm thick	-	-	-	-	0.08	m²	**-**
average 100 mm thick	-	-	-	-	0.10	m²	**-**
average 150 mm thick	-	-	-	-	0.15	m²	**-**
average 200 mm thick	-	-	-	-	0.20	m²	**-**
average 250 mm thick	-	-	-	-	0.25	m²	**-**
average 300 mm thick	-	-	-	-	0.30	m²	**-**
exceeding 300 mm thick	-	-	-	-	1.10	m³	**-**
Type 1 granular fill base; PC £18.50/tonne (£40.48/m³ compacted)							
By machine							
100 mm thick	4.05	0.03	0.59	0.45	4.05	m²	**5.08**
150 mm thick	6.07	0.03	0.53	0.61	6.07	m²	**7.20**
By hand (mechanical compaction)							
100 mm thick	4.05	0.17	3.50	0.04	4.05	m²	**7.59**
150 mm thick	6.07	0.25	5.25	0.06	6.07	m²	**11.38**
Surface treatments							
Sand blinding; to hardcore base (not included); 25 mm thick	0.99	0.03	0.70	-	0.99	m²	**1.69**

Q PAVING/PLANTING/FENCING/SITE FURNITURE

Item Excluding site overheads and profit	PC £	Labour hours	Labour £	Plant £	Material £	Unit	**Total rate £**
Sand blinding; to hardcore base (not included);							
50 mm thick	1.98	0.05	1.05	-	1.98	m²	**3.03**
Filter fabrics; to hardcore base (not included)	0.45	0.01	0.21	-	0.47	m²	**0.68**
Herbicides; Scotts							
"Casoron G" (residual) herbicide; treating substrate before laying base							
Total; 1.25 kg/100 m²	-	0.50	10.50	-	5.65	100 m²	**16.15**
Q21 IN SITU CONCRETE ROADS/PAVINGS							
Unreinforced concrete; on prepared sub-base (not included)							
Paths and roads; 21.00 N/mm² - 20 aggregate (1:2:4) mechanically mixed on site							
100 mm thick	10.73	0.13	2.63	-	11.26	m²	**13.89**
150 mm thick	16.09	0.17	3.50	-	16.89	m²	**20.40**
Formwork for in situ concrete							
Sides of foundations							
height not exceeding 250 mm	0.37	0.05	1.05	-	0.60	m	**1.65**
Extra over formwork for curved work 6 m radius	-	0.25	5.25	-	-	m	**5.25**
Steel road forms; to edges of beds or faces of foundations							
150 mm wide	-	0.20	4.20	1.10	-	m	**5.30**
Reinforcement; fabric; BS 4483; side laps 150 mm; head laps 300 mm; mesh 200 x 200 mm; in roads, footpaths or pavings							
Fabric							
ref A142 (2.22 kg/m²)	2.01	0.08	1.74	-	2.22	m²	**3.96**
ref A193 (3.02 kg/m²)	3.93	0.08	1.74	-	4.32	m²	**6.07**
Reinforced in situ concrete; mechanically mixed on site; normal Portland cement; on hardcore base (not included); reinforcement (not included)							
Roads; 11.50 N/mm² - 40 aggregate (1:3:6)							
100 mm thick	9.62	0.40	8.40	-	9.86	m²	**18.26**
150 mm thick	14.44	0.60	12.60	-	14.80	m²	**27.40**
200 mm thick	19.25	0.80	16.80	-	19.73	m²	**36.53**
Roads; 21.00 N/mm² - 20 aggregate (1:2:4)							
100 mm thick	10.73	0.40	8.40	-	11.00	m²	**19.40**
150 mm thick	16.09	0.60	12.60	-	16.50	m²	**29.10**
200 mm thick	21.45	0.80	16.80	-	21.99	m²	**38.79**
Roads; 25.00 N/mm² - 20 aggregate GEN 4 ready mixed							
100 mm thick	8.43	0.40	8.40	-	8.64	m²	**17.04**
150 mm thick	12.64	0.60	12.60	-	12.96	m²	**25.56**
200 mm thick	16.85	0.80	16.80	-	17.27	m²	**34.07**
250 mm thick	21.06	1.00	21.00	0.20	21.58	m²	**42.78**
Reinforced in situ concrete; ready mixed; discharged directly into location from supply lorry; normal Portland cement; on hardcore base (not included); reinforcement (not included)							
Roads; 11.50 N/mm² - 40 aggregate (1:3:6)							
100 mm thick	8.31	0.16	3.36	-	8.31	m²	**11.67**
150 mm thick	12.47	0.24	5.04	-	12.47	m²	**17.51**
200 mm thick	16.63	0.36	7.56	-	16.63	m²	**24.19**
250 mm thick	20.79	0.54	11.34	0.20	20.79	m²	**32.32**
300 mm thick	24.94	0.66	13.86	0.20	24.94	m²	**39.00**

Q PAVING/PLANTING/FENCING/SITE FURNITURE

Item Excluding site overheads and profit	PC £	Labour hours	Labour £	Plant £	Material £	Unit	Total rate £
Q21 IN SITU CONCRETE ROADS/PAVINGS - cont'd							
Reinforced in situ concrete - cont'd							
Roads; 21.00 N/mm² - 20 aggregate (1:2:4)							
100 mm thick	8.36	0.16	3.36	-	8.36	m²	**11.72**
150 mm thick	12.54	0.24	5.04	-	12.54	m²	**17.58**
200 mm thick	16.72	0.36	7.56	-	16.72	m²	**24.28**
250 mm thick	20.90	0.54	11.34	0.20	20.90	m²	**32.43**
300 mm thick	25.08	0.66	13.86	0.20	25.08	m²	**39.13**
Roads; 26.00 N/mm² - 20 aggregate (1:1:5:3)							
100 mm thick	8.76	0.16	3.36	-	8.76	m²	**12.12**
150 mm thick	13.13	0.24	5.04	-	13.13	m²	**18.17**
200 mm thick	17.51	0.36	7.56	-	17.51	m²	**25.07**
250 mm thick	21.89	0.54	11.34	0.20	21.89	m²	**33.43**
300 mm thick	26.27	0.66	13.86	0.20	26.27	m²	**40.32**
Concrete sundries							
Treating surfaces of unset concrete; grading to cambers, tamping with 75 mm thick steel shod tamper or similar	-	0.13	2.80	-	-	m²	**2.80**
Expansion joints							
13 mm thick joint filler; formwork							
width or depth not exceeding 150 mm	1.67	0.20	4.20	-	2.25	m	**6.45**
width or depth 150 - 300 mm	1.21	0.25	5.25	-	2.37	m	**7.62**
width or depth 300 - 450 mm	0.54	0.30	6.30	-	2.28	m	**8.58**
25 mm thick joint filler; formwork							
width or depth not exceeding 150 mm	2.19	0.20	4.20	-	2.77	m	**6.97**
width or depth 150 - 300 mm	2.19	0.25	5.25	-	3.35	m	**8.60**
width or depth 300 - 450 mm	2.19	0.30	6.30	-	3.93	m	**10.23**
Sealants; sealing top 25 mm of joint with rubberized bituminous compound	1.16	0.25	5.25	-	1.16	m	**6.41**
Q22 COATED MACADAM/ASPHALT ROADS/PAVINGS							
Coated macadam/asphalt roads/pavings - General							
Preamble: The prices for all in situ finishings to roads and footpaths include for work to falls, crossfalls or slopes not exceeding 15 degrees from horizontal; for laying on prepared bases (not included) and for rolling with an appropriate roller.							
Users should note the new terminology for the surfaces described below which is to European standard descriptions. The now redundant descriptions for each course are shown in brackets.							
Note: The rates below are for smaller areas up to 200 m².							
Macadam surfacing; Spadeoak Construction Co Ltd; surface (wearing) course; 20 mm of 6 mm dense bitumen macadam to BS4987-1 2001 ref 7.5							
Hand lay							
limestone aggregate	-	-	-	-	-	m²	**9.04**
granite aggregate	-	-	-	-	-	m²	**9.02**
red	-	-	-	-	-	m²	**11.12**

Q PAVING/PLANTING/FENCING/SITE FURNITURE

Item Excluding site overheads and profit	PC £	Labour hours	Labour £	Plant £	Material £	Unit	Total rate £
Macadam surfacing; Spadeoak Construction Co Ltd; surface (wearing) course; 30 mm of 10 mm dense bitumen macadam to BS4987-1 2001 ref 7.4							
Hand lay							
limestone aggregate	-	-	-	-	-	m^2	**10.47**
granite aggregate	-	-	-	-	-	m^2	**10.60**
red	-	-	-	-	-	m^2	**-**
Macadam surfacing; Spadeoak Construction Co Ltd; surface (wearing) course; 40 mm of 10 mm dense bitumen macadam to BS4987-1 2001 ref 7.4							
Hand lay							
limestone aggregate	-	-	-	-	-	m^2	**12.95**
granite aggregate	-	-	-	-	-	m^2	**13.11**
red	-	-	-	-	-	m^2	**22.89**
Macadam surfacing; Spadeoak Construction Co Ltd; binder (base) course; 50 mm of 20 mm dense bitumen macadam to BS4987-1 2001 ref 6.5							
Hand lay							
limestone aggregate	-	-	-	-	-	m^2	**12.98**
granite aggregate	-	-	-	-	-	m^2	**13.16**
Macadam surfacing; Spadeoak Construction Co Ltd; binder (base) course; 60 mm of 20 mm dense bitumen macadam to BS4987-1 2001 ref 6.5							
Hand lay							
limestone aggregate	-	-	-	-	-	m^2	**14.12**
granite aggregate	-	-	-	-	-	m^2	**14.33**
Macadam surfacing; Spadeoak Construction Co Ltd; base (roadbase) course; 75 mm of 28 mm dense bitumen macadam to BS4987-1 2001 ref 5.2							
Hand lay							
limestone aggregate	-	-	-	-	-	m^2	**17.01**
granite aggregate	-	-	-	-	-	m^2	**17.29**
Areas up to 300 m^2							
Macadam surfacing; Spadeoak Construction Co Ltd; base (roadbase) course; 100 mm of 28 mm dense bitumen macadam to BS4987-1 2001 ref 5.2							
Hand lay							
limestone aggregate	-	-	-	-	-	m^2	**36.23**
granite aggregate	-	-	-	-	-	m^2	**21.10**
Macadam surfacing; Spadeoak Construction Co Ltd; base (roadbase) course; 150 mm of 28 mm dense bitumen macadam in two layers to BS4987-1 2001 ref 5.2							
Hand lay							
limestone aggregate	-	-	-	-	-	m^2	**33.29**
granite aggregate	-	-	-	-	-	m^2	**33.84**
Base (roadbase) course; 200 mm of 28 mm dense bitumen macadam in two layers to BS4987-1 2001 ref 5.2							
Hand lay							
limestone aggregate	-	-	-	-	-	m^2	**28.16**
granite aggregate	-	-	-	-	-	m^2	**42.25**

Q PAVING/PLANTING/FENCING/SITE FURNITURE

Item Excluding site overheads and profit	PC £	Labour hours	Labour £	Plant £	Material £	Unit	Total rate £
Q23 GRAVEL/HOGGIN/WOODCHIP ROADS/PAVINGS							
Excavation and path preparation; excavating; 300 mm deep; to width of path; depositing excavated material at sides of excavation							
By machine							
width 1.00 m	-	-	-	4.35	-	m^2	**4.35**
width 1.50 m	-	-	-	3.62	-	m^2	**3.62**
width 2.00 m	-	-	-	3.10	-	m^2	**3.10**
width 3.00 m	-	-	-	2.61	-	m^2	**2.61**
By hand							
width 1.00 m	-	0.70	14.79	-	-	m^2	**14.79**
width 1.50 m	-	1.05	22.05	-	-	m^2	**22.05**
width 2.00 m	-	1.40	29.40	-	-	m^2	**29.40**
width 3.00 m	-	2.10	44.10	-	-	m^2	**44.10**
Excavating trenches; in centre of pathways; 100 flexible drain pipes; filling with clean broken stone or gravel rejects							
300 x 450 deep							
by machine	2.87	0.10	2.10	1.63	2.92	m	**6.65**
by hand	2.87	0.40	8.40	-	2.92	m	**11.32**
Hand trimming and compacting reduced surface of pathway; by vibrating roller							
width 1.00 m	-	0.05	1.05	0.12	-	m	**1.17**
width 1.50 m	-	0.04	0.93	0.11	-	m	**1.04**
width 2.00 m	-	0.04	0.84	0.10	-	m	**0.94**
width 3.00 m	-	0.04	0.84	0.10	-	m	**0.94**
Permeable membranes; to trimmed and compacted surface of pathway							
"Terram 1000"	0.45	0.02	0.42	-	0.47	m^2	**0.89**
Permaloc "AshphaltEdge"; RTS Ltd; extruded aluminium alloy L shaped edging with 5.33 mm exposed upper lip; edging fixed to roadway base and edge profile with 250 mm steel fixing spike; laid to straight or curvilinear road edge; subsequently filled with macadam (not included)							
Depth of macadam							
38 mm	7.72	0.02	0.35	-	7.91	m	**8.26**
51 mm	8.43	0.02	0.36	-	8.64	m	**9.00**
64 mm	9.09	0.02	0.38	-	9.32	m	**9.70**
76 mm	10.55	0.02	0.42	-	10.81	m	**11.23**
102 mm	11.92	0.02	0.44	-	12.22	m	**12.65**
Permaloc "Cleanline"; RTS Ltd; heavy duty straight profile edging; for edgings to soft landscape beds or turf areas; 3.2 mm x 102 high; 3.2 mm thick with 4.75 mm exposed upper lip; fixed to form straight or curvilinear edge with 305 mm fixing spike							
Milled aluminium							
100 deep	6.64	0.02	0.38	-	6.81	m	**7.19**
Black							
100 deep	7.31	0.02	0.38	-	7.49	m	**7.87**

Q PAVING/PLANTING/FENCING/SITE FURNITURE

Item Excluding site overheads and profit	PC £	Labour hours	Labour £	Plant £	Material £	Unit	Total rate £
Permaloc "Permastrip"; RTS Ltd; heavy duty L shaped profile maintenance strip; 3.2 mm x 89 mm high with 5.2 mm exposed top lip; for straight or gentle curves on paths or bed turf interfaces; fixed to form straight or curvilinear edge with standard 305 mm stake; other stake lengths available							
Milled aluminium							
89 deep	6.64	0.02	0.38	-	6.81	m	**7.19**
Black							
89 deep	7.31	0.02	0.38	-	7.49	m	**7.87**
Permaloc "Proline"; RTS Ltd; medium duty straight profiled maintenance strip; 3.2 mm x 102 mm high with 3.18 mm exposed top lip; for straight or gentle curves on paths or bed turf interfaces; fixed to form straight or curvilinear edge with standard 305 mm stake; other stake lengths available							
Milled aluminium							
89 deep	5.03	0.02	0.38	-	5.16	m	**5.54**
Black							
89 deep	5.80	0.02	0.38	-	5.95	m	**6.33**
Timber edging boards							
Boards; 50 x 50 x 750 mm timber pegs at 1000 mm centres (excavations and hardcore under edgings not included)							
Straight							
38 x 150 mm treated softwood edge boards	2.52	0.10	2.10	-	2.52	m	**4.62**
50 x 150 mm treated softwood edge boards	2.75	0.10	2.10	-	2.75	m	**4.86**
38 x 150 mm hardwood (iroko) edge boards	8.12	0.10	2.10	-	8.12	m	**10.22**
50 x 150 mm hardwood (iroko) edge boards	10.31	0.10	2.10	-	10.31	m	**12.41**
Curved							
38 x 150 mm treated softwood edge boards	2.52	0.20	4.20	-	2.52	m	**6.72**
50 x 150 mm treated softwood edge boards	2.75	0.25	5.25	-	2.75	m	**8.01**
38 x 150 mm hardwood (iroko) edge boards	8.12	0.20	4.20	-	8.12	m	**12.32**
50 x 150 mm hardwood (iroko) edge boards	10.31	0.25	5.25	-	10.31	m	**15.56**
Filling to make up levels							
Obtained off site; hardcore; PC £20.00/m^3							
150 mm thick	3.00	0.07	1.40	2.30	3.00	m^2	**6.70**
Obtained off site; granular fill type 1; PC £18.50/tonne (£40.48/m^3 compacted)							
100 mm thick	4.05	0.03	0.59	0.45	4.05	m^2	**5.08**
150 mm thick	6.07	0.03	0.53	0.61	6.07	m^2	**7.20**
Surface treatments							
Sand blinding; to hardcore (not included)							
50 mm thick	1.98	0.04	0.84	-	2.18	m^2	**3.02**
Filter fabric; to hardcore (not included)	0.45	0.01	0.21	-	0.47	m^2	**0.68**
Granular Pavings							
Footpath gravels; porous self binding gravel							
Breedon Special Aggregates; "Golden Gravel" or equivalent; rolling wet; on hardcore base (not included); for pavements; to falls and crossfalls and to slopes not exceeding 15 degrees from horizontal; over 300 mm wide							
50 mm thick	9.90	0.05	1.05	0.41	9.90	m^2	**11.37**
by hand; 50 mm thick	9.90	0.20	4.20	0.24	9.90	m^2	**14.35**

Q PAVING/PLANTING/FENCING/SITE FURNITURE

Item Excluding site overheads and profit	PC £	Labour hours	Labour £	Plant £	Material £	Unit	Total rate £
Q23 GRAVEL/HOGGIN/WOODCHIP ROADS/PAVINGS - cont'd							
Footpath gravels - cont'd							
Breedon Special Aggregates; 'Wayfarer' specially formulated fine gravel for use on golf course pathways							
50 mm thick	8.81	0.03	0.60	0.31	8.81	m²	**9.72**
by hand; 50 mm thick	8.81	0.10	2.10	0.14	8.81	m²	**11.05**
Hoggin (stabilized); PC £31.10/m³ on hardcore base (not included); to falls and crossfalls and to slopes not exceeding 15 degrees from horizontal; over 300 mm wide							
100 mm thick	3.11	0.02	0.42	0.29	4.67	m²	**5.37**
by hand; 100 mm thick	3.11	0.10	2.10	0.10	4.67	m²	**6.86**
150 mm thick	4.67	0.03	0.53	0.36	7.00	m²	**7.88**
Ballast; as dug; watering; rolling; on hardcore base (not included)							
100 mm thick	3.83	0.13	2.80	0.97	4.79	m²	**8.55**
150 mm thick	5.74	0.03	0.53	0.34	7.18	m²	**8.04**
CED Ltd; Cedec gravel; self-binding; laid to inert (non-limestone) base measured separately; compacting							
red, gold or silver; 50 mm thick	10.75	0.03	0.60	0.32	10.75	m²	**11.68**
by hand; red, gold or silver; 50 mm thick	10.75	0.07	1.40	0.05	10.75	m²	**12.20**
Grundon Ltd; Coxwell self-binding path gravels laid and compacted to excavation or base measured separately							
50 mm thick	1.33	0.08	1.75	0.10	1.33	m²	**3.17**
Footpath gravels; porous loose gravels							
Washed shingle; on prepared base (not included)							
25 - 50 size, 25 mm thick	0.89	0.02	0.37	0.08	0.94	m²	**1.39**
25 - 50 size, 75 mm thick	2.68	0.05	1.10	0.25	2.82	m²	**4.17**
50 - 75 size, 25 mm thick	0.89	0.02	0.42	0.10	0.94	m²	**1.46**
50 - 75 size, 75 mm thick	2.68	0.07	1.40	0.30	2.82	m²	**4.52**
Pea shingle; on prepared base (not included)							
10 - 15 size, 25 mm thick	0.97	0.02	0.37	0.08	1.02	m²	**1.47**
5 - 10 size, 75 mm thick	2.91	0.05	1.10	0.25	3.05	m²	**4.41**
Breedon Special Aggregates; Breedon Buff decorative limestone chippings							
50 mm thick	4.50	0.01	0.21	0.10	4.72	m²	**5.03**
by hand; 50 mm thick	4.50	0.03	0.70	-	4.72	m²	**5.42**
Breedon Special Aggregates; Brindle or Moorland Black chippings							
50 mm thick	5.84	0.01	0.21	0.10	6.13	m²	**6.44**
by hand; 50 mm thick	5.84	0.03	0.70	-	6.13	m²	**6.83**
Breedon Special Aggregates; slate chippings; plum/blue/green							
50 mm thick	8.47	0.01	0.21	0.10	8.90	m²	**9.20**
by hand; 50 mm thick	8.47	0.03	0.70	-	8.90	m²	**9.60**
Bark surfaces; Melcourt Industries Ltd							
Bark chips; to surface of pathways by hand; ; levelling and spreading by hand (excavation and preparation not included) (items labelled FSC are Forest Stewardship Council certified)							
"Walk Chips"; FSC; 100 mm thick (25 m³ loads)	3.62	0.02	0.35	-	3.80	m²	**4.15**
"Woodfibre"; FSC; 100 mm thick	1.95	0.05	1.05	-	2.05	m²	**3.10**

Q PAVING/PLANTING/FENCING/SITE FURNITURE

Item Excluding site overheads and profit	PC £	Labour hours	Labour £	Plant £	Material £	Unit	Total rate £
Bound aggregates; Addagrip Surface Treatments UK Ltd; natural decorative resin bonded surface dressing laid to concrete, macadam or to plywood panels priced separately							
Primer coat to macadam or concrete base	-	-	-	-	-	m²	**4.60**
Golden pea gravel 1-3 mm							
buff adhesive	-	-	-	-	-	m²	**25.30**
red adhesive	-	-	-	-	-	m²	**25.30**
green adhesive	-	-	-	-	-	m²	**25.30**
Golden pea gravel 2-5 mm							
buff adhesive	-	-	-	-	-	m²	**28.75**
Chinese bauxite 1-3 mm							
buff adhesive	-	-	-	-	-	m²	**25.30**
Q24 INTERLOCKING BRICK/BLOCK ROADS/PAVINGS							
Precast concrete block edgings; PC £7.09/m²; 200 x 100 x 60 mm; on prepared base (not included); haunching one side							
Edgings; butt joints							
stretcher course	0.71	0.17	3.50	-	3.41	m	**6.91**
header course	1.42	0.27	5.60	-	4.32	m	**9.92**
Precast concrete vehicular paving blocks; Marshalls Plc; on prepared base (not included); on 50 mm compacted sharp sand bed; blocks laid in 7 mm loose sand and vibrated; joints filled with sharp sand and vibrated; level and to falls only							
"Trafica" paving blocks; 450 x 450 x 70 mm							
"Perfecta" finish; colour natural	23.71	0.50	10.50	0.09	26.48	m²	**37.07**
"Perfecta" finish; colour buff	27.46	0.50	10.50	0.09	30.32	m²	**40.91**
"Saxon" finish; colour natural	20.85	0.50	10.50	0.09	23.55	m²	**34.14**
"Saxon" finish; colour buff	23.96	0.50	10.50	0.09	26.74	m²	**37.33**
Precast concrete vehicular paving blocks; "Keyblock" Marshalls Plc; on prepared base (not included); on 50 mm compacted sharp sand bed; blocks laid in 7 mm loose sand and vibrated; joints filled with sharp sand and vibrated; level and to falls only							
Herringbone bond							
200 x 100 x 60 mm; natural grey	7.09	1.50	31.50	0.09	9.62	m²	**41.21**
200 x 100 x 60 mm; colours	7.70	1.50	31.50	0.09	10.26	m²	**41.85**
200 x 100 x 80 mm; natural grey	7.89	1.50	31.50	0.09	10.46	m²	**42.05**
200 x 100 x 80 mm; colours	8.91	1.50	31.50	0.09	11.53	m²	**43.12**
Basketweave bond							
200 x 100 x 60 mm; natural grey	7.09	1.20	25.20	0.09	9.62	m²	**34.91**
200 x 100 x 60 mm; colours	7.70	1.20	25.20	0.09	10.26	m²	**35.55**
200 x 100 x 80 mm; natural grey	7.89	1.20	25.20	0.09	10.46	m²	**35.75**
200 x 100 x 80 mm; colours	8.91	1.20	25.20	0.09	11.53	m²	**36.82**

Q PAVING/PLANTING/FENCING/SITE FURNITURE

Item Excluding site overheads and profit	PC £	Labour hours	Labour £	Plant £	Material £	Unit	**Total rate £**
Q24 INTERLOCKING BRICK/BLOCK ROADS/PAVINGS - cont'd							
Precast concrete vehicular paving blocks; Charcon Hard Landscaping; on prepared base (not included); on 50 mm compacted sharp sand bed; blocks laid in 7 mm loose sand and vibrated; joints filled with sharp sand and vibrated; level and to falls only							
"Europa" concrete blocks							
200 x 100 x 60 mm; natural grey	9.87	1.50	31.50	0.09	12.54	m^2	**44.13**
200 x 100 x 60 mm; colours	10.57	1.50	31.50	0.09	13.28	m^2	**44.87**
"Parliament" concrete blocks							
200 x 100 x 65 mm; natural grey	23.89	1.50	31.50	0.09	26.67	m^2	**58.25**
200 x 100 x 65 mm; colours	23.89	1.50	31.50	0.09	26.67	m^2	**58.25**
Recycled polyethylene grassblocks; ADP Netlon Turf Systems; interlocking units laid to prepared base or rootzone (not included)							
"ADP Netpave 50"; load bearing 150 tonnes per m^2; 500 x 500 x 50 mm deep							
5 - 49 m^2	15.00	0.20	4.20	0.18	17.59	m^2	**21.97**
50 - 199 m^2	13.50	0.20	4.20	0.18	16.09	m^2	**20.47**
"ADP Netpave 25"; load bearing: light vehicles and pedestrians; 500 x 500 x 25 mm deep; laid onto established grass surface							
5 - 99 m^2	14.00	0.10	2.10	-	14.00	m^2	**16.10**
100 - 399 m^2	12.25	0.10	2.10	-	12.25	m^2	**14.35**
"ADP Turfguard"; extruded polyethylene flexible mesh laid to existing grass surface or newly seeded areas to provide surface protection from traffic including vehicle or animal wear and tear							
ADP Turfguard Standard 30 m x 2 m; up to 300 m^2	2.57	0.01	0.17	-	2.57	m^2	**2.75**
ADP Turfguard Standard 30 m x 2 m; up to 600 m^2	2.29	0.01	0.17	-	2.29	m^2	**2.47**
ADP Turfguard Premium 30 m x 2 m; up to 300 m^2	2.74	0.01	0.17	-	2.74	m^2	**2.91**
ADP Turfguard Premium 30 m x 2 m; over 600 m^2	2.46	0.01	0.17	-	2.46	m^2	**2.63**
Q25 SLAB/BRICK/SETT/COBBLE PAVINGS							
Bricks - General							
Preamble: BS 3921 includes the following specification for bricks for paving: bricks shall be hard, well burnt, non-dusting, resistant to frost and sulphate attack and true to shape, size and sample.							
Movement of materials							
Mechanically offloading bricks; loading wheelbarrows; transporting maximum 25 m distance	-	0.20	4.20	-	-	m^2	**4.20**
Edge restraints; to brick paving; on prepared base (not included); 65 mm thick bricks; PC £300.00/1000; haunching one side							
Header course							
200 x 100 mm; butt joints	3.00	0.27	5.60	-	5.90	m	**11.50**
210 x 105 mm; mortar joints	2.67	0.50	10.50	-	5.72	m	**16.22**
Stretcher course							
200 x 100 mm; butt joints	1.50	0.17	3.50	-	4.20	m	**7.70**
210 x 105 mm; mortar joints	1.36	0.33	7.00	-	4.13	m	**11.13**

Q PAVING/PLANTING/FENCING/SITE FURNITURE

Item Excluding site overheads and profit	PC £	Labour hours	Labour £	Plant £	Material £	Unit	**Total rate £**
Variation in brick prices; add or subtract the following amounts for every £1.00/1000 difference in the PC price							
Edgings							
100 wide	-	-	-	-	0.05	10 m	-
200 wide	-	-	-	-	0.11	10 m	-
102.5 wide	-	-	-	-	0.05	10 m	-
215 wide	-	-	-	-	0.09	10 m	-
Clay brick pavings; on prepared base (not included); bedding on 50 mm sharp sand; kiln dried sand joints							
Pavings; 200 x 100 x 65 mm wirecut chamfered paviors							
brick; PC £520.00/1000	26.00	1.97	41.35	0.23	28.54	m^2	**70.11**
Clay brick pavings; 200 x 100 x 50; laid to running stretcher, or stack bond only; on prepared base (not included); bedding on cement:sand (1:4) pointing mortar as work proceeds							
PC £600.00/1000							
laid on edge	47.62	4.76	100.00	-	55.59	m^2	**155.59**
laid on edge but pavior 65 mm thick	40.00	3.81	80.00	-	47.78	m^2	**127.78**
laid flat	25.97	2.20	46.20	-	30.66	m^2	**76.86**
PC £500.00/1000							
laid on edge	39.68	4.76	100.00	-	47.45	m^2	**147.45**
laid on edge but pavior 65 mm thick	33.33	3.81	80.00	-	40.94	m^2	**120.94**
laid flat	21.64	2.20	46.20	-	26.23	m^2	**72.43**
PC £400.00/1000							
laid on edge	31.74	4.76	100.00	-	39.32	m^2	**139.32**
laid on edge but pavior 65 mm thick	26.66	3.81	80.00	-	34.11	m^2	**114.11**
laid flat	17.32	2.20	46.20	-	21.79	m^2	**67.99**
PC £300.00/1000							
laid on edge	23.81	4.76	100.00	-	31.18	m^2	**131.18**
laid on edge but pavior 65 mm thick	20.00	3.81	80.00	-	27.28	m^2	**107.28**
laid flat	12.99	2.20	46.20	-	17.35	m^2	**63.55**
Clay brick pavings; 200 x 100 x 50; butt jointed laid herringbone or basketweave pattern only; on prepared base (not included); bedding on 50 mm sharp sand							
PC £600.00/1000							
laid flat	30.00	1.44	30.32	0.29	33.15	m^2	**63.75**
PC £500.00/1000							
laid flat	25.00	1.44	30.32	0.29	28.03	m^2	**58.63**
PC £400.00/1000							
laid flat	20.00	1.44	30.32	0.29	22.90	m^2	**53.50**
PC £300.00/1000							
laid flat	15.00	1.44	30.32	0.29	17.78	m^2	**48.38**
Clay brick pavings; 215 x 102.5 x 65 mm; on prepared base (not included); bedding on cement:sand (1:4) pointing mortar as work proceeds							
Paving bricks; PC £600.00/1000; herringbone bond							
laid on edge	35.55	3.55	74.65	-	41.86	m^2	**116.51**
laid flat	23.70	2.37	49.77	-	29.12	m^2	**78.89**
Paving bricks; PC £600.00/1000; basketweave bond							
laid on edge	35.55	2.37	49.77	-	41.86	m^2	**91.63**
laid flat	23.70	1.58	33.19	-	29.12	m^2	**62.30**

Q PAVING/PLANTING/FENCING/SITE FURNITURE

Item Excluding site overheads and profit	PC £	Labour hours	Labour £	Plant £	Material £	Unit	Total rate £
Q25 SLAB/BRICK/SETT/COBBLE PAVINGS - cont'd							
Clay brick pavings; 215 x 102.5 x 65 mm - cont'd							
Paving bricks; PC £600.00/1000; running or stack bond							
laid on edge	35.55	1.90	39.83	-	41.86	m²	**81.68**
laid flat	23.70	1.26	26.55	-	29.12	m²	**55.67**
Paving bricks; PC £500.00/1000; herringbone bond							
laid on edge	29.63	3.55	74.65	-	38.59	m²	**113.24**
laid flat	19.75	2.37	49.77	-	25.17	m²	**74.94**
Paving bricks; PC £500.00/1000; basketweave bond							
laid on edge	29.63	2.37	49.77	-	33.48	m²	**83.25**
laid flat	19.75	1.58	33.19	-	25.17	m²	**58.35**
Paving bricks; PC £500.00/1000; running or stack bond							
laid on edge	29.63	1.90	39.83	-	33.48	m²	**73.30**
laid flat	19.75	1.26	26.55	-	25.17	m²	**51.72**
Paving bricks; PC £400.00/1000; herringbone bond							
laid on edge	23.70	3.55	74.65	-	27.40	m²	**102.06**
laid flat	15.80	2.37	49.77	-	21.12	m²	**70.89**
Paving bricks; PC £400.00/1000; basketweave bond							
laid on edge	23.70	2.37	49.77	-	27.40	m²	**77.18**
laid flat	15.80	1.58	33.19	-	21.12	m²	**54.30**
Paving bricks; PC £400.00/1000; running or stack bond							
laid on edge	23.70	1.90	39.83	-	27.40	m²	**67.23**
laid flat	15.80	1.26	26.55	-	21.12	m²	**47.66**
Paving bricks; PC £300.00/1000; herringbone bond							
laid on edge	17.77	3.55	74.65	-	20.89	m²	**95.54**
laid flat	11.85	2.37	49.77	-	17.27	m²	**67.04**
Paving bricks; PC £300.00/1000; basketweave bond							
laid on edge	17.77	2.37	49.77	-	20.89	m²	**70.66**
laid flat	11.85	1.58	33.19	-	17.27	m²	**50.45**
Paving bricks; PC £300.00/1000; running or stack bond							
laid on edge	17.77	1.90	39.82	-	20.89	m²	**60.70**
laid flat	11.85	1.26	26.55	-	17.27	m²	**43.82**
Cutting							
curved cutting	-	0.44	9.33	4.74	-	m	**14.07**
raking cutting	-	0.33	7.00	3.75	-	m	**10.75**
Add or subtract the following amounts for every £10.00/1000 difference in the prime cost of bricks							
Butt joints							
200 x 100	-	-	-	-	0.50	m²	**-**
215 x 102.5	-	-	-	-	0.45	m²	**-**
10 mm mortar joints							
200 x 100	-	-	-	-	0.43	m²	**-**
215 x 102.5	-	-	-	-	0.40	m²	**-**

Q PAVING/PLANTING/FENCING/SITE FURNITURE

Item Excluding site overheads and profit	PC £	Labour hours	Labour £	Plant £	Material £	Unit	Total rate £
Precast concrete pavings; Charcon Hard Landscaping; to BS 7263; on prepared sub-base (not included); bedding on 25 mm thick cement:sand mortar (1:4); butt joints; straight both ways; jointing in cement:sand (1:3) brushed in; on 50 mm thick sharp sand base							
Pavings; natural grey							
450 x 450 x 70 mm chamfered	13.83	0.44	9.33	-	18.35	m^2	**27.68**
450 x 450 x 50 mm chamfered	12.00	0.44	9.33	-	16.52	m^2	**25.85**
600 x 300 x 50 mm	10.83	0.44	9.33	-	15.35	m^2	**24.69**
400 x 400 x 65 mm chamfered	22.25	0.40	8.40	-	26.77	m^2	**35.17**
450 x 600 x 50 mm	9.63	0.44	9.33	-	14.15	m^2	**23.48**
600 x 600 x 50 mm	7.72	0.40	8.40	-	12.24	m^2	**20.64**
750 x 600 x 50 mm	7.29	0.40	8.40	-	11.81	m^2	**20.21**
900 x 600 x 50 mm	6.68	0.40	8.40	-	11.21	m^2	**19.61**
Pavings; coloured							
450 x 450 x 70 mm chamfered	16.20	0.44	9.33	-	20.72	m^2	**30.05**
450 x 600 x 50 mm	14.52	0.44	9.33	-	19.04	m^2	**28.37**
400 x 400 x 65 mm chamfered	22.25	0.40	8.40	-	26.77	m^2	**35.17**
600 x 600 x 50 mm	12.08	0.40	8.40	-	16.60	m^2	**25.00**
750 x 600 x 50 mm	11.16	0.40	8.40	-	15.68	m^2	**24.08**
900 x 600 x 50 mm	9.90	0.40	8.40	-	14.42	m^2	**22.82**
Precast concrete pavings; Charcon Hard Landscaping; to BS 7263; on prepared sub-base (not included); bedding on 25 mm thick cement:sand mortar (1:4); butt joints; straight both ways; jointing in cement:sand (1:3) brushed in; on 50 mm thick sharp sand base							
"Appalacian" rough textured exposed aggregate pebble paving							
600 mm x 600 mm x 65 mm	28.76	0.50	10.50	-	32.02	m^2	**42.52**
Precast concrete pavings; Marshalls Plc; "Heritage" imitation riven yorkstone paving; on prepared sub-base measured separately; bedding on 25 mm thick cement:sand mortar (1:4); pointed straight both ways cement:sand (1:3)							
Square and rectangular paving							
450 x 300 x 38 mm	33.28	1.00	21.00	-	36.87	m^2	**57.87**
450 x 450 x 38 mm	20.04	0.75	15.75	-	23.63	m^2	**39.38**
600 x 300 x 38 mm	21.94	0.80	16.80	-	25.53	m^2	**42.33**
600 x 450 x 38 mm	22.24	0.75	15.75	-	25.82	m^2	**41.57**
600 x 600 x 38 mm	21.48	0.50	10.50	-	25.13	m^2	**35.63**
Extra labours for laying the a selection of the above sizes to random rectangular pattern	-	0.33	7.00	-	-	m^2	**7.00**
Radial paving for circles							
circle with centre stone and first ring (8 slabs), 450 x 230/560 x 38 mm; diameter 1.54 m (total area 1.86 m^2)	58.00	1.50	31.50	-	62.70	nr	**94.20**
circle with second ring (16 slabs), 450 x 300/460 x 38 mm; diameter 2.48 m (total area 4.83 m^2)	149.52	4.00	84.00	-	161.59	nr	**245.59**
circle with third ring (16 slabs), 450 x 470/625 x 38 mm; diameter 3.42 m (total area 9.18 m^2)	272.08	8.00	168.00	-	294.81	nr	**462.81**
Stepping stones							
380 dia x 38 mm	4.13	0.20	4.20	-	9.21	nr	**13.41**
asymmetrical 560 x 420 x 38 mm	6.16	0.20	4.20	-	11.24	nr	**15.44**

Q PAVING/PLANTING/FENCING/SITE FURNITURE

Item Excluding site overheads and profit	PC £	Labour hours	Labour £	Plant £	Material £	Unit	Total rate £
Q25 SLAB/BRICK/SETT/COBBLE PAVINGS - cont'd							
Precast concrete pavings; Marshalls Plc; "Chancery" imitation reclaimed riven yorkstone paving; on prepared sub-base measured separately; bedding on 25 mm thick cement:sand mortar (1:4); pointed straight both ways cement:sand (1:3)							
Square and rectangular paving							
300 x 300 x 45 mm	22.98	1.00	21.00	-	26.57	m^2	**47.57**
450 x 300 x 45 mm	21.67	0.90	18.90	-	25.26	m^2	**44.16**
600 x 300 x 45 mm	21.04	0.80	16.80	-	24.63	m^2	**41.43**
600 x 450 x 45 mm	21.09	0.75	15.75	-	24.68	m^2	**40.43**
450 x 450 x 45 mm	19.06	0.75	15.75	-	22.65	m^2	**38.40**
600 x 600 x 45 mm	21.62	0.50	10.50	-	25.20	m^2	**35.70**
Extra labours for laying the a selection of the above sizes to random rectangular pattern	-	0.33	7.00	-	-	m^2	**7.00**
Radial paving for circles							
circle with centre stone and first ring (8 slabs), 450 x 230/560 x 38 mm; diameter 1.54 m (total area 1.86 m^2)	69.58	1.50	31.50	-	74.28	nr	**105.78**
circle with second ring (16 slabs), 450 x 300/460 x 38 mm; diameter 2.48 m (total area 4.83 m^2)	179.50	4.00	84.00	-	191.57	nr	**275.57**
circle with third ring (16 slabs), 450 x 470/625 x 38 mm; diameter 3.42 m (total area 9.18 m^2)	326.70	8.00	168.00	-	349.43	nr	**517.43**
Squaring off set for 2 ring circle							
16 slabs; 2.72 m^2	191.14	1.00	21.00	-	191.14	nr	**212.14**
Edge restraints; to block paving; on prepared base (not included); 200 x 100 x 80 mm; PC £7.89/m^2; haunching one side							
Header course							
200 x 100 mm; butt joints	1.58	0.27	5.60	-	4.48	m	**10.08**
Stretcher course							
200 x 100 mm; butt joints	0.79	0.17	3.50	-	3.49	m	**6.99**
Concrete paviors; Marshalls Plc; on prepared base (not included); bedding on 50 mm sand; kiln dried sand joints swept in							
"Keyblock" paviors							
200 x 100 x 60 mm; grey	7.09	0.40	8.40	0.09	9.27	m^2	**17.76**
200 x 100 x 60 mm; colours	7.70	0.40	8.40	0.09	9.88	m^2	**18.37**
200 x 100 x 80 mm; grey	7.89	0.44	9.33	0.09	10.07	m^2	**19.49**
200 x 100 x 80 mm; colours	8.91	0.44	9.33	0.09	11.09	m^2	**20.51**
Concrete cobble paviors; Charcon Hard Landscaping; Concrete Products; on prepared base (not included); bedding on 50 mm sand; kiln dried sand joints swept in							
Paviors							
"Woburn" blocks; 100 - 201 x 134 x 80 mm; random sizes	25.11	0.67	14.00	0.09	27.92	m^2	**42.01**
"Woburn" blocks; 100 - 201 x 134 x 80 mm; single size	25.11	0.50	10.50	0.09	27.92	m^2	**38.51**
"Woburn" blocks; 100 - 201 x 134 x 60 mm; random sizes	20.33	0.67	14.00	0.09	23.02	m^2	**37.11**
"Woburn" blocks; 100 - 201 x 134 x 60 mm; single size	20.33	0.50	10.50	0.09	23.00	m^2	**33.59**

Q PAVING/PLANTING/FENCING/SITE FURNITURE

Item Excluding site overheads and profit	PC £	Labour hours	Labour £	Plant £	Material £	Unit	Total rate £
Concrete setts; on 25 mm sand; compacted; vibrated; joints filled with sand; natural or coloured; well rammed hardcore base (not included)							
Marshalls Plc; Tegula cobble paving							
60 mm thick; random sizes	19.07	0.57	12.00	0.09	22.20	m^2	**34.29**
60 mm thick; single size	19.07	0.45	9.54	0.09	22.40	m^2	**32.04**
80 mm thick; random sizes	21.97	0.57	12.00	0.09	25.45	m^2	**37.54**
80 mm thick; single size	21.97	0.45	9.54	0.09	25.45	m^2	**35.08**
cobbles 80 x 80 x 60 mm thick; traditional	33.30	0.56	11.67	0.09	36.51	m^2	**48.27**
Cobbles							
Charcon Hard Landscaping; Country setts							
100 mm thick; random sizes	31.01	1.00	21.00	0.09	33.39	m^2	**54.48**
100 mm thick; single size	31.01	0.67	14.00	0.09	34.16	m^2	**48.25**
Natural stone, slab or granite paving - General							
Preamble: Provide paving slabs of the specified thickness in random sizes but not less than 25 slabs per 10 m^2 of surface area, to be laid in parallel courses with joints alternately broken and laid to falls.							
Reconstituted yorkstone aggregate pavings; Marshalls Plc; "Saxon" on prepared sub-base measured separately; bedding on 25 mm thick cement:sand mortar (1:4) ;on 50 mm thick sharp sand base							
Square and rectangular paving in various colours; butt joints straight both ways							
300 x 300 x 35 mm	30.64	0.80	16.80	-	35.78	m^2	**52.58**
600 x 300 x 35 mm	20.10	0.65	13.65	-	25.25	m^2	**38.90**
450 x 450 x 50 mm	21.98	0.75	15.75	-	27.12	m^2	**42.87**
600 x 600 x 35 mm	15.06	0.50	10.50	-	20.21	m^2	**30.71**
600 x 600 x 50 mm	18.93	0.55	11.55	-	24.07	m^2	**35.62**
Square and rectangular paving in natural; butt joints straight both ways							
300 x 300 x 35 mm	25.53	0.80	16.80	-	30.67	m^2	**47.47**
450 x 450 x 50 mm	18.72	0.70	14.70	-	23.86	m^2	**38.56**
600 x 300 x 35 mm	17.70	0.75	15.75	-	22.84	m^2	**38.59**
600 x 600 x 35 mm	12.99	0.50	10.50	-	18.13	m^2	**28.63**
600 x 600 x 50 mm	15.76	0.60	12.60	-	20.91	m^2	**33.51**
Radial paving for circles; 20 mm joints							
circle with centre stone and first ring (8 slabs), 450 x 230/560 x 35 mm; diameter 1.54 m (total area 1.86 m^2)	59.92	1.50	31.50	-	64.88	nr	**96.38**
circle with second ring (16 slabs), 450 x 300/460 x 35 mm; diameter 2.48 m (total area 4.83 m^2)	146.00	4.00	84.00	-	158.81	nr	**242.81**
circle with third ring (24 slabs), 450 x 310/430 x 35 mm; diameter 3.42 m (total area 9.18 m^2)	275.12	8.00	168.00	-	299.62	nr	**467.62**
Granite setts; bedding on 25 mm cement:sand (1:3)							
Natural granite setts; 100 x 100 mm to 125 x 150 mm; x 150 to 250 mm length; riven surface; silver grey							
new; standard grade	21.15	2.00	42.00	-	28.73	m^2	**70.73**
new; high grade	27.38	2.00	42.00	-	34.95	m^2	**76.95**
reclaimed; cleaned	44.80	2.00	42.00	-	53.26	m^2	**95.26**

Q PAVING/PLANTING/FENCING/SITE FURNITURE

Item Excluding site overheads and profit	PC £	Labour hours	Labour £	Plant £	Material £	Unit	Total rate £
Q25 SLAB/BRICK/SETT/COBBLE PAVINGS - cont'd							
Natural stone, slate or granite flag pavings; CED Ltd; on prepared base (not included); bedding on 25 mm cement:sand (1:3); cement:sand (1:3) joints							
Yorkstone; riven laid random rectangular							
new slabs; 40 - 60 mm thick	63.84	1.71	36.00	-	73.54	m^2	**109.54**
reclaimed slabs, Cathedral grade; 50 - 75 mm thick	77.84	2.80	58.80	-	88.94	m^2	**147.74**
Donegal quartzite slabs; standard tiles							
200 x random lengths x 15-25 mm	63.28	4.20	88.24	-	66.60	m^2	**154.84**
250 x random lengths x 15-25 mm	63.28	3.82	80.18	-	66.60	m^2	**146.79**
300 x random lengths x 15-25 mm	63.28	3.47	72.92	-	66.60	m^2	**139.52**
350 x random lengths x 15-25 mm	63.28	3.16	66.27	-	66.60	m^2	**132.87**
400 x random lengths x 15-25 mm	63.28	2.87	60.26	-	66.60	m^2	**126.86**
450 x random lengths x 15-25 mm	63.28	2.61	54.77	-	69.76	m^2	**124.54**
Natural Yorkstone, pavings or edgings; Johnsons Wellfield Quarries; sawn 6 sides; 50 mm thick; on prepared base measured separately; bedding on 25 mm cement:sand (1:3); cement:sand (1:3) joints							
Paving							
laid to random rectangular pattern	59.52	1.71	36.00	-	66.25	m^2	**102.24**
laid to coursed laying pattern; 3 sizes	64.55	1.72	36.21	-	70.93	m^2	**107.14**
Paving; single size							
600 x 600 mm	71.47	0.85	17.85	-	78.20	m^2	**96.05**
600 x 400 mm	70.32	1.00	21.00	-	76.99	m^2	**97.99**
300 x 200 mm	81.51	2.00	42.00	-	88.74	m^2	**130.74**
215 x 102.5 mm	85.28	2.50	52.50	-	92.70	m^2	**145.20**
Paving; cut to template off site; 600 x 600; radius:							
1.00 m	204.00	3.33	70.00	-	207.16	m^2	**277.16**
2.50 m	196.67	2.00	42.00	-	199.83	m^2	**241.83**
5.00 m	191.43	2.00	42.00	-	194.59	m^2	**236.59**
Edgings							
100 mm wide x random lengths	7.85	0.50	10.50	-	8.56	m	**19.06**
100 mm x 100 mm	8.15	0.50	10.50	-	11.72	m	**22.22**
100 mm x 200 mm	16.30	0.50	10.50	-	20.27	m	**30.77**
250 mm wide x random lengths	15.39	0.40	8.40	-	19.32	m	**27.72**
500 mm wide x random lengths	30.79	0.33	7.00	-	35.49	m	**42.49**
Yorkstone edgings; 600 mm long x 250 mm wide; cut to radius							
1.00 m to 3.00	53.35	0.50	10.50	-	56.34	m	**66.84**
3.00 m to 5.00 m	51.78	0.44	9.33	-	54.69	m	**64.02**
exceeding 5.00 m	50.21	0.40	8.40	-	53.04	m	**61.44**
Natural Yorkstone, pavings or edgings; Johnsons Wellfield Quarries; sawn 6 sides; 75 mm thick; on prepared base measured separately; bedding on 25 mm cement:sand (1:3); cement:sand (1:3) joints							
Paving							
laid to random rectangular pattern	65.96	0.95	19.95	-	72.41	m^2	**92.36**
laid to coursed laying pattern; 3 sizes	74.12	0.95	19.95	-	80.98	m^2	**100.93**
Paving; single size							
600 x 600 mm	82.60	0.95	19.95	-	89.89	m^2	**109.84**
600 x 400 mm	82.60	0.95	19.95	-	89.89	m^2	**109.84**
300 x 200 mm	94.85	0.75	15.75	-	102.75	m^2	**118.50**
215 x 102.5 mm	99.56	2.50	52.50	-	107.69	m^2	**160.19**

Q PAVING/PLANTING/FENCING/SITE FURNITURE

Item Excluding site overheads and profit	PC £	Labour hours	Labour £	Plant £	Material £	Unit	Total rate £
Paving; cut to template off site; 600 x 600; radius							
1.00 m	238.68	4.00	84.00	-	241.84	m²	**325.84**
2.50 m	232.40	2.50	52.50	-	235.56	m²	**288.06**
5.00 m	226.14	2.50	52.50	-	229.30	m²	**281.80**
Edgings							
100 mm wide x random lengths	10.05	0.60	12.60	-	10.87	m	**23.47**
100 mm x 100 mm	99.56	0.60	12.60	-	107.69	m	**120.29**
100 mm x 200 mm	19.93	0.50	10.50	-	24.09	m	**34.59**
250 mm wide x random lengths	18.53	0.50	10.50	-	22.61	m	**33.11**
500 mm wide x random lengths	37.06	0.40	8.40	-	42.07	m	**50.47**
Edgings; 600 mm long x 250 mm wide; cut to radius							
1.00 m to 3.00	62.45	0.60	12.60	-	65.89	m	**78.49**
3.00 m to 5.00 m	60.71	0.50	10.50	-	64.06	m	**74.56**
exceeding 5.00 m	59.91	0.44	9.33	-	63.22	m	**72.55**
CED Ltd; Indian sandstone, riven pavings or edgings; 25-35 mm thick; on prepared base measured separately; bedding on 25 mm cement:sand (1:3); cement:sand (1:3) joints							
Paving							
laid to random rectangular pattern	23.30	2.40	50.41	-	27.62	m²	**78.03**
laid to coursed laying pattern; 3 sizes	23.30	2.00	42.00	-	27.62	m²	**69.62**
Paving; single size							
600 x 600 mm	23.30	1.00	21.00	-	27.62	m²	**48.62**
600 x 400 mm	23.30	1.25	26.25	-	27.62	m²	**53.87**
400 x 400 mm	23.30	1.67	35.00	-	27.62	m²	**62.62**
Natural stone, slate or granite flag pavings; CED Ltd; on prepared base (not included); bedding on 25 mm cement:sand (1:3); cement:sand (1:3) joints							
Granite paving; sawn 6 sides; textured top							
new slabs; silver grey, blue grey or yellow; 50 mm thick	38.98	1.71	36.00	-	44.25	m²	**80.25**
new slabs; black; 50 mm thick	54.88	1.71	36.00	-	60.94	m²	**96.94**
Cobble pavings - General							
Cobbles should be embedded by hand, tight-butted, endwise to a depth of 60% of their length. A dry grout of rapid-hardening cement:sand (1:2) shall be brushed over the cobbles until the interstices are filled to the level of the adjoining paving. Surplus grout shall then be brushed off and a light, fine spray of water applied over the area.							
Cobble pavings							
Cobbles; to present a uniform colour in panels; or varied in colour as required							
Scottish beach cobbles; 200 - 100 mm	18.76	2.00	42.00	-	25.68	m²	**67.68**
Scottish beach cobbles; 100 - 75 mm	13.27	2.50	52.50	-	20.19	m²	**72.69**
Scottish beach cobbles; 75 - 50 mm	7.73	3.33	70.00	-	14.65	m²	**84.65**

Q PAVING/PLANTING/FENCING/SITE FURNITURE

Item Excluding site overheads and profit	PC £	Labour hours	Labour £	Plant £	Material £	Unit	Total rate £
Q26 SPECIAL SURFACINGS/PAVINGS FOR SPORT/GENERAL AMENITY							
Market prices of surfacing materials							
Surfacings; Melcourt Industries Ltd (items labelled FSC are Forest Stewardship Council certified)							
Playbark 10/50®; per 25 m^3 load	-	-	-	-	51.60	m^3	-
Playbark 8/25®; per 25 m^3 load	-	-	-	-	50.10	m^3	-
Playchips®; per 25 m^3 load; FSC	-	-	-	-	34.50	m^3	-
Kushyfall; per 25 m^3 load; FSC	-	-	-	-	32.70	m^3	-
Softfall; per 25 m^3 load	-	-	-	-	20.40	m^3	-
Playsand; per 10 t load	-	-	-	-	81.36	m^3	-
Woodfibre; per 25 m^3 load; FSC	-	-	-	-	29.95	m^3	-
Playgrounds; Wicksteed Leisure Ltd							
Safety tiles; on prepared base (not included)							
1000 x 1000 x 60 mm; red or green	47.00	0.20	4.20	-	47.00	m^2	**51.20**
1000 x 1000 x 60 mm; black	44.00	0.20	4.20	-	44.00	m^2	**48.20**
1000 x 1000 x 43 mm; red or green	44.00	0.20	4.20	-	44.00	m^2	**48.20**
1000 x 1000 x 43 mm; black	41.00	0.20	4.20	-	41.00	m^2	**45.20**
Playgrounds; SMP Playgrounds Ltd							
Tiles; on prepared base (not included)							
Premier 25; 1000 x 1000 x 25 mm; black; for general use	38.00	0.20	4.20	-	44.16	m^2	**48.36**
Premier 70; 1000 x 1000 x 70 mm; black; for higher equipment	64.00	0.25	5.25	-	70.16	m^2	**75.41**
Playgrounds; Melcourt Industries Ltd							
"Playbark"; on drainage layer (not included); to BS 5696; minimum 300 mm settled depth							
"Playbark 8/25", 8 - 25 mm particles; red/brown	15.03	0.35	7.37	-	15.03	m^2	**22.40**
"Playbark 10/50"; 10 - 50 mm particles; red/brown	17.20	0.55	10.69	-	17.20	m^2	**27.89**
Playgrounds; timber edgings							
Timber edging boards; 50 x 50 x 750 mm timber pegs at 1000 mm centres; excavations and hardcore under edgings (not included)							
50 x 150 mm; hardwood (iroko) edge boards	9.60	0.10	2.10	-	10.31	m	**12.41**
38 x 150 mm; hardwood (iroko) edge boards	7.41	0.10	2.10	-	8.12	m	**10.22**
50 x 150 mm; treated softwood edge boards	2.42	0.10	2.10	-	2.42	m	**4.52**
38 x 150 mm; treated softwood edge boards	1.80	0.10	2.10	-	2.52	m	**4.62**
Q30 SEEDING/TURFING							
Seeding/turfing - General							
Market prices of seeding materials							
Please see market prices in the Major Works section of this book.							
Market prices of chemicals and application rates							
Please see market prices in the Major Works section of this book.							
Cultivation							
Breaking up existing ground; using pedestrian operated tine cultivator or rotavator							
100 mm deep	-	0.50	10.50	4.75	-	100 m^2	**15.25**
150 mm deep	-	0.57	12.00	5.42	-	100 m^2	**17.42**
200 mm deep	-	0.67	14.00	6.33	-	100 m^2	**20.33**

Q PAVING/PLANTING/FENCING/SITE FURNITURE

Item Excluding site overheads and profit	PC £	Labour hours	Labour £	Plant £	Material £	Unit	Total rate £
As above but in heavy clay or wet soils							
100 mm deep	-	0.67	14.00	6.33	-	100 m²	**20.33**
150 mm deep	-	0.80	16.80	7.59	-	100 m²	**24.39**
200 mm deep	-	1.00	21.00	9.49	-	100 m²	**30.49**
Rolling cultivated ground lightly; using self-propelled agricultural roller	-	0.06	1.17	0.60	-	100 m²	**1.77**
Importing and storing selected and approved topsoil; to BS 3882; inclusive of settlement							
small quantities, less than 15 m³	31.05	-	-	-	37.26	m³	-
over 15 m³	18.50	-	-	-	22.20	m³	-
Spreading and lightly consolidating approved topsoil (imported or from spoil heaps); in layers not exceeding 150 mm; travel distance from spoil heaps not exceeding 25 m							
By machine (imported topsoil not included)							
minimum depth 100 mm	-	-	-	0.96	-	m²	**0.96**
minimum depth 200 mm	-	-	-	1.92	-	m²	**1.92**
minimum depth 300 mm	-	-	-	2.88	-	m²	**2.88**
minimum depth 500 mm	-	-	-	4.80	-	m²	**4.80**
over 500 mm	-	-	-	9.61	-	m³	**9.61**
By hand (imported topsoil not included)							
minimum depth 100 mm	-	0.20	4.20	-	-	m²	**4.20**
minimum depth 150 mm	-	0.30	6.31	-	-	m²	**6.31**
minimum depth 300 mm	-	0.60	12.57	-	-	m²	**12.57**
minimum depth 450 mm	-	0.90	18.92	-	-	m²	**18.92**
over 450 mm deep	-	2.22	46.62	-	-	m³	**46.62**
Extra over for spreading topsoil to slopes 15 - 30 degrees by machine or hand	-	-	-	-	-	10%	-
Extra over for spreading topsoil to slopes over 30 degrees by machine or hand	-	-	-	-	-	25%	-
Extra over for spreading topsoil from spoil heaps travel exceeding 100 m; by machine							
100 - 150 m	-	0.01	0.26	0.07	-	m³	**0.33**
150 - 200 m	-	0.02	0.39	0.11	-	m³	**0.50**
200 - 300 m	-	0.03	0.58	0.17	-	m³	**0.75**
Extra over spreading topsoil for travel exceeding 100 m; by hand							
100 m	-	0.83	17.50	-	-	m³	**17.50**
200 m	-	1.67	35.00	-	-	m³	**35.00**
300 m	-	2.50	52.50	-	-	m³	**52.50**
Evenly grading; to general surfaces to bring to finished levels							
by pedestrian operated rotavator	-	-	0.08	0.04	-	m²	**0.12**
by hand	-	0.01	0.21	-	-	m²	**0.21**
Extra over grading for slopes 15 - 30 degrees by machine or hand	-	-	-	-	-	10%	-
Extra over grading for slopes over 30 degrees by machine or hand	-	-	-	-	-	25%	-
Apply screened topdressing to grass surfaces; spread using Tru-Lute							
Sand soil mixes 90/10 to 50/50	0.12	-	0.04	0.03	0.12	m²	**0.20**
Spread only existing cultivated soil to final levels using Tru-Lute							
Cultivated soil	-	-	0.04	0.03	-	m²	**0.07**
Clearing stones; disposing off site; to distance not exceeding 13 km							
by hand; stones not exceeding 50 mm in any direction; loading to skip 4.6 m³	-	0.01	0.21	0.03	-	m²	**0.24**
Lightly cultivating; weeding; to fallow areas; disposing debris off site; to distance not exceeding 13 km							
by hand	-	0.01	0.30	-	0.07	m²	**0.37**

Item Excluding site overheads and profit	PC £	Labour hours	Labour £	Plant £	Material £	Unit	Total rate £
Q30 SEEDING/TURFING - cont'd							
Surface applications and soil additives; pre-seeding; material delivered to a maximum of 25 m from area of application; applied; by hand							
Soil conditioners; to cultivated ground; ground limestone; PC £35.00/tonne; including turning in							
0.25 kg/m^2 = 2.50 tonnes/ha	-	1.20	25.20	-	0.88	100 m^2	**26.07**
0.50 kg/m^2 = 5.00 tonnes/ha	1.75	1.33	28.00	-	1.75	100 m^2	**29.75**
0.75 kg/m^2 = 7.50 tonnes/ha	2.63	1.50	31.50	-	2.63	100 m^2	**34.13**
1.00 kg/m^2 = 10.00 tonnes/ha	3.50	1.71	36.00	-	3.50	100 m^2	**39.50**
Soil conditioners; to cultivated ground; medium bark; based on deliveries of 25 m^3 loads; PC £37.55/m^3; including turning in							
1 m^3 per 40 m^2 = 25 mm thick	0.94	0.02	0.47	-	0.94	m^2	**1.40**
1 m^3 per 20 m^2 = 50 mm thick	1.88	0.04	0.93	-	1.88	m^2	**2.81**
1 m^3 per 13.33 m^2 = 75 mm thick	2.82	0.07	1.40	-	2.82	m^2	**4.21**
1 m^3 per 10 m^2 = 100 mm thick	3.75	0.08	1.68	-	3.75	m^2	**5.43**
Soil conditioners; to cultivated ground; mushroom compost; delivered in 25 m^3 loads; PC £22.08/m^3; including turning in							
1 m^3 per 40 m^2 = 25 mm thick	0.55	0.02	0.47	-	0.55	m^2	**1.02**
1 m^3 per 20 m^2 = 50 mm thick	1.10	0.04	0.93	-	1.10	m^2	**2.04**
1 m^3 per 13.33 m^2 = 75 mm thick	1.66	0.07	1.40	-	1.66	m^2	**3.05**
1 m^3 per 10 m^2 = 100 mm thick	2.21	0.08	1.68	-	2.21	m^2	**3.89**
Soil conditioners; to cultivated ground; mushroom compost; delivered in 60 m^3 loads; PC £8.62/m^3; including turning in							
1 m^3 per 40 m^2 = 25 mm thick	0.22	0.02	0.47	-	0.22	m^2	**0.68**
1 m^3 per 20 m^2 = 50 mm thick	0.38	0.04	0.93	-	0.38	m^2	**1.32**
1 m^3 per 13.33 m^2 = 75 mm thick	0.65	0.07	1.40	-	0.65	m^2	**2.05**
1 m^3 per 10 m^2 = 100 mm thick	0.86	0.08	1.68	-	0.86	m^2	**2.54**
Preparation of seedbeds - General Preamble: for preliminary operations see "Cultivation" section.							
Preparation of seedbeds; soil preparation							
Lifting selected and approved topsoil from spoil heaps; passing through 6 mm screen; removing debris	-	0.08	1.75	4.50	0.01	m^3	**6.26**
Topsoil; supply only; PC £18.50/m^3; allowing for 20% settlement; 20 tonne loads							
25 mm	0.46	-	-	-	0.56	m^2	-
50 mm	0.93	-	-	-	1.11	m^2	-
100 mm	1.85	-	-	-	2.22	m^2	-
150 mm	2.77	-	-	-	3.33	m^2	-
200 mm	3.70	-	-	-	4.44	m^2	-
250 mm	4.63	-	-	-	5.55	m^2	-
300 mm	5.55	-	-	-	6.66	m^2	-
400 mm	7.40	-	-	-	8.88	m^2	-
450 mm	8.32	-	-	-	9.99	m^2	-
Topsoil; supply only; PC £31.90/m^3; allowing for 20% settlement; 10 tonne loads							
25 mm	0.80	-	-	-	0.96	m^2	-
50 mm	1.59	-	-	-	1.91	m^2	-
100 mm	3.19	-	-	-	3.83	m^2	-
150 mm	4.79	-	-	-	5.74	m^2	-
200 mm	6.38	-	-	-	7.66	m^2	-
250 mm	7.97	-	-	-	9.57	m^2	-
300 mm	9.57	-	-	-	11.48	m^2	-
400 mm	12.76	-	-	-	15.31	m^2	-
450 mm	14.36	-	-	-	17.23	m^2	-

Q PAVING/PLANTING/FENCING/SITE FURNITURE

Item Excluding site overheads and profit	PC £	Labour hours	Labour £	Plant £	Material £	Unit	Total rate £
Spreading topsoil to form seedbeds average thickness not exceeding 200 mm (topsoil not included); by machine; grading and cultivation not included							
Excavated material; from spoil heaps							
average 25 m distance	-	0.10	1.85	5.75	-	m^3	**7.60**
average 50 m distance	-	0.10	1.85	6.27	-	m^3	**8.12**
average 100 m distance	-	0.10	1.85	6.84	-	m^3	**8.69**
average 200 m distance	-	0.10	1.85	8.55	-	m^3	**10.40**
Excavated material; spreading on site							
25 mm deep	-	-	-	0.21	-	m^2	**0.21**
50 mm deep	-	-	-	0.42	-	m^2	**0.42**
75 mm deep	-	-	-	0.62	-	m^2	**0.62**
100 mm deep	-	-	-	0.83	-	m^2	**0.83**
150 mm deep	-	-	-	1.00	-	m^2	**1.00**
Spreading only topsoil to form seedbeds (topsoil not included); by hand							
25 mm deep	-	0.03	0.53	-	-	m^2	**0.53**
50 mm deep	-	0.03	0.70	-	-	m^2	**0.70**
75 mm deep	-	0.04	0.90	-	-	m^2	**0.90**
100 mm deep	-	0.05	1.05	-	-	m^2	**1.05**
150 mm deep	-	0.08	1.57	-	-	m^2	**1.57**
Bringing existing topsoil to a fine tilth for seeding; by raking or harrowing; stones not to exceed 6 mm; by machine	-	-	0.08	0.04	-	m^2	**0.13**
Bringing existing topsoil to a fine tilth for seeding; by raking or harrowing; stones not to exceed 6 mm; by hand	-	0.01	0.18	-	-	m^2	**0.18**
Preparation of seedbeds; soil treatments							
For the following topsoil improvement and seeding operations add or subtract the following amounts for every £0.10 difference in the material cost price							
35 g/m^2	-	-	-	-	0.35	100 m^2	-
50 g/m^2	-	-	-	-	0.50	100 m^2	-
70 g/m^2	-	-	-	-	0.70	100 m^2	-
100 g/m^2	-	-	-	-	1.00	100 m^2	-
Pre-seeding fertilizers (6:9:6); PC £0.39/kg; to seedbeds; by hand							
35 g/m^2	1.36	0.17	3.50	-	1.36	100 m^2	**4.86**
50 g/m^2	1.94	0.17	3.50	-	1.94	100 m^2	**5.44**
70 g/m^2	2.72	0.17	3.50	-	2.72	100 m^2	**6.22**
100 g/m^2	3.89	0.20	4.20	-	3.89	100 m^2	**8.09**
125 g/m^2	4.86	0.20	4.20	-	4.86	100 m^2	**9.06**
Seeding grass areas - General							
Preamble: the British Standard recommendations for seed and seeding of grass areas are contained in BS 4428: 1989. The prices given in this section are based on compliance with the standard.							
Seeding							
Grass seed; spreading in two operations; PC £3.00/kg (for changes in material prices please refer to table above); by hand							
35 g/m^2	-	0.17	3.50	-	10.50	100 m^2	**14.00**
50 g/m^2	-	0.17	3.50	-	15.00	100 m^2	**18.50**
70 g/m^2	-	0.17	3.50	-	21.00	100 m^2	**24.50**
100 g/m^2	-	0.17	3.50	-	30.00	100 m^2	**33.50**
125 g/m^2	-	0.20	4.20	-	37.50	100 m^2	**41.70**

Q PAVING/PLANTING/FENCING/SITE FURNITURE

Item Excluding site overheads and profit	PC £	Labour hours	Labour £	Plant £	Material £	Unit	Total rate £
Q30 SEEDING/TURFING - cont'd							
Seeding - cont'd							
Extra over seeding by hand for slopes over 30 degrees (allowing for the actual area but measured in plan)							
35 g/m^2	1.56	-	0.08	-	1.56	100 m^2	**1.64**
50 g/m^2	2.25	-	0.08	-	2.25	100 m^2	**2.33**
70 g/m^2	3.15	-	0.08	-	3.15	100 m^2	**3.23**
100 g/m^2	4.50	-	0.09	-	4.50	100 m^2	**4.59**
125 g/m^2	5.61	-	0.09	-	5.61	100 m^2	**5.70**
Harrowing seeded areas; light chain harrow	-	-	-	0.09	-	100 m^2	**0.09**
Raking over seeded areas							
by hand	-	0.80	16.80	-	-	100 m^2	**16.80**
Rolling seeded areas; light roller							
by pedestrian operated mechanical roller	-	0.08	1.75	0.41	-	100 m^2	**2.16**
by hand drawn roller	-	0.17	3.50	-	-	100 m^2	**3.50**
Extra over harrowing, raking or rolling seeded areas for slopes over 30 degrees; by machine or hand	-	-	-	-	-	25%	**-**
Turf edging; to seeded areas; 300 mm wide	2.20	0.13	2.80	-	2.20	m^2	**5.00**
Preparation of turf beds							
Rolling turf to be lifted; lifting by hand or mechanical turf stripper; stacks to be not more than 1 m high							
cutting only preparing to lift; pedestrian turf cutter	-	0.75	15.75	9.48	-	100 m^2	**25.23**
lifting and stacking; by hand	-	8.33	175.00	-	-	100 m^2	**175.00**
Rolling up; moving to stacks							
distance not exceeding 100 m	-	2.50	52.50	-	-	100 m^2	**52.50**
extra over rolling and moving turf to stacks to transport per additional 100 m	-	0.83	17.50	-	-	100 m^2	**17.50**
Lifting selected and approved topsoil from spoil heaps							
passing through 6 mm screen; removing debris	-	0.17	3.50	9.00	-	m^3	**12.50**
Extra over lifting topsoil and passing through screen for imported topsoil; plus 20% allowance for settlement	18.50	-	-	-	22.20	m^3	**-**
Spreading topsoil to form turfbeds (topsoil not included); by machine							
25 mm deep	-	-	-	0.24	-	m^2	**0.24**
50 mm deep	-	-	-	0.48	-	m^2	**0.48**
75 mm deep	-	-	-	0.72	-	m^2	**0.72**
100 mm deep	-	-	-	0.96	-	m^2	**0.96**
150 mm deep	-	-	-	1.44	-	m^2	**1.44**
Bringing existing topsoil to a fine tilth for turfing by raking or harrowing; stones not to exceed 6 mm; by hand	-	0.01	0.28	-	-	m^2	**0.28**
Spreading topsoil to form turfbeds (topsoil not included); by hand							
25 mm deep	-	0.04	0.84	-	-	m^2	**0.84**
50 mm deep	-	0.08	1.68	-	-	m^2	**1.68**
75 mm deep	-	0.12	2.52	-	-	m^2	**2.52**
100 mm deep	-	0.15	3.15	-	-	m^2	**3.15**
150 mm deep	-	0.23	4.73	-	-	m^2	**4.73**
Operations after spreading of topsoil; by hand							
Bringing existing topsoil to a fine tilth for turfing by raking or harrowing; stones not to exceed 6 mm							
topsoil spread by machine (soil compacted)	-	0.03	0.53	-	-	m^2	**0.53**
topsoil spread by hand	-	0.02	0.35	-	-	m^2	**0.35**

Q PAVING/PLANTING/FENCING/SITE FURNITURE

Item Excluding site overheads and profit	PC £	Labour hours	Labour £	Plant £	Material £	Unit	Total rate £
Turfing							
Turfing; laying only; to stretcher bond; butt joints; including providing and working from barrow plank runs where necessary to surfaces not exceeding 30 degrees from horizontal; average size of turves 0.50 yd² (0.418 m²); 2.39 turves per m²							
specially selected lawn turves from previously lifted stockpile	-	0.20	4.20	-	-	m²	**4.20**
cultivated lawn turves; to larger open areas	-	0.11	2.24	-	-	m²	**2.24**
cultivated lawn turves; to domestic or garden areas	-	0.13	2.80	-	-	m²	**2.80**
Industrially grown turf; PC prices listed represent the general range of industrial turf prices for sportsfields and amenity purposes; prices will vary with quantity and site location							
"Rolawn"							
ref RB Medallion; sports fields, domestic lawns, general landscape	2.20	0.13	2.80	-	2.20	m²	**5.00**
"Inturf"							
ref Inturf 1; fine lawns, golf greens, bowling greens	3.03	0.13	2.80	-	3.03	m²	**5.83**
ref Inturf 2; football grounds, parks, hardwearing areas	1.83	0.13	2.80	-	1.83	m²	**4.63**
ref Inturf 2 Bargold; fine turf ; football grounds, parks, hardwearing areas	2.13	0.13	2.80	-	2.13	m²	**4.93**
ref Inturf 3; hockey grounds, polo, medium wearing areas	2.13	0.13	2.80	-	2.13	m²	**4.93**
Firming turves with wooden beater	-	0.01	0.21	-	-	m²	**0.21**
Rolling turfed areas; light roller							
by pedestrian operated mechanical roller	-	0.08	1.75	0.41	-	100 m²	**2.16**
by hand drawn roller	-	0.17	3.50	-	-	100 m²	**3.50**
Dressing with finely sifted topsoil; brushing into joints	0.02	0.05	1.05	-	0.02	m²	**1.07**
Turfing; laying only							
to slopes over 30 degrees; to diagonal bond (measured as plan area - add 15% to these rates for the incline area of 30 degree slopes)	-	0.12	2.52	-	-	m²	**2.52**
Extra over laying turfing for pegging down turves							
wooden or galvanized wire pegs; 200 mm long; 2 pegs per 0.50 m²	1.44	0.01	0.28	-	1.44	m²	**1.71**
Artificial grass; Artificial Lawn Company; laid to sharp sand bed priced separately 15 kg kiln sand brushed in per m²							
"Leisure"; 24 mm thick artificial sports turf; sand filled	-	-	-	-	-	m²	**26.17**
"Greenfield"; for general use; budget surface; sand filled	-	-	-	-	-	m²	**20.94**
"Garden"; patios conservatories and pool surrounds; sand filled	-	-	-	-	-	m²	**21.90**
"Premier"; lawns and patios	-	-	-	-	-	m²	**28.08**
"Elite"; grass/sand and rubber filled	-	-	-	-	-	m²	**31.45**
"Grassflex"; safety surfacing for play areas	-	-	-	-	-	m²	**44.97**
Maintenance operations (Note: the following rates apply to aftercare maintenance executed as part of a landscaping contract only)							
Initial cutting; to turfed areas							
20 mm high; using pedestrian guided power driven cylinder mower; including boxing off cuttings (stone picking and rolling not included)	-	0.18	3.78	0.28	-	100 m²	**4.06**

Q PAVING/PLANTING/FENCING/SITE FURNITURE

Item Excluding site overheads and profit	PC £	Labour hours	Labour £	Plant £	Material £	Unit	Total rate £
Q30 SEEDING/TURFING - cont'd							
Maintenance operations - cont'd							
Repairing damaged grass areas							
scraping out; removing slurry; from ruts and holes; average 100 mm deep	-	0.13	2.80	-	-	m^2	**2.80**
100 mm topsoil	-	0.13	2.80	-	2.22	m^2	**5.02**
Repairing damaged grass areas; sowing grass seed to match existing or as specified; to individually prepared worn patches							
35 g/m^2	0.11	0.01	0.21	-	0.12	m^2	**0.33**
50 g/m^2	0.15	0.01	0.21	-	0.17	m^2	**0.38**
Using pedestrian operated mechanical equipment and blowers							
grassed areas with perimeters of mature trees such as sports fields and amenity areas	-	0.04	0.84	0.05	-	100 m^2	**0.89**
grassed areas containing ornamental trees and shrub beds	-	0.10	2.10	0.13	-	100 m^2	**2.23**
verges	-	0.07	1.40	0.08	-	100 m^2	**1.49**
By hand							
grassed areas with perimeters of mature trees such as sports fields and amenity areas	-	0.05	1.05	0.10	-	100 m^2	**1.15**
grassed areas containing ornamental trees and shrub beds	-	0.08	1.75	0.17	-	100 m^2	**1.92**
verges	-	1.00	21.00	1.99	-	100 m^2	**22.99**
Removal of arisings							
areas with perimeters of mature trees	-	0.01	0.14	0.09	0.80	100 m^2	**1.03**
areas containing ornamental trees and shrub beds	-	0.02	0.42	0.34	2.00	100 m^2	**2.76**
Cutting grass to specified height; per cut							
ride-on triple cylinder mower	-	0.01	0.29	0.13	-	100 m^2	**0.42**
ride-on triple rotary mower	-	0.01	0.29	-	-	100 m^2	**0.29**
pedestrian mower (open areas)	-	0.18	3.78	0.35	-	100 m^2	**4.13**
pedestrian mower (small areas or areas with obstacles)	-	0.33	7.00	0.42	-	100 m^2	**7.42**
Cutting rough grass; per cut							
power flail or scythe cutter	-	0.04	0.73	-	-	100 m^2	**0.73**
Extra over cutting grass for slopes not exceeding 30 degrees	-	-	-	-	-	10%	-
Extra over cutting grass for slopes exceeding 30 degrees	-	-	-	-	-	40%	-
Cutting fine sward							
pedestrian operated seven-blade cylinder lawn mower	-	0.14	2.94	0.21	-	100 m^2	**3.15**
Extra over cutting fine sward for boxing off cuttings							
pedestrian mower	-	0.03	0.59	0.04	-	100 m^2	**0.63**
Cutting areas of rough grass							
scythe	-	1.00	21.00	-	-	100 m^2	**21.00**
sickle	-	2.00	42.00	-	-	100 m^2	**42.00**
petrol operated strimmer	-	0.30	6.31	0.42	-	100 m^2	**6.73**
Cutting areas of rough grass which contain trees or whips							
petrol operated strimmer	-	0.40	8.40	0.56	-	100 m^2	**8.96**
Extra over cutting rough grass for on site raking up and dumping	-	0.33	7.00	-	-	100 m^2	**7.00**
Trimming edge of grass areas; edging tool							
with petrol powered strimmer	-	0.13	2.80	0.19	-	100 m	**2.99**
by hand	-	0.67	14.00	-	-	100 m	**14.00**
Rolling grass areas; light roller							
by pedestrian operated mechanical roller	-	0.08	1.75	0.41	-	100 m^2	**2.16**
by hand drawn roller	-	0.17	3.50	-	-	100 m^2	**3.50**

Q PAVING/PLANTING/FENCING/SITE FURNITURE

Item Excluding site overheads and profit	PC £	Labour hours	Labour £	Plant £	Material £	Unit	Total rate £
Aerating grass areas; to a depth of 100 mm							
using pedestrian-guided motor powered solid or slitting tine turf aerator	-	0.18	3.68	2.82	-	100 m^2	**6.50**
using hollow tine aerator; including sweeping up and dumping corings	-	0.50	10.50	5.64	-	100 m^2	**16.14**
using hand aerator or fork	-	1.67	35.00	-	-	100 m^2	**35.00**
Extra over aerating grass areas for on site sweeping up and dumping corings	-	0.17	3.50	-	-	100 m^2	**3.50**
Switching off dew; from fine turf areas	-	0.20	4.20	-	-	100 m^2	**4.20**
Scarifying grass areas to break up thatch; removing dead grass							
using self-propelled scarifier; including removing and disposing of grass on site	-	0.33	7.00	0.14	-	100 m^2	**7.14**
by hand	-	3.03	63.64	-	-	100 m^2	**63.64**
For the following topsoil improvement and seeding operations add or subtract the following amounts for every £0.10 difference in the material cost price							
35 g/m^2	-	-	-	-	0.35	100 m^2	-
50 g/m^2	-	-	-	-	0.50	100 m^2	-
70 g/m^2	-	-	-	-	0.70	100 m^2	-
100 g/m^2	-	-	-	-	1.00	100 m^2	-
Top dressing fertilizers (7:7:7); PC £0.62/kg; to seedbeds; by hand							
35 g/m^2	2.27	0.17	3.50	-	2.27	100 m^2	**5.78**
50 g/m^2	3.25	0.17	3.50	-	3.25	100 m^2	**6.75**
70 g/m^2	4.55	0.17	3.50	-	4.55	100 m^2	**8.05**
Watering turf; evenly; at a rate of 5 litre/m^2							
using sprinkler equipment and with sufficient water pressure to run 1 nr 15 m radius sprinkler	-	0.02	0.42	-	-	100 m^2	**0.42**
using hand-held watering equipment	-	0.25	5.25	-	-	100 m^2	**5.25**
Q31 PLANTING							
Site protection; temporary protective fencing							
Cleft chestnut rolled fencing; to 100 mm diameter chestnut posts; driving into firm ground at 3 m centres; pales at 50 mm centres							
900 mm high	3.82	0.11	2.24	-	7.06	m	**9.30**
1100 mm high	4.48	0.11	2.24	-	7.54	m	**9.78**
1500 mm high	7.11	0.11	2.24	-	10.17	m	**12.41**
Extra over temporary protective fencing for removing and making good (no allowance for re-use of material)	-	0.07	1.40	0.20	-	m	**1.60**
Cultivation							
Treating soil with "Paraquat-Diquat" weedkiller at rate of 5 litre/ha; PC £9.13/litre; in accordance with manufacturer's instructions; including all safety precautions							
by hand	-	0.33	7.00	-	0.46	100 m^2	**7.46**
Breaking up existing ground; using pedestrian operated tine cultivator or rotavator							
100 mm deep	-	0.55	11.55	4.75	-	100 m^2	**16.30**
150 mm deep	-	0.63	13.20	5.42	-	100 m^2	**18.62**
200 mm deep	-	0.73	15.40	6.33	-	100 m^2	**21.73**
As above but in heavy clay or wet soils							
100 mm deep	-	0.73	15.40	6.33	-	100 m^2	**21.73**
150 mm deep	-	0.88	18.48	7.59	-	100 m^2	**26.07**
200 mm deep	-	1.10	23.10	9.49	-	100 m^2	**32.59**
Importing only selected and approved topsoil							
1 - 14 m^3	18.50	-	-	-	55.50	m^3	-
over 15 m^3	18.50	-	-	-	18.50	m^3	-

Q PAVING/PLANTING/FENCING/SITE FURNITURE

Item Excluding site overheads and profit	PC £	Labour hours	Labour £	Plant £	Material £	Unit	Total rate £
Q31 PLANTING - cont'd							
Imported topsoil; spreading and lightly consolidating approved topsoil in layers not exceeding 150 mm; travel distance from offloading point not exceeding 25 m							
By machine							
minimum depth 100 mm	-	-	-	0.96	2.59	m^2	**3.55**
minimum depth 200 mm	-	-	-	1.92	5.17	m^2	**7.10**
minimum depth 300 mm	-	-	-	2.88	7.76	m^2	**10.65**
minimum depth 500 mm	-	-	-	4.80	12.94	m^2	**17.74**
over 500 mm	-	-	-	9.61	25.88	m^3	**35.48**
By hand							
minimum depth 100 mm	-	0.20	4.20	-	2.59	m^2	**6.79**
minimum depth 150 mm	-	0.30	6.31	-	3.88	m^2	**10.19**
minimum depth 300 mm	-	0.60	12.57	-	7.76	m^2	**20.34**
minimum depth 500 mm	-	0.90	18.92	-	12.94	m^2	**31.86**
over 500 mm deep	-	2.22	46.62	-	25.88	m^3	**72.50**
Spreading and lightly consolidating approved topsoil (imported or from spoil heaps); in layers not exceeding 150 mm; travel distance from spoil heaps not exceeding 25 m							
By machine (imported topsoil not included)							
minimum depth 100 mm	-	-	-	0.96	-	m^2	**0.96**
minimum depth 200 mm	-	-	-	1.92	-	m^2	**1.92**
minimum depth 300 mm	-	-	-	2.88	-	m^2	**2.88**
minimum depth 500 mm	-	-	-	4.80	-	m^2	**4.80**
over 500 mm	-	-	-	9.61	-	m^3	**9.61**
By hand (imported topsoil not included)							
minimum depth 100 mm	-	0.20	4.20	-	-	m^2	**4.20**
minimum depth 150 mm	-	0.30	6.31	-	-	m^2	**6.31**
minimum depth 300 mm	-	0.60	12.57	-	-	m^2	**12.57**
minimum depth 500 mm	-	0.90	18.92	-	-	m^2	**18.92**
over 500 mm deep	-	2.22	46.62	-	-	m^3	**46.62**
Extra over for spreading topsoil to slopes 15 - 30 degrees by machine or hand	-	-	-	-	-	10%	-
Extra over for spreading topsoil to slopes over 30 degrees by machine or hand	-	-	-	-	-	25%	-
Extra over spreading topsoil for travel exceeding 100 m; by machine							
100 - 150 m	-	-	-	0.63	-	m^3	**0.63**
150 - 200 m	-	-	-	0.84	-	m^3	**0.84**
200 - 300 m	-	-	-	1.13	-	m^3	**1.13**
Extra over spreading topsoil for travel exceeding 100 m; by hand							
100 m	-	2.50	52.50	-	-	m^3	**52.50**
200 m	-	3.50	73.50	-	-	m^3	**73.50**
300 m	-	4.50	94.50	-	-	m^3	**94.50**
Evenly grading; to general surfaces to bring to finished levels							
by pedestrian operated rotavator	-	-	0.08	0.04	-	m^2	**0.12**
by hand	-	0.01	0.21	-	-	m^2	**0.21**
Extra over grading for slopes 15 - 30 degrees by machine or hand	-	-	-	-	-	10%	-
Extra over grading for slopes over 30 degrees by machine or hand	-	-	-	-	-	25%	-
Clearing stones; disposing off site							
by hand; stones not exceeding 50 mm in any direction; loading to skip 4.6 m^3	-	0.01	0.21	0.03	-	m^2	**0.24**
Lightly cultivating; weeding; to fallow areas; disposing debris off site							
by hand	-	0.01	0.30	-	0.07	m^2	**0.37**

Q PAVING/PLANTING/FENCING/SITE FURNITURE

Item Excluding site overheads and profit	PC £	Labour hours	Labour £	Plant £	Material £	Unit	Total rate £
Preparation of planting operations; herbicides and granular additives							
For the following topsoil improvement and planting operations add or subtract the following amounts for every £0.10 difference in the material cost price							
35 g/m^2	-	-	-	-	0.35	100 m^2	-
50 g/m^2	-	-	-	-	0.50	100 m^2	-
70 g/m^2	-	-	-	-	0.70	100 m^2	-
100 g/m^2	-	-	-	-	1.00	100 m^2	-
General herbicides; in accordance with manufacturer's instructions; "Knapsack" spray application							
Dextrone X" at 50 ml/100 m^2	-	0.33	7.00	-	0.46	100 m^2	**7.46**
"Roundup Pro - Bi Active" at 50 ml/100 m^2	0.52	0.33	7.00	-	0.52	100 m^2	**7.52**
"Casoron G (residual)" at 1 kg/125 m^2	6.75	0.33	7.00	-	7.42	100 m^2	**14.42**
Fertilizers; in top 150 mm of topsoil; at 35 g/m^2							
fertilizer (18:0:0+Mg+Fe)	7.58	0.12	2.57	-	7.58	100 m^2	**10.15**
"Enmag"	6.30	0.12	2.57	-	6.61	100 m^2	**9.18**
fertilizer (7:7:7)	2.27	0.12	2.57	-	2.27	100 m^2	**4.85**
"Super Phosphate Powder"	2.23	0.12	2.57	-	2.34	100 m^2	**4.92**
fertilizer (20:10:10)	3.06	0.12	2.57	-	3.06	100 m^2	**5.63**
"Bone meal"	3.65	0.12	2.57	-	3.83	100 m^2	**6.40**
Fertilizers; in top 150 mm of topsoil at 70 g/m^2							
fertilizer (18:0:0+Mg+Fe)	15.15	0.12	2.57	-	15.15	100 m^2	**17.73**
"Enmag"	12.59	0.12	2.57	-	13.22	100 m^2	**15.80**
fertilizer (7:7:7)	4.55	0.12	2.57	-	4.55	100 m^2	**7.12**
"Super Phosphate Powder"	4.46	0.12	2.57	-	4.69	100 m^2	**7.26**
fertilizer (20:10:10)	6.13	0.12	2.57	-	6.13	100 m^2	**8.70**
"Bone meal"	7.30	0.12	2.57	-	7.66	100 m^2	**10.23**
Preparation of planting areas; movement of materials to location maximum 25 m from offload location							
By machine	-	-	-	6.66	-	m^3	**6.66**
By hand	-	1.00	21.00	-	-	m^3	**21.00**
Preparation of planting operations; spreading only; movement of material to planting beds not included							
Composted bark and manure soil conditioner (20 m^3 loads); from not further than 25 m from location; cultivating into topsoil by pedestrian operated cultivator							
50 mm thick	157.00	2.86	60.00	1.58	164.85	100 m^2	**226.43**
100 mm thick	314.00	6.05	127.03	1.58	329.70	100 m^2	**458.31**
150 mm thick	471.00	8.90	187.00	1.58	494.55	100 m^2	**683.13**
200 mm thick	628.00	12.90	271.00	1.58	659.40	100 m^2	**931.98**
Mushroom compost (20 m^3 loads); from not further than 25 m from location; cultivating into topsoil by pedestrian operated cultivator							
50 mm thick	110.40	2.86	60.00	1.58	110.40	100 m^2	**171.98**
100 mm thick	220.80	6.05	127.03	1.58	220.80	100 m^2	**349.41**
150 mm thick	331.20	8.90	187.00	1.58	331.20	100 m^2	**519.78**
200 mm thick	441.60	12.90	271.00	1.58	441.60	100 m^2	**714.18**
Tree planting; pre-planting operations							
Excavating tree pits; depositing soil alongside pits; by machine							
600 mm x 600 mm x 600 mm deep	-	0.15	3.09	0.71	-	nr	**3.80**
900 mm x 900 mm x 600 mm deep	-	0.33	6.93	1.58	-	nr	**8.51**
1.00 m x 1.00 m x 600 mm deep	-	0.61	12.89	1.96	-	nr	**14.85**
1.25 m x 1.25 m x 600 mm deep	-	0.96	20.17	3.07	-	nr	**23.24**

Q PAVING/PLANTING/FENCING/SITE FURNITURE

Item Excluding site overheads and profit	PC £	Labour hours	Labour £	Plant £	Material £	Unit	Total rate £
Q31 PLANTING - cont'd							
Tree planting; pre-planting operations - cont'd							
1.00 m x 1.00 m x 1.00 m deep	-	1.02	21.48	3.27	-	nr	**24.75**
1.50 m x 1.50 m x 750 mm deep	-	1.73	36.24	5.52	-	nr	**41.77**
1.50 m x 1.50 m x 1.00 m deep	-	2.30	48.20	7.34	-	nr	**55.55**
1.75 m x 1.75 m x 1.00 mm deep	-	3.13	65.78	10.02	-	nr	**75.80**
2.00 m x 2.00 m x 1.00 mm deep	-	4.09	85.90	13.09	-	nr	**99.00**
Excavating tree pits; depositing soil alongside pits; by hand							
600 mm x 600 mm x 600 mm deep	-	0.44	9.24	-	-	nr	**9.24**
900 mm x 900 mm x 600 mm deep	-	1.00	21.00	-	-	nr	**21.00**
1.00 m x 1.00 m x 600 mm deep	-	1.13	23.63	-	-	nr	**23.63**
1.25 m x 1.25 m x 600 mm deep	-	1.93	40.53	-	-	nr	**40.53**
1.00 m x 1.00 m x 1.00 m deep	-	2.06	43.26	-	-	nr	**43.26**
1.50 m x 1.50 m x 750 mm deep	-	3.47	72.87	-	-	nr	**72.87**
1.75 m x 1.50 m x 750 mm deep	-	4.05	85.05	-	-	nr	**85.05**
1.50 m x 1.50 m x 1.00 m deep	-	4.63	97.23	-	-	nr	**97.23**
2.00 m x 2.00 m x 750 mm deep	-	6.17	129.57	-	-	nr	**129.57**
2.00 m x 2.00 m x 1.00 m deep	-	8.23	172.84	-	-	nr	**172.84**
Breaking up subsoil in tree pits; to a depth of 200 mm	-	0.03	0.70	-	-	m^2	**0.70**
Spreading and lightly consolidating approved topsoil (imported or from spoil heaps); in layers not exceeding 150 mm; distance from spoil heaps not exceeding 50 m (imported topsoil not included); by machine							
minimum depth 100 mm	-	-	-	1.22	-	m^2	**1.22**
minimum depth 150 mm	-	-	-	1.83	-	m^2	**1.83**
minimum depth 300 mm	-	-	-	3.38	-	m^2	**3.38**
minimum depth 450 mm	-	-	-	4.84	-	m^2	**4.84**
Spreading and lightly consolidating approved topsoil (imported or from spoil heaps); in layers not exceeding 150 mm; distance from spoil heaps not exceeding 100 m (imported topsoil not included); by hand							
minimum depth 100 mm	-	0.13	2.63	-	-	m^2	**2.63**
minimum depth 150 mm	-	0.19	3.94	-	-	m^2	**3.94**
minimum depth 300 mm	-	0.38	7.89	-	-	m^2	**7.89**
minimum depth 450 mm	-	0.56	11.87	-	-	m^2	**11.87**
Extra for filling tree pits with imported topsoil; PC £18.50/m^3; plus allowance for 20% settlement							
depth 100 mm	-	-	-	-	2.22	m^2	-
depth 150 mm	-	-	-	-	3.33	m^2	-
depth 200 mm	-	-	-	-	4.44	m^2	-
depth 300 mm	-	-	-	-	6.66	m^2	-
depth 400 mm	-	-	-	-	8.88	m^2	-
depth 450 mm	-	-	-	-	9.99	m^2	-
depth 500 mm	-	-	-	-	11.10	m^2	-
depth 600 mm	-	-	-	-	13.32	m^2	-
Add or deduct the following amounts for every £0.50 change in the material price of topsoil							
depth 100 mm	-	-	-	-	0.06	m^2	-
depth 150 mm	-	-	-	-	0.09	m^2	-
depth 200 mm	-	-	-	-	0.12	m^2	-
depth 300 mm	-	-	-	-	0.18	m^2	-
depth 400 mm	-	-	-	-	0.24	m^2	-
depth 450 mm	-	-	-	-	0.27	m^2	-
depth 500 mm	-	-	-	-	0.30	m^2	-
depth 600 mm	-	-	-	-	0.36	m^2	-

Q PAVING/PLANTING/FENCING/SITE FURNITURE

Item Excluding site overheads and profit	PC £	Labour hours	Labour £	Plant £	Material £	Unit	Total rate £
Tree staking							
J Toms Ltd; extra over trees for tree stake(s); driving 500 mm into firm ground; trimming to approved height; including two tree ties to approved pattern							
one stake; 75 mm diameter x 2.40 m long	4.01	0.20	4.20	-	4.01	nr	**8.21**
two stakes; 60 mm diameter x 1.65 m long	5.56	0.30	6.30	-	5.56	nr	**11.86**
two stakes; 75 mm diameter x 2.40 m long	7.26	0.30	6.30	-	7.26	nr	**13.56**
three stakes; 75 mm diameter x 2.40 m long	10.88	0.36	7.56	-	10.88	nr	**18.44**
Tree anchors							
Platipus Anchors Ltd; extra over trees for tree anchors							
ref RF1 rootball kit; for 75 to 220 mm girth, 2 to 4.5 m high inclusive of "Plati-Mat" PM1	32.09	1.00	21.00	-	32.09	nr	**53.09**
ref RF2; rootball kit; for 220 to 450 mm girth, 4.5 to 7.5 m high inclusive of "Plati-Mat" Pm^2	52.58	1.33	27.93	-	52.58	nr	**80.51**
ref RF3; rootball kit; for 450 to 750 mm girth, 7.5 to 12 m high "Plati-Mat" Pm^3	111.10	1.50	31.50	-	111.10	nr	**142.60**
ref CG1; guy fixing kit; 75 to 220 mm girth, 2 to 4.5 m high	20.90	1.67	35.00	-	20.90	nr	**55.90**
ref CG2; guy fixing kit; 220 to 450 mm girth, 4.5 to 7.5 m high	36.75	2.00	42.00	-	36.75	nr	**78.75**
installation tools; drive rod for RF1/CG1 kits	-	-	-	-	67.75	nr	**-**
installation tools; drive rod for RF2/CG2 kits	-	-	-	-	102.07	nr	**-**
Extra over trees for land drain to tree pits; 100 mm diameter perforated flexible agricultural drain; including excavating drain trench; laying pipe; backfilling	0.97	1.00	21.00	-	1.07	m	**22.07**
Tree planting; tree pit additives							
Melcourt Industries Ltd; "Topgrow"; incorporating into topsoil at 1 part "Topgrow" to 3 parts excavated topsoil; supplied in 75 L bags; pit size							
600 mm x 600 mm x 600 mm	1.41	0.02	0.42	-	1.41	nr	**1.83**
900 mm x 900 mm x 900 mm	4.76	0.06	1.26	-	4.76	nr	**6.02**
1.00 m x 1.00 m x 1.00 m	6.53	0.24	5.04	-	6.53	nr	**11.57**
1.25 m x 1.25 m x 1.25 m	12.76	0.40	8.40	-	12.76	nr	**21.16**
1.50 m x 1.50 m x 1.50 m	22.05	0.90	18.90	-	22.05	nr	**40.95**
Melcourt Industries Ltd; "Topgrow"; incorporating into topsoil at 1 part "Topgrow" to 3 parts excavated topsoil; supplied in 65 m^3 loose loads; pit size							
600 mm x 600 mm x 600 mm	1.07	0.02	0.35	-	1.07	nr	**1.43**
900 mm x 900 mm x 900 mm	3.63	0.05	1.05	-	3.63	nr	**4.68**
1.00 m x 1.00 m x 1.00 m	4.97	0.20	4.20	-	4.97	nr	**9.18**
1.25 m x 1.25 m x 1.25 m	9.72	0.33	7.00	-	9.72	nr	**16.72**
1.50 m x 1.50 m x 1.50 m	16.79	0.75	15.75	-	16.79	nr	**32.54**
Alginure Products; "Alginure Root Dip"; to bare rooted plants at 1 part "Alginure Root Dip" to 3 parts water							
transplants; at 3000/15 kg bucket	0.88	0.07	1.40	-	0.88	100 nr	**2.28**
standard trees; at 112/15 kg bucket of dip	0.23	0.03	0.70	-	0.23	nr	**0.93**
medium shrubs; at 600/15 kg bucket	0.04	0.02	0.35	-	0.04	nr	**0.39**
Mulching of tree pits; Melcourt Industries Ltd (items labelled FSC are Forest Stewardship Council certified)							
Spreading mulch; to individual trees; maximum distance 25 m (mulch not included)							
50 mm thick	-	0.05	1.02	-	-	m^2	**1.02**
75 mm thick	-	0.07	1.53	-	-	m^2	**1.53**
100 mm thick	-	0.10	2.04	-	-	m^2	**2.04**

Q PAVING/PLANTING/FENCING/SITE FURNITURE

Item Excluding site overheads and profit	PC £	Labour hours	Labour £	Plant £	Material £	Unit	Total rate £
Q31 PLANTING - cont'd							
Mulching of tree pits - cont'd							
Mulch; "Bark Nuggets®"; to individual trees; delivered in 25 m^3 loads; maximum distance 25 m							
50 mm thick	2.46	0.05	1.02	-	2.58	m^2	**3.60**
75 mm thick	3.68	0.05	1.05	-	3.87	m^2	**4.92**
100 mm thick	4.91	0.07	1.40	-	5.16	m^2	**6.56**
Mulch; "Amenity Bark Mulch"; FSC; to individual trees; delivered in 25 m^3 loads; maximum distance 25 m							
50 mm thick	1.88	0.05	1.02	-	1.97	m^2	**2.99**
75 mm thick	2.82	0.07	1.53	-	2.96	m^2	**4.49**
100 mm thick	3.75	0.07	1.40	-	3.94	m^2	**5.34**
Trees; planting labours only							
Bare root trees; including backfilling with previously excavated material (all other operations and materials not included)							
light standard; 6 - 8 cm girth	-	0.35	7.35	-	-	nr	**7.35**
standard; 8 - 10 cm girth	-	0.40	8.40	-	-	nr	**8.40**
selected standard; 10 - 12 cm girth	-	0.58	12.18	-	-	nr	**12.18**
heavy standard; 12 - 14 cm girth	-	0.83	17.50	-	-	nr	**17.50**
extra heavy standard; 14 - 16 cm girth	-	1.00	21.00	-	-	nr	**21.00**
Root balled trees; including backfilling with previously excavated material (all other operations and materials not included)							
standard; 8 - 10 cm girth	-	0.50	10.50	-	-	nr	**10.50**
selected standard; 10 - 12 cm girth	-	0.60	12.60	-	-	nr	**12.60**
heavy standard; 12 - 14 cm girth	-	0.80	16.80	-	-	nr	**16.80**
extra heavy standard; 14 - 16 cm girth	-	1.50	31.50	-	-	nr	**31.50**
16 - 18 cm girth	-	1.30	27.22	21.38	-	nr	**48.60**
18 - 20 cm girth	-	1.60	33.60	26.40	-	nr	**60.00**
20 - 25 cm girth	-	4.50	94.50	86.67	-	nr	**181.17**
25 - 30 cm girth	-	6.00	126.00	113.90	-	nr	**239.90**
30 - 35 cm girth	-	11.00	231.00	200.13	-	nr	**431.13**
Tree planting; root balled trees; advanced nursery stock and semi-mature - General							
Preamble: the cost of planting semi-mature trees will depend on the size and species, and on the access to the site for tree handling machines. Prices should be obtained for individual trees and planting.							
Tree planting; bare root trees; nursery stock; James Coles & Sons (Nurseries) Ltd							
"Acer platanoides"; including backfillling with excavated material (other operations not included)							
light standard; 6 - 8 cm girth	8.40	0.35	7.35	-	8.40	nr	**15.75**
standard; 8 - 10 cm girth	10.75	0.40	8.40	-	10.75	nr	**19.15**
selected standard; 10 - 12 cm girth	16.90	0.58	12.18	-	16.90	nr	**29.08**
heavy standard; 12 - 14 cm girth	28.00	0.83	17.50	-	28.00	nr	**45.50**
extra heavy standard; 14 - 16 cm girth	16.50	1.00	21.00	-	16.50	nr	**37.50**
"Carpinus betulus"; including backfillling with excavated material (other operations not included)							
light standard; 6 - 8 cm girth	11.75	0.35	7.35	-	11.75	nr	**19.10**
standard; 8 - 10 cm girth	16.10	0.40	8.40	-	16.10	nr	**24.50**
selected standard; 10 - 12 cm girth	26.70	0.58	12.18	-	26.70	nr	**38.88**
heavy standard; 12 - 14 cm girth	26.50	0.83	17.51	-	26.50	nr	**44.01**
extra heavy standard; 14 - 16 cm girth	40.00	1.00	21.00	-	40.00	nr	**61.00**

Q PAVING/PLANTING/FENCING/SITE FURNITURE

Item Excluding site overheads and profit	PC £	Labour hours	Labour £	Plant £	Material £	Unit	Total rate £
"Fraxinus excelsior"; including backfillling with excavated material (other operations not included)							
light standard; 6 - 8 cm girth	7.00	0.35	7.35	-	7.00	nr	**14.35**
standard; 8 - 10 cm girth	9.00	0.40	8.40	-	9.00	nr	**17.40**
selected standard; 10 - 12 cm girth	14.75	0.58	12.18	-	14.75	nr	**26.93**
heavy standard; 12 - 14 cm girth	26.50	0.83	17.43	-	26.50	nr	**43.93**
extra heavy standard; 14 - 16 cm girth	33.50	1.00	21.00	-	33.50	nr	**54.50**
"Prunus avium Plena"; including backfillling with excavated material (other operations not included)							
light standard; 6 - 8 cm girth	7.00	0.36	7.64	-	7.00	nr	**14.64**
standard; 8 - 10 cm girth	9.00	0.40	8.40	-	9.00	nr	**17.40**
selected standard; 10 - 12 cm girth	25.20	0.58	12.18	-	25.20	nr	**37.38**
heavy standard; 12 - 14 cm girth	42.00	0.83	17.50	-	42.00	nr	**59.50**
extra heavy standard; 14 - 16 cm girth	50.40	1.00	21.00	-	50.40	nr	**71.40**
"Quercus robur"; including backfillling with excavated material (other operations not included)							
light standard; 6 - 8 cm girth	11.20	0.35	7.35	-	11.20	nr	**18.55**
standard; 8 - 10 cm girth	17.50	0.40	8.40	-	17.50	nr	**25.90**
selected standard; 10 - 12 cm girth	24.00	0.58	12.18	-	24.00	nr	**36.18**
heavy standard; 12 - 14 cm girth	39.20	0.83	17.50	-	39.20	nr	**56.70**
extra heavy standard; 14 - 16 cm girth	44.80	1.00	21.00	-	44.80	nr	**65.80**
"Robinia pseudoacacia Frisia"; including backfillling with excavated material (other operations not included)							
light standard; 6 - 8 cm girth	19.00	0.35	7.35	-	19.00	nr	**26.35**
standard; 8 - 10 cm girth	25.75	0.40	8.40	-	25.75	nr	**34.15**
selected standard; 10 - 12 cm girth	37.50	0.58	12.18	-	37.50	nr	**49.68**
Tree planting; root balled trees; nursery stock; James Coles & Sons (Nurseries) Ltd							
"Acer platanoides"; including backfillling with excavated material (other operations not included)							
standard; 8 - 10 cm girth	25.75	0.48	10.08	-	25.75	nr	**35.83**
selected standard; 10 - 12 cm girth	31.90	0.56	11.77	-	31.90	nr	**43.67**
heavy standard; 12 - 14 cm girth	43.00	0.76	16.05	-	43.00	nr	**59.05**
extra heavy standard; 14 - 16 cm girth	51.50	1.20	25.20	-	51.50	nr	**76.70**
"Carpinus betulus"; including backfillling with excavated material (other operations not included)							
standard; 8 - 10 cm girth	31.10	0.48	10.08	-	31.10	nr	**41.18**
selected standard; 10 - 12 cm girth	41.70	0.56	11.77	-	41.70	nr	**53.47**
heavy standard; 12 - 14 cm girth	54.25	0.76	16.05	-	54.25	nr	**70.30**
extra heavy standard; 14 - 16 cm girth	64.00	1.20	25.20	-	64.00	nr	**89.20**
Tree planting; "Airpot" container grown trees; advanced nursery stock and semi-mature; Deepdale Trees Ltd							
"Acer platanoides Emerald Queen"; including backfilling with excavated material (other operations not included)							
16 - 18 cm girth	95.00	1.98	41.58	27.72	95.00	nr	**164.30**
18 - 20 cm girth	130.00	2.18	45.74	30.49	130.00	nr	**206.23**
20 - 25 cm girth	190.00	2.38	49.90	33.26	190.00	nr	**273.16**
25 - 30 cm girth	250.00	2.97	62.37	55.44	250.00	nr	**367.81**
30 - 35 cm girth	450.00	3.96	83.16	80.27	450.00	nr	**613.43**

Q PAVING/PLANTING/FENCING/SITE FURNITURE

Item Excluding site overheads and profit	PC £	Labour hours	Labour £	Plant £	Material £	Unit	Total rate £
Q31 PLANTING - cont'd							
Tree planting; "Airpot" container grown trees - cont'd							
"Aesculus briotti"; including backfilling with excavated material (other operations not included)							
16 - 18 cm girth	90.00	1.98	41.58	27.72	90.00	nr	**159.30**
18 - 20 cm girth	130.00	1.60	33.60	30.49	130.00	nr	**194.09**
20 - 25 cm girth	180.00	2.38	49.90	33.26	180.00	nr	**263.16**
25 - 30 cm girth	240.00	2.97	62.37	72.37	240.00	nr	**374.74**
30 - 35 cm girth	330.00	3.96	83.16	80.27	330.00	nr	**493.43**
"Betula pendula multistem"; including backfilling with excavated material (other operations not included)							
3.0 - 3.5 m high	140.00	1.98	41.58	27.72	140.00	nr	**209.30**
3.5 - 4.0 m high	175.00	1.60	33.60	30.49	175.00	nr	**239.09**
4.0 - 4.5 m high	225.00	2.38	49.90	48.16	225.00	nr	**323.06**
4.5 - 5.0 m high	275.00	2.97	62.37	72.37	275.00	nr	**409.74**
5.0 - 6.0 m high	375.00	3.96	83.16	80.27	375.00	nr	**538.43**
6.0 - 7.0 m high	495.00	4.50	94.50	96.05	495.00	nr	**685.55**
"Pinus sylvestris"; including backfilling with excavated material (other operations not included)							
3.0 - 3.5 m high	300.00	1.98	41.58	27.72	300.00	nr	**369.30**
3.5 - 4.0 m high	375.00	1.60	33.60	30.49	375.00	nr	**439.09**
4.0 - 4.5 m high	450.00	2.38	49.90	33.26	450.00	nr	**533.16**
4.5 - 5.0 m high	500.00	2.97	62.37	72.37	500.00	nr	**634.74**
5.0 - 6.0 m high	650.00	3.96	83.16	80.27	650.00	nr	**813.43**
6.0 - 7.0 m high	950.00	4.50	94.50	98.85	950.00	nr	**1143.35**
Tree planting; "Airpot" container grown trees; semi-mature and mature trees; Deepdale Trees Ltd; planting and back filling; planted by telehandler or by crane; delivery included; all other operations priced separately							
Semi-mature trees indicative prices							
40 - 45 cm girth	550.00	4.00	84.00	48.00	550.00	nr	**682.00**
45 - 50 cm girth	750.00	4.00	84.00	48.00	750.00	nr	**882.00**
55 - 60 cm girth	1350.00	6.00	126.00	48.00	1350.00	nr	**1524.00**
60 - 70 cm girth	2500.00	7.00	147.00	65.00	2500.00	nr	**2712.00**
70 - 80 cm girth	3500.00	7.50	157.50	82.00	3500.00	nr	**3739.50**
80 - 90 cm girth	4500.00	8.00	168.00	96.00	4500.00	nr	**4764.00**
Tree planting; root balled trees; advanced nursery stock and semi-mature; Lorenz von Ehren							
"Acer platanoides Emerald Queen"; including backfilling with excavated material (other operations not included)							
16 - 18 cm girth	80.00	1.30	27.22	3.75	80.00	nr	**110.97**
18 - 20 cm girth	93.00	1.60	33.60	3.75	93.00	nr	**130.35**
20 - 25 cm girth	115.00	4.50	94.50	18.04	115.00	nr	**227.54**
25 - 30 cm girth	210.00	6.00	126.00	22.40	210.00	nr	**358.40**
30 - 35 cm girth	340.00	11.00	231.00	29.88	340.00	nr	**600.88**
"Quercus palustris - Pin Oak"; including backfilling with excavated material (other operations not included)							
16 - 18 cm girth	82.00	1.30	27.22	3.75	82.00	nr	**112.97**
18 - 20 cm girth	110.00	1.60	33.60	3.75	110.00	nr	**147.35**
20 - 25 cm girth	150.00	4.50	94.50	18.04	150.00	nr	**262.54**
25 - 30 cm girth	220.00	6.00	126.00	22.40	220.00	nr	**368.40**
30 - 35 cm girth	320.00	11.00	231.00	29.88	320.00	nr	**580.88**

Q PAVING/PLANTING/FENCING/SITE FURNITURE

Item Excluding site overheads and profit	PC £	Labour hours	Labour £	Plant £	Material £	Unit	**Total rate £**
"Tilia cordata Green Spire"; including backfilling with excavated material (other operations not included)							
16 - 18 cm girth	68.00	1.30	27.22	3.75	68.00	nr	**98.97**
18 - 20 cm girth	86.00	1.60	33.60	3.75	86.00	nr	**123.35**
20 - 25 cm girth	104.00	4.50	94.50	18.04	104.00	nr	**216.54**
25 - 30 cm girth; 5 x transplanted 4.0 - 5.0 m tall	160.00	6.00	126.00	22.40	160.00	nr	**308.40**
30 - 35 cm girth; 5 x transplanted 5.0 - 7.0 m tall	260.00	11.00	231.00	29.88	260.00	nr	**520.88**
Hedges							
Excavating trench for hedges; depositing soil alongside trench; by machine							
300 mm deep x 300 mm wide	-	0.03	0.63	0.35	-	m	**0.98**
300 mm deep x 450 mm wide	-	0.05	0.94	0.52	-	m	**1.46**
Excavating trench for hedges; depositing soil alongside trench; by hand							
300 mm deep x 300 mm wide	-	0.12	2.52	-	-	m	**2.52**
300 mm deep x 450 mm wide	-	0.23	4.73	-	-	m	**4.73**
Setting out; notching out; excavating trench; breaking up subsoil to minimum depth 300 mm							
minimum 400 mm deep	-	0.25	5.25	-	-	m	**5.25**
Hedge planting; including backfill with excavated topsoil; PC £0.32/nr							
single row; 200 mm centres	1.65	0.06	1.31	-	1.65	m	**2.96**
single row; 300 mm centres	1.10	0.06	1.17	-	1.10	m	**2.27**
single row; 400 mm centres	0.82	0.04	0.88	-	0.82	m	**1.70**
single row; 500 mm centres	0.66	0.03	0.70	-	0.66	m	**1.36**
double row; 200 mm centres	3.30	0.17	3.50	-	3.30	m	**6.80**
double row; 300 mm centres	2.20	0.13	2.80	-	2.20	m	**5.00**
double row; 400 mm centres	1.65	0.08	1.75	-	1.65	m	**3.40**
double row; 500 mm centres	1.32	0.07	1.40	-	1.32	m	**2.72**
Extra over hedges for incorporating manure; at 1 m^3 per 30 m	0.71	0.03	0.53	-	0.71	m	**1.23**
Shrub planting - General							
Preamble: for preparation of planting areas see "Cultivation" at the beginning of the planting section.							
Shrub planting							
Setting out; selecting planting from holding area; loading to wheelbarrows; planting as plan or as directed; distance from holding area maximum 50 m; plants 2 - 3 litre containers							
plants in groups of 100 nr minimum	-	0.01	0.24	-	-	nr	**0.24**
plants in groups of 10 - 100 nr	-	0.02	0.35	-	-	nr	**0.35**
plants in groups of 3 - 5 nr	-	0.03	0.53	-	-	nr	**0.53**
single plants not grouped	-	0.04	0.84	-	-	nr	**0.84**
Forming planting holes; in cultivated ground (cultivating not included); by mechanical auger; trimming holes by hand; depositing excavated material alongside holes							
250 diameter	-	0.03	0.70	0.04	-	nr	**0.74**
250 x 250 mm	-	0.04	0.84	0.08	-	nr	**0.92**
300 x 300 mm	-	0.08	1.57	0.10	-	nr	**1.67**

Q PAVING/PLANTING/FENCING/SITE FURNITURE

Item Excluding site overheads and profit	PC £	Labour hours	Labour £	Plant £	Material £	Unit	Total rate £
Q31 PLANTING - cont'd							
Shrub planting - cont'd							
Hand excavation; forming planting holes; in cultivated ground (cultivating not included); depositing excavated material alongside holes							
100 mm x 100 mm x 100 mm deep; with mattock or hoe	-	0.01	0.14	-	-	nr	**0.14**
250 mm x 250 mm x 300 mm deep	-	0.04	0.84	-	-	nr	**0.84**
300 mm x 300 mm x 300 mm deep	-	0.06	1.17	-	-	nr	**1.17**
400 mm x 400 mm x 400 mm deep	-	0.13	2.63	-	-	nr	**2.63**
500 mm x 500 mm x 500 mm deep	-	0.25	5.25	-	-	nr	**5.25**
600 mm x 600 mm x 600 mm deep	-	0.43	9.09	-	-	nr	**9.09**
900 mm x 900 mm x 600 mm deep	-	1.00	21.00	-	-	nr	**21.00**
1.00 m x 1.00 m x 600 mm deep	-	1.23	25.83	-	-	nr	**25.83**
1.25 m x 1.25 m x 600 mm deep	-	1.93	40.53	-	-	nr	**40.53**
Hand excavation; forming planting holes; in uncultivated ground; depositing excavated material alongside holes							
100 mm x 100 mm x 100 mm deep with mattock or hoe	-	0.03	0.53	-	-	nr	**0.53**
250 mm x 250 mm x 300 mm deep	-	0.06	1.17	-	-	nr	**1.17**
300 mm x 300 mm x 300 mm deep	-	0.06	1.31	-	-	nr	**1.31**
400 mm x 400 mm x 400 mm deep	-	0.25	5.25	-	-	nr	**5.25**
500 mm x 500 mm x 500 mm deep	-	0.33	6.84	-	-	nr	**6.84**
600 mm x 600 mm x 600 mm deep	-	0.55	11.55	-	-	nr	**11.55**
900 mm x 900 mm x 600 mm deep	-	1.25	26.25	-	-	nr	**26.25**
1.00 m x 1.00 m x 600 mm deep	-	1.54	32.29	-	-	nr	**32.29**
1.25 m x 1.25 m x 600 mm deep	-	2.41	50.66	-	-	nr	**50.66**
Bare root planting; to planting holes (forming holes not included); including backfilling with excavated material (bare root plants not included)							
bare root 1+1; 30 - 90 mm high	-	0.02	0.35	-	-	nr	**0.35**
bare root 1+2; 90 - 120 mm high	-	0.02	0.35	-	-	nr	**0.35**
Containerised planting; to planting holes (forming holes not included); including backfilling with excavated material (shrub or ground cover not included)							
9 cm pot	-	0.01	0.21	-	-	nr	**0.21**
2 litre container	-	0.02	0.42	-	-	nr	**0.42**
3 litre container	-	0.02	0.47	-	-	nr	**0.47**
5 litre container	-	0.03	0.70	-	-	nr	**0.70**
10 litre container	-	0.05	1.05	-	-	nr	**1.05**
15 litre container	-	0.07	1.40	-	-	nr	**1.40**
20 litre container	-	0.08	1.75	-	-	nr	**1.75**
Shrub planting; 2 litre containerised plants; in cultivated ground (cultivating not included); PC £2.69/nr							
average 2 plants per m^2	-	0.06	1.18	-	5.38	m^2	**6.56**
average 3 plants per m^2	-	0.08	1.76	-	8.07	m^2	**9.83**
average 4 plants per m^2	-	0.11	2.35	-	10.76	m^2	**13.11**
average 6 plants per m^2	-	0.17	3.53	-	16.14	m^2	**19.67**
Extra over shrubs for stakes	0.50	0.02	0.35	-	0.50	nr	**0.85**
Composted bark soil conditioners; 20 m^3 loads; on beds by mechanical loader; spreading and rotavating into topsoil; by machine							
50 mm thick	157.00	-	-	6.11	157.00	100 m^2	**163.11**
100 mm thick	314.00	-	-	10.36	329.70	100 m^2	**340.06**
150 mm thick	471.00	-	-	14.78	494.55	100 m^2	**509.33**
200 mm thick	628.00	-	-	19.11	659.40	100 m^2	**678.51**

Q PAVING/PLANTING/FENCING/SITE FURNITURE

Item Excluding site overheads and profit	PC £	Labour hours	Labour £	Plant £	Material £	Unit	Total rate £
Mushroom compost; 25 m^3 loads; delivered not further than 25 m from location; cultivating into topsoil by pedestrian operated machine							
50 mm thick	110.40	2.86	60.00	1.58	110.40	100 m^2	**171.98**
100 mm thick	220.80	6.05	127.03	1.58	220.80	100 m^2	**349.41**
150 mm thick	331.20	8.90	187.00	1.58	331.20	100 m^2	**519.78**
200 mm thick	441.60	12.90	271.00	1.58	441.60	100 m^2	**714.18**
Manure; 20 m^3 loads; delivered not further than 25 m from location; cultivating into topsoil by pedestrian operated machine							
50 mm thick	212.50	2.86	60.00	1.58	212.50	100 m^2	**274.08**
100 mm thick	425.00	6.05	127.03	1.58	446.25	100 m^2	**574.86**
150 mm thick	637.50	8.90	187.00	1.58	669.38	100 m^2	**857.96**
200 mm thick	850.00	12.90	271.00	1.58	892.50	100 m^2	**1165.08**
Fertilizers (7:7:7); PC £0.65/kg; to beds; by hand							
35 g/m^2	2.27	0.17	3.50	-	2.27	100 m^2	**5.78**
50 g/m^2	3.25	0.17	3.50	-	3.25	100 m^2	**6.75**
70 g/m^2	4.55	0.17	3.50	-	4.55	100 m^2	**8.05**
Fertilizers; "Enmag" PC £1.80/kg slow release fertilizer; to beds; by hand							
35 g/m^2	6.30	0.17	3.50	-	6.30	100 m^2	**9.80**
50 g/m^2	9.00	0.17	3.50	-	9.00	100 m^2	**12.50**
70 g/m^2	13.49	0.17	3.50	-	13.49	100 m^2	**16.99**
Note: for machine incorporation of fertilizers and soil conditioners see "Cultivation".							
Herbaceous and groundcover planting							
Herbaceous plants; PC £1.20/nr; including forming planting holes in cultivated ground (cultivating not included); backfilling with excavated material; 1 litre containers							
average 4 plants per m^2 - 500 mm centres	-	0.09	1.96	-	4.80	m^2	**6.76**
average 6 plants per m^2 - 408 mm centres	-	0.14	2.94	-	7.20	m^2	**10.14**
average 8 plants per m^2 - 354 mm centres	-	0.19	3.92	-	9.60	m^2	**13.52**
Note: for machine incorporation of fertilizers and soil conditioners see "Cultivation".							
Bulb planting							
Bulbs; including forming planting holes in cultivated area (cultivating not included); backfilling with excavated material							
small	13.00	0.83	17.50	-	13.00	100 nr	**30.50**
medium	22.00	0.83	17.50	-	22.00	100 nr	**39.50**
large	25.00	0.91	19.09	-	25.00	100 nr	**44.09**
Bulbs; in grassed area; using bulb planter; including backfilling with screened topsoil or peat and cut turf plug							
small	13.00	1.67	35.00	-	13.00	100 nr	**48.00**
medium	22.00	1.67	35.00	-	22.00	100 nr	**57.00**
large	25.00	2.00	42.00	-	25.00	100 nr	**67.00**
Aquatic planting							
Aquatic plants; in prepared growing medium in pool; plant size 2 - 3 litre containerised (plants not included)	-	0.04	0.84	-	-	nr	**0.84**
Operations after planting							
Initial cutting back to shrubs and hedge plants; including disposal of all cuttings	-	1.00	21.00	-	-	100 m^2	**21.00**
Mulch; Melcourt Industries Ltd; "Bark Nuggets®"; to plant beds; delivered in 25 m^3 loads; maximum distance 25 m							
50 mm thick	2.18	0.03	0.61	-	2.29	m^2	**2.90**
75 mm thick	3.68	0.07	1.40	-	3.87	m^2	**5.27**
100 mm thick	4.91	0.09	1.87	-	5.16	m^2	**7.02**

Q PAVING/PLANTING/FENCING/SITE FURNITURE

Item Excluding site overheads and profit	PC £	Labour hours	Labour £	Plant £	Material £	Unit	Total rate £
Q31 PLANTING - cont'd							
Operations after planting - cont'd							
Mulch; Melcourt Industries Ltd; "Amenity Bark Mulch"; FSC; to plant beds; delivered in 25 m^3 loads; maximum distance 25 m							
50 mm thick	1.88	0.04	0.93	-	1.97	m^2	**2.90**
75 mm thick	2.82	0.07	1.40	-	2.96	m^2	**4.36**
100 mm thick	3.75	0.09	1.87	-	3.94	m^2	**5.81**
Maintenance operations (Note: the following rates apply to aftercare maintenance executed as part of a landscaping contract only)							
Weeding and hand forking planted areas; including disposing weeds and debris on site; areas maintained weekly	-	-	0.08	-	-	m^2	**0.08**
Weeding and hand forking planted areas; including disposing weeds and debris on site; areas maintained monthly	-	0.01	0.21	-	-	m^2	**0.21**
Mulch; Melcourt Industries Ltd; "Bark Nuggets®"; to plant beds; delivered in 25 m^3 loads; maximum distance 25 m							
50 mm thick	2.46	0.04	0.93	-	2.58	m^2	**3.51**
75 mm thick	3.68	0.07	1.40	-	3.87	m^2	**5.27**
Mulch; Melcourt Industries Ltd; "Amenity Bark Mulch"; FSC; to plant beds; delivered in 25 m^3 loads; maximum distance 25 m							
50 mm thick	1.88	0.04	0.93	-	1.97	m^2	**2.90**
75 mm thick	2.82	0.07	1.40	-	2.96	m^2	**4.36**
Watering planting; evenly; at a rate of 5 litre/m^2							
using hand-held watering equipment	-	0.25	5.25	-	-	100 m^2	**5.25**
using sprinkler equipment and with sufficient water pressure to run 1 nr 15 m radius sprinkler	-	0.14	2.92	-	-	100 m^2	**2.92**
Work to existing planting							
Cutting and trimming ornamental hedges; to specified profiles; including cleaning out hedge bottoms; hedge cut 2 occasions per annum; by hand							
up to 2.00 m high	-	0.03	0.70	-	0.25	m	**0.95**
2.00 - 4.00 m high	-	0.05	1.05	1.58	0.50	m	**3.13**
Q35 LANDSCAPE MAINTENANCE							
Grass cutting - pedestrian operated equipment							
Using cylinder lawn mower fitted with not less than five cutting blades, front and rear rollers; on surface not exceeding 30 deg from horizontal; arisings let fly; width of cut							
51 cm	-	0.06	1.28	0.19	-	100 m	**1.47**
61 cm	-	0.05	1.07	0.17	-	100 m	**1.24**
Using rotary self-propelled mower; width of cut							
45 cm	-	0.04	0.83	0.06	-	100 m^2	**0.89**
Add for using grass box for collecting and depositing arisings							
removing and depositing arisings	-	0.05	1.05	-	-	100 m^2	**1.05**
Add for 30 to 50 deg from horizontal	-	-	-	-	-	33%	**-**
Add for slopes exceeding 50 deg	-	-	-	-	-	100%	**-**
Cutting grass or light woody undergrowth; using trimmer with nylon cord or metal disc cutter; on surface							
not exceeding 30 deg from horizontal	-	0.20	4.20	0.28	-	100 m^2	**4.48**
30 - 50 deg from horizontal	-	0.40	8.40	0.56	-	100 m^2	**8.96**
exceeding 50 deg from horizontal	-	0.50	10.50	0.70	-	100 m^2	**11.20**

Q PAVING/PLANTING/FENCING/SITE FURNITURE

Item Excluding site overheads and profit	PC £	Labour hours	Labour £	Plant £	Material £	Unit	Total rate £
Grass cutting - collecting arisings							
Extra over for tractor drawn and self-propelled machinery using attached grass boxes; depositing arisings							
22 cuts per year	-	0.05	1.05	-	-	100 m^2	**1.05**
18 cuts per year	-	0.08	1.57	-	-	100 m^2	**1.57**
12 cuts per year	-	0.10	2.10	-	-	100 m^2	**2.10**
4 cuts per year	-	0.25	5.25	-	-	100 m^2	**5.25**
Disposing arisings							
22 cuts per year	-	0.01	0.17	0.02	0.06	100 m^2	**0.25**
18 cuts per year	-	0.01	0.20	0.03	0.07	100 m^2	**0.30**
12 cuts per year	-	0.01	0.30	0.04	0.10	100 m^2	**0.45**
4 cuts per year	-	0.04	0.92	0.13	0.32	100 m^2	**1.36**
Scarifying by hand							
hand implement	-	0.50	10.50	-	-	100 m^2	**10.50**
add for disposal of arisings	-	0.03	0.53	1.02	10.00	100 m^2	**11.55**
Rolling							
Rolling grassed area; equipment towed by tractor; once over, using							
smooth roller	-	0.01	0.30	0.15	-	100 m^2	**0.45**
Turf aeration							
By hand							
hand fork; to effect a minimum penetration of 100 mm and spaced 150 mm apart	-	1.33	28.00	-	-	100 m^2	**28.00**
hollow tine hand implement; to effect a minimum penetration of 100 mm and spaced 150 mm apart	-	2.00	42.00	-	-	100 m^2	**42.00**
collection of arisings by hand	-	3.00	63.00	-	-	100 m^2	**63.00**
Turf areas; surface treatments and top dressing; Boughton Loam Ltd							
Apply screened topdressing to grass surfaces; spread using Tru-Lute							
Sand soil mixes 90/10 to 50/50	0.12	-	0.04	0.03	0.12	m^2	**0.20**
Using pedestrian operated mechanical equipment and blowers							
Grassed areas with perimeters of mature trees such as sports fields and amenity areas	-	0.04	0.84	0.05	-	100 m^2	**0.89**
Grassed areas containing ornamental trees and shrub beds	-	0.10	2.10	0.13	-	100 m^2	**2.23**
verges	-	0.07	1.40	0.08	-	100 m^2	**1.49**
By hand							
Grassed areas with perimeters of mature trees such as sports fields and amenity areas	-	0.05	1.05	0.10	-	100 m^2	**1.15**
grassed areas containing ornamental trees and shrub beds	-	0.08	1.75	0.17	-	100 m^2	**1.92**
verges	-	1.00	21.00	1.99	-	100 m^2	**22.99**
Removal of arisings							
Areas with perimeters of mature trees	-	0.01	0.14	0.09	0.80	100 m^2	**1.03**
Areas containing ornamental trees and shrub beds	-	0.02	0.42	0.34	2.00	100 m^2	**2.76**
Litter clearance							
Collection and disposal of litter from grassed area	-	0.01	0.21	-	0.05	100 m^2	**0.26**
Collection and disposal of litter from isolated grassed area not exceeding 1000 m^2	-	0.04	0.84	-	0.05	100 m^2	**0.89**
Edge maintenance							
Maintain edges where lawn abuts pathway or hard surface using							
Strimmer	-	0.01	0.11	0.01	-	m	**0.11**
Shears	-	0.02	0.35	-	-	m	**0.05**

Q PAVING/PLANTING/FENCING/SITE FURNITURE

Item Excluding site overheads and profit	PC £	Labour hours	Labour £	Plant £	Material £	Unit	Total rate £
Q35 LANDSCAPE MAINTENANCE - cont'd							
Edge maintenance - cont'd							
Maintain edges where lawn abuts plant bed using							
Mechanical edging tool	-	0.01	0.14	0.03	-	m	**0.17**
Shears	-	0.01	0.23	-	-	m	**0.23**
Half moon edging tool	-	0.02	0.42	-	-	m	**0.42**
Tree guards, stakes and ties, etc.							
Adjusting existing tree tie	-	0.03	0.70	-	-	nr	**0.70**
Taking up single or double tree stake and ties; removing and disposing	-	0.05	1.05	-	-	nr	**1.05**
Pruning shrubs							
Trimming ground cover planting							
Soft groundcover; vinca ivy and the like	-	1.00	21.00	-	-	100 m^2	**21.00**
Woody groundcover; cotoneaster and the like	-	1.50	31.50	-	-	100 m^2	**31.50**
Pruning massed shrub border (measure ground area)							
Shrub beds pruned annually	-	0.01	0.21	-	-	m^2	**0.21**
Shrub beds pruned hard every 3 years	-	0.03	0.58	-	-	m^2	**0.58**
Cutting off dead heads							
Bush or standard rose	-	0.05	1.05	-	-	nr	**1.05**
Climbing rose	-	0.08	1.75	-	-	nr	**1.75**
Pruning roses							
Bush or standard rose	-	0.05	1.05	-	-	nr	**1.05**
Climbing rose or rambling rose; tying in as required	-	0.07	1.40	-	-	nr	**1.40**
Pruning ornamental shrubs; height before pruning (increase these rates by 50% if pruning work has not been executed during the previous two years)							
Not exceeding 1 m	-	0.04	0.84	-	-	nr	**0.84**
1 to 2 m	-	0.06	1.17	-	-	nr	**1.17**
Exceeding 2 m	-	0.13	2.63	-	-	nr	**2.63**
Removing excess growth etc from face of building etc.; height before pruning							
Not exceeding 2 m	-	0.03	0.60	-	-	nr	**0.60**
2 to 4 m	-	0.05	1.05	-	-	nr	**1.05**
4 to 6 m	-	0.08	1.75	-	-	nr	**1.75**
6 to 8 m	-	0.13	2.63	-	-	nr	**2.63**
8 to 10 m	-	0.14	3.00	-	-	nr	**3.00**
Removing epicormic growth from base of shrub or trunk and base of tree: any height; any diameter; number of growths							
Not exceeding 10	-	0.05	1.05	-	-	nr	**1.05**
10 to 20	-	0.07	1.40	-	-	nr	**1.40**
Beds, borders and planters							
Lifting							
bulbs	-	0.50	10.50	-	-	100 nr	**10.50**
tubers or corms	-	0.40	8.40	-	-	100 nr	**8.40**
established herbaceous plants; hoeing and depositing for replanting	-	2.00	42.00	-	-	100 nr	**42.00**
Temporary staking and tying in herbaceous plant	-	0.03	0.70	-	0.10	nr	**0.80**
Cutting down spent growth of herbaceous plant; clearing arisings							
unstaked	-	0.02	0.42	-	-	nr	**0.42**
staked; not exceeding 4 stakes per plant; removing stakes and putting into store	-	0.03	0.53	-	-	nr	**0.53**
Hand weeding							
newly planted areas	-	2.00	42.00	-	-	100 m^2	**42.00**
established areas	-	0.50	10.50	-	-	100 m^2	**10.50**

Q PAVING/PLANTING/FENCING/SITE FURNITURE

Item Excluding site overheads and profit	PC £	Labour hours	Labour £	Plant £	Material £	Unit	Total rate £
Removing grasses from groundcover areas	-	3.00	63.06	-	-	100 m^2	**63.06**
Hand digging with fork; not exceeding 150 mm deep; breaking down lumps; leaving surface with a medium tilth	-	1.33	27.99	-	-	100 m^2	**27.99**
Hand digging with fork or spade to an average depth of 230 mm; breaking down lumps; leaving surface with a medium tilth	-	2.00	42.00	-	-	100 m^2	**42.00**
Hand hoeing; not exceeding 50 mm deep; leaving surface with a medium tilth	-	0.40	8.40	-	-	100 m^2	**8.40**
Hand raking to remove stones etc.; breaking down lumps: leaving surface with a fine tilth prior to planting	-	0.67	14.00	-	-	100 m^2	**14.00**
Hand weeding; planter, window box; not exceeding 1.00 m^2							
ground level box	-	0.05	1.05	-	-	nr	**1.05**
box accessed by stepladder	-	0.08	1.75	-	-	nr	**1.75**
Spreading only compost, mulch or processed bark to a depth of 75 mm							
on shrub bed with existing mature planting	-	0.09	1.91	-	-	m^2	**1.91**
recently planted areas	-	0.07	1.40	-	-	m^2	**1.40**
groundcover and herbaceous areas	-	0.08	1.57	-	-	m^2	**1.57**
Clearing cultivated area of leaves, litter and other extraneous debris; using hand implement							
weekly maintenance	-	0.13	2.63	-	-	100 m^2	**2.63**
daily maintenance	-	0.02	0.35	-	-	100 m^2	**0.35**
Bedding							
Lifting							
bedding plants; hoeing and depositing for disposal	-	3.00	63.00	-	-	100 m^2	**63.00**
Hand digging with fork; not exceeding 150 mm deep: breaking down lumps; leaving surface with a medium tilth	-	0.75	15.75	-	-	100 m^2	**15.75**
Hand weeding							
newly planted areas	-	2.00	42.00	-	-	100 m^2	**42.00**
established areas	-	0.50	10.50	-	-	100 m^2	**10.50**
Hand digging with fork or spade to an average depth of 230 mm; breaking down lumps; leaving surface with a medium tilth	-	0.50	10.50	-	-	100 m^2	**10.50**
Hand hoeing: not exceeding 50 mm deep; leaving surface with a medium tilth	-	0.40	8.40	-	-	100 m^2	**8.40**
Hand raking to remove stones etc.; breaking down lumps; leaving surface with a fine tilth prior to planting	-	0.67	14.00	-	-	100 m^2	**14.00**
Hand weeding; planter, window box; not exceeding 1.00 m^2							
ground level box	-	0.05	1.05	-	-	nr	**1.05**
box accessed by stepladder	-	0.08	1.75	-	-	nr	**1.75**
Spreading only; compost, mulch or processed bark to a depth of 75 mm							
on shrub bed with existing mature planting	-	0.09	1.91	-	-	m^2	**1.91**
recently planted areas	-	0.07	1.40	-	-	m^2	**1.40**
groundcover and herbaceous areas	-	0.08	1.57	-	-	m^2	**1.57**
Collecting bedding from nursery	-	3.00	63.00	13.64	-	100 m^2	**76.64**
Setting out							
mass planting single variety	-	0.13	2.63	-	-	m^2	**2.63**
pattern	-	0.33	7.00	-	-	m^2	**7.00**
Planting only							
massed bedding plants	-	0.20	4.20	-	-	m^2	**4.20**
Clearing cultivated area of leaves, litter and other extraneous debris; using hand implement							
weekly maintenance	-	0.13	2.63	-	-	100 m^2	**2.63**
daily maintenance	-	0.02	0.35	-	-	100 m^2	**0.35**

Q PAVING/PLANTING/FENCING/SITE FURNITURE

Item Excluding site overheads and profit	PC £	Labour hours	Labour £	Plant £	Material £	Unit	Total rate £
Q35 LANDSCAPE MAINTENANCE - cont'd							
Irrigation and watering							
Hand held hosepipe; flow rate 25 litres per minute; irrigation requirement							
10 litres/m^2	-	0.74	15.48	-	-	100 m^2	**15.48**
15 litres/m^2	-	1.10	23.10	-	-	100 m^2	**23.10**
20 litres/m^2	-	1.46	30.72	-	-	100 m^2	**30.72**
25 litres/m^2	-	1.84	38.58	-	-	100 m^2	**38.58**
Hand held hosepipe; flow rate 40 litres per minute; irrigation requirement							
10 litres/m^2	-	0.46	9.70	-	-	100 m^2	**9.70**
15 litres/m^2	-	0.69	14.55	-	-	100 m^2	**14.55**
20 litres/m^2	-	0.91	19.17	-	-	100 m^2	**19.17**
25 litres/m^2	-	1.15	24.07	-	-	100 m^2	**24.07**
Hedge cutting; field hedges cut once or twice annually							
Trimming sides and top using hand tool or hand held mechanical tools							
not exceeding 2 m high	-	0.10	2.10	0.14	-	10 m^2	**2.24**
2 to 4 m high	-	0.33	7.00	0.47	-	10 m^2	**7.47**
Hedge cutting; ornamental							
Trimming sides and top using hand tool or hand held mechanical tools							
not exceeding 2 m high	-	0.13	2.63	0.18	-	10 m^2	**2.80**
2 to 4 m high	-	0.50	10.50	0.70	-	10 m^2	**11.20**
Hedge cutting; reducing width; hand tool or hand held mechanical tools							
Not exceeding 2 m high							
Average depth of cut not exceeding 300 mm	-	0.10	2.10	0.14	-	10 m^2	**2.24**
Average depth of cut 300 to 600 mm	-	0.83	17.50	1.17	-	10 m^2	**18.67**
Average depth of cut 600 to 900 mm	-	1.25	26.25	1.75	-	10 m^2	**28.00**
2 to 4 m high							
Average depth of cut not exceeding 300 mm	-	0.03	0.53	0.04	-	10 m^2	**0.56**
Average depth of cut 300 to 600 mm	-	0.13	2.63	0.18	-	10 m^2	**2.80**
Average depth of cut 600 to 900 mm	-	2.50	52.50	3.50	-	10 m^2	**56.00**
4 to 6 m high							
Average depth of cut not exceeding 300 mm	-	0.10	2.10	0.14	-	m^2	**2.24**
Average depth of cut 300 to 600 mm	-	0.17	3.50	0.23	-	m^2	**3.73**
Average depth of cut 600 to 900 mm	-	0.50	10.50	0.70	-	m^2	**11.20**
Hedge cutting; reducing height; hand tool or hand held mechanical tools							
Not exceeding 2 m high							
Average depth of cut not exceeding 300 mm	-	0.07	1.40	0.09	-	10 m^2	**1.49**
Average depth of cut 300 to 600 mm	-	0.13	2.80	0.19	-	10 m^2	**2.99**
Average depth of cut 600 to 900 mm	-	0.40	8.40	0.56	-	10 m^2	**8.96**
2 to 4 m high							
Average depth of cut not exceeding 300 mm	-	0.03	0.70	0.05	-	m^2	**0.75**
Average depth of cut 300 to 600 mm	-	0.07	1.40	0.09	-	m^2	**1.49**
Average depth of cut 600 to 900 mm	-	0.20	4.20	0.28	-	m^2	**4.48**
4 to 6 m high							
Average depth of cut not exceeding 300 mm	-	0.07	1.40	0.09	-	m^2	**1.49**
Average depth of cut 300 to 600 mm	-	0.13	2.63	0.18	-	m^2	**2.80**
Average depth of cut 600 to 900 mm	-	0.25	5.25	0.35	-	m^2	**5.60**
Hedge cutting; removal and disposal of arisings							
Sweeping up and depositing arisings							
300 mm cut	-	0.05	1.05	-	-	10 m^2	**1.05**
600 mm cut	-	0.20	4.20	-	-	10 m^2	**4.20**
900 mm cut	-	0.40	8.40	-	-	10 m^2	**8.40**

Q PAVING/PLANTING/FENCING/SITE FURNITURE

Item Excluding site overheads and profit	PC £	Labour hours	Labour £	Plant £	Material £	Unit	Total rate £
Chipping arisings							
300 mm cut	-	0.02	0.42	0.11	-	10 m^2	**0.53**
600 mm cut	-	0.08	1.75	0.48	-	10 m^2	**2.23**
900 mm cut	-	0.20	4.20	1.14	-	10 m^2	**5.34**
Disposal of unchipped arisings							
300 mm cut	-	0.02	0.35	0.51	1.33	10 m^2	**2.20**
600 mm cut	-	0.03	0.70	1.02	2.00	10 m^2	**3.72**
900 mm cut	-	0.08	1.75	2.56	5.00	10 m^2	**9.31**
Disposal of chipped arisings							
300 mm cut	-	-	0.07	0.17	2.00	10 m^2	**2.24**
600 mm cut	-	0.02	0.35	0.17	4.00	10 m^2	**4.52**
900 mm cut	-	0.03	0.70	0.34	10.00	10 m^2	**11.04**
Spraying; labour rates only; for chemical rates please use the tables at the beginning of the section Q30 Seeding/Turfing							
Herbicide applications; standard backpack spray applicators; application to maintain 1.00 m diameter clear circles (0.79 m^2) around new planting							
plants at 1.50 m centres; 4444 nr/ha	-	0.25	5.18	-	-	100 m^2	**5.18**
plants at 1.75 m centres; 3265 nr/ha	-	0.18	3.81	-	-	100 m^2	**3.81**
plants at 2.00 m centres; 2500 nr/ha	-	0.14	2.92	-	-	100 m^2	**2.92**
understory of mature planting or trees	-	0.13	2.63	-	-	100 m^2	**2.63**
mass spraying low vegetation or hard surfaces	-	0.10	2.10	-	-	100 m^2	**2.10**
Q40 FENCING							
Protective fencing							
Cleft chestnut rolled fencing; fixing to 100 mm diameter chestnut posts; driving into firm ground at 3 m centres							
900 mm high	3.82	0.11	2.24	-	7.06	m	**9.30**
1200 mm high	4.48	0.16	3.36	-	7.77	m	**11.13**
1500 mm high; 3 strand	7.11	0.21	4.49	-	10.17	m	**14.66**
Boundary fencing; strained wire and wire mesh; AVS Fencing Supplies Ltd							
Strained wire fencing; concrete posts only at 2750 mm centres, 610 mm below ground; excavating holes; filling with concrete; replacing topsoil; disposing surplus soil off site							
900 mm high	3.20	0.56	11.67	0.60	4.65	m	**16.92**
1200 mm high	4.81	0.56	11.67	0.60	6.26	m	**18.53**
1400 mm high	5.10	0.56	11.67	0.60	6.54	m	**18.81**
1800 mm high	5.49	0.78	16.33	1.49	6.94	m	**24.77**
2400 mm high	7.88	1.67	35.00	0.60	9.33	m	**44.93**
Extra over strained wire fencing for concrete straining posts with one strut; posts and struts 610 mm below ground; struts, cleats, stretchers, winders, bolts, and eye bolts; excavating holes; filling to within 150 mm of ground level with concrete (1:12) - 40 mm aggregate; replacing topsoil; disposing surplus soil off site							
900 mm high	25.92	0.67	14.07	1.34	57.58	nr	**72.99**
1200 mm high	28.83	0.67	14.00	1.35	61.21	nr	**76.56**
1400 mm high	33.11	0.67	14.00	1.35	88.20	nr	**103.55**
1800 mm high	41.02	0.67	14.00	1.35	75.52	nr	**90.87**
2400 mm high	56.38	0.83	17.49	1.59	100.01	nr	**119.09**

Q PAVING/PLANTING/FENCING/SITE FURNITURE

Item Excluding site overheads and profit	PC £	Labour hours	Labour £	Plant £	Material £	Unit	Total rate £
Q40 FENCING - cont'd							
Boundary fencing - cont'd							
Extra over strained wire fencing for concrete straining posts with two struts; posts and struts 610 mm below ground; excavating holes; filling to within 150 mm of ground level with concrete (1:12) - 40 mm aggregate; replacing topsoil; disposing surplus soil off site							
900 mm high	37.38	0.91	19.11	1.73	84.33	nr	**105.17**
1200 mm high	41.68	0.85	17.85	1.35	95.94	nr	**115.14**
1400 mm high	48.07	0.91	19.07	1.35	110.08	nr	**130.49**
1800 mm high	59.56	0.85	17.85	1.35	116.37	nr	**135.57**
2400 mm high	80.86	0.67	14.00	1.59	155.13	nr	**170.72**
Strained wire fencing; galvanised steel angle posts only at 2750 mm centres; 610 mm below ground; driving in							
900 mm high; 40 mm x 40 mm x 5 mm	5.47	0.03	0.64	-	5.47	m	**6.11**
1200 mm high; 40 mm x 40 mm x 5 mm	6.41	0.03	0.70	-	6.41	m	**7.11**
1400 mm high; 40 mm x 40 mm x 5 mm	7.65	0.06	1.27	-	7.65	m	**8.92**
1800 mm high; 40 mm x 40 mm x 5 mm	9.31	0.07	1.53	-	9.31	m	**10.83**
2400 mm high; 45 mm x 45 mm x 5 mm	11.29	0.12	2.55	-	11.29	m	**13.83**
Galvanised steel angle straining posts with one strut for strained wire fencing; setting in concrete							
900 mm high; 50 mm x 50 mm x 6 mm	55.98	0.67	14.00	-	55.98	nr	**69.98**
1200 mm high; 50 mm x 50 mm x 6 mm	78.30	0.67	14.00	-	78.30	nr	**92.30**
1400 mm high; 50 mm x 50 mm x 6 mm	91.01	0.67	14.00	-	91.01	nr	**105.01**
1800 mm high; 50 mm x 50 mm x 6 mm	103.49	0.67	14.00	-	103.49	nr	**117.49**
2400 mm high; 60 mm x 60 mm x 6 mm	123.55	0.67	14.00	-	123.55	nr	**137.55**
Galvanised steel straining posts with two struts for strained wire fencing; setting in concrete							
900 mm high; 50 mm x 50 mm x 6 mm	83.77	1.00	21.00	-	83.77	nr	**104.77**
1200 mm high; 50 mm x 50 mm x 6 mm	115.26	1.00	21.00	-	115.26	nr	**136.26**
1400 mm high; 50 mm x 50 mm x 6 mm	136.47	1.00	21.00	-	136.47	nr	**157.47**
1800 mm high; 50 mm x 50 mm x 6 mm	165.91	1.00	21.00	-	165.91	nr	**186.91**
2400 mm high; 60 mm x 60 mm x 6 mm	205.92	1.00	21.00	-	205.92	nr	**226.92**
Strained wire; to posts (posts not included); 3 mm galvanized wire; fixing with galvanized stirrups							
900 mm high; 2 wire	0.28	0.03	0.70	-	0.40	m	**1.10**
1200 mm high; 3 wire	0.42	0.05	0.98	-	0.54	m	**1.52**
1400 mm high; 3 wire	0.42	0.05	0.98	-	0.54	m	**1.52**
1800 mm high; 3 wire	0.42	0.05	0.98	-	0.54	m	**1.52**
Barbed wire; to posts (posts not included); 3 mm galvanised wire; fixing with galvanised stirrups							
900 mm high; 2 wire	0.26	0.07	1.40	-	0.38	m	**1.78**
1200 mm high; 3 wire	0.40	0.09	1.96	-	0.52	m	**2.47**
1400 mm high; 3 wire	0.40	0.09	1.96	-	0.52	m	**2.47**
1800 mm high; 3 wire	0.40	0.09	1.96	-	0.52	m	**2.47**
Chain link fencing; AVS Fencing Supplies Ltd; to strained wire and posts priced separately; 3 mm galvanised wire; 51 mm mesh; galvanized steel components; fixing to line wires threaded through posts and strained with eye-bolts; posts (not included)							
900 mm high	4.84	0.07	1.40	-	4.96	m	**6.36**
1200 mm high	6.04	0.07	1.40	-	6.16	m	**7.56**
1800 mm high	6.82	0.10	2.10	-	6.94	m	**9.04**
2400 mm high	10.56	0.10	2.10	-	10.68	m	**12.78**

Q PAVING/PLANTING/FENCING/SITE FURNITURE

Item Excluding site overheads and profit	PC £	Labour hours	Labour £	Plant £	Material £	Unit	Total rate £
Chain link fencing; to strained wire and posts priced separately; 3.15 mm plastic coated galvanized wire (wire only 2.50 mm); 51 mm mesh; galvanised steel components; fencing with line wires threaded through posts and strained with eye-bolts; posts (not included) (Note: plastic coated fencing can be cheaper than galvanised finish as wire of a smaller cross-sectional area can be used)							
900 mm high	2.86	0.07	1.40	-	3.10	m	**4.50**
1200 mm high	3.96	0.07	1.40	-	4.20	m	**5.60**
1400 mm high	4.84	0.07	1.40	-	5.08	m	**6.48**
1800 mm high	5.46	0.13	2.63	-	6.42	m	**9.04**
Extra over strained wire fencing for cranked arms and galvanized barbed wire							
1 row	3.07	0.02	0.35	-	3.07	m	**3.42**
2 row	3.20	0.05	1.05	-	3.20	m	**4.25**
3 row	3.33	0.05	1.05	-	3.33	m	**4.38**
Field fencing; Jacksons Fencing; welded wire mesh; fixed to posts and straining wires measured separately							
Cattle fence; 1100 m high 114 x 300 mm at bottom to 230 x 300 mm at top	1.07	0.10	2.10	-	1.30	m	**3.40**
Sheep fence; 900 mm high; 140 x 300 mm at bottom to 230 x 300 mm at top	0.94	0.10	2.10	-	1.17	m	**3.27**
Deer fence; 1900 mm high; 89 x 150 mm at bottom to 267 x 300 mm at top	3.54	0.13	2.63	-	3.77	m	**6.40**
Extra for concreting in posts	-	0.50	10.50	-	4.62	nr	**15.12**
Extra for straining post	9.70	0.75	15.75	-	9.70	nr	**25.45**
Rabbit netting; Jacksons Fencing; timber stakes; peeled kiln dried pressure treated; pointed; 1.8 m posts driven 900 mm into ground at 3 m centres (line wires and netting priced separately)							
75 - 100 mm stakes	1.78	0.25	5.25	-	1.78	m	**7.03**
Corner posts or straining posts 150 mm diameter 2.3 m high set in concrete; centres to suit local conditions or changes of direction							
1 strut	11.64	1.00	21.00	3.84	18.50	each	**43.34**
2 strut	13.96	1.00	21.00	3.84	20.82	each	**45.66**
Strained wire; to posts (posts not included); 3 mm galvanized wire; fixing with galvanized stirrups							
900 mm high; 2 wire	0.28	0.03	0.70	-	0.40	m	**1.10**
1200 mm high; 3 wire	0.42	0.05	0.98	-	0.54	m	**1.52**
Rabbit netting; 31 mm 19 gauge 1050 high netting fixed to posts line wires and straining posts or corner posts all priced separately							
900 high turned in	0.77	0.04	0.84	-	0.78	m	**1.62**
900 high buried 150 mm in trench	0.77	0.08	1.75	-	0.78	m	**2.53**
Boundary fencing; strained wire and wire mesh; Jacksons Fencing							
Tubular chain link fencing; galvanized; plastic coated; 60.3 mm diameter posts at 3.0 m centres; setting 700 mm into ground; choice of ten mesh colours; including excavating holes; backfilling and removing surplus soil; with top rail only							
900 mm high	-	-	-	-	-	m	**29.93**
1200 mm high	-	-	-	-	-	m	**31.55**
1800 mm high	-	-	-	-	-	m	**35.00**
2000 mm high	-	-	-	-	-	m	**37.11**

Q PAVING/PLANTING/FENCING/SITE FURNITURE

Item Excluding site overheads and profit	PC £	Labour hours	Labour £	Plant £	Material £	Unit	Total rate £
Q40 FENCING - cont'd							
Boundary fencing - cont'd							
Tubular chain link fencing; galvanised; plastic coated; 60.3 mm diameter posts at 3.0 m centres; cranked arms and 3 lines barbed wire; setting 700 mm into ground; including excavating holes; backfilling and removing surplus soil; with top rail only							
1800 mm high	-	-	-	-	-	m	**38.10**
2000 mm high	-	-	-	-	-	m	**38.86**
Timber fencing; AVS Fencing Supplies Ltd							
Palisade fencing; 22 mm x 75 mm softwood vertical palings with flat tops; nailing to 3 nr 50 mm x 100 mm horizontal softwood rails; housing into 100 mm x 100 mm softwood posts with weathered tops at 3.00 m centres; setting in concrete; all treated timber							
1800 mm high	23.75	0.47	9.81	-	28.18	m	**37.99**
Palisade fencing; 22 mm x 75 mm oak vertical palings with flat tops; nailing to 3 nr 50 mm x 125 mm horizontal oak rails; housing into 100 mm x 100 mm oak posts with weathered tops at 3.00 m centres; setting in concrete							
1800 mm high	28.28	0.47	9.80	-	32.70	m	**42.50**
Post-and-rail fencing; 3 nr 90 mm x 38 mm softwood horizontal rails; fixing with galvanized nails to 150 mm x 75 mm softwood posts; including excavating and backfilling into firm ground at 1.80 m centres; all treated timber							
1200 mm high	16.24	0.35	7.35	-	16.32	m	**23.67**
Post-and-rail fencing; 3 nr 90 mm x 38 mm oak horizontal rails; fixing with galvanised nails to 150 mm x 75 mm oak posts; including excavating and backfilling into firm ground at 1.80 m centres							
1200 mm high	32.62	0.35	7.35	-	32.70	m	**40.06**
Morticed post-and-rail fencing; 3 nr horizontal 90 mm x 38 mm softwood rails; fixing with galvanized nails; 90 mm x 38 mm softwood centre prick posts; to 150 mm x 75 mm softwood posts; including excavating and backfilling into firm ground at 2.85 m centres; all treated timber							
1200 mm high	13.81	0.35	7.35	-	13.89	m	**21.25**
1350 mm high five rails	16.55	0.40	8.40	-	16.63	m	**25.03**
Cleft rail fencing; oak or chestnut adze tapered rails 2.80 m long; morticed into joints; to 125 mm x 100 mm softwood posts 1.95 m long; including excavating and backfilling into firm ground at 2.80 m centres							
two rails	18.40	0.25	5.25	-	18.40	m	**23.65**
three rails	21.32	0.28	5.88	-	21.32	m	**27.20**
four rails	24.24	0.35	7.35	-	24.24	m	**31.59**
Close boarded fencing; 2 nr softwood rails; 150 x 25 mm gravel boards; 2 ex 100 x 22 softwood pales lapped 13 mm; to concrete posts; including excavating and backfilling into firm ground at 3.00 m centres							
900 mm high	11.11	0.47	9.81	-	17.39	m	**27.21**
1050 mm high	12.31	0.47	9.81	-	19.82	m	**29.63**
1200 mm high	13.39	0.47	9.81	-	20.90	m	**30.71**

Q PAVING/PLANTING/FENCING/SITE FURNITURE

Item Excluding site overheads and profit	PC £	Labour hours	Labour £	Plant £	Material £	Unit	Total rate £
Close boarded fencing; 3 nr softwood rails; 150 x 25 mm gravel boards; 89 mm x 19 mm softwood pales lapped 13 mm; to concrete posts; including excavating and backfilling into firm ground at 3.00 m centres							
1350 mm high	22.04	0.56	11.76	-	23.49	m	**35.26**
1650 mm high	24.51	0.56	11.76	-	25.96	m	**37.72**
1800 mm high	24.99	0.56	11.76	-	26.44	m	**38.20**
Close boarded fencing; 3 nr softwood rails; 150 x 25 mm gravel boards; 89 mm x 19 mm softwood pales lapped 13 mm; to softwood posts 100 x 100 mm; including excavating and backfilling into firm ground at 3.00 m centres							
1350 mm high	19.00	0.56	11.76	-	20.45	m	**32.21**
1650 mm high	20.72	0.56	11.76	-	22.17	m	**33.93**
1800 mm high	21.52	0.56	11.76	-	22.97	m	**34.74**
Close boarded fencing; 3 nr softwood rails; 150 x 25 mm gravel boards; 89 mm x 19 mm softwood pales lapped 13 mm; to oak posts; including excavating and backfilling into firm ground at 3.00 m centres							
1350 mm high	22.17	0.56	11.76	-	23.62	m	**35.38**
1650 mm high	24.60	0.56	11.76	-	26.05	m	**37.81**
1800 mm high	25.75	0.56	11.76	-	27.20	m	**38.96**
Close boarded fencing; 3 nr softwood rails; 150 x 25 mm gravel boards; 89 mm x 19 mm softwood pales lapped 13 mm; to oak posts; including excavating and backfilling into firm ground at 2.47 m centres							
1800 mm high	27.48	0.68	14.23	-	28.82	m	**43.05**
Close boarded fencing - all oak; 2 nr 76 mm x 38 mm rectangular rails; 152 mm x 25 mm gravel boards; 89 mm x 19 mm oak featheredge boards lapped 13 mm; to oak posts; including excavating and backfilling into firm ground at 2.47 m centres							
1070 mm high	10.36	0.56	11.76	-	44.77	m	**56.54**
1800 mm high	21.03	0.56	11.76	-	52.38	m	**64.14**
Extra over close boarded fencing for 65 mm x 38 mm oak cappings	6.59	0.05	1.05	-	6.59	nr	**7.64**
"Hit and miss" horizontal rail fencing; 87 mm x 38 mm top and bottom rails; 100 mm x 22 mm vertical boards arranged alternately on opposite side of rails; to 100 mm x 100 mm posts; including excavating and backfilling into firm ground; setting in concrete at 1.8 m centres							
treated softwood; 1800 mm high	34.15	1.33	28.00	-	37.17	m	**65.16**
treated softwood; 2000 mm high	35.53	1.33	28.00	-	43.17	m	**71.17**
primed softwood; 1800 mm high	34.15	1.67	35.00	-	37.17	m	**72.17**
primed softwood; 2000 mm high	35.53	1.67	35.00	-	43.17	m	**78.17**
Screen fencing in horizontal wavy edge overlap panels; 1.83 m wide; to 100 mm x 100 mm preservative treated softwood posts; including excavating and backfilling into firm ground							
900 mm high	9.17	0.47	9.81	-	17.14	m	**26.95**
1200 mm high	9.17	0.47	9.81	-	18.00	m	**27.81**
1500 mm high	9.32	0.56	11.80	-	20.48	m	**32.28**
1800 mm high	9.35	0.56	11.80	-	21.31	m	**33.11**
Screen fencing for oak in lieu of softwood posts							
900 mm high	16.07	0.47	9.81	-	20.78	m	**30.59**
1200 mm high	17.90	0.47	9.81	-	22.61	m	**32.42**
1500 mm high	19.68	0.56	11.80	-	25.90	m	**37.70**
1800 mm high	21.37	0.56	11.80	-	27.60	m	**39.39**
Extra over screen fencing for 300 mm high trellis tops, slats at 100 mm centres; including additional length of posts	4.36	0.10	2.10	-	4.36	m	**6.46**

Q PAVING/PLANTING/FENCING/SITE FURNITURE

Item Excluding site overheads and profit	PC £	Labour hours	Labour £	Plant £	Material £	Unit	Total rate £
Q40 FENCING - cont'd							
Traditional trellis panels; The Garden Trellis Company; bespoke trellis panels for decorative, screening or security applications; timber planed all round; height of trellis 1800 mm							
Freestanding panels; posts 70 x 70 set in concrete; timber frames mitred and grooved 45 x 34 mm; heights and widths to suit; slats 32 mm x 10 mm joinery quality tanalised timber at 100 mm ccs; elements fixed by galvanised staples; capping rail 70 x 34							
HV68 horizontal and vertical slats; softwood	69.30	1.00	21.00	-	75.46	m	**96.46**
D68 diagonal slats; softwood	79.20	1.00	21.00	-	85.36	m	**106.36**
HV68 horizontal and vertical slats; hardwood iroko	138.60	1.00	21.00	-	144.76	m	**165.76**
D68 diagonal slats; hardwood iroko or Western Red Cedar	158.40	1.00	21.00	-	164.56	m	**185.56**
Trellis panels fixed to face of existing wall or railings; timber frames mitred and grooved 45 x 34 mm; heights and widths to suit; slats 32 mm x 10 mm joinery quality tanalised timber at 100 mm ccs; elements fixed by galvanised staples; capping rail 70 x 34							
HV68 horizontal and vertical slats; softwood	63.36	0.50	10.50	-	65.47	m	**75.97**
D68 diagonal slats; softwood	75.24	0.50	10.50	-	77.35	m	**87.85**
HV68 horizontal and vertical slats; iroko or Western Red Cedar	128.70	0.50	10.50	-	130.81	m	**141.31**
D68 Diagonal slats; iroko or Western Red Cedar	148.50	0.50	10.50	-	150.61	m	**161.11**
Contemporary style trellis panels; The Garden Trellis Company; bespoke trellis panels for decorative, screening or security applications; timber planed all round							
Freestanding panels 30/15; posts 70 x 70 set in concrete with 90 x 30 top capping; slats 30 x 14 mm with 15 mm gaps; vertical support at 450 mm ccs							
joinery treated softwood	89.10	1.00	21.00	-	95.26	m	**116.26**
hardwood iroko or Western Red Cedar	178.20	1.00	21.00	-	184.36	m	**205.36**
Panels 30/15 face fixed to existing wall or fence; posts 70 x 70 set in concrete with 90 x 30 top capping; slats 30 x 14 mm with 15 mm gaps; vertical support at 450 mm ccs							
joinery treated softwood	79.20	0.50	10.50	-	81.31	m	**91.81**
hardwood iroko or Western Red Cedar	168.30	0.50	10.50	-	170.41	m	**180.91**
Integral arches to ornamental trellis panels; The Garden Trellis Company							
Arches to trellis panels in 45 mm x 34 mm grooved timbers to match framing; fixed to posts							
R450 1/4 circle; joinery treated softwood	38.50	1.50	31.50	-	44.66	nr	**76.16**
R450 1/4 circle; hardwood iroko or Western Red Cedar	44.00	1.50	31.50	-	50.16	nr	**81.66**
Arches to span 1800 wide							
joinery treated softwood	49.50	1.50	31.50	-	55.66	nr	**87.16**
hardwood iroko or Western Red Cedar	66.00	1.50	31.50	-	72.16	nr	**103.66**
Painting or staining of trellis panels; high quality coatings							
microporous opaque paint or spirit based stain	-	-	-	-	16.50	m^2	-

Q PAVING/PLANTING/FENCING/SITE FURNITURE

Item Excluding site overheads and profit	PC £	Labour hours	Labour £	Plant £	Material £	Unit	Total rate £
Concrete fencing							
Panel fencing; to precast concrete posts; in 2 m bays; setting posts 600 mm into ground; sandfaced finish							
900 mm high	10.94	0.25	5.25	-	12.76	m	**18.01**
1200 mm high	15.10	0.25	5.25	-	16.92	m	**22.17**
Panel fencing; to precast concrete posts; in 2 m bays; setting posts 750 mm into ground; sandfaced finish							
1500 mm high	19.81	0.33	7.00	-	21.63	m	**28.63**
1800 mm high	24.23	0.36	7.64	-	26.05	m	**33.68**
2100 mm high	28.20	0.40	8.40	-	30.02	m	**38.42**
2400 mm high	32.57	0.40	8.40	-	34.39	m	**42.79**
extra over capping panel to all of the above	5.57	-	-	-	5.57	m	**-**
Windbreak fencing							
Fencing; English Woodlands; "Shade and Shelter Netting" windbreak fencing; green; to 100 mm diameter treated softwood posts; setting 450 mm into ground; fixing with 50 mm x 25 mm treated softwood battens nailed to posts, including excavating and backfilling into firm ground; setting in concrete at 3 m centres							
1200 mm high	1.36	0.16	3.27	-	6.34	m	**9.61**
1800 mm high	1.94	0.16	3.27	-	6.91	m	**10.18**
Gates - General							
Preamble: gates in fences; see specification for fencing, as gates in traditional or proprietary fencing systems are usually constructed of the same materials and finished as the fencing itself.							
Gates; hardwood; Longlyf Timber Products Ltd							
Hardwood entrance gate, five bar diamond braced, curved hanging stile, planed iroko; fixed to 150 mm x 150 mm softwood posts; inclusive of hinges and furniture							
ref 1100 040; 0.9 m wide	219.53	5.00	105.00	-	279.04	nr	**384.04**
ref 1100 041; 1.2 m wide	234.37	5.00	105.00	-	293.88	nr	**398.88**
ref 1100 042; 1.5 m wide	303.60	5.00	105.00	-	363.11	nr	**468.11**
ref 1100 043; 1.8 m wide	321.63	5.00	105.00	-	381.14	nr	**486.14**
ref 1100 044; 2.1 m wide	352.77	5.00	105.00	-	412.28	nr	**517.28**
ref 1100 045; 2.4 m wide	373.49	5.00	105.00	-	433.00	nr	**538.00**
ref 1100 047; 3.0 m wide	406.85	5.00	105.00	-	466.36	nr	**571.36**
ref 1100 048; 3.3 m wide	425.38	5.00	105.00	-	484.89	nr	**589.89**
ref 1100 049; 3.6 m wide	440.18	5.00	105.00	-	499.69	nr	**604.69**
Hardwood field gate, five bar diamond braced, planed iroko; fixed to 150 mm x 150 mm softwood posts; inclusive of hinges and furniture							
ref 1100 100; 0.9 m wide	158.40	4.00	84.00	-	217.91	nr	**301.91**
ref 1100 101; 1.2 m wide	172.37	4.00	84.00	-	231.88	nr	**315.88**
ref 1100 102; 1.5 m wide	205.52	4.00	84.00	-	265.03	nr	**349.03**
ref 1100 103; 1.8 m wide	221.20	4.00	84.00	-	280.71	nr	**364.71**
ref 1100 104; 2.1 m wide	272.18	4.00	84.00	-	331.69	nr	**415.69**
ref 1100 105; 2.4 m wide	288.67	4.00	84.00	-	348.18	nr	**432.18**
ref 1100 106; 2.7 m wide	305.17	4.00	84.00	-	364.68	nr	**448.68**
ref 1100 108; 3.3 m wide	338.15	4.00	84.00	-	397.66	nr	**481.66**
ref 1100 109; 3.6 m wide	355.52	4.00	84.00	-	415.03	nr	**499.03**

Q PAVING/PLANTING/FENCING/SITE FURNITURE

Item Excluding site overheads and profit	PC £	Labour hours	Labour £	Plant £	Material £	Unit	Total rate £
Q40 FENCING - cont'd							
Gates; softwood; Jacksons Fencing							
Timber field gates, including wrought iron ironmongery; five bar type; diamond braced; 1.80 m high; to 200 mm x 200 mm posts; setting 750 mm into firm ground							
treated softwood; width 2400 mm	79.85	10.00	210.00	-	173.56	nr	**383.56**
treated softwood; width 2700 mm	87.53	10.00	210.00	-	181.24	nr	**391.24**
treated softwood; width 3000 mm	91.18	10.00	210.00	-	184.89	nr	**394.89**
treated softwood; width 3300 mm	97.84	10.00	210.00	-	191.55	nr	**401.55**
Featherboard garden gates, including ironmongery; to 100 mm x 120 mm posts; 1 nr diagonal brace							
treated softwood; 1.0 m x 1.2 m high	16.52	3.00	63.00	-	96.49	nr	**159.49**
treated softwood; 1.0 m x 1.5 m high	21.83	3.00	63.00	-	103.76	nr	**166.76**
treated softwood; 1.0 m x 1.8 m high	24.43	3.00	63.00	-	110.71	nr	**173.71**
Picket garden gates, including ironmongery; to match picket fence; width 1000 mm; to 100 mm x 120 mm posts; 1 nr diagonal brace							
treated softwood; 950 mm high	66.47	3.00	63.00	-	91.68	nr	**154.68**
treated softwood; 1200 mm high	69.52	3.00	63.00	-	94.73	nr	**157.73**
treated softwood; 1800 mm high	80.28	3.00	63.00	-	113.39	nr	**176.39**
Gates; tubular steel; Jacksons Fencing							
Tubular mild steel field gates, including ironmongery; diamond braced; 1.80 m high; to tubular steel posts; setting in concrete							
galvanized; width 3000 mm	120.42	5.00	105.00	-	221.02	nr	**326.02**
galvanized; width 3300 mm	127.42	5.00	105.00	-	228.02	nr	**333.02**
galvanized; width 3600 mm	134.70	5.00	105.00	-	235.30	nr	**340.30**
galvanized; width 4200 mm	149.22	5.00	105.00	-	249.82	nr	**354.82**
Stiles and kissing gates							
Stiles; Jacksons Fencing; "Jacksons Nr 2007" stiles; 2 nr posts; setting into firm ground; 3 nr rails; 2 nr treads	71.68	3.00	63.00	-	83.99	nr	**146.99**
Kissing gates; Jacksons Fencing; in galvanised metal bar; fixing to fencing posts (posts not included); 1.65 m x 1.30 m x 1.00 m high	204.19	5.00	105.00	-	204.19	nr	**309.19**

R DISPOSAL SYSTEMS

Item Excluding site overheads and profit	PC £	Labour hours	Labour £	Plant £	Material £	Unit	Total rate £
R12 DRAINAGE BELOW GROUND							
Silt pits and inspection chambers							
Excavating pits; starting from ground level; by machine							
maximum depth not exceeding 1.00 m	-	-	-	3.11	-	m³	**3.11**
maximum depth not exceeding 2.00 m	-	-	-	4.14	-	m³	**4.14**
Disposal of excavated material; depositing on site in permanent spoil heaps; average 50 m	-	-	-	6.66	-	m³	**6.66**
Surface treatments; compacting; bottoms of excavations	-	0.05	1.05	-	-	m²	**1.05**
Earthwork support; distance between opposing faces not exceeding 2.00 m							
maximum depth not exceeding 1.00 m	-	0.20	4.20	-	4.28	m³	**8.48**
maximum depth not exceeding 2.00 m	-	0.30	6.30	-	4.28	m³	**10.58**
maximum depth not exceeding 4.00 m	-	0.67	13.99	-	1.69	m³	**15.67**
Silt pits and inspection chambers; in situ concrete							
Beds; plain in situ concrete; 11.50 N/mm² - 40 mm aggregate							
thickness not exceeding 150 mm	-	1.00	21.00	-	107.27	m³	**128.27**
thickness 150 mm - 450 mm	-	0.67	14.00	-	107.27	m³	**121.27**
Benchings in bottoms; plain in situ concrete; 25.50 N/mm² - 20 mm aggregate							
thickness 150 mm - 450 mm	-	2.00	42.00	-	107.27	m³	**149.27**
Isolated cover slabs; reinforced In situ concrete; 21.00 N/mm² - 20 mm aggregate							
thickness not exceeding 150 mm	-	4.00	84.00	-	107.27	m³	**191.27**
Fabric reinforcement; BS 4483; A193 (3.02 kg/m²) in cover slabs	3.93	0.06	1.31	-	4.32	m²	**5.64**
Formwork to reinforced in situ concrete; isolated cover slabs							
soffits; horizontal	-	3.28	68.88	-	4.40	m²	**73.28**
height not exceeding 250 mm	-	0.97	20.37	-	1.87	m	**22.24**
Brickwork							
Walls to manholes; bricks; PC £300.00/1000; in cement mortar (1:3)							
one brick thick	36.00	3.00	63.00	-	45.38	m²	**108.38**
one and a half brick thick	54.00	4.00	84.00	-	68.06	m²	**152.06**
two brick thick projection of footing or the like	72.00	4.80	100.80	-	90.75	m²	**191.55**
Walls to manholes; engineering bricks; PC £260.00/1000; in cement mortar (1:3)							
one brick thick	31.20	3.00	63.00	-	40.34	m²	**103.34**
one and a half brick thick	46.80	3.00	63.00	-	56.72	m²	**119.72**
two brick thick projection of footing or the like	62.40	3.00	63.00	-	73.10	m²	**136.10**
Extra over common or engineering bricks in any mortar for fair face; flush pointing as work proceeds; English bond walls or the like	-	0.13	2.80	-	-	m²	**2.80**
In situ finishings; cement:sand mortar (1:3); steel trowelled; 13 mm one coat work to manhole walls; to brickwork or blockwork base; over 300 mm wide	-	0.80	16.80	-	2.53	m²	**19.33**
Building into brickwork; ends of pipes; making good facings or renderings							
small	-	0.20	4.20	-	-	nr	**4.20**
large	-	0.30	6.30	-	-	nr	**6.30**
extra large	-	0.40	8.40	-	-	nr	**8.40**
extra large; including forming ring arch cover	-	0.50	10.50	-	-	nr	**10.50**

R DISPOSAL SYSTEMS

Item Excluding site overheads and profit	PC £	Labour hours	Labour £	Plant £	Material £	Unit	Total rate £
R12 DRAINAGE BELOW GROUND - cont'd							
Inspection chambers; Hepworth Plc							
Polypropylene up to 600 mm deep							
300 mm diameter; 960 mm deep; heavy duty round covers and frames; double seal recessed with 6 nr 110 mm outlets/inlets	216.70	3.00	63.00	-	219.59	nr	**282.59**
Polypropylene up to 1200 mm deep							
475 mm diameter; 1030 mm deep; heavy duty round covers and frames; double seal recessed with 5 nr 110 mm outlets/inlets	71.81	4.00	84.00	-	288.90	nr	**372.90**
Step irons; to BS 1247; Ashworth Ltd drainage systems; malleable cast iron; galvanized; building into joints							
General purpose pattern; for one brick walls	4.82	0.17	3.57	-	4.82	nr	**8.39**
Best quality vitrified clay half section channels; Hepworth Plc; bedding and jointing in cement:mortar (1:2)							
Channels; straight							
100 mm	6.18	0.80	16.80	-	9.15	m	**25.95**
150 mm	10.29	1.00	21.00	-	13.25	m	**34.25**
225 mm	19.26	1.35	28.35	-	23.71	m	**52.06**
300 mm	47.46	1.80	37.80	-	50.42	m	**88.22**
Bends; 15, 30, 45 or 90 degrees							
100 mm bends	5.32	0.75	15.75	-	6.80	nr	**22.55**
150 mm bends	9.19	0.90	18.90	-	12.15	nr	**31.05**
225 mm bends	35.67	1.20	25.20	-	38.63	nr	**63.83**
300 mm bends	72.72	1.10	23.10	-	76.43	nr	**99.53**
Intercepting traps							
Vitrified clay; inspection arms; brass stoppers; iron levers; chains and staples; galvanized; staples cut and pinned to brickwork; cement:mortar (1:2) joints to vitrified clay pipes and channels; bedding and surrounding in concrete; 11.50 N/mm^2 - 40 mm aggregate; cutting and fitting brickwork; making good facings							
100 mm inlet; 100 mm outlet	65.79	3.00	55.50	-	95.24	nr	**150.74**
150 mm inlet; 150 mm outlet	94.86	2.00	37.00	-	130.16	nr	**167.16**
Excavating trenches; using 3 tonne tracked excavator; to receive pipes; grading bottoms; earthwork support; filling with excavated material to within 150 mm of finished surfaces and compacting; completing fill with topsoil; disposal of surplus soil							
Services not exceeding 200 mm nominal size							
average depth of run not exceeding 0.50 m	-	0.02	0.42	3.19	1.72	m	**5.33**
average depth of run not exceeding 0.75 m	-	0.03	0.70	4.76	1.72	m	**7.18**
average depth of run not exceeding 1.00 m	-	0.05	1.05	8.25	1.72	m	**11.02**
average depth of run not exceeding 1.25 m	-	0.10	2.10	10.96	1.72	m	**14.79**
Granular beds to trenches; lay granular material, to trenches excavated separately, to receive pipes (not included)							
300 wide x 100 thick							
reject sand	-	0.03	0.53	0.77	0.97	m	**2.27**
shingle 20 mm aggregate	-	0.03	0.53	0.77	1.07	m	**2.37**
sharp sand	-	0.03	0.53	0.77	1.19	m	**2.49**
300 wide x 150 thick							
reject sand	-	0.04	0.79	1.16	0.49	m	**2.43**
shingle 20 mm aggregate	-	0.04	0.79	1.16	0.54	m	**2.48**
sharp sand	-	0.04	0.79	1.16	0.59	m	**2.54**

R DISPOSAL SYSTEMS

Item Excluding site overheads and profit	PC £	Labour hours	Labour £	Plant £	Material £	Unit	Total rate £
Excavating trenches; using 3 tonne tracked excavator; to receive pipes; grading bottoms; earthwork support; filling with imported granular material type 2 and compacting; disposal of surplus soil							
Services not exceeding 200 mm nominal size							
average depth of run not exceeding 0.50 m	4.86	0.02	0.42	2.16	8.31	m	**10.89**
average depth of run not exceeding 0.75 m	7.29	0.03	0.53	2.57	12.46	m	**15.56**
average depth of run not exceeding 1.00 m	9.72	0.03	0.60	3.58	16.62	m	**20.80**
average depth of run not exceeding 1.25 m	12.15	0.03	0.60	6.33	20.78	m	**27.70**
Excavating trenches, using 3 tonne tracked excavator, to receive pipes; grading bottoms; earthwork support; filling with concrete, ready mixed ST2; disposal of surplus soil							
Services not exceeding 200 mm nominal size							
average depth of run not exceeding 0.50 m	12.47	0.02	0.42	2.06	15.92	m	**18.40**
average depth of run not exceeding 0.75 m	18.71	0.03	0.53	2.57	23.88	m	**26.98**
average depth of run not exceeding 1.00 m	24.94	0.03	0.60	3.58	31.84	m	**36.02**
average depth of run not exceeding 1.25 m	31.18	0.03	0.60	6.33	39.81	m	**46.73**
Clay pipes and fittings: to BS EN295:1:1991; Hepworth Plc; Supersleeve							
100 mm clay pipes; polypropylene slip coupling; in trenches (trenches not included)							
laid straight	2.90	0.25	5.25	-	5.04	m	**10.29**
short runs under 3.00 m	2.90	0.31	6.56	-	5.04	m	**11.61**
Extra over 100 mm clay pipes for							
bends; 15-90 degree; single socket	9.66	0.25	5.25	-	9.66	nr	**14.91**
junction; 45 or 90 degree; double socket	20.36	0.25	5.25	-	20.36	nr	**25.61**
slip couplings polypropylene	3.42	0.08	1.75	-	3.42	nr	**5.17**
gully with "P" trap; 100 mm; 154 mm x 154 mm plastic grating	59.14	1.00	21.00	-	61.06	nr	**82.06**
PVC-u pipes and fittings; to BS EN1401; Wavin Plastics Ltd; OsmaDrain							
110 mm PVC-u pipes; in trenches (trenches not included)							
laid straight	6.45	0.08	1.68	-	6.45	m	**8.13**
short runs under 3.00 m	6.45	0.12	2.52	-	6.45	m	**8.97**
Extra over 110 mm PVC-u pipes for							
bends; short radius	12.61	0.25	5.25	-	12.61	nr	**17.86**
bends; long radius	22.43	0.25	5.25	-	22.43	nr	**27.68**
junctions; equal; double socket	15.05	0.25	5.25	-	15.05	nr	**20.30**
slip couplings	-	0.10	2.10	-	7.30	nr	**9.40**
adaptors to clay	13.49	0.50	10.50	-	13.64	nr	**24.14**
Kerbs to gullies							
Kerbs; to gullies; in one course Class B engineering bricks; to 4 nr sides; rendering in cement:mortar (1:3); dished to gully gratings	2.40	1.00	21.00	-	3.16	nr	**24.16**
Gullies vitrified clay; Hepworth Plc; bedding in concrete; 11.50 N/mm^2 - 40 mm aggregate							
Vitrified clay yard gullies (mud); trapped; domestic duty (up to 1 tonne)							
RGP5; 100 mm outlet; 100 mm dia; 225 mm internal width 585 mm internal depth	112.49	3.50	73.50	-	113.12	nr	**186.62**
RGP7; 150 mm outlet; 100 mm dia; 225 mm internal width 585 mm internal depth	112.49	3.50	73.50	-	113.12	nr	**186.62**

R DISPOSAL SYSTEMS

Item Excluding site overheads and profit	PC £	Labour hours	Labour £	Plant £	Material £	Unit	Total rate £
R12 DRAINAGE BELOW GROUND - cont'd							
Gullies vitrified clay - cont'd							
Vitrified clay yard gullies (mud); trapped; medium duty (up to 5 tonnes)							
100 mm outlet; 100 mm dia; 225 mm internal width 585 mm internal depth	112.49	3.50	73.50	-	113.12	nr	**186.62**
150 mm outlet; 100 mm dia; 225 mm internal width 585 mm internal depth	158.94	3.50	73.50	-	159.57	nr	**233.07**
Combined filter and silt bucket for yard gullies							
225 mm wide	36.97	-	-	-	36.97	nr	**-**
Vitrified clay road gullies; trapped with rodding eye							
100 mm outlet; 300 mm internal dia, 600 mm internal depth	98.75	3.50	73.50	-	99.38	nr	**172.88**
150 mm outlet; 300 mm internal dia, 600 mm internal depth	101.12	3.50	73.50	-	101.75	nr	**175.25**
150 mm outlet; 400 mm internal dia, 750 mm internal depth	117.28	3.50	73.50	-	117.91	nr	**191.41**
150 mm outlet; 450 mm internal dia, 900 mm internal depth	158.68	3.50	73.50	-	159.32	nr	**232.81**
Hinged gratings and frames for gullies; alloy							
120 mm x 120 mm	-	-	-	-	8.32	nr	**-**
135 mm for 100 mm dia gully	-	-	-	-	13.55	nr	**-**
150 mm x 150 mm	-	-	-	-	15.07	nr	**-**
193 mm for 150 mm dia gully	-	-	-	-	23.43	nr	**-**
230 mm x 230 mm	-	-	-	-	27.56	nr	**-**
316 mm x 316 mm	-	-	-	-	73.07	nr	**-**
Hinged gratings and frames for gullies; cast iron							
265 mm for 225 mm dia gully	-	-	-	-	46.82	nr	**-**
150 mm x 150 mm	-	-	-	-	15.07	nr	**-**
230 mm x 230 mm	-	-	-	-	27.56	nr	**-**
316 mm x 316 mm	-	-	-	-	73.07	nr	**-**
Universal gully trap PVC-u; Wavin Plastics Ltd; OsmaDrain system; bedding in concrete; 11.50 N/mm^2 - 40 mm aggregate							
Universal gully fitting; comprising gully trap only							
110 mm outlet; 110 mm dia; 205 mm internal depth	10.70	3.50	73.50	-	11.33	nr	**84.83**
Vertical inlet hopper c\w plastic grate							
272 x 183 mm	14.38	0.25	5.25	-	14.38	nr	**19.63**
Sealed access hopper							
110 x 110 mm	31.24	0.25	5.25	-	31.24	nr	**36.49**
Universal gully PVC-u; Wavin Plastics Ltd; OsmaDrain system; accessories to Universal gully trap							
Hoppers; backfilling with clean granular material; tamping; surrounding in lean mix concrete							
plain hopper with 110 spigot 150 mm long	11.75	0.40	8.40	-	13.23	nr	**21.63**
vertical inlet hopper with 110 spigot 150 mm long	14.38	0.40	8.40	-	15.86	nr	**24.26**
sealed access hopper with 110 spigot 150 mm long	31.24	0.40	8.40	-	32.72	nr	**41.12**
plain hopper; solvent weld to trap	8.03	0.40	8.40	-	9.51	nr	**17.91**
vertical inlet hopper; solvent weld to trap	13.08	0.40	8.40	-	14.56	nr	**22.96**
sealed access cover; PVC-u	15.91	0.10	2.10	-	15.91	nr	**18.01**
Gullies PVC-u; Wavin Plastics Ltd; OsmaDrain system; bedding in concrete; 11.50 N/mm^2 - 40 mm aggregate							
Bottle gully; providing access to the drainage system for cleaning							
bottle gully; 228 x 228 x 317 mm deep	26.38	0.50	10.50	-	26.96	nr	**37.46**
sealed access cover; PVC-u 217 x 217 mm	20.51	0.10	2.10	-	20.51	nr	**22.61**
grating; ductile iron 215 x 215 mm	15.81	0.10	2.10	-	15.81	nr	**17.91**
bottle gully riser; 325 mm	3.49	0.50	10.50	-	4.74	nr	**15.24**

R DISPOSAL SYSTEMS

Item Excluding site overheads and profit	PC £	Labour hours	Labour £	Plant £	Material £	Unit	Total rate £
Yard gully; trapped 300 mm diameter 600 mm deep; including catchment bucket and ductile iron cover and frame, medium duty loading							
305 mm diameter 600 mm deep	163.27	2.50	52.50	-	166.84	nr	**219.34**
Linear Drainage							
Loadings for slot drains							
A15 - 1.5 tonne - pedestrian							
B125 - 12.5 tonne domestic use							
C250 - 25 tonne; car parks, supermarkets, industrial units							
D400 - 40 tonne; highways							
E600 - 60 tonne; forklifts							
Linear drainage; channels; Aco Building Products; laid to concrete bed C25 on compacted granular base on 200 mm deep concrete bed; haunched with 200 mm concrete surround; all in 750 x 430 wide trench with compacted 200 granular base surround (excavation and sub-base not included)							
Slot drains							
ACO Brickslot; load class C250; discreet slot drainage system ideal for visually sensitive paved areas	79.43	0.33	7.00	-	81.12	m	**88.12**
Extra over for laying of paving (PC £300.00/1000) adjacent to channel on epoxy mortar bed; stretcher bond	-	0.33	7.00	-	2.05	m	**9.05**
Channel footpath drain							
Aco Channel drain to pedestrian footpaths; connection to existing kerbdrain; laying on 150 mm concrete base and surround; excavation sub-base not included	27.09	2.00	42.00	-	37.56	m	**79.56**
NK100; to C250; 100 mm bore; polymer concrete drain units; tapered to falls; on 150 mm concrete with 150 concrete surround							
Constant or pre-sloped invert	52.55	0.80	16.80	-	63.02	m	**79.82**
Accessories for NK100							
Sump unit - complete with sediment bucket	94.56	1.00	21.00	-	94.56	nr	**115.56**
End cap - closing piece	-	-	-	-	8.67	nr	-
End cap - inlet/outlet	-	-	-	-	18.49	nr	-
Step connectors	-	-	-	-	14.17	nr	-
Aco Building Products; grating covers for NK100							
Load class A15 (pedestrian areas)							
Slotted galvanised steel grating; Ref: 423QL/424QL	-	-	-	-	37.31	m	-
Slotted stainless steel grating; Ref: 470QL	-	-	-	-	108.93	m	-
1433 load class C250 (cars and light vans)							
Slotted galvanised steel grating; Ref: 405/406	-	-	-	-	37.31	m	-
Galvanised steel mesh grating; Ref: 413/414	-	-	-	-	27.97	m	-
Heelguard ™' composite 522QL	-	-	-	-	36.48	m	-
'Heelguard™' ductile iron grating; Ref: 508QL	-	-	-	-	38.49	m	-
Access covers and frames							
Loading information; Note: groups 5 - 6 are not covered within the scope of this book, please refer to the Spon's Civil Engineering Price Book							
Group 1 - Class A15; pedestrian access only - 1.5 tonne maximum weight							
Group 2 - Class B125; car parks and pedestrian areas with occasional vehicle use - 12.5 tonne maximum weight limit							

R DISPOSAL SYSTEMS

Item Excluding site overheads and profit	PC £	Labour hours	Labour £	Plant £	Material £	Unit	Total rate £
R12 DRAINAGE BELOW GROUND - cont'd							
Access covers and frames; BSEN124; St Gobain Pipelines Plc; bedding frame in cement mortar (1:3); cover in grease and sand; light duty; clear opening sizes; base size shown in brackets							
Group 1; solid top single seal; ductile iron; plastic frame							
Pedestrian; 450 mm diameter (525) x 35 deep	17.01	3.00	63.00	-	18.90	nr	**81.90**
Driveway; 450 mm diameter (550) x 35 deep	26.52	3.00	63.00	-	28.41	nr	**91.41**
Group 1; solid top single seal; ductile iron; steel frame							
450 mm x 450 mm (565 x 565) x 35 deep	82.52	3.00	63.00	-	84.41	nr	**147.41**
450 mm x 600 mm (710 x 510) x 35 deep	122.93	3.00	63.00	-	124.82	nr	**187.82**
600 mm x 600 mm (710 x 710) x 35 deep	127.00	3.00	63.00	-	128.89	nr	**191.89**
Group 1; coated; double seal solid top; fabricated steel							
450 mm x 450 mm	44.10	3.00	63.00	-	45.99	nr	**108.99**
600 mm x 450 mm	57.00	3.00	63.00	-	58.89	nr	**121.89**
600 mm x 600 mm	56.70	3.00	63.00	-	58.59	nr	**121.59**
Group 2; single seal solid top							
450 diameter x 41 deep	46.87	3.00	63.00	-	48.76	nr	**111.76**
600 diameter x 75 deep	-	3.00	63.00	-	91.14	nr	**154.14**
600 x 450 (710x 560) x 41 deep	62.93	3.00	63.00	-	64.82	nr	**127.82**
600 x 600 (711 x 711) x 44 deep	67.43	3.00	63.00	-	69.32	nr	**132.32**
Recessed covers and frames; Bripave; galvanized; single seal; clear opening sizes shown in brackets							
450 x 450 x 93 (596 x 596) 1.5 tonne load	77.46	1.50	31.50	-	79.35	nr	**110.85**
450 x 450 x 93 (596 x 596) 5 tonne load	97.00	1.50	31.50	-	98.89	nr	**130.39**
450 x 450 x 93 (596 x 596) 11 tonne load	117.00	2.00	42.00	-	118.89	nr	**160.89**
600 x 450 x 93 (746 x 596) 1.5 tonne load	235.21	2.00	42.00	-	237.10	nr	**279.10**
600 x 450 x 93 (746 x 596) 5 tonne load	252.82	2.00	42.00	-	254.71	nr	**296.71**
600 x 450 x 93 (746 x 596) 11 tonne load	260.19	3.00	63.00	-	262.08	nr	**325.08**
600 x 600 x 93 (746 x 596) 1.5 tonne load	245.65	2.33	48.93	-	247.54	nr	**296.47**
600 x 600 x 93 (746 x 596) 5 tonne load	257.59	2.33	48.93	-	259.48	nr	**308.41**
600 x 600 x 93 (746 x 596) 11 tonne load	270.49	2.75	57.75	-	272.38	nr	**330.13**
750 x 600 x 93 (896 x 746) 1.5 tonne load	279.94	2.33	48.93	-	281.83	nr	**330.76**
750 x 600 x 93 (896 x 746) 5 tonne load	311.30	2.75	57.75	-	313.19	nr	**370.94**
750 x 600 x 93 (896 x 746) 11 tonne load	340.11	3.00	63.00	-	342.00	nr	**405.00**
Extra over to the above for double seal to covers							
450 x 450 x 93 (596 x 596)	-	-	-	-	56.00	nr	-
600 x 450 x 93 (746 x 596)	-	-	-	-	60.00	nr	-
600 x 600 x 93 (746 x 596)	-	-	-	-	64.00	nr	-
750 x 600 x 93 (896 x 746)	-	-	-	-	69.00	nr	-
Extra over manhole frames and covers for							
filling recessed manhole covers with brick paviors; PC £305.00/1000	12.20	1.00	21.00	-	13.95	m^2	**34.95**
filling recessed manhole covers with vehicular paving blocks; PC £7.70/m^2	7.70	0.75	15.75	-	8.22	m^2	**23.97**
filling recessed manhole covers with concrete paving flags; PC £10.20/m^2	10.21	0.35	7.35	-	10.52	m^2	**17.87**

R DISPOSAL SYSTEMS

Item Excluding site overheads and profit	PC £	Labour hours	Labour £	Plant £	Material £	Unit	**Total rate £**
R13 LAND DRAINAGE							
Agricultural drainage; excavating for drains; by 3 tonne tracked excavator; including disposing spoil to spoil heaps not exceeding 100 m							
Width 225 mm							
depth 450 mm	-	-	-	344.03	-	100 m	**344.03**
depth 600 mm	-	-	-	353.59	-	100 m	**353.59**
depth 700 mm	-	-	-	363.97	-	100 m	**363.97**
depth 900 mm	-	-	-	387.72	-	100 m	**387.72**
depth 1000 mm	-	-	-	401.37	-	100 m	**401.37**
depth 1200 mm	-	-	-	416.46	-	100 m	**416.46**
Agricultural drainage; excavating for drains; by hand, including disposing spoil to spoil heaps not exceeding 100 m							
Width 150 mm							
depth 450 mm	-	22.03	462.63	-	-	100 m	**462.63**
depth 600 mm	-	29.38	616.98	-	-	100 m	**616.98**
depth 700 mm	-	34.27	719.67	-	-	100 m	**719.67**
depth 900 mm	-	44.06	925.26	-	-	100 m	**925.26**
Width 225 mm							
depth 450 mm	-	33.05	694.05	-	-	100 m	**694.05**
depth 600 mm	-	44.06	925.26	-	-	100 m	**925.26**
depth 700 mm	-	51.41	1079.61	-	-	100 m	**1079.61**
depth 900 mm	-	66.10	1388.10	-	-	100 m	**1388.10**
depth 1000 mm	-	146.88	2900.88	-	-	100 m	**2900.88**
Wavin Plastics Ltd; flexible plastic perforated pipes in trenches (not included); to a minimum depth of 450 mm (couplings not included)							
"OsmaDrain"; flexible plastic perforated pipes in trenches (not included); to a minimum depth of 450 mm (couplings not included)							
80 mm diameter; available in 100 m coil	56.50	2.00	42.00	-	56.50	100 m	**98.50**
100 mm diameter; available in 100 m coil	97.00	2.00	42.00	-	97.00	100 m	**139.00**
160 mm diameter; available in 35 m coil	235.00	2.00	42.00	-	240.88	100 m	**282.88**
"WavinCoil"; plastic pipe junctions							
80 mm x 80 mm	2.50	0.05	1.05	-	2.50	nr	**3.55**
100 mm x 100 mm	2.79	0.05	1.05	-	2.79	nr	**3.84**
100 mm x 60 mm	2.75	0.05	1.05	-	2.75	nr	**3.80**
100 mm x 80 mm	2.66	0.05	1.05	-	2.66	nr	**3.71**
160 mm x 160 mm	6.83	0.05	1.05	-	6.83	nr	**7.88**
"WavinCoil"; couplings for flexible pipes							
80 mm diameter	0.95	0.03	0.62	-	0.95	nr	**1.57**
100 mm diameter	1.05	0.03	0.62	-	1.05	nr	**1.67**
160 mm diameter	1.42	0.03	0.62	-	1.42	nr	**2.04**
Market prices of backfilling materials							
Sand	15.25	-	-	-	18.30	m^3	-
Gravel rejects	15.00	-	-	-	16.50	m^3	-
Topsoil; allowing for 20% settlement	18.50	-	-	-	22.20	m^3	-
Agricultural drainage; backfilling trench after laying pipes with gravel rejects or similar; blind filling with ash or sand; topping with 150 mm topsoil from dumps not exceeding 100 m; by machine							
Width 150 mm							
depth 450 mm	-	3.20	67.20	44.96	129.24	100 m	**241.40**
depth 600 mm	-	4.20	88.20	59.95	168.70	100 m	**316.86**
depth 750 mm	-	4.86	102.06	74.94	195.30	100 m	**372.30**
depth 900 mm	-	6.20	130.20	89.93	230.15	100 m	**450.27**

R DISPOSAL SYSTEMS

Item Excluding site overheads and profit	PC £	Labour hours	Labour £	Plant £	Material £	Unit	Total rate £
R13 LAND DRAINAGE - cont'd							
Agricultural drainage - cont'd							
Width 225 mm							
depth 450 mm	-	4.80	100.80	68.94	183.70	100 m	**353.45**
depth 600 mm	-	6.30	132.30	89.93	253.22	100 m	**475.45**
depth 750 mm	-	7.80	163.80	112.37	183.70	100 m	**459.88**
depth 900 mm	-	9.30	195.30	137.89	183.70	100 m	**516.89**
Agricultural drainage; backfilling trench after laying pipes with gravel rejects or similar, blind filling with ash or sand, topping with 150 mm topsoil from dumps not exceeding 100 m; by hand							
Width 150 mm							
depth 450 mm	-	18.63	391.23	-	106.49	100 m	**497.72**
depth 600 mm	-	24.84	521.64	-	151.31	100 m	**672.95**
depth 750 mm	-	31.05	652.05	-	129.44	100 m	**781.49**
depth 900 mm	-	37.26	782.46	-	216.07	100 m	**998.53**

SPON'S PRICEBOOKS 2008

Spon's Architects' and Builders' Price Book 2008

Davis Langdon

The most detailed, professionally relevant source of construction price information currently available anywhere.

The 2008 edition provides an extensive overhaul of prices in line with buoyant market conditions. It includes an expanded Approximate Estimates section and a new section on Sustainability. It also includes the following new items: Additional high quality balustrades, PVC wall linings, Rainscreen cladding, Resin-bonded gravel, SAS suspended ceilings, Self-levelling screeds, 'Speedfit' pipes and fittings, Tapered insulation boards, Waterproofing membranes, Window refurbishment

HB & electronic package * £135
978 0-415-42448-6 1000pp approx.

Spon's External Works and Landscape Price Book 2008

Davis Langdon

The only comprehensive source of information for detailed external works and landscape costs.

The 2008 edition gives an expanded Approximate Estimates section, and new items including Professional and amenity sports pitch construction, Costs of tender production, Small unit concrete kerbs, Ornamental trellises, Borehole drilling, irrigation sensors, and an irrigation water recycling system. And more on excavation and filling, landscape chemicals prices and applications, and linear drainage.

HB & electronic package * £105
978 0-415-42444-8 424pp approx.

Spon's Mechanical and Electrical Services Price Book 2008

Davis Langdon, Mott Green Wall

The only comprehensive and best annual services engineering price book available for the UK.

As well as an extensive overhaul of prices, updates to the 2008 edition include an expanded Approximate Estimate section, updated coverage of sustainability and of communication and data, along with additions to both the ductwork and lifts sections. High efficiency plant has been refreshed and the CHP and Biomass plant has been reviewed.

HB & electronic package * £135
978 0-415-42442-4 640pp approx.

Spon's Civil Engineering and Highway Works Price Book 2008

Davis Langdon

The 21st edition, in its easy-to-read format, incorporates a general review throughout, updated in line with the latest CIJC wage agreement, and includes a restructuring of the highway works unit costs section to align with the latest revisions to the *Manual of Contract Documents for Highway Works.*

HB & electronic package * £145
978 0-415-42446-2 800pp approx.

Receive a free electronic version when you order any Spon 2008 Price Book with estimating software to help you produce tender documents, customise data, perform word searches and simple calculations.
Visit www.pricebooks.co.uk

To Order: Tel: +44 (0) 8700 768853, or +44 (0) 1264 343071 Fax: +44 (0) 1264 343005, or
Post: Taylor and Francis Customer Services, Thomson Publishing Services, Cheriton House, Andover, Hants, SP10 5BE, UK Email: book.orders@tandf.co.uk

For a complete listing of all our titles visit www.tandf.co.uk

Taylor & Francis
Taylor & Francis Group

ESSENTIAL READING FROM TAYLOR AND FRANCIS

Spon's Estimating Costs Guide to Small Groundworks, Landscaping and Gardening

Second Edition

Especially written for contractors and small businesses carrying out small works, Spon's Estimating Costs Guide to Small Groundworks, Landscaping and Gardening contains accurate information on thousands of rates, each broken down to labour, material overheads and profit.

This is the first book to focus on garden maintenance work on blocks of flats and individual houses, schools and sports fields, garden makeovers, laying patios and paths, small land drainage schemes on farmland, and small-scale local authority maintenance work.

More than just a price book, it gives easy-to-read, professional advice on setting up and running a business including help on producing estimates faster, and keeping estimates accurate and competitive.

Suitable for any size firm from one-man-band to established business, this book contains valuable commercial and cost information that contractors can't afford to be without.

December 2007: 216x138 mm: 304 pages
PB: 978-0-415-43442-3: £27.99

To Order: Tel: +44 (0) 1264 343071 Fax: +44 (0) 1264 343005, or
Post: Taylor and Francis Customer Services, Thomson Publishing Services, Cheriton House, Andover, Hants, SP10 5BE, UK Email: book.orders@tandf.co.uk

For a complete listing of all our titles visit:
www.sponpress.com

Taylor & Francis
Taylor & Francis Group

ESSENTIAL READING FROM TAYLOR AND FRANCIS

Project Management Demystified

Third Edition

Geoff Reiss

Concise, practical and entertaining to read, this excellent introduction to project management is an indispensable book for both professionals and students working in or studying project management in business, engineering or the public sector.

Approachable and written in an easy-to-use style, it shows readers how, where and when to use the various project management techniques, demonstrating how to achieve efficient management of human, material and financial resources to make major contributions to projects and be an appreciated and successful project manager.

This new edition contains expanded sections on programme management, portfolio management, and the public sector. An entirely new chapter covers the evaluation, analysis and management of risks and issues. A much expanded section explores the rise and utilisation of methodologies like Prince2.

Contents: Introduction. Setting the Stage. Getting the Words in the Right Order. Nine Steps to a Successful Project. The Scope of the Project and its Objectives. Project Planning. A Fly on the Wall. Resource Management. Progress Monitoring and Control. Advanced Critical-Path Topics. The People Issues. Risk and Issue Management. Terminology

June 2007: 234x156mm: 240 pages
PB: 978-0-415-42163-8: £19.99

To Order: Tel: +44 (0) 1264 343071 Fax: +44 (0) 1264 343005, or
Post: Taylor and Francis Customer Services, Thomson Publishing Services, Cheriton House, Andover, Hants, SP10 5BE, UK Email: book.orders@tandf.co.uk

For a complete listing of all our titles visit:
www.tandf.co.uk

Taylor & Francis
Taylor & Francis Group

ESSENTIAL READING FROM TAYLOR AND FRANCIS

2nd Edition

Spon's Irish Construction Price Book

Franklin + Andrews

This new edition of *Spon's Irish Construction Price Book*, edited by Franklin + Andrews, is the only complete and up-to-date source of cost data for this important market.

- **All the materials costs, labour rates, labour constants and cost per square metre are based on current conditions in Ireland**
- **Structured according to the new Agreed Rules of Measurement (second edition)**
- **30 pages of Approximate Estimating Rates for quick pricing**

This price book is an essential aid to profitable contracting for all those operating in Ireland's booming construction industry.

Franklin + Andrews, Construction Economists, have offices in 100 countries and in-depth experience and expertise in all sectors of the construction industry.

April 2004: 246x174 mm: 448 pages
HB: 0-415-34409-3: £135.00

To Order: Tel: +44 (0) 1264 343071 Fax: +44 (0) 1264 343005, or Post: Taylor and Francis Customer Services, Thomson Publishing Services, Cheriton House, Andover, Hants, SP10 5BE, UK Email: book.orders@tandf.co.uk

For a complete listing of all our titles visit: www.sponpress.com

Taylor & Francis
Taylor & Francis Group

PART 6

Fees for Professional Services

LANDSCAPE ARCHITECTS' FEES

The Landscape Institute no longer sets any fee scales for its members. A publication which offers guidance in determining fees for different types of project is available from:

The Landscape Institute
33 Great Portland Street
London
W1W 8QG

Telephone: 020 7299 4500

The publication is entitled "Engaging a Landscape Consultant - Guidance for Clients on Fees 2002"

The document refers to suggested fee systems on the following basis:

- Time Charged Fee Basis
- Lump Sum Fee Basis
- Percentage Fee Basis
- Retainer Fee Basis

In regard to the 'Percentage Fee Basis' the publication lists various project types and relates them to complexity ratings for works valued at £22,500 and above. The suggested fee scales are dependent on the complexity rating of the project.

The following are samples of the Complexity Categories:

Category 1: Golf Courses, Country Parks and Estates, Planting Schemes

Category 2: Coastal, River and Agricultural Works, Roads, Rural Recreation Schemes

Category 3: Hospital Grounds, Sport Stadia, Urban Offices and Commercial Properties, Housing

Category 4: Domestic/Historic Garden Design, Urban Rehabilitation and Environmental Improvements

Table Showing Suggested Percentage Fees for Various Categories of Complexity of Landscape Design or Consultancy

	Complexity			
Project Value £	**1**	**2**	**3**	**4**
22,500	15.00%	16.50%	18.00%	21.00%
30,000	13.75	15.00	16.50	19.25
50,000	10.50	12.75	14.00	16.25
100,000	9.50	10.50	11.50	13.50
150,000	8.75	9.50	10.50	12.50
200,000	8.00	8.75	9.75	11.25
300,000	7.50	8.25	9.00	10.50
500,000	6.75	7.50	8.25	9.75
750,000	6.50	7.25	7.75	9.25
1,000,000	6.25	7.00	7.50	8.75
10,000,000	6.25	6.75	7.25	8.25

(extrapolated from the graph/curve chart in the aforementioned publication)

Guide to Stage Payments of Fees, Relevant Fee Basis and Proportion of Fee Applicable to Lump Sum and Percentage Fee Basis. *Details of Preliminary, Standard and Other Services are set out in detail in the Landscape Consultant's Appointment.*

Work Stage	Relevant Fee Basis			Proportion of Fee	
	Time	Lump	%	Proportion of fee	Total
Preliminary Services					
A Inception	✓	✓	n/a	n/a	n/a
B Feasibility	✓	✓	n/a	n/a	n/a
Standard Services					
C Outline Proposals	✓	✓	✓	15%	15%
D Sketch Scheme Proposals	✓	✓	✓	15%	30%
E Detailed Proposals	✓	✓	✓	15%	45%
FG Production Information	✓	✓	✓	20%	65%
HJ Tender Action & Contract Preparation	✓	✓	✓	5%	70%
K Operations on Site	✓	✓	✓	25%	95%
L Completion	✓	✓	✓	5%	100%
Other Services	✓	✓	n/a		
Timing of Fee Payments Percentage fees are normally paid at the end of each work stage. Time based fees are normally paid at monthly intervals. Lump sum fees are normally paid at intervals by agreement. Retainer or term commission fees are normally paid in advance, for predetermined periods of service.					

WORKED EXAMPLES OF PERCENTAGE FEE CALCULATIONS

Worked Example 1

Project Type	Caravan Site
Services Required	To Detailed Proposals - Work Stages C to E
Budget	£120,000

Step 1 Decide on Work Type and therefore Complexity Rating - Complexity Rating 2
Step 2 Decide on Services required and Proportion of Fee - To Detailed Proposals, 45%
Step 3 Read off Graph, Complexity Rating 2, the % fee of £120,000 - Graph Fee 9.9%
Step 4 Multiply the Proportion of Fee (45%) by the Graph Fee (9.9%) - Adjusted Fee - 4.46%
Step 5 Calculate the Guide Fee (4.46% of £120,000) - Guide Fee - £5,352
Step 6 Agree fee with Client, complete Memorandum of Agreement & Schedule of Services & Fees

Worked Example 2

Project Type	New Housing
Services Required	Full Standard Services - Work Stages C to L
Budget	£350,000

Step 1 Decide on Work Type and therefore Complexity Rating - Complexity Rating 3
Step 2 Decide on Services required and Proportion of Fee - To Completion, 100%
Step 3 Read off Graph, Complexity Rating 3 the % fee of £350,000 - Graph Fee 8.8%
Step 4 Multiply the Proportion of Fee (100%) by the Graph Fee (8.8%) - Adjusted Fee - 8.8%
Step 5 Calculate the Guide Fee (8.8% of £350,000) - Guide Fee - £30,800
Step 6 Agree fee with Client, complete Memorandum of Agreement & Schedule of Services & Fees

Worked Example 3

Project Type	Urban Environmental Improvements
Services Required	To Production Information - Work Stages C to G
Budget	£1,250,000

Step 1 Decide on Work Type and therefore Complexity Rating - Complexity Rating 4
Step 2 Decide on Services required and Proportion of Fee - To Production Information, 65%
Step 3 Read off Graph, Complexity Rating 4 the % fee of £1,250,000 - Graph Fee 8.6%
Step 4 Multiply the Proportion of Fee (65%) by the Graph Fee (8.6%) - Adjusted Fee - 5.59%
Step 5 Calculate the Guide Fee (5.59% of £1,250,000) - Guide Fee - £69,875
Step 6 Agree fee with Client, complete Memorandum of Agreement & Schedule of Services & Fees

PART 7

Rates of Wages

BUILDING INDUSTRY WAGES

Authorized rates of wages, etc., in the building industry in England, Wales and Scotland agreed by the Building and Civil Engineering Joint Negotiating Committee. Effective from 25 June 2007. Basic pay. Weekly rate based on 39 hours.

Craft Rate	£379.08
Skill Rate 1	£361.53
Skill Rate 2	£347.88
Skill Rate 3	£325.65
Skill Rate 4	£306.93
General Operative	£285.09

The Prices for Measured Works used in this book are based upon the commercial wage rates currently used in the landscaping industry and are typical of the London area.
However, it is recognized that some contractors involved principally in soft landscaping and planting works may base their wages on the rates determined by the Agricultural Wages Board (England and Wales) and therefore the following information is given to assist readers to adjust the prices if necessary.

AGRICULTURAL WAGES, ENGLAND AND WALES

Minimum payments agreed by the Agricultural Wages Board for standard 39 hour, five day week. Effective 1 December 2006.

Minimum rates of pay are specified for 6 grades and 4 categories of workers:

Grade 1 - Initial Grade
Grade 2 - Standard Worker
Grade 3 - Lead Worker
Grade 4 - Craft Grade
Grade 5 - Supervisory Grade
Grade 6 - Farm Management Grade
Full Time Flexible Worker
Part Time Flexible Worker
Apprentice
Trainee

The following rates apply to workers who meet the conditions for the applicable grade who are not Full or Part Time Flexible Workers.

Grade	Weekly Pay (£)	Hourly Pay (£)	Overtime (£ per hour)
1 (of compulsory school age)	-	2.68	4.02
1 (above compulsory school age)	208.65	5.35	8.03
2	223.86	5.74	8.61

AGRICULTURAL WAGES, ENGLAND AND WALES

Grade	Weekly Pay (£)	Hourly Pay (£)	Overtime (£ per hour)
3	246.09	6.31	9.47
4	264.03	6.77	10.16
5	280.02	7.18	10.77
6	302.25	7.75	11.63

Weekly rates apply to workers contractually required to work 39 basic hours per week; hourly rates apply to workers contractually required to work less than a 39 basic hours per week.

Overtime rates apply to all overtime hours worked by workers other than Flexible Workers -

- more than 39 basic hours in any week
- or more than 8 hours on any day
- or any hours beyond the normal working hours specified in a contract
- or on a Public Holiday
- or on a Sunday (applies only to workers engaged under a contract which started before 1 October 2006)

The rate of holiday pay should be calculated using the following principle:

$$\text{Holiday pay per day} = \frac{\text{Normal or average weekly pay}}{\text{Number of days worked weekly}}$$

Copies of the Agricultural Wages Order 2006 (Number 1) may be obtained from:

The Secretary to the Agricultural Wages Board for England and Wales
Area 3A Ergon House
Horseferry Road
London SW1P 2AL
Telephone: 020 7238 6523
Fax: 020 7238 6553
Email: m.agriwages@arp.defra.gsi.gov.uk, or online from www.defra.gov.uk

AGRICULTURAL WAGES, SCOTLAND

New rules relating to agricultural wages and other conditions of service took effect from 1 January 2006. The new rules are contained in the Agricultural Wages (Scotland) Order (No. 53) 2005. The Order contains the detailed legal requirements for the calculations of minimum pay etc, but the Scottish Office has produced 'A Guide for Workers and Employers' which attempts to explain the new rules in simpler terms.

Both 'The Agricultural Wages (Scotland) Order (No. 53) 2005' and 'A Guide for Workers and Employers' may be obtained from:

Scottish Agricultural Wages Board
Pentland House
47 Robb's Loan
Edinburgh EH14 1TY
Telephone: 0131 244 6397
Fax: 0131 244 6551
Email: sawb@scotland.gsi.gov.uk

The new Order applies equally to all workers employed in agriculture in Scotland. It makes no distinction between full-time employees, part-time employees and students.

The old classifications of general farm worker, shepherd, stockworker, tractorman and supervisory grade have been removed and replaced by a single category covering all agricultural workers. The one exception to this is that special arrangements have been made for hill-shepherds.

Instead of minimum weekly rates of pay, the new Order works on minimum hourly rates of pay (with the exception of hill-shepherds). It is very important to note that although wages are to be calculated on an hourly basis, a worker is still entitled to be paid for the full number of hours for which he is contracted (unless he is unavailable for work). In particular, this means that an employer cannot reduce the number of hours for which he is to be paid simply by e.g. sending him home early.

Minimum rates of wages
There are two main pay scales: one for workers who have been with their present employer for not more than 10 weeks and the other for workers who have been with their present employer for more than 10 weeks.

Within these two scales, there are varying minimum hourly rates of pay depending on the age of the worker.

The minimum hourly rates of pay are as follows:

	Minimum Hourly Rates (£)	
Age	Up to 10 weeks continuous employment	More than 10 weeks continuous employment
Under 16	2.03	2.71
16 and under 17	3.12	3.28
17 and under 18	3.12	3.81
18 and under 19	4.26	4.64
19 and over	5.05	5.46

Workers who have been with the same employer for more than 10 weeks and hold either:

(a) a Scottish, or National, Vocational Qualification in an agricultural subject at Level III or above, or
(b) an apprenticeship certificate approved by Lantra, NTO (formerly ATB Landbase), or a certificate of acquired experience issued by the ATB Landbase

shall be paid, for each hour worked, an additional sum of not less than £0.82.

Minimum hourly overtime rate
The minimum hourly rate of wages payable to a worker:

(a) for each hour worked in excess of 8 hours on any day, and
(b) for each hour worked in excess of 39 hours in any week

shall be calculated in accordance with the following formula:

M x 1.5 (where M = the minimum hourly rate of pay to which the worker is entitled)

ESSENTIAL READING FROM TAYLOR AND FRANCIS

Spon's Estimating Cost Guide to Minor Works, Alterations and Repairs to Fire, Flood, Gale and Theft Damage

Third Edition

Bryan Spain

All the cost data you need to keep your estimating accurate, competitive and profitable.

Do you work on jobs between £50 and £50,000? Then this book is for you.

Specially written for contractors, quantity surveyors and clients carrying out small works, this book contains accurate information on thousands of rates each broken down to labour, material overheads and profit.More than just a price book, it gives easy-to-read professional advice on setting up and running a business.

December 2005: 216x138 mm: 320 pages
PB: 0-415-38213-0: £29.99

To Order: Tel: +44 (0) 1264 343071 Fax: +44 (0) 1264 343005, or
Post: Taylor and Francis Customer Services, Thomson Publishing Services, Cheriton House, Andover, Hants, SP10 5BE, UK Email: book.orders@tandf.co.uk

For a complete listing of all our titles visit:
www.tandf.co.uk

Taylor & Francis
Taylor & Francis Group

PART 8

Daywork and Prime Cost

When work is carried out which cannot be valued in any other way it is customary to assess the value on a cost basis with an allowance to cover overheads and profit. The basis of costing is a matter for agreement between the parties concerned, but definitions of prime cost for the building industry have been prepared and published jointly by the Royal Institution of Chartered Surveyors and the National Federation of Building Trades Employers (now the Construction Confederation) for the convenience of those who wish to use them. These documents are reproduced with the permission of the Royal Institution of Chartered Surveyors, which owns the copyright.

The daywork schedule published by the Civil Engineering Contractors Association is included in the External Works and Landscape Book's companion title, *"Spon's Civil Engineering and Highway Works Price Book"*.

For larger Prime Cost contracts the reader is referred to the form of contract issued by the Royal Institute of British Architects.

BUILDING INDUSTRY

DEFINITION OF PRIME COST OF DAYWORK CARRIED OUT UNDER A BUILDING CONTRACT (JUNE 2007 - THIRD EDITION)

This definition of Prime Cost is published by the Royal Institution of Chartered Surveyors and the Construction Confederation, for convenience and for use by people who choose to use it. Members of the Construction Confederation are not in any way debarred from defining Prime Cost and rendering their accounts for work carried out on that basis in any way they choose. Building owners are advised to reach agreement with contractors on the Definition of Prime Cost to be used prior to issuing instructions.

INTRODUCTION

This new edition of the Definition includes two options for dealing with the prime cost of labour:

Option 'A' – Percentage Addition, is based upon the traditional method of pricing labour in daywork, and allows for a percentage addition to be made for incidental costs, overheads and profit, to the prime cost of labour applicable at the time the daywork is carried out.

Option 'B' – All inclusive Rates, includes not only the prime cost of labour but also includes an allowance for incidental costs, overheads and profit. The all-inclusive rates are deemed to be fixed for the period of the contract. However, where a fluctuating price contract is used, or where the rates in the contract are to be index-linked, the all-inclusive rates shall be adjusted by a suitable index in accordance with the contract conditions.

Model documentation, intended for inclusion in a building contract, is included in Appendix A, which illustrates how the Definition of Prime Cost may be applied in practice.
Example calculations of the Prime Cost of Labour in Daywork are given in Appendix B, *and shown on page 383*

The publication of this new edition for the Prime Cost of Dayworks carried out under a building contract has coincided with the final proofing of this edition of Spon's External Works and Landscape Price Book . Due to the last minute publication of this information, the authors have only been able to include the Introduction and the examples from APPENDIX B in this year's publication. Readers are advised to purchase the original document from the RICS for a further explanation and model documentation of this latest definition.

ESSENTIAL READING FROM TAYLOR AND FRANCIS

Spon's Estimating Costs Guide to Electrical Works

Third Edition

Bryan Spain

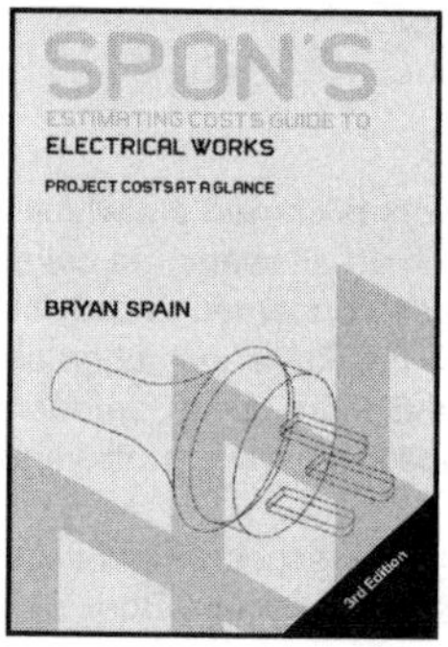

All the cost data you need to keep your estimating accurate, competitive and profitable.

Do you work on jobs between £50 and £25,000? Then this book is for you.

Specially written for contractors and small businesses carrying out small works, Spon's Estimating Costs Guide to Electrical Works provides accurate information on thousands of rates, each broken down to labour, material overheads and profit for residential, retail and light industrial premises. It is the first book to include typical project costs for new installations, stripping out, rewiring and upgrading for flats and houses.

In addition, vital information and advice is given on setting up and running a business, employing staff, tax, VAT and CIS4's.

For the cost of approximately two hours of your charge-out rate (or less), this book will help you to:

produce estimates faster
keep your estimates accurate and competitive
run your business more effectively
save time.
No matter how big your firm is - from one-man-band to an established business - this book contains valuable commercial and cost information that you can't afford to be without.

December 2005: 216x138 mm: 304 pages
PB: 0-415-38614-4: £29.99

To Order: Tel: +44 (0) 1264 343071 Fax: +44 (0) 1264 343005, or
Post: Taylor and Francis Customer Services, Thomson Publishing Services, Cheriton House, Andover, Hants, SP10 5BE, UK Email: book.orders@tandf.co.uk

For a complete listing of all our titles visit:
www.tandf.co.uk

Taylor & Francis
Taylor & Francis Group

Daywork Rates – Building Operatives

APPENDIX B *(Extracted from the full text issued by the RICS)*
Example 1 – Option A – Example Calculations of Prime Cost of Labour in Daywork

Example of calculation of typical standard hourly base rate (as defined in Section 3) for CIJC Building Craft operative and General Operative based upon rates applicable 6th April 2007.

		Rate (£)	Craft Operative	Rate (£)	General Operative
Basic Wages:	46.2 weeks	363.48	£16,792.78	273.39	£12,630.62
Extra Payments:	Where applicable	-	-		-
Sub Total:			£16,792.78		£12,630.62
National Insurance:	12.80% above ET (46.2 wks @£100.00pw)		£1,558.12		£1,025.36
Holidays with Pay:	226 hours	9.32	£2,106.32	7.01	£1,584.26
Welfare Benefit:	52 stamps	10.90	£566.80	10.90	£566.80
CITB Levy:	0.5% of	18,899.10	£94.50	14,214.88	£71.07
Annual labour cost:			**£21,118.52**		**£15,878.11**
Hourly Base Rate:	Divide by 1802 hours		**£11.72**		**£8.81**

Note:

(1) Standard working hours per annum calculated as follows:

52 weeks @ 39 hours	2028
Less \	
hours annual holiday	163
hours public holiday	63
Standard working hours per year	1802

(2) It has been assumed that employers who follow the CIJC Working Rules Agreement will match the employee pension contributions (part of welfare benefit) between £3.00 and £10.00 per week. Furthermore it has been assumed that employees have contributed £10.00 per week to the pension scheme.

(3) It should be noted that all labour costs incurred by the Contractor in his capacity as an employer other than those contained in the hourly base rate, are to be taken into account under section 6.

(4) The above example is for the convenience of users only and does not form part of the Definition; all the basic costs are subject to re-examination according to the time when and in the area where the daywork is executed.

Example 2 – Option A – Example Calculations of Non productive overtime in Daywork, is similar to Example 1 above, except that the Holidays with Pay and Welfare Benefit lines do not apply, with the calculation proceeding as follows:

		Rate (£)	Craft Operative	Rate (£)	General Operative
Calculations, as above	\Less HWP & WB		£18,350.90		£13,655.98
CITB Levy:	0.5% of	16,792.78	£83.96	12,630.62	£63.15
Annual labour cost:			**£18,434.86**		**£13,719.13**
Hourly Base Rate:	Divide by 1802 hours		**£10.23**		**£7.61**

For example 2, only Notes 1,3 and 4 (re-numbered) apply from above.

ESSENTIAL READING FROM TAYLOR AND FRANCIS

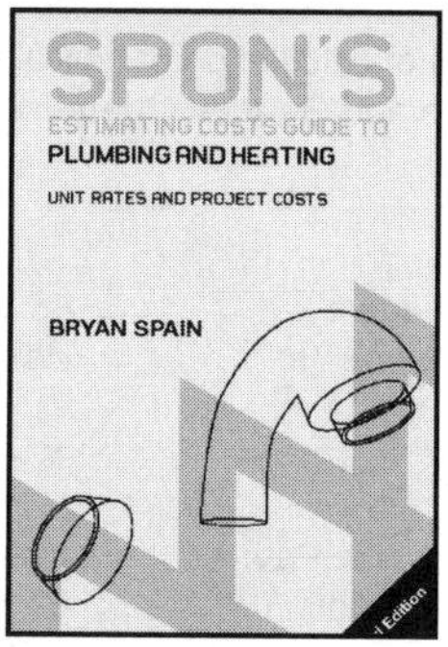

Spon's Estimating Cost Guide to Plumbing and Heating

Unit Rates and Project Costs

Third Edition

Bryan Spain

All the cost data you need to keep your estimating accurate, competitive and profitable.
Do you work on jobs between £50 and £25,000? Then this book is for you.
Specially written for contractors and small businesses carrying out small works, Spon's Estimating Costs Guide to Plumbing and Heating contains accurate information on thousands of rates, each broken down into labour, material overheads and profit.
The first book to include typical project costs for:
* rainwater goods installations
* bathrooms
* external waste systems
* central heating systems
* hot and cold water systems
More than just a price book, it gives easy-to-read professional advice on setting up and running a business including:
* taxation
* book-keeping, VAT and CIS4's
* legal obligations
For the cost of approximately two hours of your charge-out rate (or less), this book will help you to:
* produce estimates faster
* keep your estimates accurate and competitive
* run your business more effectively
* help save you time
No matter how big your firm is - from one-man-band to an established business - this book contains valuable commercial and cost information that you can't afford to be without.

December 2005: 216x138 mm: 264 pages
PB: 0-415-38618-7: £29.99

To Order: Tel: +44 (0) 1264 343071 Fax: +44 (0) 1264 343005, or
Post: Taylor and Francis Customer Services, Thomson Publishing Services, Cheriton House, Andover, Hants, SP10 5BE, UK Email: book.orders@tandf.co.uk

For a complete listing of all our titles visit:
www.tandf.co.uk

Taylor & Francis
Taylor & Francis Group

ESSENTIAL READING FROM TAYLOR AND FRANCIS

The Dynamic Landscape

Naturalistic Planting in an Urban Context

Edited by
Nigel Dunnett and James Hitchmough

The last quarter of the twentieth-century witnessed a burgeoning of interest in ecological or naturally inspired use of vegetation in the designed landscape. More recently a strong aesthetic element has been added to what was formerly a movement aimed at creating nature-like landscapes.

This book advances a fusion of scientific and ecological planting design philosophy that can address the need for more sustainable designed landscapes. It is a major statement on the design, implementation and management of ecologically-inspired landscape vegetation. With contributions from people at the forefront of developments in this field, in both Europe and North America, it provides a valuable synthesis of current thinking.

Contents: 1. Introduction 2. The Historical Development of Naturalistic Planting 3. A Contemporary Overview of Naturalistic Planting 4. The Dynamic Nature of Plant Communities 5. A Naturalistic Design Process 6. Herbaceous Plantings 7. Exploring Woodland Design 8. Wetlands and Water Bodies 9. Communicating Naturalistic Plantings: Plans and Specifications 10. The Creative Management of Naturalistic Plantings 11. The Social and Cultural Context of Naturalistic Plantings. Index.

June 2004: 238x225mm: 336 pages
20 tables, 40 line drawings, 67 b+w photos and 85 colour photos
HB: 0-415-25620-8: £63.00

To Order: Tel: +44 (0) 1264 343071 Fax: +44 (0) 1264 343005, or
Post: Taylor and Francis Customer Services, Thomson Publishing Services, Cheriton House, Andover, Hants, SP10 5BE, UK Email: book.orders@tandf.co.uk

For a complete listing of all our titles visit:
www.sponpress.com

Taylor & Francis
Taylor & Francis Group

ESSENTIAL READING FROM TAYLOR AND FRANCIS

Visualization in Landscape and Environmental Planning
Technology and Applications

Ian Bishop and Eckart Lange

This major reference presents the challenges, issues and directions of computer-based visualization of the natural and built environment and the role of such visualization in landscape and environmental planning. It offers a uniquely systematic approach to the potential of visualization and the writers are acknowledged experts in their field of specialization. Case studies are presented to illustrate many aspects of landscape management including forestry, agriculture, ecology, mining and urban development.

Foreword: Stephen Ervin. Part One. 1. Communication, perception and visualization. 2. Visualization Classified. 3. Data sources for three-dimensional models. 4. Visualization Technology. 5. Validity, reliability, and ethics in visualization. Part Two - Applications. 6 - Application in the Forest Landscape. 7. Applications in the agricultural landscape. 8. Applications in Energy, Industry and Infrastructure. 9 Applications in the urban landscape Part Three - Prospects. 10. Visualization Prospects. Index.

May 2005: 246x189 mm: 340 pages
HB: 0-415-30510-1: £58.00

To Order: Tel: +44 (0) 1264 343071 Fax: +44 (0) 1264 343005, or
Post: Taylor and Francis Customer Services, Thomson Publishing Services, Cheriton House, Andover, Hants, SP10 5BE, UK Email: book.orders@tandf.co.uk

For a complete listing of all our titles visit:
www.sponpress.com

Taylor & Francis
Taylor & Francis Group

PART 9

Tables and Memoranda

This part of the book contains the following sections:

ESSENTIAL READING FROM TAYLOR AND FRANCIS

Spon's First Stage Estimating Handbook
2nd Edition

Bryan Spain

The ultimate paradox of construction project management is that the earlier a decision has to be made, the more difficult the task is of estimating the cost implications. Yet the most important decisions are taken earliest - when the cost estimates are the most difficult to obtain.

Spon's First Stage Estimating Handbook is the only comprehensive and reliable source of first stage estimating costs. Covering the whole spectrum of building, civil engineering, mechanical and electrical work, and landscaping work, vital cost data is presented as costs per square metre, elemental cost analyses, principal rates and composite rates. Compact and clear, Spon's First Stage Estimating Handbook is ideal for meetings with clients when the inevitable question 'just an idea ... a ballpark figure ...' needs an accurate response.

With additional sections on whole life costing and general information, this is an essential reference for all construction professionals and clients making early judgements on the viability of new projects.

April 2006: 216x138 mm: 240 pages
PB: 0-415-38619-5: £42.00

To Order: Tel: +44 (0) 1264 343071 Fax: +44 (0) 1264 343005, or
Post: Taylor and Francis Customer Services, Thomson Publishing Services, Cheriton House, Andover, Hants, SP10 5BE, UK Email: book.orders@tandf.co.uk

For a complete listing of all our titles visit:
www.tandf.co.uk

Taylor & Francis
Taylor & Francis Group

CONVERSION TABLES

Conversion Factors

	Imperial		Metric	
1. LINEAR				
0.039	in	1	mm	25.4
3.281	ft	1	metre	0.305
1.094	yd	1	metre	0.914
2. WEIGHT				
0.020	cwt	1	kg	50.802
0.984	ton	1	tonne	1.016
2.205	lb	1	kg	0.454
3. CAPACITY				
1.760	pint	1	litre	0.568
0.220	gal	1	litre	4.546
4. AREA				
0.002	in^2	1	mm^2	645.16
10.764	ft^2	1	m^2	0.093
1.196	yd^2	1	m^2	0.836
2.471	acre	1	ha	0.405
0.386	$mile^2$	1	km^2	2.59
5. VOLUME				
0.061	in^3	1	cm^3	16.387
35.315	ft^3	1	m^3	0.028
1.308	yd^3	1	m^3	0.765
6. POWER				
1.310	HP	1	kW	0.746

Conversion Factors – Metric to Imperial

Multiply Metric	Unit	By	To Obtain Imperial	Unit
Length				
kilometre	km	0.6214	statute mile	ml
metre	m	1.0936	yard	yd
centimetre	cm	0.0328	foot	ft
millimetre	mm	0.0394	inch	in
Area				
hectare	ha	2.471	acre	
square kilometre	km^2	0.3861	square mile	sq ml
square metre	m^2	10.764	square foot	sq ft
square metre	m^2	1550	square inch	sq in
square centimetre	cm^2	0.155	square inch	sq in
Volume				
cubic metre	m^3	35.336	cubic foot	cu ft
cubic metre	m^3	1.308	cubic yard	cu yd
cubic centimetre	cm^3	0.061	cubic inch	cu in
cubic centimetre	cm^3	0.0338	fluid ounce	fl oz
Liquid volume				
litre	l	0.0013	cubic yard	cu yd
litre	l	61.02	cubic inch	cu in
litre	l	0.22	Imperial gallon	gal
litre	l	0.2642	US gallon	US gal
litre	l	1.7596	pint	pt
Mass				
metric tonne	t	0.984	long ton	lg ton
metric tonne	t	1.102	short ton	sh ton
kilogram	kg	2.205	pound, avoirdupois	lb
gram	g or gr	0.0353	ounce, avoirdupois	oz
Unit mass				
kilograms/cubic metre	kg/m^3	0.062	pounds/cubic foot	lbs/cu ft
kilograms/cubic metre	kg/m^3	1.686	pounds/cubic yard	lbs/cu yd
tonnes/cubic metre	t/m^3	1692	pound/cubic yard	lbs/cu yd
kilograms/sq centimetre	kg/cm^2	14.225	pounds/square inch	lbs/sq in
kilogram-metre	kg.m	7.233	foot-pound	ft-lb

Conversion Factors – Metric to Imperial

Multiply Metric	Unit	By	To Obtain Imperial	Unit
Force				
meganewton	MN	9.3197	tons force	tonf
kilonewton	kN	225	pounds force	lbf
newton	N	0.225	pounds force	lbf
Pressure and stress				
meganewton per square metre	MN/m^2	9.3197	tons force/square foot	$tonf/ft^2$
kilopascal	kPa	0.145	pounds/square inch	psi
bar		14.5	pounds/square inch	psi
kilogram metre	kgm	7.2307	foot-pound	ft-lb
Energy				
kilocalorie	kcal	3.968	British thermal unit	Btu
metric horsepower	CV	0.9863	horse power	hp
kilowatt	kW	1.341	horse power	hp
Speed				
kilometres/hour	km/h	0.621	miles/hour	mph

Conversion Factors – Imperial to Metric

Multiply Imperial	Unit	By	To Obtain Metric	Unit
Length				
statute mile	ml	1.609	kilometre	km
yard	yd	0.9144	metre	m
foot	ft	30.48	centimetre	cm
inch	in	25.4	millimetre	mm
Area				
acre	acre	0.4047	hectare	ha
square mile	sq ml	2.59	square kilometre	km^2
square foot	sq ft	0.0929	square metre	m^2
square inch	sq in	0.0006	square metre	m^2
square inch	sq in	6.4516	square centimetre	cm^2
Volume				
cubic foot	cu ft	0.0283	cubic metre	m^3
cubic yard	cu yd	0.7645	cubic metre	m^3
cubic inch	cu in	16.387	cubic centimetre	cm^3
fluid ounce	fl oz	29.57	cubic centimetre	cm^3
Liquid volume				
cubic yard	cu yd	764.55	litre	l
cubic inch	cu in	0.0164	litre	l
Imperial gallon	gal	4.5464	litre	l
US gallon	US gal	3.785	litre	l
US gallon	US gal	0.833	Imperial gallon	gal
pint	pt	0.5683	litre	l
Mass				
long ton	lg ton	1.016	metric tonne	tonne
short ton	sh ton	0.907	metric tonne	tonne
pound	lb	0.4536	kilogram	kg
ounce	oz	28.35	gram	g
Unit mass				
pounds/cubic foot	lb/ cu ft	16.018	kilogram's/cubic metre	kg/m^3
pounds/cubic yard	lb/cu yd	0.5933	cubic/cubic metre	kg/m^3
pounds/cubic yard	lb/cu yd	0.0006	tonnes/cubic metre	t/m^3
foot-pound	ft-lb	0.1383	kilogram-metre	kg.m

Conversion Factors – Imperial to Metric

Multiply Imperial	Unit	By	To Obtain Metric	Unit
Force				
tons force	tonf	0.1073	meganewton	MN
pounds force	lbf	0.0045	kilonewton	kN
pounds force	lbf	4.45	newton	N
Pressure and stress				
pounds/square inch	psi	0.1073	kilogram/sq. centimetre	kg/cm^2
pounds/square inch	psi	6.89	kilopascal	kPa
pounds/square inch	psi	0.0689	bar	
foot-pound	ft-lb	0.1383	kilogram metre	kgm
Energy				
British Thermal Unit	Btu	0.252	kilocalorie	kcal
horsepower (hp)	hp	1.014	metric horsepower	CV
horsepower (hp)	hp	0.7457	kilowatt	kW
Speed				
miles/hour	mph	1.61	kilometres/hour	km/h

Conversion Table

Length

Millimetre	mm	1 in	=	25.4 mm	1 mm	=	0.0394 in
Centimetre	cm	1 in	=	2.54 cm	1 cm	=	0.3937 in
Metre	m	1 ft	=	0.3048 m	1 m	=	3.2808 ft
		1 yd	=	0.9144 m	1 m	=	1.0936 yd
Kilometre	km	1 mile	=	1.6093 km	1 km	=	0.6214 mile

Note:	1 cm =	10 mm	1 ft	=	12 in
	1 m =	100 cm	1 yd	=	3 ft
	1 km =	1,000 m	1 mile	=	1,760 yd

Area

Square millimetre	mm^2	1 in^2	=	645.2 mm^2	1 mm^2	=	0.0016 in^2
Square centimetre	cm^2	1 in^2	=	6.4516 cm^2	1 cm^2	=	1.1550 in^2
Square metre	m^2	1 ft^2	=	0.0929 m^2	1 m^2	=	10.764 ft^2
		1 yd^2	=	0.8361 m^2	1 m^2	=	1.1960 yd^2
Square Kilometre	km^2	1 mile2	=	2.590 km^2	1 km^2	=	0.3861 mile2

Note:	1 cm^2	=	100 m^2	1 ft^2	=	144 in^2
	1 m^2	=	10,000 cm^2	1 yd^2	=	9 ft^2
	1 km^2	=	100 hectares	1 mile^2	=	640 acres
				1 acre	=	4, 840 yd^2

Volume

Cubic Centimetre	cm^3	1 cm^3	=	0.0610 in^3	1 in^3	=	16.387 cm^3
Cubic Decimetre	dm^3	1 dm^3	=	0.0353 ft^3	1 ft^3	=	28.329 dm^3
Cubic metre	m^3	1 m^3	=	35.3147 ft^3	1 ft^3	=	0.0283 m^3
		1 m^3	=	1.3080 yd^3	1 yd^3	=	0.7646 m^3
Litre	L	1 L	=	1.76 pint	1 pint	=	0.5683 L
			=	2.113 US pt		=	0.4733 US L

Note:	1 dm^3	=	1,000 cm^3	1 ft^3	=	1.728 in^3
	1 m^3	=	1,000 dm^3	1 yd^3	=	27 ft^3
	1 L	=	1 dm^3	1 pint	=	20 fl oz
	1 HL	=	100 L	1 gal	=	8 pints

Conversion Table

Mass

Milligram	mg	1 mg	=	0.0154 grain	1 grain	=	64.935 mg
Gram	g	1 g	=	0.0353 oz	1 oz	=	28.35 g
Kilogram	kg	1 kg	=	2.2046 lb	1 lb	=	0.4536 kg
Tonne	t	1 t	=	0.9842 ton	1 ton	=	1.016 t

Note:	1 g	=	1,000 mg	1 oz	=	437.5 grains
	1 kg	=	1000 g	1 lb	=	16 oz
	1 t	=	1,000 kg	1 stone	=	14 lb
				1 cwt	=	112 lb
				1 ton	=	20 cwt

Force

Newton	N	1lbf	=	4.448 N	1 kgf	=	9.807 N
Kilonewton	kN	1lbf	=	0.00448 kN	1 ton f	=	9.964 kN
Meganewton	MN	100 tonf	=	0.9964 MN			

Pressure and stress

Kilonewton per		1 lbf/in²	=	6.895 kN/m²
square metre	kN/m²	1 bar	=	100 kN/m²
Meganewton per		1 tonf/ft²	=	107.3 kN/m² = 0.1073 MN/m²
square metre	MN/m²	1 kgf/cm²	=	98.07 kN/m²
		1 lbf/ft²	=	0.04788 kN/m²

Temperature

Degree Celsius ^{o}C

$$^{o}C = \frac{5 \times (^{o}F - 32)}{9} \qquad ^{o}F = \frac{(9 \times ^{o}C) + 32}{5}$$

Metric Equivalents

1 km	=	1000 m
1 m	=	100 cm
1 cm	=	10 mm
1 km^2	=	100 ha
1 ha	=	10,000 m^2
1 m^2	=	10,000 cm^2
1 cm^2	=	100 mm^2
1 m^3	=	1,000 litres
1 litre	=	1,000 cm^3
1 metric tonne	=	1,000 kg
1 quintal	=	100 kg
1 N	=	0.10197 kg
1 kg	=	1000 g
1 g	=	1000 mg
1 bar	=	14.504 psi
1 cal	=	427 kg.m
1 cal	=	0.0016 cv.h
torque unit	=	0.00116 kw.h
1 CV	=	75 kg.m/s
1 kg/cm^2	=	0.97 atmosph

Imperial Unit Equivalents

1 mile	=	1760 yd
1 yd	=	3 ft
1 ft	=	12 in
1 sq mile	=	640 acres
1 acre	=	43,560 sq ft
1 sq ft	=	144 sq in
1 cu ft	=	7.48 gal liq
1 gal	=	231 cu in
	=	4 quarts liq
1 quart	=	32 fl oz
1 fl oz	=	1.80 cu in
	=	437.5 grains
1 stone	=	14 lb
1 cwt	=	112 lb
1 sh ton	=	2000 lb
1 lg ton	=	2240 lb
	=	20 cwt
1 lb	=	16 oz, avdp
1 Btu	=	778 ft lb
	=	0.000393 hph
	=	0.000293 kwh
1 hp	=	550 ft-lb/sec
1 atmosph	=	14.7 lb/in^2

Power Units

kW	=	Kilowatt
HP	=	Horsepower
CV	=	Cheval Vapeur (Steam Horsepower)
	=	French designation for Metric Horsepower
PS	=	Pferderstarke (Horsepower)
	=	German designation of Metric Horsepower
1 HP	=	1.014 CV = 1.014 PS
	=	0.7457 kW
1 PS	=	1 CV = 0.9863 HP
	=	0.7355 kW
1 kW	=	1.341 HP
	=	1.359 CV
	=	1.359 PS

Speed Conversion

km/h	m/min	mph	fpm
1	16.7	0.6	54.7
2	33.3	1.2	109.4
3	50.0	1.9	164.0
4	66.7	2.5	218.7
5	83.3	3.1	273.4
6	100.0	3.7	328.1
7	116.7	4.3	382.8
8	133.3	5.0	437.4
9	150.0	5.6	492.1
10	166.7	6.2	546.8
11	183.3	6.8	601.5
12	200.0	7.5	656.2
13	216.7	8.1	710.8
14	233.3	8.7	765.5
15	250.0	9.3	820.2
16	266.7	9.9	874.9
17	283.3	10.6	929.6
18	300.0	11.2	984.3
19	316.7	11.8	1038.9
20	333.3	12.4	1093.6
21	350.0	13.0	1148.3
22	366.7	13.7	1203.0
23	383.3	14.3	1257.7
24	400.0	14.9	1312.3
25	416.7	15.5	1367.0
26	433.3	16.2	1421.7
27	450.0	16.8	1476.4
28	466.7	17.4	1531.1
29	483.3	18.0	1585.7
30	500.0	18.6	1640.4
31	516.7	19.3	1695.1
32	533.3	19.9	1749.8
33	550.0	20.5	1804.5
34	566.7	21.1	1859.1
35	583.3	21.7	1913.8

Speed Conversion – cont'd

km/h	m/min	mph	fpm
36	600.0	22.4	1968.5
37	616.7	23.0	2023.2
38	633.3	23.6	2077.9
39	650.0	24.2	2132.5
40	666.7	24.9	2187.2
41	683.3	25.5	2241.9
42	700.0	26.1	2296.6
43	716.7	26.7	2351.3
44	733.3	27.3	2405.9
45	750.0	28.0	2460.6
46	766.7	28.6	2515.3
47	783.3	29.2	2570.0
48	800.0	29.8	2624.7
49	816.7	30.4	2679.4
50	833.3	31.1	2734.0

FORMULAE

Two Dimensional Figures

Figure	Diagram of figure	Surface area	Perimeter
Square	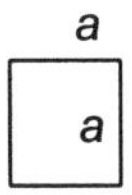	a^2	$4a$
Rectangle	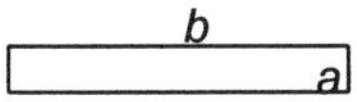	ab	$2(a + b)$
Triangle	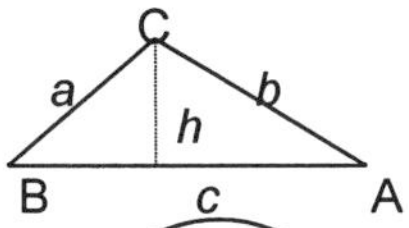	$\frac{1}{2}ch$ $\frac{1}{2}ab \sin C$ $\sqrt{\{s(s-a)(s-b)(s-c)\}}$ where $s = \frac{1}{2}(a + b + c)$	$a + b + c = 2s$
Circle	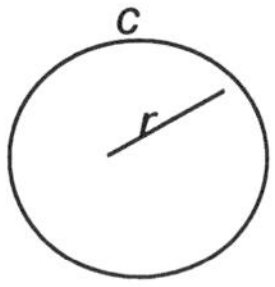	πr^2 $\frac{1}{4}\pi d^2$ where $2r = d$	$2\pi r$ πd
Parallelogram	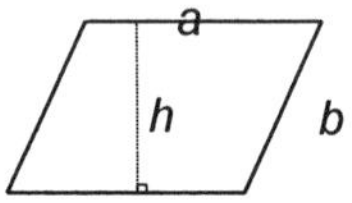	ah	$2(a + b)$
Trapezium	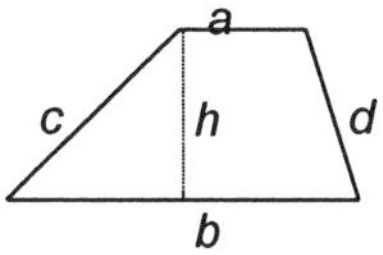	$\frac{1}{2}h(a + b)$	$a + b + c + d$
Ellipse	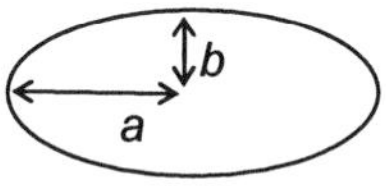	Approximately πab	$\pi(a + b)$
Hexagon	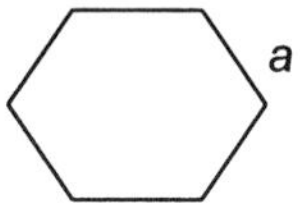	$2.6 \times a^2$	
Octagon	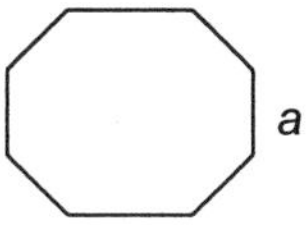	$4.83 \times a^2$	
Sector of circle	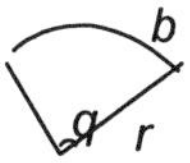	$\frac{1}{2} rb$ or $\frac{q}{360}\pi r^2$ note: b = angle $\frac{q}{360} \times \pi 2r$	
Segment of a circle		S - T where S = area of sector T = area of triangle	
Bellmouth	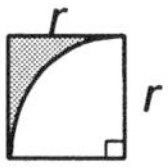	$\frac{3}{14} \times r^2$	

Three Dimensional Figures

Figure	Diagram of figure	Surface area	Volume
Cube	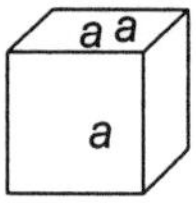	$6a^2$	a^3
Cuboid/ rectangular block	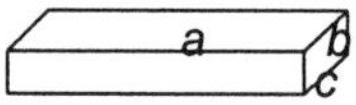	$2(ab + ac + bc)$	abc
Prism/ triangular block	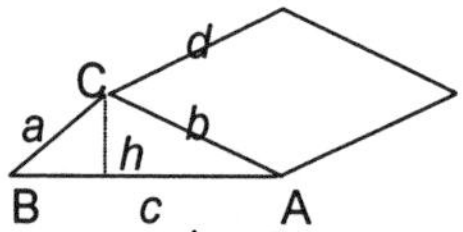	$bd + hc + dc + ad$	$½hcd$ $½ab \sin C\, d$ $d\,\sqrt{\{s(s-a)(s-b)(s-c)}$ where $s = ½(a + b + c)$
Cylinder	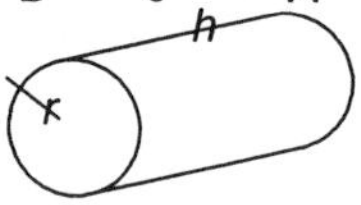	$2\ rh + 2\ r^2$ $dh + ½\ d^2$	r^2h $¼\ d^2h$
Sphere	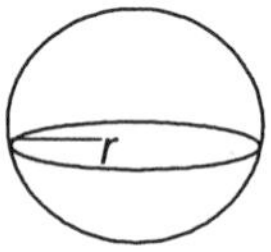	$4\ r^2$	$^4/_3$□r^3
Segment of sphere	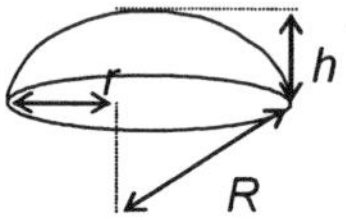	$2\ Rh$	$^1/_6\ h(3r^2 + h^2)$ $^1/_3\ h^2(3R - h)$
Pyramid	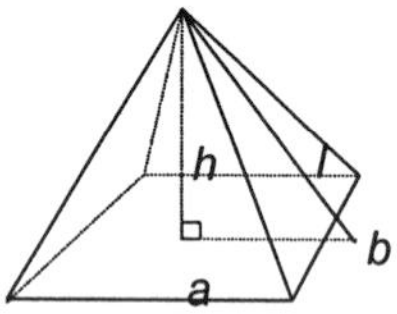	$(a + b)l + ab$	$^1/_3abh$
Frustrum of a pyramid	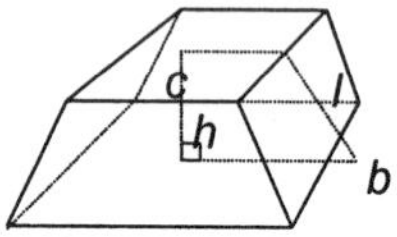	$l(a+b+c+d) + \sqrt{(ab+cd)}$ [regular figure only]	$^h/_3(ab + cd + \sqrt{abcd})$
Cone	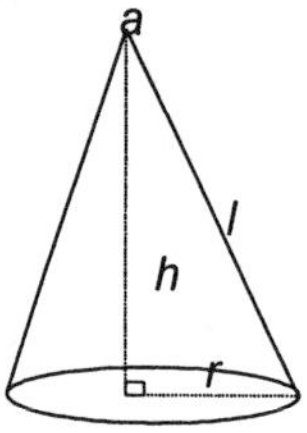	$rl + r^2$ $½\ dh + ¼\ d^2$	$^1/_3\ r^2h$ $^1/_{12}\ d^2h$
Frustrum of a cone	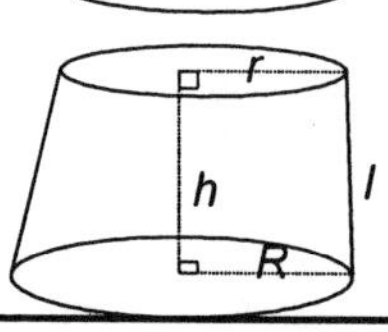	$r^2 + R^2 + h(R + r)$	$^1/_3\ h(R^2 + Rr + r^2)$

Geometric Formulae

Formula	**Description**
Pythagoras theorem	A2 = B2 + C2 where A is the hypotenuse of a right-angled triangle and B and C are the two adjacent sides
Simpson's Rule	The Area is divided into an even number of strips of equal width, and therefore has an odd number of ordinates at the division points area = $\frac{S\ (A + 2B + 4C)}{3}$ where S = common interval (strip width) A = sum of first and last ordinates B = sum of remaining odd ordinates C = sum of the even ordinates The Volume can be calculated by the same formula, but by substituting the area of each co-ordinate rather than its length.
Trapezoidal Rule	A given trench is divided into two equal sections, giving three ordinates, the first, the middle and the last. volume = $\frac{S \times (A + B + 2C)}{2}$ where S = width of the strips A = area of the first section B = area of the last section C = area of the rest of the sections
Prismoidal Rule	A given trench is divided into two equal sections, giving three ordinates, the first, the middle and the last. volume = $\frac{L \times (A + 4B + C)}{6}$ where L = total length of trench A = area of the first section B = area of the middle section C = area of the last section

EARTHWORK

Weights of Typical Materials Handled by Excavators

The weight of the material is that of the state in its natural bed and includes moisture. Adjustments should be made to allow for loose or compacted states.

Material	kg/m³	lb/cu yd
Adobe	1914	3230
Ashes	610	1030
Asphalt, rock	2400	4050
Basalt	2933	4950
Bauxite: alum ore	2619	4420
Borax	1730	2920
Caliche	1440	2430
Carnotite	2459	4150
Cement	1600	2700
Chalk (hard)	2406	4060
Cinders	759	1280
Clay: dry	1908	3220
Clay: wet	1985	3350
Coal: bituminous	1351	2280
Coke	510	860
Conglomerate	2204	3720
Dolomite	2886	4870
Earth: dry	1796	3030
Earth: moist	1997	3370
Earth: wet	1742	2940
Feldspar	2613	4410
Felsite	2495	4210
Fluorite	3093	5220
Gabbro	3093	5220
Gneiss	2696	4550
Granite	2690	4540
Gravel, dry	1790	3020
Gypsum	2418	4080
Hardcore (consolidated)	1928	120
Lignite broken	1244	2100

Weights of Typical Materials Handled by Excavators

Material	kg/m³	lb/cu yd
Limestone	2596	4380
Magnesite, magnesium ore	2993	5050
Marble	2679	4520
Marl	2216	3740
Peat	700	1180
Potash	2193	3700
Pumice	640	1080
Quarry waste	1438	90
Quartz	2584	4360
Rhyolite	2400	4050
Sand: dry	1707	2880
Sand: wet	1831	3090
Sand and gravel - dry	1790	3020
- wet	2092	3530
Sandstone	2412	4070
Schist	2684	4530
Shale	2637	4450
Slag (blast)	2868	4840
Slate	2667	4500
Snow - dry	130	220
- wet	510	860
Taconite	3182	5370
Topsoil	1440	2430
Trachyte	2400	4050
Traprock	2791	4710
Water	1000	62

Transport Capacities

Type of vehicle	Capacity of vehicle Payload	Heaped capacity
Wheelbarrow	150	0.10
1 tonne dumper	1250	1.00
2.5 tonne dumper	4000	2.50
Articulated dump truck (Volvo A20 6x4)	18500	11.00
Articulated dump truck (Volvo A35 6x6)	32000	19.00
Large capacity rear dumper (Euclid R35)	35000	22.00
Large capacity rear dumper (Euclid R85)	85000	50.00

Machine Volumes for Excavating and Filling

Machine type	Cycles per minute	Volume per minute (m^3)
1.5 tonne excavator	1	0.04
	2	0.08
	3	0.12
3 tonne excavator	1	0.13
	2	0.26
	3	0.39
5 tonne excavator	1	0.28
	2	0.56
	3	0.84
7 tonne excavator	1	0.28
	2	0.56
	3	0.84
21 tonne excavator	1	1.21
	2	2.42
	3	3.63
Backhoe loader JCB3CX excavating Rear bucket capacity 0.28m3	1	0.28
	2	0.56
	3	0.84
Backhoe loader JCB3CX loading Front bucket capacity 1.00m3	1	1.00
	2	2.00

Machine Volumes for Excavating and Filling

Machine type	Loads per hour	Volume per hour (m^3)
1 tonne high tip skip loader Volume 0.485m3	5 7 10	2.43 3.40 4.85
3 tonne dumper Max. volume 2.40m3 Available volume 1.9m3	4 5 7 10	7.60 9.50 13.30 19.00
6 tonne dumper Max. volume 3.40m3 Available volume 3.77m3	4 5 7 10	15.08 18.85 26.39 37.70

Bulkage of Soils (After excavation)

Type of soil	Approximate bulking of 1 m^3 after excavation
Vegetable soil and loam	25 - 30%
Soft clay	30 - 40%
Stiff clay	10 - 20%
Gravel	20 - 25%
Sand	40 - 50%
Chalk	40 - 50%
Rock, weathered	30 - 40%
Rock, unweathered	50 - 60%

Shrinkage of Materials (On being deposited)

Type of soil	Approximate bulking of 1 m^3 after excavation
Clay	10%
Gravel	8%
Gravel and sand	9%
Loam and light sandy soils	12%
Loose vegetable soils	15%

Voids in Material Used as Sub-bases or Beddings

Material	m^3 of voids/m^3
Alluvium	0.37
River grit	0.29
Quarry sand	0.24
Shingle	0.37
Gravel	0.39
Broken stone	0.45
Broken bricks	0.42

Angles of Repose

Type of soil		Degrees
Clay	- dry	30
	- damp, well drained	45
	- wet	15 - 20
Earth	- dry	30
	- damp	45
Gravel	- moist	48
Sand	- dry or moist	35
	- wet	25
Loam		40

Slopes and Angles

Ratio of base to height	Angle in degrees
5 : 1	11
4 : 1	14
3 : 1	18
2 : 1	27
1½ : 1	34
1 : 1	45
1 : 1½	56
1 : 2	63
1 : 3	72
1 : 4	76
1 : 5	79

Grades (In Degrees and Percents)

Degrees	Percent	Degrees	Percent
1	1.8	24	44.5
2	3.5	25	46.6
3	5.2	26	48.8
4	7.0	27	51.0
5	8.8	28	53.2
6	10.5	29	55.4
7	12.3	30	57.7
8	14.0	31	60.0
9	15.8	32	62.5
10	17.6	33	64.9
11	19.4	34	67.4
12	21.3	35	70.0
13	23.1	36	72.7
14	24.9	37	75.4
15	26.8	38	78.1
16	28.7	39	81.0
17	30.6	40	83.9
18	32.5	41	86.9
19	34.4	42	90.0
20	36.4	43	93.3
21	38.4	44	96.6
22	40.4	45	100.0

Bearing Powers

Ground conditions	Bearing power		
	km/mý	lb/iný	metric t/m²
Rock (broken)	483	70	50
Rock (solid)	2,415	350	240
Clay, dry or hard	380	55	40
medium dry	190	27	20
soft or wet	100	14	10
Gravel, cemented	760	110	80
Sand, compacted	380	55	40
clean dry	190	27	20
Swamp and alluvial soils	48	7	5

Earthwork Support

Maximum depth of excavation in various soils without the use of earthwork support:

Ground conditions	Feet (ft)	Metres (m)
Compact soil	12	3.66
Drained loam	6	1.83
Dry sand	1	0.3
Gravelly earth	2	0.61
Ordinary earth	3	0.91
Stiff clay	10	3.05

It is important to note that the above table should only be used as a guide.
Each case must be taken on its merits and, as the limited distances given above are approached, careful watch must be kept for the slightest signs of caving in.

CONCRETE WORK

Weights of Concrete and Concrete Elements

Type of material			kg/m³	lb/cu ft
<u>Ordinary concrete (dense aggregates)</u>				
Non-reinforced plain or mass concrete				
	Nominal weight		2305	144
	Aggregate	- limestone	2162 to 2407	135 to 150
		- gravel	2244 to 2407	140 to 150
		- broken brick	2000 (av)	125 (av)
		- other crushed stone	2326 to 2489	145 to 155
Reinforced concrete				
	Nominal weight		2407	150
	Reinforcement	- 1%	2305 to 2468	144 to 154
		- 2%	2356 to 2519	147 to 157
		- 4%	2448 to 2703	153 to 163
<u>Special concretes</u>				
Heavy concrete				
	Aggregates -	barytes, magnetite	3210 (min)	200 (min)
		steel shot, punchings	5280	330
Lean mixes				
	Dry-lean (gravel aggregate)		2244	140
	Soil-cement (normal mix)		1601	100

Weights of Concrete and Concrete Elements - cont'd

Type of material		kg/m² per mm thick	lb/sq ft per inch thick
Ordinary concrete (dense aggregates)			
Solid slabs (floors, walls etc.)			
Thickness:	75 mm or 3 in	184	37.5
	100 mm or 4 in	245	50
	150 mm or 6 in	378	75
	250 mm or 10 in	612	125
	300 mm or 12 in	734	150
Ribbed slabs			
Thickness:	125 mm or 5 in	204	42
	150 mm or 6 in	219	45
	225 mm or 9 in	281	57
	300 mm or 12 in	342	70
Special concretes			
Finishes etc			
Rendering, screed etc Granolithic, terrazzo		1928 to 2401	10 to 12.5
Glass-block (hollow) concrete		1734 (approx)	9 (approx)
Pre-stressed concrete		Weights as for reinforced concrete (upper limits)	
Air-entrained concrete		Weights as for plain or reinforced concrete	

Average Weight of Aggregates

Materials	**Voids %**	**Weight kg/m³**
Sand	39	1660
Gravel 10 - 20 mm	45	1440
Gravel 35 - 75 mm	42	1555
Crushed stone	50	1330
Crushed granite (over 15 mm)	50	1345
(n.e. 15 mm)	47	1440
'All-in' ballast	32	1800 - 2000

Material	**kg/m³**	**lb/cu yd**
Vermiculite (aggregate)	64-80	108-135

Material	**kg/m³**	**lb/cu ft**
All-in aggregate	1999	125

Common Mixes (per m³)

Recommended mix	Class of work suitable for: -	Cement (kg)	Sand (kg)	Coarse aggregate (kg)	No 50kg bags cement per m³ of combined aggregate
1:3:6	Roughest type of mass concrete such as footings, road haunching over 300 mm thick	208	905	1509	4.00
1:2.5:5	Mass concrete of better class than 1:3:6 such as bases for machinery, walls below ground etc.	249	881	1474	5.00
1:2:4	Most ordinary uses of concrete, such as mass walls above ground, road slabs etc. and general reinforced concrete work	304	889	1431	6.00
1:1.5:3	Watertight floors, pavements and walls, tanks, pits, steps, paths, surface of 2 course roads, reinforced concrete where extra strength is required	371	801	1336	7.50
1:1:2	Work of thin section such as fence posts and small precast work	511	720	1206	10.50

Prescribed Mixes for Ordinary Structural Concrete

Weights of cement and total dry aggregates in kg to produce approximately one cubic metre of fully compacted concrete together with the percentages by weight of fine aggregate in total dry aggregates.

Conc. grade	Nominal max. size of aggregate (mm)	40		20		14		10	
	Workability	Med.	High	Med.	High	Med.	High	Med.	High
	Limits to slump that may be expected (mm)	50-100	100-150	25-75	75-125	10-50	50-100	10-25	25-50
7	Cement (kg)	180	200	210	230	-	-	-	-
	Total aggregate (kg)	1950	1850	1900	1800	-	-	-	-
	Fine aggregate (%)	30-45	30-45	35-50	35-50	-	-	-	-
10	Cement (kg)	210	230	240	260	-	-	-	-
	Total aggregate (kg)	1900	1850	1850	1800	-	-	-	-
	Fine aggregate (%)	30-45	30-45	35-50	35-50	-	-	-	-
15	Cement (kg)	250	270	280	310	-	-	-	-
	Total aggregate (kg)	1850	1800	1800	1750	-	-	-	-
	Fine aggregate (%)	30-45	30-45	35-50	35-50	-	-	-	-
20	Cement (kg)	300	320	320	350	340	380	360	410
	Total aggregate (kg)	1850	1750	1800	1750	1750	1700	1750	1650
	Sand								
	Zone 1 (%)	35	40	40	45	45	50	50	55
	Zone 2 (%)	30	35	35	40	40	45	45	50
	Zone 3 (%)	30	30	30	35	35	40	40	45
25	Cement (kg)	340	360	360	390	380	420	400	450
	Total aggregate (kg)	1800	1750	1750	1700	1700	1650	1700	1600
	Sand								
	Zone 1 (%)	35	40	40	45	45	50	50	55
	Zone 2 (%)	30	35	35	40	40	45	45	50
	Zone 3 (%)	30	30	30	35	35	40	40	45
30	Cement (kg)	370	390	400	430	430	470	460	510
	Total aggregate (kg)	1750	1700	1700	1650	1700	1600	1650	1550
	Sand								
	Zone 1 (%)	35	40	40	45	45	50	50	55
	Zone 2 (%)	30	35	35	40	40	45	45	50
	Zone 3 (%)	30	30	30	35	35	40	40	45

Weights of Bar Reinforcement

Nominal sizes (mm)	Cross-sectional area (mm^2)	Mass kg/m	Length of bar m/tonne
6	28.27	0.222	4505
8	50.27	0.395	2534
10	78.54	0.617	1622
12	113.1	0.888	1126
16	201.06	1.578	634
20	314.16	2.466	405
25	490.87	3.853	260
32	804.25	6.313	158
40	1265.64	9.865	101
50	1963.5	15.413	65

Weights of Bars at Specific Spacings

Weights of metric bars in kilograms per square metre.

Size	Spacing of bars in millimetres									
(mm)	75	100	125	150	175	200	225	250	275	300
6	2.96	2.220	1.776	1.480	1.27	1.110	0.99	0.89	0.81	0.74
8	5.26	3.95	3.16	2.63	2.26	1.97	1.75	1.58	1.44	1.32
10	8.22	6.17	4.93	4.11	3.52	3.08	2.74	2.47	2.24	2.06
12	11.84	8.88	7.10	5.92	5.07	4.44	3.95	3.55	3.23	2.96
16	21.04	15.78	12.63	10.52	9.02	7.89	7.02	6.31	5.74	5.26
20	32.88	24.66	19.73	16.44	14.09	12.33	10.96	9.87	8.97	8.22
25	51.38	38.53	30.83	25.69	22.02	19.27	17.13	15.41	14.01	12.84
32	84.18	63.13	50.51	42.09	36.08	31.57	28.06	25.25	22.96	21.04
40	131.53	98.65	78.92	65.76	56.37	49.32	43.84	39.46	35.87	32.88
50	205.51	154.13	123.31	102.76	88.08	77.07	68.50	61.65	56.05	51.38

Basic weight of steelwork taken as 7850 kg/m^3

Basic weight of bar reinforcement per metre run = 0.00785 kg/mm^2

The value of PI has been taken as 3.141592654

Fabric Reinforcement

Fabric reference	Longitudinal wires			Cross wires			Mass
	Nominal wire size (mm)	Pitch (mm)	Area (mm/m²)	Nominal wire size (mm)	Pitch (mm)	Area (mm/m²)	(kg/m²)
Square mesh							
A393	10	200	393	10	200	393	6.16
A252	8	200	252	8	200	252	3.95
A193	7	200	193	7	200	193	3.02
A142	6	200	142	6	200	142	2.22
A98	5	200	98	5	200	98	1.54
Structural mesh							
B1131	12	100	1131	8	200	252	10.90
B785	10	100	785	8	200	252	8.14
B503	8	100	503	8	200	252	5.93
B385	7	100	385	7	200	193	4.53
B283	6	100	283	7	200	193	3.73
B196	5	100	196	7	200	193	3.05
Long mesh							
C785	10	100	785	6	400	70.8	6.72
C636	9	100	636	6	400	70.8	5.55
C503	8	100	503	5	400	49.00	4.34
C385	7	100	385	5	400	49.00	3.41
C283	6	100	283	5	400	49.00	2.61
Wrapping mesh							
D98	5	200	98	5	200	98	1.54
D49	2.5	100	49	2.5	100	49	0.77
Stock sheet size	**Length 4.8 m**			**Width 2.4 m**		**Sheet area 11.52m²**	

Wire

SWG	6g	5g	4g	3g	2g	1g	1/0g	2/0g	3/0g	4/0g	5/0g
diameter											
in	0.192	0.212	0.232	0.252	0.276	0.300	0.324	0.348	0.372	0.400	0.432
mm	4.9	5.4	5.9	6.4	7.0	7.6	8.2	8.8	9.5	0.2	1.0
area											
in^2	0.029	0.035	0.042	0.050	0.060	0.071	0.082	0.095	0.109	0.126	0.146
mm^2	19	23	27	32	39	46	53	61	70	81	95

Average Weight (kg/m^3) of Steelwork Reinforcement in Concrete for Various Building Elements

	kg/m^3 concrete
Substructure	
Pile caps	110 - 150
Tie beams	130 - 170
Ground beams	230 - 330
Bases	90 - 130
Footings	70 - 110
Retaining walls	110 - 150
Superstructure	
Slabs - one way	75 - 125
Slabs - two way	65 - 135
Plate slab	95 - 135
Cantilevered slab	90 - 130
Ribbed floors	80 - 120
Columns	200 - 300
Beams	250 - 350
Stairs	130 - 170
Walls - normal	30 - 70
Walls - wind	50 - 90

Note: For exposed elements add the following %: Walls 50%, Beams 100%, Columns 15%

Formwork Stripping Times – Normal Curing Periods

Conditions under which concrete is maturing	Minimum periods of protection for different types of cement					
	Number of days (where the average surface temperature of the concrete exceeds 10°C during the whole period)			Equivalent maturity (degree hours) calculated as the age of the concrete in hours multiplied by the number of degrees Celsius by which the average surface temperature of the concrete exceeds - 10°C		
	Other	SRPC	OPC or RHPC	Other	SRPC	OPC or RHPC
1. Hot weather or drying winds	7	4	3	3500	2000	1500
2. Conditions not covered by 1	4	3	2	2000	1500	1000

KEY

OPC - Ordinary Portland Cement

RHPC - Rapid-hardening Portland cement

SRPC - Sulphate-resisting Portland cement

Minimum period before striking formwork

	Minimum period before striking		
	Surface temperature of concrete		
	16 °C	**17 °C**	**t °C(0-25)**
Vertical formwork to columns, walls and large beams	12 hours	18 hours	$\frac{300}{t+10}$ hours
Soffit formwork to slabs	4 days	6 days	$\frac{100}{t+10}$ days
Props to slabs	10 days	15 days	$\frac{250}{t+10}$ days
Soffit formwork to beams	9 days	14 days	$\frac{230}{t+10}$ days
Props to beams	14 days	21 days	$\frac{360}{t+10}$ days

MASONRY

Weights of Bricks and Blocks

Walls and components of walls	kg/m² per mm thick	lb/sq ft per inch thick
Blockwork		
Hollow clay blocks; average)	1.15	6
Common clay blocks	1.90	10
Brickwork		
Engineering clay bricks	2.30	12
Refractory bricks	1.15	6
Sand-lime (and similar) bricks	2.02	10.5

Weights of Stones

Type of stone		kg/m3	lb/cu ft
Natural stone (solid)			
Granite		2560 to 2927	160 to 183
Limestone	- Bath stone	2081	130
	- Marble	2723	170
	- Portland stone	2244	140
Sandstone		2244 to 2407	140 to 150
Slate		2880	180
Stone rubble (packed)		2244	140

Quantities of Bricks and Mortar

Materials per m² of wall:		
Thickness	**No. of Bricks**	**Mortar m³**
Half brick (112.5 mm)	58	0.022
One brick (225 mm)	116	0.055
Cavity, both skins (275 mm)	116	0.045
1.5 brick (337 mm)	174	0.074
Mass brickwork per m³	464	0.36

Mortar Mixes: Quantities of Dry Materials

	Imperial cu yd			Metric m³		
Mix	Cement cwts	Lime cwts	Sand cu yds	Cement tonnes	Lime tonnes	Sand cu m
1:3	7.0	-	1.04	0.54	-	1.10
1:4	6.3	-	1.10	0.40	-	1.20
1:1:6	3.9	1.6	1.10	0.27	0.13	1.10
1:2:9	2.6	2.1	1.10	0.20	0.15	1.20
0:1:3	-	3.3	1.10	-	0.27	1.00

Mortar Mixes for Various Uses

Mix	Use
1:3	Construction designed to withstand heavy loads in all seasons.
1:1:6	Normal construction not designed for heavy loads. Sheltered and moderate conditions in spring and summer. Work above d:p:c - sand, lime bricks, clay blocks, etc.
1:2:9	Internal partitions with blocks which have high drying shrinkage, pumice blocks, etc. any periods.
0:1:3	Hydraulic lime only should be used in this mix and may be used for construction not designed for heavy loads and above d:p:c spring and summer.

Quantities of Bricks and Mortar Required per m² of Walling

Description	Unit	Nr of bricks required	Mortar required (m³)		
			No frogs	Single frogs	Double frogs
Standard bricks					
Brick size					
215 x 102.5 x 50 mm					
half brick wall (103 mm)					
(103 mm)	m²	72	0.022	0.027	0.032
2 x half brick cavity wall					
(270 mm)	m²	144	0.044	0.054	0.064
one brick wall					
(215 mm)	m²	144	0.052	0.064	0.076
one and a half brick wall					
(328 mm)	m²	216	0.073	0.091	0.108
mass brickwork	m³	576	0.347	0.413	0.480
Brick size					
215 x 102.5 x 65 mm					
half brick wall	m²	58	0.019	0.022	0.026
(103 mm)					
2 x half brick cavity wall					
(270 mm)	m²	116	0.038	0.045	0.055
one brick wall					
(215 mm)	m²	116	0.046	0.055	0.064
one and a half brick wall					
(328 mm)	m²	174	0.063	0.074	0.088
mass brickwork	m³	464	0.307	0.360	0.413
Metric modular bricks - Perforated					
Brick co-ordinating size					
200 x 100 x 75 mm					
90 mm thick	m²	67	0.016	0.019	
190 mm thick	m²	133	0.042	0.048	
290 mm thick	m²	200	0.068	0.078	
Brick co-ordinating size					
200 x 100 x 100 mm					
90 mm thick	m²	50	0.013	0.016	
190 mm thick	m²	100	0.036	0.041	
290 mm thick	m²	150	0.059	0.067	
Brick co-ordinating size					
300 x 100 x 75 mm					
90 mm thick	m²	33	-	0.015	
Brick co-ordinating size					
00 x 100 x 100 mm					
90 mm thick	m²	44	0.015	0.018	

Note: Assuming 10 mm deep joints

Mortar Required per m^2 Blockwork (9.88 blocks/m^2)

Wall thickness	75	90	100	125	140	190	215
Mortar m^3/m^2	0.005	0.006	0.007	0.008	0.009	0.013	0.014

Mortar Mixes

Mortar group	Cement: lime: sand	Masonry cement:sand	Cement: sand with plasticizer
1	1 : 0-0.25:3		
2	1 : 0.5 :4-4.5	1 : 2.5-3.5	1 : 3-4
3	1 : 1:5-6	1 : 4-5	1 : 5-6
4	1 : 2:8-9	1 : 5.5-6.5	1 : 7-8
5	1 : 3:10-12	1 : 6.5-7	1 : 8

Group 1: strong inflexible mortar
Group 5: weak but flexible

All mixes within a group are of approximately similar strength.
Frost resistance increases with the use of plasticisers.
Cement:lime:sand mixes give the strongest bond and greatest resistance to rain penetration.
Masonry cement equals ordinary Portland cement plus a fine neutral mineral filler and an air entraining agent.

Calcium Silicate Bricks

Type	Strength	Location
Class 2 crushing strength	14.0N/mm2	not suitable for walls
Class 3	20.5N/mm2	walls above dpc
Class 4	27.5N/mm2	cappings and copings
Class 5	34.5N/mm2	retaining walls
Class 6	41.5N/mm2	walls below ground
Class 7	48.5N/mm2	walls below ground

The Class 7 calcium silicate bricks are therefore equal in strength to Class B bricks.
Calcium silicate bricks are not suitable for DPCs.

Durability of Bricks

FL	Frost resistant with low salt content
FN	Frost resistant with normal salt content
ML	Moderately frost resistant with low salt content
MN	Moderately frost resistant with normal salt content

Brickwork Dimensions

No. of horizontal bricks	Dimensions mm	No. of vertical courses	No. of vertical courses
1/2	112.5	1	75
1	225.0	2	150
1 1/2	337.5	3	225
2	450.0	4	300
2 1/2	562.5	5	375
3	675.0	6	450
3 1/2	787.5	7	525
4	900.0	8	600
4 1/2	1012.5	9	675
5	1125.0	10	750
5 1/2	1237.5	11	825
6	1350.0	12	900
6 1/2	1462.5	13	975
7	1575.0	14	1050
7 1/2	1687.5	15	1125
8	1800.0	16	1200
8 1/2	1912.5	17	1275
9	2025.0	18	1350
9 1/2	2137.5	19	1425
10	2250.0	20	1500
20	4500.0	24	1575
40	9000.0	28	2100
50	11250.0	32	2400
60	13500.0	36	2700
75	16875.0	40	3000

Standard Available Block Sizes

Block	Co-ordinating size Length x height (mm)	Work size Length x height (mm)	Thicknesses (work size) (mm)
A	400 x 100	390 x 90	75, 90, 100, 140, 190
	400 x 200	440 x 190	75, 90, 100, 140, 190
	450 x 225	440 x 215	75, 90, 100, 140, 190, 215
B	400 x 100	390 x 90	75, 90, 100, 140, 190
	400 x 200	390 x 190	75, 90, 100, 140, 190
	450 x 200	440 x 190	75, 90, 100, 140, 190, 215
	450 x 225	440 x 215	75, 90, 100, 140, 190, 215
	450 x 300	440 x 290	75, 90, 100, 140, 190, 215
	600 x 200	590 x 190	75, 90, 100, 140, 190, 215
	600 x 225	590 x 215	75, 90, 100, 140, 190, 215
C	400 x 200	390 x 190	60, 75
	450 x 200	440 x 190	60, 75
	450 x 225	440 x 215	60, 75
	450 x 300	440 x 290	60, 75
	600 x 200	590 x 190	60, 75
	600 x 225	590 x 215	60, 75

TIMBER

Weights of Timber

Material	kg/m³	lb/cu ft
General	806 (avg)	50 (avg)
Douglas fir	479	30
Yellow pine, spruce	479	30
Pitch pine	673	42
Larch, elm	561	35
Oak (English)	724 to 959	45 to 60
Teak	643 to 877	40 to 55
Jarrah	959	60
Greenheart	1040 to 1204	65 to 75
Quebracho	1285	80
Material	**kg/m² per mm thickness**	**lb/sq ft per inch thickness**
Wooden boarding and blocks		
Softwood	0.48	2.5
Hardwood	0.76	4
Hardboard	1.06	5.5
Chipboard	0.76	4
Plywood	0.62	3.25
Blockboard	0.48	2.5
Fibreboard	0.29	1.5
Wood-wool	0.58	3
Plasterboard	0.96	5
Weather boarding	0.35	1.8

Conversion Tables (for sawn timber only)

Inches	>	Millimetres	Feet	>	Metres
1		25	1		0.300
2		50	2		0.600
3		75	3		0.900
4		100	4		1.200
5		125	5		1.500
6		150	6		1.800
7		175	7		2.100
8		200	8		2.400
9		225	9		2.700
10		250	10		3.000
11		275	11		3.300
12		300	12		3.600
13		325	13		3.900
14		350	14		4.200
15		375	15		4.500
16		400	16		4.800
17		425	17		5.100
18		450	18		5.400
19		475	19		5.700
20		500	20		6.000
21		525	21		6.300
22		550	22		6.600
23		575	23		6.900
24		600	24		7.200

Planed Softwood

The finished end section size of planed timber is usually 3/16" less than the original size from which it is produced. This however varies slightly dependant upon availability of material and origin of species used.

Standard (timber) to Cubic Metres and Cubic Metres to Standard (timber)

m^3	m^3/Standards	Standard
4.672	1	0.214
9.344	2	0.428
14.017	3	0.642
18.689	4	0.856
23.361	5	1.070
28.033	6	1.284
32.706	7	1.498
37.378	8	1.712
42.05	9	1.926
46.722	10	2.140
93.445	20	4.281
140.167	30	6.421
186.890	40	8.561
233.612	50	10.702
280.335	60	12.842
327.057	70	14.982
373.779	80	17.122
420.502	90	19.263
467.224	100	21.403

1 cu metre = 35.3148 cu ft
1 cu ft = 0.028317 cu metres
1 std = 4.67227 cu metres

Standards (timber) to Cubic Metres and Cubic Metres to Standards (timber)

1 cu metre	=	35.3148 cu ft	=	0.21403 std
1 cu ft	=	0.028317 cu metres		
1 std	=	4.67227 cu metres		

Basic Sizes of Sawn Softwood Available (cross sectional areas)

Thickness (mm)	Width (mm)								
	75	100	125	150	175	200	225	250	300
16	x	x	x	x					
19	x	x	x	x					
22	x	x	x	x					
25	x	x	x	x	x	x	x	x	x
32	x	x	x	x	x	x	x	x	x
36	x	x	x	x					
38	x	x	x	x	x	x	x		
44	x	x	x	x	x	x	x	x	x
47*	x	x	x	x	x	x	x	x	x
50	x	x	x	x	x	x	x	x	x
63	x	x	x	x	x	x	x		
75		x	x	x	x	x	x	x	x
100		x		x		x		x	x
150				x		x			x
200						x			
250								x	
300									x

* This range of widths for 47 mm thickness will usually be found to be available in construction quality only.

Note: The smaller sizes below 100 mm thick and 250 mm width are normally but not exclusively of European origin. Sizes beyond this are usually of North and South American origin.

Basic Lengths of Sawn Softwood Available (metres)

1.80	2.10	3.00	4.20	5.10	6.00	7.20
	2.40	3.30	4.50	5.40	6.30	
	2.70	3.60	4.80	5.70	6.60	
		3.90			6.90	

Note: Lengths of 6.00 m and over will generally only be available from North American species and may have to be re-cut from larger sizes.

Reductions From Basic Size to Finished Size of Timber By Planing of Two Opposed Faces

Purpose	15 – 35 mm	36 – 100 mm	101 – 150 mm	Over 150 mm
a) constructional timber	3 mm	3 mm	5 mm	6 mm
b) matching interlocking boards	4 mm	4 mm	6 mm	6 mm
c) wood trim not specified in BS 584	5 mm	7 mm	7 mm	9 mm
d) joinery and cabinet work	7 mm	9 mm	11 mm	13 mm

Note: The reduction of width or depth is overall the extreme size and is exclusive of any reduction of the face by the machining of a tongue or lap joints.

METAL

Weights of Metals

Material	kg/m³	lb/cu ft
Metals, steel construction, etc.		
Iron		
- cast	7207	450
- wrought	7687	480
- ore - general	2407	150
- (crushed) Swedish	3682	230
Steel	7854	490
Copper		
- cast	8731	545
- wrought	8945	558
Brass	8497	530
Bronze	8945	558
Aluminium	2774	173
Lead	11322	707
Zinc (rolled)	7140	446
	g/mm² per metre	**lb/sq ft per foot**
Steel bars	7.85	3.4
Structural steelwork	Net weight of member @ 7854 kg/m³	
riveted	+ 10% for cleats, rivets, bolts, etc	
welded	+ 1.25% to 2.5% for welds, etc	
Rolled sections		
beams	+ 2.5%	
stanchions	+ 5% (extra for caps and bases)	
Plate		
web girders	+ 10% for rivets or welds, stiffeners, etc	
	kg/m	**lb/ft**
Steel stairs : industrial type		
1 m or 3 ft wide	84	56
Steel tubes		
50 mm or 2 in bore	5 to 6	3 to 4
Gas piping		
20 mm or 3/4 in	2	1¼

KERBS/EDGINGS/CHANNELS

Precast Concrete Kerbs to BS 7263
Straight kerb units: length from 450 to 915 mm

150 mm high x 125 mm thick			
bullnosed		type BN	
half battered		type HB3	
255 mm high x 125 mm thick			
45 degree splayed		type SP	
half battered		type HB2	
305 mm high x 150 mm thick			
half battered		type HB1	
Quadrant kerb units			
150 mm high x 305 and 455 mm radius to match		type BN	type QBN
150 mm high x 305 and 455 mm radius to match		type HB2, HB3	type QHB
150 mm high x 305 and 455 mm radius to match		type SP	type QSP
255 mm high x 305 and 455 mm radius to match		type BN	type QBN
255 mm high x 305 and 455 mm radius to match		type HB2, HB3	type QHB
225 mm high x 305 and 455 mm radius to match		type SP	type QSP
Angle kerb units			
305 x 305 x 225 mm high x 125 mm thick			
bullnosed external angle		type XA	
splayed external angle to match type SP		type XA	
bullnosed internal angle		type IA	
splayed internal angle to match type SP		type IA	
Channels			
255 mm wide x 125 mm high flat		type CS1	
150 mm wide x 125 mm high flat type		CS2	
255 mm wide x 125 mm high dished		type CD	
Transition kerb units			
from kerb type SP to HB	left handed	type TL	
	right handed	type TR	
from kerb type BN to HB	left handed	type DL1	
	right handed	type DR1	
from kerb type BN to SP	left handed	type DL2	
	right handed	type DR2	

Radial Kerbs and Channels

All profiles of kerbs and channels

External radius (mm)	Internal radius (mm)
1000	3000
2000	4500
3000	6000
4500	7500
6000	9000
7500	1050
9000	1200
1050	
1200	

Precast Concrete Edgings to BS 7263

Round top type ER	Flat top type EF	Bullnosed top type EBN
150 x 50 mm	150 x 50 mm	150 x 50 mm
200 x 50	200 x 50	200 x 50
250 x 50	250 x 50	250 x 50

BASES

Cement Bound Material for Bases and Sub-bases

CBM1:	very carefully graded aggregate from 37.5 - 75 ym, with a 7-day strength of 4.5N/mm2
CBM2:	same range of aggregate as CBM1 but with more tolerance in each size of aggregate with a 7-day strength of 7.0N/mm2
CBM3:	crushed natural aggregate or blast furnace slag, graded from 37.5 mm - 150 ym for 40 mm aggregate, and from 20 - 75 ym for 20 mm aggregate, with a 7-day strength of 10N/mm2
CBM4:	crushed natural aggregate or blast furnace slag, graded from 37.5 mm - 150 ym for 40 mm aggregate, and from 20 - 75 ym for 20 mm aggregate, with a 7-day strength of 15N/mm2

INTERLOCKING BRICK/BLOCK ROADS/PAVINGS

Sizes of Precast Concrete Paving Blocks to BS 6717: Part 1

Type R blocks
200 x 100 x 60 mm
200 x 100 x 65
200 x 100 x 80
200 x 100 x 100

Type S
Any shape within a 295 mm space

Sizes of Clay Brick Pavers to BS 6677: Part 1
200 x 100 x 50 mm thick
200 x 100 x 65
210 x 105 x 50
210 x 105 x 65
215 x 102.5 x 50
215 x 102.5 x 65

Type PA: 3 kN
Footpaths and pedestrian areas, private driveways, car parks, light vehicle traffic and over-run.

Type PB: 7 kN
Residential roads, lorry parks, factory yards, docks, petrol station forecourts, hardstandings, bus stations.

PAVING AND SURFACING

Weights and Sizes of Paving and Surfacing

Description of Item	Size	Quantity per tonne
Paving 50 mm thick	900 x 600 mm	15
Paving 50 mm thick	750 x 600 mm	18
Paving 50 mm thick	600 x 600 mm	23
Paving 50 mm thick	450 x 600 mm	30
Paving 38 mm thick	600 x 600 mm	30
Path edging	914 x 50 x 150 mm	60
Kerb (including radius and tapers)	125 x 254 x 914 mm	15
Kerb (including radius and tapers)	125 x 150 x 914 mm	25
Square channel	125 x 254 x 914 mm	15
Dished channel	125 x 254 x 914 mm	15
Quadrants	300 x 300 x 254 mm	19
Quadrants	450 x 450 x 254 mm	12
Quadrants	300 x 300 x 150 mm	30
Internal angles	300 x 300 x 254 mm	30
Fluted pavement channel	255 x 75 x 914 mm	25
Corner stones	300 x 300 mm	80
Corner stones	360 x 360 mm	60
Cable covers	914 x 175 mm	55
Gulley kerbs	220 x 220 x 150 mm	60
Gulley kerbs	220 x 200 x 75 mm	120

Weights and Sizes of Paving and Surfacing

Material	kg/m³	lb/cu yd
Tarmacadam	2306	3891
Macadam (waterbound)	2563	4325
Vermiculite (aggregate)	64-80	108-135
Terracotta	2114	3568
Cork - compressed	388	24
	kg/m²	**lb/sq ft**
Clay floor tiles, 12.7 mm	27.3	5.6
Pavement lights	122	25
Damp proof course	5	1
	kg/m² per mm thickness	**lb/sq ft per inch thickness**
Paving Slabs (stone)	2.3	12
Granite setts	2.88	15
Asphalt	2.30	12
Rubber flooring	1.68	9
Poly-vinylchloride	1.94 (avg)	10 (avg)

Coverage (m²) Per Cubic Metre of Materials Used as Sub-bases or Capping Layers

Consolidated thickness laid in (mm)	Square metre coverage		
	Gravel	Sand	Hardcore
50	15.80	16.50	-
75	10.50	11.00	-
100	7.92	8.20	7.42
125	6.34	6.60	5.90
150	5.28	5.50	4.95
175	-	-	4.23
200	-	-	3.71
225	-	-	3.30
300	-	-	2.47

Approximate Rate of Spreads

Average thickness of course	Description	Approximate rate of spread			
		Open Textured		Dense, Medium & Fine Textured	
mm		kg/m²	m²/t	kg/m²	m²/t
35	14 mm open textured or dense wearing course	60-75	13-17	70-85	12-14
40	20 mm open textured or dense base course	70-85	12-14	80-100	10-12
45	20 mm open textured or dense base course	80-100	10-12	95-100	9-10
50	20 mm open textured or dense, or 28 mm dense base course	85-110	9-12	110-120	8-9
60	28 mm dense base course, 40 mm open textured of dense base course or 40 mm single course as base course		8-10	130-150	7-8
65	28 mm dense base course, 40 mm open textured or dense base course or 40 mm single course	100-135	7-10	140-160	6-7
75	40 mm single course, 40 mm open textured or dense base course, 40 mm dense roadbase	120-150	7-8	165-185	5-6
100	40 mm dense base course or roadbase	-	-	220-240	4-4.5

Surface Dressing Roads: Coverage (m²) per Tonne of Material

Size in mm	Sand	Granite chips	Gravel	Limestone chips
Sand	168	-	-	-
3	-	148	152	165
6	-	130	133	144
9	-	111	114	123
13	-	85	87	95
19	-	68	71	78

Sizes of Flags to BS 7263

Reference	Nominal size (mm)	Thickness (mm)
A	600 x 450	50 and 63
B	600 x 600	50 and 63
C	600 x 750	50 and 63
D	600 x 900	50 and 63
E	450 x 450	50 and 70 chamfered top surface
F	400 x 400	50 and 65 chamfered top surface
G	300 x 300	50 and 60 chamfered top surface

Sizes of Natural Stone Setts to BS 435

Width (mm)		Length (mm)		Depth (mm)
100	x	100	x	100
75	x	150 to 250	x	125
75	x	150 to 250	x	150
100	x	150 to 250	x	100
100	x	150 to 250	x	150

SPORTS

Sizes of Sports Areas
Sizes in metres given include clearances:

Association football	Senior		114	x	72
	Junior		108	x	58
	International		100 - 110	x	64 - 75
Football	American	Pitch	109.80	x	48.80
		Overall	118.94	x	57.94
	Australian Rules	Overall	135 - 185	x	110 - 155
	Canadian	Overall	145.74	x	59.47
	Gaelic		128 - 146.4	x	76.8 - 91.50
Handball			91 - 110	x	55 - 65
Hurling			137	x	82
Rugby	Union pitch		56	x	81
	League pitch		134	x	80
Hockey pitch			100.5	x	61
Men's lacrosse pitch			106	x	61
Women's lacrosse pitch			110	x	60
Target archery ground			150	x	50
Archery (Clout)			7.3m	firing area	
			Range 109.728 (Women), 146.304 (Men).		
			182.88 (Normal range)		
400m Running Track			115.61	bend length	x 2
6 lanes			84.39	straight length	x 2
		Overall	176.91 long	x	92.52 wide
Baseball		Overall	60m	x	70
Basketball			14.0	x	26.0

Sizes of Sports Areas

Badminton			6.10	x	13.40
Camogie			91 - 110	x	54 - 68
Discus and Hammer		Safety cage 2.74m square Landing area 45 arc (65° safety) 70 m radius			
Javelin		Runway	36.5	x	4.27
		Landing area	80 - 95	x	48
Jump	High	Running area	38.8	x	19
		Landing area	5	x	4
	Long	Runway	45	x	1.22
		Landing area	9	x	2.750
	Triple	Runway	45	x	1.22
		Landing area	7.3	x	2.75
Korfball			90	x	40
Netball			15.25	x	30.48
Pole Vault		Runway	45	x	1.22
		Landing area	5	x	5
Polo			275	x	183
Rounders		Overall	19	x	17
Shot Putt		Base	2.135	dia	
		Landing area	65° arc,	25m radius	
Shinty			128 -183	x	64 - 91.5
Tennis		Court	23.77	x	10.97
		Overall minimum	36.27	x	18.29
Tug-of-war			46	x	5

SEEDING/TURFING AND PLANTING

BS 3882: 1994 Topsoil Quality

Topsoil grade	Properties
Premium	natural topsoil, high fertility, loamy texture, good soil structure, suitable for intensive cultivation
General Purpose	natural or manufactured topsoil of lesser quality than Premium, suitable for agriculture or amenity landscape, may need fertilizer or soil structure improvement.
Economy	selected subsoil, natural mineral deposit such as river silt or greensand. The grade comprises two sub-grades; "Low clay" and "High clay" which is more liable to compaction in handling. This grade is suitable for low production agricultural land and amenity woodland or conservation planting areas.

Forms of Trees to BS 3936: 1992

Standards:	shall be clear with substantially straight stems. Grafted and budded trees shall have no more than a slight bend at the union. Standards shall be designated as Half, Extra light, Light, Standard, Selected standard, Heavy, and Extra heavy.
Sizes of Standards	
Heavy standard	12-14 cm girth x 3.50 to 5.00 m high
Extra Heavy standard	14-16 cm girth x 4.25 to 5.00 m high
Extra Heavy standard	16-18 cm girth x 4.25 to 6.00 m high
Extra Heavy standard	18-20 cm girth x 5.00 to 6.00 m high
Semi-mature trees:	between 6.0 m and 12.0 m tall with a girth of 20 to 75 cm at 1.0 m above ground.
Feathered trees:	shall have a defined upright central leader, with stem furnished with evenly spread and balanced lateral shoots down to or near the ground.
Whips:	shall be without significant feather growth as determined by visual inspection.
Multi-stemmed trees:	shall have two or more main stems at, near, above or below ground.

Seedlings grown from seed and not transplanted shall be specified when ordered for sale as:

1+0 one year old seedling		
2+0 two year old seedling		
1+1 one year seed bed,	one year transplanted	= two year old seedling
1+2 one year seed bed,	two years transplanted	= three year old seedling
2+1 two year seed bed,	one year transplanted	= three year old seedling
1u1 two years seed bed,	undercut after 1 year	= two year old seedling
2u2 four years seed bed,	undercut after 2 years	= four year old seedling

Cuttings

The age of cuttings (plants grown from shoots, stems, or roots of the mother plant) shall be specified when ordered for sale. The height of transplants and undercut seedlings/cuttings (which have been transplanted or undercut at least once) shall be stated in centimetres. The number of growing seasons before and after transplanting or undercutting shall be stated.

0+1	one year cutting
0+2	two year cutting
0+1+1	one year cutting bed, one year transplanted = two year old seedling
0+1+2	one year cutting bed, two years transplanted = three year old seedling

Grass Cutting Capacities in m2 per Hour

Speed mph	Width Of Cut in metres												
	0.5	0.7	1.0	1.2	1.5	1.7	2.0	2.0	2.1	2.5	2.8	3.0	3.4
1.0	724	1127	1529	1931	2334	2736	3138	3219	3380	4023	4506	4828	5472
1.5	1086	1690	2293	2897	3500	4104	4707	4828	5069	6035	6759	7242	8208
2.0	1448	2253	3058	3862	4667	5472	6276	6437	6759	8047	9012	9656	10944
2.5	1811	2816	3822	4828	5834	6840	7846	8047	8449	10058	11265	12070	13679
3.0	2173	3380	4587	5794	7001	8208	9415	9656	10139	12070	13518	14484	16415
3.5	2535	3943	5351	6759	8167	9576	10984	11265	11829	14082	15772	16898	19151
4.0	2897	4506	6115	7725	9334	10944	12553	12875	13518	16093	18025	19312	21887
4.5	3259	5069	6880	8690	10501	12311	14122	14484	15208	18105	20278	21726	24623
5.0	3621	5633	7644	9656	11668	13679	15691	16093	16898	20117	22531	24140	27359
5.5	3983	6196	8409	10622	12834	15047	17260	17703	18588	22128	24784	26554	30095
6.0	4345	6759	9173	11587	14001	16415	18829	19312	20278	24140	27037	28968	32831
6.5	4707	7322	9938	12553	15168	17783	20398	20921	21967	26152	29290	31382	35566
7.0	5069	7886	10702	13518	16335	19151	21967	22531	23657	28163	31543	33796	38302

Number of Plants per m2: For Plants Planted on an Evenly Spaced Grid

Planting distances

mm	0.10	0.15	0.20	0.25	0.35	0.40	0.45	0.50	0.60	0.75	0.90	1.00	1.20	1.50
0.10	100.00	66.67	50.00	40.00	28.57	25.00	22.22	20.00	16.67	13.33	11.11	10.00	8.33	6.67
0.15	66.67	44.44	33.33	26.67	19.05	16.67	14.81	13.33	11.11	8.89	7.41	6.67	5.56	4.44
0.20	50.00	33.33	25.00	20.00	14.29	12.50	11.11	10.00	8.33	6.67	5.56	5.00	4.17	3.33
0.25	40.00	26.67	20.00	16.00	11.43	10.00	8.89	8.00	6.67	5.33	4.44	4.00	3.33	2.67
0.35	28.57	19.05	14.29	11.43	8.16	7.14	6.35	5.71	4.76	3.81	3.17	2.86	2.38	1.90
0.40	25.00	16.67	12.50	10.00	7.14	6.25	5.56	5.00	4.17	3.33	2.78	2.50	2.08	1.67
0.45	22.22	14.81	11.11	8.89	6.35	5.56	4.94	4.44	3.70	2.96	2.47	2.22	1.85	1.48
0.50	20.00	13.33	10.00	8.00	5.71	5.00	4.44	4.00	3.33	2.67	2.22	2.00	1.67	1.33
0.60	16.67	11.11	8.33	6.67	4.76	4.17	3.70	3.33	2.78	2.22	1.85	1.67	1.39	1.11
0.75	13.33	8.89	6.67	5.33	3.81	3.33	2.96	2.67	2.22	1.78	1.48	1.33	1.11	0.89
0.90	11.11	7.41	5.56	4.44	3.17	2.78	2.47	2.22	1.85	1.48	1.23	1.11	0.93	0.74
1.00	10.00	6.67	5.00	4.00	2.86	2.50	2.22	2.00	1.67	1.33	1.11	1.00	0.83	0.67
1.20	8.33	5.56	4.17	3.33	2.38	2.08	1.85	1.67	1.39	1.11	0.93	0.83	0.69	0.56
1.50	6.67	4.44	3.33	2.67	1.90	1.67	1.48	1.33	1.11	0.89	0.74	0.67	0.56	0.44

Grass Clippings Wet: Based on 3.5 m3 /tonne

Annual kg/100 m2	Average 20 cuts kg/100m2	m2/tonne	m2/m3
32.0	1.6	61162.1	214067.3

Nr of cuts		**22**	**20**	**18**	**16**	**12**	**4**
Kg/cut		1.45	1.60	1.78	2.00	2.67	8.00
Area capacity of 3 tonne vehicle per load							
m2		206250	187500	168750	150000	112500	37500
Load m3		**100 m2 units/m3 of vehicle space**					
	1	196.4	178.6	160.7	142.9	107.1	35.7
	2	392.9	357.1	321.4	285.7	214.3	71.4
	3	589.3	535.7	482.1	428.6	321.4	107.1
	4	785.7	714.3	642.9	571.4	428.6	142.9
	5	982.1	892.9	803.6	714.3	535.7	178.6

Transportation of Trees

To unload large trees a machine with the necessary lifting strength is required. The weight of the trees must therefore be known in advance. The following table gives a rough overview. The additional columns with root ball dimensions and the number of plants per trailer provide additional information for example about preparing planting holes and calculating unloading times.

Girth in cm	Root ball Diameter in cm	Ball height in cm	Weight in kg	Numbers of trees per trailer
16 - 18	50 - 60	40	150	100 - 120
18 - 20	60 - 70	40 - 50	200	80 - 100
20 - 25	60 - 70	40 - 50	270	50 - 70
25 - 30	80	50 - 60	350	50
30 - 35	90 - 100	60 - 70	500	12 - 18
35 - 40	100 - 110	60 - 70	650	10 - 15
40 - 45	110 - 120	60 - 70	850	8 - 12
45 - 50	110 - 120	60 - 70	1100	5 - 7
50 - 60	130 - 140	60 - 70	1600	1 - 3
60 - 70	150 - 160	60 - 70	2500	1
70 - 80	180 - 200	70	4000	1
80 - 90	200 - 220	70 - 80	5500	1
90 - 100	230 - 250	80 - 90	7500	1
100 - 120	250 - 270	80 - 90	9500	1

Data supplied by Lorenz von Ehren GmbH

The information in the table is approximate; deviations depend on soil type, genus and weather.

FENCING AND GATES

Types of Preservative to BS 5589:1989

Creosote (tar oil) can be "factory" applied	
	by pressure to BS 144: pts 1&2
	by immersion to BS 144: pt 1
	by hot and cold open tank to BS 144: pts 1&2
Copper/chromium/arsenic (CCA)	by full cell process to BS 4072 pts 1&2
Organic solvent (OS)	by double vacuum (vacvac) to BS 5707 pts 1&3
	by immersion to BS 5057 pts 1&3
Pentachlorophenol (PCP)	by heavy oil double vacuum to BS 5705 pts 2&3
Boron diffusion process (treated with disodium octaborate to BWPA Manual 1986.	

Note: Boron is used on green timber at source and the timber is supplied dry.

Cleft Chestnut Pale Fences to BS 1722: Part 4:1986

Pales	Pale spacing	Wire lines	
900 mm long	75 mm	2	temporary protection
1050	75 or 100	2	light protective fences
1200	75	3	perimeter fences
1350	75	3	perimeter fences
1500	50	3	narrow perimeter fences
1800	50	3	light security fences

Close-boarded Fences to BS 1722 :Pt 5: 1986

Close-boarded fences 1.05 to 1.8m high
Type BCR (recessed) or BCM (morticed) with concrete posts 140 x 115 mm tapered and Type BW with timber posts.

Palisade Fences to BS 1722: pt 6: 1986

Wooden palisade fences
Type WPC with concrete posts 140 x 115 mm tapered and Type WPW with timber posts.

For both types of fence:

Height of fence 1050 mm:	two rails
Height of fence 1200 mm:	two rails
Height of fence 1500 mm:	three rails
Height of fence 1650 mm:	three rails
Height of fence 1800 mm:	three rails

Post and Rail Fences to BS 1722: part 7

Wooden post and rail fences
Type MPR 11/3 morticed rails and Type SPR 11/3 nailed rails
Height to top of rail 1100 mm
Rails: three rails 87 mm 38 mm

Type MPR 11/4 morticed rails and Type SPR 11/4 nailed rails
Height to top of rail 1100 mm
Rails: four rails 87 mm 38 mm.

Type MPR 13/4 morticed rails and Type SPR 13/4 nailed rails
Height to top of rail 1300 mm
Rail spacing 250 mm, 250 mm, and 225 mm from top
Rails: four rails 87 mm 38 mm.

Steel Posts to BS 1722: Part 1

Rolled steel angle iron posts for chain link fencing:

Posts	Fence height	Strut	Straining post
1500 x 40 x 40 x 5 mm	900 mm	1500 x 40 x 40 x 5 mm	1500 x 50 x 50 x 6 mm
1800 x 40 x 40 x 5 mm	1200 mm	1800 x 40 x 40 x 5 mm	1800 x 50 x 50 x 6 mm
2000 x 45 x 45 x 5 mm	1400 mm	2000 x 45 x 45 x 5 mm	2000 x 60 x 60 x 6 mm
2600 x 45 x 45 x 5 mm	1800 mm	2600 x 45 x 45 x 5 mm	2600 x 60 x 60 x 6 mm
3000 x 50 x 50 x 6 mm with arms	1800 mm	2600 x 45 x 45 x 5 mm	3000 x 60 x 60 x 6 mm

Concrete Posts to BS 1722: Part 1

Concrete posts for chain link fencing:

Posts and straining posts	Fence height	Strut
1570 mm 100 x 100 mm	900 mm	1500 mm x 75 x 75 mm
1870 mm 125 x 125 mm	1200 mm	1830 mm x 100 x 75 mm
2070 mm 125 x 125 mm	1400 mm	1980 mm x 100 x 75 mm
2620 mm 125 x 125 mm	1800 mm	2590 mm x 100 x 85 mm
3040 mm 125 x 125 mm	1800 mm	2590 mm x 100 x 85 mm (with arms)

Rolled Steel Angle Posts to BS 1722: Part 2

Rolled steel angle posts for rectangular wire mesh (field) fencing

Posts	Fence height	Strut	Straining post
1200 x 40 x 40 x 5 mm	600 mm	1200 x 75 x 75 mm	1350 x 100 x 100 mm
1400 x 40 x 40 x 5 mm	800 mm	1400 x 75 x 75 mm	1550 x 100 x 100 mm
1500 x 40 x 40 x 5 mm	900 mm	1500 x 75 x 75 mm	1650 x 100 x 100 mm
1600 x 40 x 40 x 5 mm	1000 mm	1600 x 75 x 75 mm	1750 x 100 x 100 mm
1750 x 40 x 40 x 5 mm	1150 mm	1750 x 75 x 100 mm	1900 x 125 x 125 mm

Concrete Posts to BS 1722: Part 2

Concrete posts for rectangular wire mesh (field) fencing

Posts	Fence height	Strut	Straining post
1270 x 100 x 100 mm	600 mm	1200 x 75 x 75 mm	1420 x 100 x 100 mm
1470 x 100 x 100 mm	800 mm	1350 x 75 x 75 mm	1620 x 100 x 100 mm
1570 x 100 x 100 mm	900 mm	1500 x 75 x 75 mm	1720 x 100 x 100 mm
1670 x 100 x 100 mm	600 mm	1650 x 75 x 75 mm	1820 x 100 x 100 mm
1820 x 125 x 125 mm	1150 mm	1830 x 75 x 100 mm	1970 x 125 x 125 mm

Cleft Chestnut Pale Fences to BS 1722: part 4: 1986
Timber Posts to BS 1722: Part 2

Timber posts for wire mesh and hexagonal wire netting fences
Round timber for general fences

Posts	Fence height	Strut	Straining post
1300 x 65 mm dia.	600 mm	1200 x 80 mm dia	1450 x 100 mm dia
1500 x 65 mm dia	800 mm	1400 x 80 mm dia	1650 x 100 mm dia
1600 x 65 mm dia.	900 mm	1500 x 80 mm dia	1750 x 100 mm dia
1700 x 65 mm dia.	1050 mm	1600 x 80 mm dia	1850 x 100 mm dia
1800 x 65 mm dia.	1150 mm	1750 x 80 mm dia	2000 x 120 mm dia

Squared timber for general fences

Posts	Fence height	Strut	Straining post
1300 x 75 x 75 mm	600 mm	1200 x 75 x 75 mm	1450 x 100 x 100 mm
1500 x 75 x 75 mm	800 mm	1400 x 75 x 75 mm	1650 x 100 x 100 mm
1600 x 75 x 75 mm	900 mm	1500 x 75 x 75 mm	1750 x 100 x 100 mm
1700 x 75 x 75 mm	1050 mm	1600 x 75 x 75 mm	1850 x 100 x 100 mm
1800 x 75 x 75 mm	1150 mm	1750 x 75 x 75 mm	2000 x 125 x 100 mm

Steel Fences to BS 1722: Pt 9: 1992

	Fence height	Top/bottom rails and flat posts	Vertical bars
Light	1000 mm	40 x 10 mm 450 mm in ground	12 mm dia at 115 mm cs
	1200 mm	40 x 10 mm 550 mm in ground	12 mm dia at 115 mm cs
	1400 mm	40 x 10 mm 550 mm in ground	12 mm dia at 115 mm cs
Light	1000 mm	40 x 10 mm 450 mm in ground	16 mm dia at 120 mm cs
	1200 mm	40 x 10 mm 550 mm in ground	16 mm dia at 120 mm cs
	1400 mm	40 x 10 mm 550 mm in ground	16 mm dia at 120 mm cs
Medium	1200 mm	50 x 10 mm 550 mm in ground	20 mm dia at 125 mm cs
	1400 mm	50 x 10 mm 550 mm in ground	20 mm dia at 125 mm cs
	1600 mm	50 x 10 mm 600 mm in ground	22 mm dia at 145 mm cs
		50 x 10 mm 600 mm in ground	
	1800 mm	50 x 10 mm 600 mm in ground	22 mm dia at 145 mm cs
Heavy	1600 mm	50 x 10 mm 600 mm in ground	22 mm dia at 145 mm cs
		50 x 10 mm 600 mm in ground	
	1800 mm	50 x 10 mm 600 mm in ground	22 mm dia at 145 mm cs
	2000 mm	50 x 10 mm 600 mm in ground	22 mm dia at 145 mm cs
		50 x 10 mm 600 mm in ground	
	2200 mm	50 x 10 mm 600 mm in ground	22 mm dia at 145 mm cs

Notes:
Mild steel fences: round or square verticals; flat standards and horizontals.
Tops of vertical bars may be bow-top, blunt, or pointed.
Round or square bar railings.

Timber Field Gates to BS 3470: 1975

Gates made to this standard are designed to open one way only.
All timber gates are 1100 mm high.
Width over stiles 2400, 2700, 3000, 3300, 3600, and 4200 mm.
Gates over 4200 mm should be made in two leaves.

Steel Field Gates to BS 3470: 1975

All steel gates are 1100 mm high.

Heavy duty:	width over stiles 2400, 3000, 3600 and 4500 mm
Light duty:	width over stiles 2400, 3000, and 3600 mm

Domestic Front Entrance Gates to BS 4092: part 1: 1966

Metal gates:	Single gates are 900 mm high minimum, 900 mm, 1000 mm and 1100 mm wide

Domestic Front Entrance Gates to BS 4092: part 2: 1966

Wooden gates:	All rails shall be tenoned into the stiles Single gates are 840 mm high minimum, 801 mm and 1020 mm wide Double gates are 840 mm high minimum, 2130, 2340 and 2640 mm wide

Timber Bridle Gates to BS 5709:1979 (Horse Or Hunting Gates)

Gates open one way only	
Minimum width between posts	1525 mm
Minimum height	1100 mm

Timber Kissing Gates to BS 5709:1979

Minimum width	700 mm
Minimum height	1000 mm
Minimum distance between shutting posts	600 mm
Minimum clearance at mid-point	600 mm

Metal Kissing Gates to BS 5709:1979

Sizes are the same as those for timber kissing gates.
Maximum gaps between rails 120 mm.

Categories of Pedestrian Guard Rail to BS 3049:1976

Class A for normal use.
Class B where vandalism is expected.
Class C where crowd pressure is likely.

DRAINAGE

Weights and Dimensions - Vitrified Clay Pipes

Product	Nominal diameter (mm)	Effective length (mm)	BS 65 limits of tolerance		Crushing Strength (kN/m)	Weight	
			min (mm)	max (mm)		kg/pipe	kg/m
Supersleve	100	1600	96	105	35.00	14.71	9.19
	150	1750	146	158	35.00	29.24	16.71
Hepsleve	225	1850	221	236	28.00	84.03	45.42
	300	2500	295	313	34.00	193.05	77.22
	150	1500	146	158	22.00	37.04	24.69
Hepseal	225	1750	221	236	28.00	85.47	48.84
	300	2500	295	313	34.00	204.08	81.63
	400	2500	394	414	44.00	357.14	142.86
	450	2500	444	464	44.00	454.55	181.63
	500	2500	494	514	48.00	555.56	222.22
	600	2500	591	615	57.00	796.23	307.69
	700	3000	689	719	67.00	1111.11	370.45
	800	3000	788	822	72.00	1351.35	450.45
Hepline	100	1600	95	107	22.00	14.71	9.19
	150	1750	145	160	22.00	29.24	16.71
	225	1850	219	239	28.00	84.03	45.42
	300	1850	292	317	34.00	142.86	77.22
Hepduct	90	1500	-	-	28.00	12.05	8.03
(Conduit)	100	1600	-	-	28.00	14.71	9.19
	125	1750	-	-	28.00	20.73	11.84
	150	1750	-	-	28.00	29.24	16.71
	225	1850	-	-	28.00	84.03	45.42
	300	1850	-	-	34.00	142.86	77.22

Weights and Dimensions - Vitrified Clay Pipes

Nominal internal diameter (mm)	Nominal wall thickness (mm)	Approximate weight kg/m
150	25	45
225	29	71
300	32	122
375	35	162
450	38	191
600	48	317
750	54	454
900	60	616
1200	76	912
1500	89	1458
1800	102	1884
2100	127	2619

Wall thickness, weights and pipe lengths vary, depending on type of pipe required.

The particulars shown above represent a selection of available diameters and are applicable to strength class 1 pipes with flexible rubber ring joints.

Tubes with Ogee joints are also available.

Weights and Dimensions - PVC-U Pipes

	Nominal size	Mean outside diameter (mm)		Wall thickness (mm)	Weight (kg/m)
		min	max		
Standard pipes					
	82.4	82.4	82.7	3.2	1.2
	110.0	110.0	110.4	3.2	1.6
	160.0	160.0	160.6	4.1	3.0
	200.0	200.0	200.6	4.9	4.6
	250.0	250.0	250.7	6.1	7.2
Perforated pipes					
- heavy grade	As above	As above	As above	As above	As above
- thin wall	82.4	82.4	82.7	1.7	-
	110.0	110.0	110.4	2.2	-
	160.0	160.0	160.6	3.2	-

Width of Trenches Required for Various Diameters of Pipes

Pipe diameter (mm)	Trench n.e. 1.5 m deep (mm)	Trench over 1.5 m deep (mm)
n.e. 100	450	600
100-150	500	650
150-225	600	750
225-300	650	800
300-400	750	900
400-450	900	1050
450-600	1100	1300

DRAINAGE BELOW GROUND AND LAND DRAINAGE

Flow of Water Which Can Be Carried by Various Sizes of Pipe

Clay or concrete pipes

	Gradient of pipeline							
	1:10	1:20	1:30	1:40	1:50	1:60	1:80	1:100
Pipe size	**Flow in litres per second**							
DN 100 15.0	8.5	6.8	5.8	5.2	4.7	4.0	3.5	
DN 150 28.0	19.0	16.0	14.0	12.0	11.0	9.1	8.0	
DN 225 140.0	95.0	76.0	66.0	58.0	53.0	46.0	40.0	

Plastic pipes

	Gradient of pipeline							
	1:10	1:20	1:30	1:40	1:50	1:60	1:80	1:100
Pipe size	**Flow in litres per second**							
82.4 mm i/dia	12.0	8.5	6.8	5.8	5.2	4.7	4.0	3.5
110 mm i/dia	28.0	19.0	16.0	14.0	12.0	11.0	9.1	8.0
160 mm i/dia	76.0	53.0	43.0	37.0	33.0	29.0	25.0	22.0
200 mm i/dia	140.0	95.0	76.0	66.0	58.0	53.0	46.0	40.0

Vitrified (Perforated) Clay Pipes and Fittings to BS En 295-5 1994

Length not specified		
75 mm bore	**250 mm bore**	**600 mm bore**
100	300	700
125	350	800
150	400	1000
200	450	1200
225	500	

Pre-cast Concrete Pipes: Pre-stressed Non-pressure Pipes and Fittings: Flexible Joints to BS 5911: Pt. 103: 1994

Rationalized metric nominal sizes: 450, 500			
Length:	500	-	1000 by 100 increments
	1000	-	2200 by 200 increments
	2200	-	2800 by 300 increments
Angles: length:	450 - 600 angles 45, 22.5,11.25 °		
	600 or more angles 22.5, 11.25 °		

Precast Concrete Pipes: Un-reinforced and Circular Manholes and Soakaways to BS 5911: Pt. 200: 1994

Nominal sizes:	
Shafts:	675, 900 mm
Chambers:	900, 1050, 1200, 1350, 1500, 1800, 2100, 2400, 2700, 3000 mm.
Large chambers:	To have either tapered reducing rings or a flat reducing slab in order to accept the standard cover.
Ring depths:	1. 300 - 1200 mm by 300 mm increments except for bottom slab and rings below cover slab, these are by 150 mm increments.
	2. 250 - 1000 mm by 250 mm increments except for bottom slab and rings below cover slab, these are by 125 mm increments.
Access hole:	750 x 750 mm for DN 1050 chamber 1200 x 675 mm for DN 1350 chamber

Calculation of Soakaway Depth

The following formula determines the depth of concrete ring soakaway that would be required for draining given amounts of water.

$$h = \frac{4ar}{3\pi D^2}$$

h	=	depth of the chamber below the invert pipe
A	=	The are to be drained
r	=	The hourly rate of rainfall (50 mm per hour)
π	=	pi
D	=	internal diameter of the soakaway

This table shows the depth of chambers in each ring size which would be required to contain the volume of water specified. These allow a recommended storage capacity of 1/3 (one third of the hourly rain fall figure).

Table Showing Required Depth of Concrete Ring Chambers in Metres

Area m²	50	100	150	200	300	400	500
Ring size							
0.9	1.31	2.62	3.93	5.24	7.86	10.48	13.10
1.1	0.96	1.92	2.89	3.85	5.77	7.70	9.62
1.2	0.74	1.47	2.21	2.95	4.42	5.89	7.37
1.4	0.58	1.16	1.75	2.33	3.49	4.66	5.82
1.5	0.47	0.94	1.41	1.89	2.83	3.77	4.72
1.8	0.33	0.65	0.98	1.31	1.96	2.62	3.27
2.1	0.24	0.48	0.72	0.96	1.44	1.92	2.41
2.4	0.18	0.37	0.55	0.74	1.11	1.47	1.84
2.7	0.15	0.29	0.44	0.58	0.87	1.16	1.46
3.0	0.12	0.24	0.35	0.47	0.71	0.94	1.18

Precast Concrete Inspection Chambers and Gullies to BS 5911: Pt. 230: 1994

Nominal sizes:	375 diameter, 750, 900 mm deep 450 diameter, 750, 900, 1050, 1200 mm deep
Depths:	from the top for trapped or un-trapped units: centre of outlet 300 mm invert (bottom) of the outlet pipe 400 mm
Depth of water seal for trapped gullies:	85 mm, rodding eye int. diam. 100 mm
Cover slab:	65 mm min.

Ductile Iron Pipes to BS En 598: 1995

Type K9 with flexible joints should be used for surface water drainage.
5500 mm or 8000 mm long

80 mm bore	400 mm bore	1000 mm bore
100	450	1100
150	500	1200
200	600	1400
250	700	1600
300	800	
350	900	

Bedding Flexible Pipes: PVC-U Or Ductile Iron

Type 1	=	100 mm fill below pipe, 300 mm above pipe: single size material
Type 2	=	100 mm fill below pipe, 300 mm above pipe: single size or graded material
Type 3	=	100 mm fill below pipe, 75 mm above pipe with concrete protective slab over
Type 4	=	100 mm fill below pipe, fill laid level with top of pipe
Type 5	=	200 mm fill below pipe, fill laid level with top of pipe
Concrete	=	25 mm sand blinding to bottom of trench, pipe supported on chocks, 100 mm concrete under the pipe, 150 mm concrete over the pipe.

Bedding Rigid Pipes: Clay or Concrete
(for vitrified clay pipes the manufacturer should be consulted)

Class D:	Pipe laid on natural ground with cut-outs for joints, soil screened to remove stones over 40 mm and returned over pipe to 150 mm min depth. Suitable for firm ground with trenches trimmed by hand.
Class N:	Pipe laid on 50 mm granular material of graded aggregate to Table 4 of BS 882, or 10 mm aggregate to Table 6 of BS 882, or as dug light soil (not clay) screened to remove stones over 10 mm. Suitable for machine dug trenches.
Class B:	As Class N, but with granular bedding extending half way up the pipe diameter.
Class F:	Pipe laid on 100 mm granular fill to BS 882 below pipe, minimum 150 mm granular fill above pipe: single size material. Suitable for machine dug trenches.
Class A:	Concrete 100 mm thick under the pipe extending half way up the pipe, backfilled with the appropriate class of fill. Used where there is only a very shallow fall to the drain. Class A bedding allows the pipes to be laid to an exact gradient.
Concrete surround:	25 mm sand blinding to bottom of trench, pipe supported on chocks, 100 mm concrete under the pipe, 150 mm concrete over the pipe. It is preferable to bed pipes under slabs or wall in granular material.

PIPED SUPPLY SYSTEMS

Identification of Service Tubes From Utility to Dwellings

Utility	Colour	Size	Depth
British Telecom	grey	54 mm od	450 mm
Electricity	black	38 mm od	450 mm
Gas	yellow	42 mm od rigid 60 mm od convoluted	450 mm
Water	may be blue	(normally untubed)	750 mm

ELECTRICAL SUPPLY/POWER/LIGHTING SYSTEMS

Electrical Insulation Class En 60.598 BS 4533

Class 1:	luminaires comply with class 1 (I) earthed electrical requirements
Class 2:	luminaires comply with class 2 (II) double insulated electrical requirements
Class 3:	luminaires comply with class 3 (III) electrical requirements

Protection to Light Fittings

BS EN 60529:1992 Classification for degrees of protection provided by enclosures.
(IP Code - International or ingress Protection)

1st characteristic: against ingress of solid foreign objects		
The figure	2	indicates that fingers cannot enter
	3	that a 2.5 mm diameter probe cannot enter
	4	that a 1.0 mm diameter probe cannot enter
	5	the fitting is dust proof (no dust around live parts)
	6	the fitting is dust tight (no dust entry)
2nd characteristic: ingress of water with harmful effects		
The figure	0	indicates unprotected
	1	vertically dripping water cannot enter
	2	water dripping 15° (tilt) cannot enter
	3	spraying water cannot enter
	4	splashing water cannot enter
	5	jetting water cannot enter
	6	powerful jetting water cannot enter
	7	proof against temporary immersion
	8	proof against continuous immersion
Optional additional codes:		A-D protects against access to hazardous parts
	H	High voltage apparatus
	M	fitting was in motion during water test
	S	fitting was static during water test
	W	protects against weather
Marking code arrangement:		(example) IPX5S = IP (International or Ingress Protection) X (denotes omission of first characteristic); 5 = jetting; S = static during water test.

ESSENTIAL READING FROM TAYLOR AND FRANCIS

4th Edition

Rock Slope Engineering

Civil and Mining

Duncan C. Wyllie and
Christopher W. Mah

The stability of rock slopes is an important issue in both civil and mining engineering. On civil projects, rock cuts must be safe from rock falls and large-scale slope instability during both construction and operation. In open pit mining, where slope heights can be many hundreds of meters, the economics of the operation are closely related to the steepest stable slope angle that can be mined.

This extensively updated version of the classic text, *Rock Slope Engineering* by Hoek and Bray, deals comprehensively with the investigation, design and operation of rock slopes. New material contained in this book includes the latest developments in earthquake engineering related to slope stability, probabilistic analysis, numerical analysis, blasting, slope movement monitoring and stabilization methods.

Rock Slopes Engineering: Civil and Mining contains both worked examples illustrating data interpretation and design methods, and chapters on civil and mining case studies. It also includes an introduction by Dr. Evert Hoek.

August 2004: 246x174 mm: 480 pages
242 line drawings, 32 b+w photos and 42 tables
HB: 0-415-28000-1: £112.00
PB: 0-415-28001-X: £40.00

To Order: Tel: +44 (0) 1264 343071 Fax: +44 (0) 1264 343005, or
Post: Taylor and Francis Customer Services, Thomson Publishing Services, Cheriton House, Andover, Hants, SP10 5BE, UK Email: book.orders@tandf.co.uk

For a complete listing of all our titles visit:
www.sponpress.com

Taylor & Francis
Taylor & Francis Group

SPON'S PRICEBOOKS 2008

Spon's Architects' and Builders' Price Book 2008

Davis Langdon

The most detailed, professionally relevant source of construction price information currently available anywhere.

The 2008 edition provides an extensive overhaul of prices in line with buoyant market conditions. It includes an expanded Approximate Estimates section and a new section on Sustainability. It also includes the following new items: Additional high quality balustrades, PVC wall linings, Rainscreen cladding, Resin-bonded gravel, SAS suspended ceilings, Self-levelling screeds, 'Speedfit' pipes and fittings, Tapered insulation boards, Waterproofing membranes, Window refurbishment

HB & electronic package * £135
978 0-415-42448-6 1000pp approx.

Spon's External Works and Landscape Price Book 2008

Davis Langdon

The only comprehensive source of information for detailed external works and landscape costs.

The 2008 edition gives an expanded Approximate Estimates section, and new items including Professional and amenity sports pitch construction, Costs of tender production, Small unit concrete kerbs, Ornamental trellises, Borehole drilling, irrigation sensors, and an irrigation water recycling system. And more on excavation and filling, landscape chemicals prices and applications, and linear drainage.

HB & electronic package * £105
978 0-415-42444-8 424pp approx.

Spon's Mechanical and Electrical Services Price Book 2008

Davis Langdon, Mott Green Wall

The only comprehensive and best annual services engineering price book available for the UK.

As well as an extensive overhaul of prices, updates to the 2008 edition include an expanded Approximate Estimate section, updated coverage of sustainability and of communication and data, along with additions to both the ductwork and lifts sections. High efficiency plant has been refreshed and the CHP and Biomass plant has been reviewed.

HB & electronic package * £135
978 0-415-42442-4 640pp approx.

Spon's Civil Engineering and Highway Works Price Book 2008

Davis Langdon

The 21st edition, in its easy-to-read format, incorporates a general review throughout, updated in line with the latest CIJC wage agreement, and includes a restructuring of the highway works unit costs section to align with the latest revisions to the *Manual of Contract Documents for Highway Works.*

HB & electronic package * £145
978 0-415-42446-2 800pp approx.

Receive a free electronic version when you order any Spon 2008 Price Book with estimating software to help you produce tender documents, customise data, perform word searches and simple calculations.
Visit www.pricebooks.co.uk

To Order: Tel: +44 (0) 8700 768853, or +44 (0) 1264 343071 Fax: +44 (0) 1264 343005, or
Post: Taylor and Francis Customer Services, Thomson Publishing Services, Cheriton House, Andover, Hants, SP10 5BE, UK Email: book.orders@tandf.co.uk

For a complete listing of all our titles visit
www.tandf.co.uk

Taylor & Francis
Taylor & Francis Group

ESSENTIAL READING FROM TAYLOR AND FRANCIS

Spon's Estimating Costs Guide to Small Groundworks, Landscaping and Gardening

Second Edition

Especially written for contractors and small businesses carrying out small works, Spon's Estimating Costs Guide to Small Groundworks, Landscaping and Gardening contains accurate information on thousands of rates, each broken down to labour, material overheads and profit.

This is the first book to focus on garden maintenance work on blocks of flats and individual houses, schools and sports fields, garden makeovers, laying patios and paths, small land drainage schemes on farmland, and small-scale local authority maintenance work.

More than just a price book, it gives easy-to-read, professional advice on setting up and running a business including help on producing estimates faster, and keeping estimates accurate and competitive.

Suitable for any size firm from one-man-band to established business, this book contains valuable commercial and cost information that contractors can't afford to be without.

December 2007: 216x138 mm: 304 pages
PB: 978-0-415-43442-3: £27.99

To Order: Tel: +44 (0) 1264 343071 Fax: +44 (0) 1264 343005, or
Post: Taylor and Francis Customer Services, Thomson Publishing Services, Cheriton House, Andover, Hants, SP10 5BE, UK Email: book.orders@tandf.co.uk

For a complete listing of all our titles visit:
www.sponpress.com

Taylor & Francis
Taylor & Francis Group

Index

Software and eBook Single-User Licence Agreement

We welcome you as a user of this Taylor & Francis Software and eBook and hope that you find it a useful and valuable tool. Please read this document carefully. **This is a legal agreement** between you (hereinafter referred to as the "Licensee") and Taylor and Francis Books Ltd. (the "Publisher"), which defines the terms under which you may use the Product. **By breaking the seal and opening the document inside the back cover of the book containing the access code you agree to these terms and conditions outlined herein. If you do not agree to these terms you must return the Product to your supplier intact, with the seal on the document unbroken.**

1. Definition of the Product

The product which is the subject of this Agreement, *Spon's External Works and Landscape Price Book 2008* Software and eBook (the "Product") consists of:

1.1 Underlying data comprised in the product (the "Data")
1.2 A compilation of the Data (the "Database")
1.3 Software (the "Software") for accessing and using the Database
1.4 An electronic book containing the data in the price book (the "eBook")

2. Commencement and licence

2.1 This Agreement commences upon the breaking open of the document containing the access code by the Licensee (the "Commencement Date").

2.2 This is a licence agreement (the "Agreement") for the use of the Product by the Licensee, and not an agreement for sale.

2.3 The Publisher licenses the Licensee on a non-exclusive and non-transferable basis to use the Product on condition that the Licensee complies with this Agreement. The Licensee acknowledges that it is only permitted to use the Product in accordance with this Agreement.

3. Multiple use

For more than one user or for a wide area network or consortium, use is only permissible with the purchase from the Publisher of a multiple-user licence and adherence to the terms and conditions of that licence.

4. Installation and Use

4.1 The Licensee may provide access to the Product for individual study in the following manner: The Licensee may install the Product on a secure local area network on a single site for use by one user.

4.2 The Licensee shall be responsible for installing the Product and for the effectiveness of such installation.

4.3 Text from the Product may be incorporated in a coursepack. Such use is only permissible with the express permission of the Publisher in writing and requires the payment of the appropriate fee as specified by the Publisher and signature of a separate licence agreement.

4.4 The Product is a free addition to the book and no technical support will be provided.

5. Permitted Activities

5.1 The Licensee shall be entitled:

5.1.1 to use the Product for its own internal purposes;

5.1.2 to download onto electronic, magnetic, optical or similar storage medium reasonable portions of the Database provided that the purpose of the Licensee is to undertake internal research or study and provided that such storage is temporary;

5.2 The Licensee acknowledges that its rights to use the Product are strictly set out in this Agreement, and all other uses (whether expressly mentioned in Clause 6 below or not) are prohibited.

6. Prohibited Activities

The following are prohibited without the express permission of the Publisher:

6.1 The commercial exploitation of any part of the Product.

6.2 The rental, loan, (free or for money or money's worth) or hire purchase of this product, save with the express consent of the Publisher.

6.3 Any activity which raises the reasonable prospect of impeding the Publisher's ability or opportunities to market the Product.

6.4 Any networking, physical or electronic distribution or dissemination of the product save as expressly permitted by this Agreement.

6.5 Any reverse engineering, decompilation, disassembly or other alteration of the Product save in accordance with applicable national laws.

6.6 The right to create any derivative product or service from the Product save as expressly provided for in this Agreement.

6.7 Any alteration, amendment, modification or deletion from the Product, whether for the purposes of error correction or otherwise.

7. General Responsibilities of the License

7.1 The Licensee will take all reasonable steps to ensure that the Product is used in accordance with the terms and conditions of this Agreement.

7.2 The Licensee acknowledges that damages may not be a sufficient remedy for the Publisher in the event of breach of this Agreement by the Licensee, and that an injunction may be appropriate.

7.3 The Licensee undertakes to keep the Product safe and to use its best endeavours to ensure that the product does not fall into the hands of third parties, whether as a result of theft or otherwise.

7.4 Where information of a confidential nature relating to the product of the business affairs of the Publisher comes into the possession of the Licensee pursuant to this Agreement (or otherwise), the Licensee agrees to use such information solely for the purposes of this Agreement, and under no circumstances to disclose any element of the information to any third party save strictly as permitted under this Agreement. For the avoidance of doubt, the Licensee's obligations under this sub-clause 7.4 shall survive the termination of this Agreement.

8. Warrant and Liability

8.1 The Publisher warrants that it has the authority to enter into this agreement and that it has secured all rights and permissions necessary to enable the Licensee to use the Product in accordance with this Agreement.

8.2 The Publisher warrants that the Product as supplied on the Commencement Date shall be free of defects in materials and workmanship, and undertakes to replace any defective Product within 28 days of notice of such defect being received provided such notice is received within 30 days of such supply. As an alternative to replacement, the Publisher agrees fully to refund the Licensee in such circumstances, if the Licensee so requests, provided that the Licensee returns this copy of *Spon's External Works and Landscape Price Book 2008* to the Publisher. The provisions of this sub-clause 8.2 do not apply where the defect results from an accident or from misuse of the product by the Licensee.

8.3 Sub-clause 8.2 sets out the sole and exclusive remedy of the Licensee in relation to defects in the Product.

8.4 The Publisher and the Licensee acknowledge that the Publisher supplies the Product on an "as is" basis. The Publisher gives no warranties:

- 8.4.1 that the Product satisfies the individual requirements of the Licensee; or
- 8.4.2 that the Product is otherwise fit for the Licensee's purpose; or
- 8.4.3 that the Data are accurate or complete or free of errors or omissions; or
- 8.4.4 that the Product is compatible with the Licensee's hardware equipment and software operating environment.

8.5 The Publisher hereby disclaims all warranties and conditions, express or implied, which are not stated above.

8.6 Nothing in this Clause 8 limits the Publisher's liability to the Licensee in the event of death or personal injury resulting from the Publisher's negligence.

8.7 The Publisher hereby excludes liability for loss of revenue, reputation, business, profits, or for indirect or consequential losses, irrespective of whether the Publisher was advised by the Licensee of the potential of such losses.

8.8 The Licensee acknowledges the merit of independently verifying Data prior to taking any decisions of material significance (commercial or otherwise) based on such data. It is agreed that the Publisher shall not be liable for any losses which result from the Licensee placing reliance on the Data or on the Database, under any circumstances.

8.9 Subject to sub-clause 8.6 above, the Publisher's liability under this Agreement shall be limited to the purchase price.

9. Intellectual Property Rights

9.1 Nothing in this Agreement affects the ownership of copyright or other intellectual property rights in the Data, the Database of the Software.

9.2 The Licensee agrees to display the Publishers' copyright notice in the manner described in the Product.

9.3 The Licensee hereby agrees to abide by copyright and similar notice requirements required by the Publisher, details of which are as follows:

"© 2008 Taylor & Francis. All rights reserved. All materials in *Spon's External Works and Landscape Price Book 2008* are copyright protected. All rights reserved. No such materials may be used, displayed, modified, adapted, distributed, transmitted, transferred, published or otherwise reproduced in any form or by any means now or hereafter developed other than strictly in accordance with the terms of the licence agreement enclosed with *Spon's External Works and Landscape Price Book 2008*. However, text and images may be printed and copied for research and private study within the preset program limitations. Please note the copyright notice above, and that any text or images printed or copied must credit the source."

9.4 This Product contains material proprietary to and copyedited by the Publisher and others. Except for the licence granted herein, all rights, title and interest in the Product, in all languages, formats and media throughout the world, including copyrights therein, are and remain the property of the Publisher or other copyright holders identified in the Product.

10. Non-assignment

This Agreement and the licence contained within it may not be assigned to any other person or entity without the written consent of the Publisher.

11. Termination and Consequences of Termination.

11.1 The Publisher shall have the right to terminate this Agreement if:

11.1.1 the Licensee is in material breach of this Agreement and fails to remedy such breach (where capable of remedy) within 14 days of a written notice from the Publisher requiring it to do so; or

11.1.2 the Licensee becomes insolvent, becomes subject to receivership, liquidation or similar external administration; or

11.1.3 the Licensee ceases to operate in business.

11.2 The Licensee shall have the right to terminate this Agreement for any reason upon two month's written notice. The Licensee shall not be entitled to any refund for payments made under this Agreement prior to termination under this sub-clause 11.2.

11.3 Termination by either of the parties is without prejudice to any other rights or remedies under the general law to which they may be entitled, or which survive such termination (including rights of the Publisher under sub-clause 7.4 above).

11.4 Upon termination of this Agreement, or expiry of its terms, the Licensee must destroy all copies and any back up copies of the product or part thereof.

12. General

12.1 **Compliance with export provisions**

The Publisher hereby agrees to comply fully with all relevant export laws and regulations of the United Kingdom to ensure that the Product is not exported, directly or indirectly, in violation of English law.

12.2 **Force majeure**

The parties accept no responsibility for breaches of this Agreement occurring as a result of circumstances beyond their control.

12.3 **No waiver**

Any failure or delay by either party to exercise or enforce any right conferred by this Agreement shall not be deemed to be a waiver of such right.

12.4 **Entire agreement**

This Agreement represents the entire agreement between the Publisher and the Licensee concerning the Product. The terms of this Agreement supersede all prior purchase orders, written terms and conditions, written or verbal representations, advertising or statements relating in any way to the Product.

12.5 **Severability**

If any provision of this Agreement is found to be invalid or unenforceable by a court of law of competent jurisdiction, such a finding shall not affect the other provisions of this Agreement and all provisions of this Agreement unaffected by such a finding shall remain in full force and effect.

12.6 **Variations**

This agreement may only be varied in writing by means of variation signed in writing by both parties.

12.7 **Notices**

All notices to be delivered to: Spon's Price Books, Taylor & Francis Books Ltd., 2 Park Square, Milton Park, Abingdon, Oxfordshire, OX14 4RN, UK.

12.8 **Governing law**

This Agreement is governed by English law and the parties hereby agree that any dispute arising under this Agreement shall be subject to the jurisdiction of the English courts.

If you have any queries about the terms of this licence, please contact:

Spon's Price Books
Taylor & Francis Books Ltd.
2 Park Square, Milton Park, Abingdon, Oxfordshire, OX14 4RN
Tel: +44 (0) 20 7017 6672
Fax: +44 (0) 20 7017 6702
www.tandfbuiltenvironment.com/

Taylor & Francis
Taylor & Francis Group

Software Installation and Use Instructions

System requirements

Minimum

- Pentium processor
- 32 MB of RAM
- 10 MB available hard disk space
- Microsoft Windows 95/98/2000/NT/ME/XP
- SVGA screen
- Internet connection

Recommended

- Pentium 266 MHz processor
- 256 MB of RAM
- 100 MB available hard disk space
- Microsoft Windows 2000/NT/XP
- XVGA screen
- Broadband Internet connection

Microsoft ® is a registered trademark and Windows™ is a trademark of the Microsoft Corporation.

Installation

Spon's External Works and Landscape Price Book 2008 Electronic Version is no longer supplied on CD-ROM but is now supplied solely by internet download.

In your internet browser type in http://ebookstore.tandf.co.uk/supplements/pricebook and follow the instructions on screen. Then type in the unique access code which is sealed inside the back cover of this book.

If the access code is successfully validated, a web page with details of the available download content will be displayed.

Please note: you will only be allowed one download of these files and onto one computer.

Click on the download links to download the files. A folder called *PriceBook* will be placed on your desktop, which will need to be unzipped. Then click on the application called *install.*

Use

- As before, the installation process will create a folder containing an icon for *Spon's External Works and Landscape Price Book 2008* and also an icon on your desktop.
- Double click the icon (from the folder or desktop) installed by the Setup program.
- Follow the instructions on screen.

Technical Support

Support for the installation is provided on http://www.ebookstore.tandf.co.uk/html/helpdesk.asp

The *Electronic Version* is a free addition to the book. For help with the running of the software please visit www.pricebooks.co.uk

---

All materials in *Spon's External Works and Landscape Price Book 2008 Electronic Version* are copyright protected. No such materials may be used, displayed, modified, adapted, distributed, transmitted, transferred, published or otherwise reproduced in any form or by any means now or hereafter developed other than strictly in accordance with the terms of the above licence agreement.

The software used in *Spon's External Works and Landscape Price Book 2008 Electronic Version* is furnished under a single user licence agreement. The software may be used only in accordance with the terms of the licence agreement, unless the additional multi-user licence agreement is purchased from Taylor & Francis, 2 Park Square, Milton Park, Abingdon, Oxon OX14 4RN Tel: +44 (0) 20 7017 6000

© COPYRIGHT ALL RIGHTS RESERVED

Multiple-user use of the Spon Press Software and eBook

To buy a licence to install your Spon Press Price Book Software and eBook on a secure local area network or a wide area network, and for the supply of network key files, for an agreed number of users please contact:

Spon's Price Books
Taylor & Francis Books Ltd.
2 Park Square, Milton Park, Abingdon, Oxfordshire, OX14 4RN
Tel: +44 (0) 207 7017 6672
Fax: +44 (0) 207 7017 6072
www.sponpress.com

Number of users	Licence cost
2–5	£420
6–10	£850
11–20	£1300
21–30	£2000
31–50	£3900
51–75	£5500
76–100	£6600
Over 100	Please contact Spon for details